ADVANCES IN HYDROLOGY AND CLIMATE CHANGE

Historical Trends and New Approaches in Water Resources Management

Cover photo credits:

Left: Canyon Okatse, Georgia. Photo by Alexandr Hovhannisya.
https://unsplash.com/photos/3DyHCJwNs-k

Top right: Godly Confluence aka Devprayaga - place where Alaknanda and Bhagirathi rivers join to form Ganga; photo by Ashwini Chaudhary.
https://unsplash.com/photos/OvwXZvQculI

Bottom right: Photo by Eberhard Grossgasteiger.
https://unsplash.com/photos/4etUEweU4WI

Middle: Vatnajokull, Iceland. Photo by Deborah Diem
https://unsplash.com/photos/ds_NPvoAzro

ADVANCES IN HYDROLOGY AND CLIMATE CHANGE

Historical Trends and New Approaches in Water Resources Management

Edited by

Surendra Kumar Chandniha, PhD, PD

Anil Kumar Lohani, PhD

Gopal Krishan, PhD

Ajay Krishna Prabhakar, PhD

AAP | APPLE ACADEMIC PRESS

First edition published 2023

Apple Academic Press Inc.
1265 Goldenrod Circle, NE,
Palm Bay, FL 32905 USA

760 Laurentian Drive, Unit 19,
Burlington, ON L7N 0A4, CANADA

CRC Press
6000 Broken Sound Parkway NW,
Suite 300, Boca Raton, FL 33487-2742 USA

4 Park Square, Milton Park,
Abingdon, Oxon, OX14 4RN UK

Library and Archives Canada Cataloguing in Publication

Title: Advances in hydrology and climate change : historical trends and new approaches in water resources management / edited by Surendra Kumar Chandniha, PhD, PD, Anil Kumar Lohani, PhD, Gopal Krishan, PhD, Ajay Krishna Prabhakar, PhD.

Names: Chandniha, Surendra Kumar, editor. | Lohani, Anil Kumar, editor. | Krishan, Gopal (Scientist), editor. | Prabhakar, Ajay Krishna, editor.

Description: First edition. | Includes bibliographical references and index.

Identifiers: Canadiana (print) 20220266476 | Canadiana (ebook) 20220266573 | ISBN 9781774910306 (hardcover) | ISBN 9781774910313 (softcover) | ISBN 9781003282365 (ebook)

Subjects: LCSH: Water-supply—Management. | LCSH: Climatic changes.

Classification: LCC TD353 .A38 2023 | DDC 333.91/15—dc23

Library of Congress Cataloging-in-Publication Data

CIP data on file with US Library of Congress

ISBN: 978-1-77491-030-6 (hbk)
ISBN: 978-1-77491-031-3 (pbk)
ISBN: 978-1-00328-236-5 (ebk)

About the Editors

Surendra Kumar Chandniha, PhD, PD
Assistant Professor/Scientist, Department of Soil and Water Engineering, BRSM College of Agricultural Engineering and Technology & Research Station, IGKV, Mungeli, Chhattisgarh, India

Surendra Kumar Chandniha, PhD, is Assistant Professor/Scientist in the Department of Soil and Water Engineering at BRSM College of Agricultural Engineering and Technology and Research Station, Mungeli, India. Dr. Chandniha has more than nine years of experience in the field of water resources, surface water hydrology, soil and water engineering, and climate change. He has published more than 65 research papers in peer-reviewed journals as well as book chapters. He has done quality research work at IIT Roorkee and NIH Roorkee and has received several national and international awards in the water sectors area. Dr. Chandniha has delivered expert talks and invited lectures in offline and online modes at government organizations throughout India. He has supervised students of the MTech and BTech programs in agricultural engineering. He is a member of several professional organizations, including The Institute of Engineers, India; Indian Association of Hydrologists, Indian Water Resources Society, Association of Agrometeorologists, and American Society of Civil Engineering. And he is an editorial board member of the journal Hydrology.

Dr. Chandniha acquired his BTech (Agricultural Engineering), MTech (Soil and Water Engineering), from Indira Gandhi Krishi Vishwavidyalaya, Raipur, Chhattisgarh, India, and PhD (Water Resources Development and Management) from the Indian Institute of Technology (IIT), Roorkee, India. In the year 2016, he was awarded the Best Water Resources Student Award in IIT, Roorkee, India. He was awarded SERB-National Post Doctoral Fellowship in 2017 and was National Post-Doctoral Fellow and Research Fellow at the National Institute of Hydrology, Roorkee.

Anil Kumar Lohani, PhD

Scientist-G, National Institute of Hydrology, Roorkee, Uttarakhand, India

Anil Kumar Lohani, PhD, is currently Scientist-G and Head of the Surface Water Hydrology Division at the National Institute of Hydrology, Roorkee, India. Dr. Lohani is Coordinator of the Central India Hydrology Centre of the National Institute of Hydrology, Bhopal, and Head and Coordinator of the Centre for Flood Management Studies, National Institute of Hydrology, Guwahati, India. Dr. Lohani has over 30 years of research experience in surface water hydrological modelling, flood management, decision support system, and soft computing techniques. He has supervised several PhD, postdoctoral, MTech, and BTech students, and is currently Training Coordinator and DSS Coordinator of the National Hydrology Project of the National Institute of Hydrology. Dr. Lohani has completed over 40 international and national funded projects and is continuing work on six projects at present. He has published more than over 225 research papers in international/national journals and conferences as well as books, book chapters, and technical reports. Dr. Lohani is a member of several professional organizations and has received awards for his work. Dr. Lohnai has been an expert committee member of various national-level committees and has coordinated and organized more than 100 training programs and conferences.

Dr. Lohani received his BTech from G.B. Pant University of Agricultural & Technology, Pantnagar, Uttarakhand, India; his MTech in Water Resources Development & Management from IIT, Kharagpur, India; and his PhD in Hydrology from IIT, Roorkee, India.

Gopal Krishan, PhD

Scientist-C, National Institute of Hydrology, Roorkee, Uttarakhand, India

Gopal Krishan, PhD, is Scientist-C at the National Institute of Hydrology, Roorkee, India, and Ex-Researcher–Indo Gangetic Basin, Groundwater Resilience Project, British Geological Survey, United Kingdom and IIRS (ISRO). Dr. Krishan has over 20 years of research experience in hydrological evaluations, surface water and groundwater hydrology project management, and field investigations. He has supervised PhD and other postgraduate students. He has completed several internationally and nationally funded projects and is continuing work on five projects right now. He has published more than 150 research papers in international/national journals, including in Nature *Geoscience, Science of Total Environment, Geophysical Research Letters, Journal of Hydrology, Environment Pollution, Water,*

Environmental Earth Sciences, and *Current Science.* He has also published papers in conferences, along with a book, book chapters, newsletter articles, and technical reports. Dr. Krishan is a member of the Society of Earth Scientists, Indian Association of Hydrologists, International Association of Hydrological Sciences, International Association of Hydro-geologists, Indian Society of Remote Sensing, and American Water Resources Association. Dr. Krishan has been an expert committee member of a national committee constituted by NGT for finalizing groundwater guidelines and has been invited by the Punjab and West Bengal governments for various meetings on hydrological perspective. Dr. Krishan has conducted training courses to officers of various states under the National Hydrology Project and has organized two international conferences.

Dr. Gopal received his PhD in Soil Science and Water Management in 2001.

Ajay Krishna Prabhakar, PhD
Assistant Professor, Department of Civil Engineering,
National Institute of Technology, Kurukshetra, Haryana, India

Ajay Krishna Prabhakar, PhD, is Assistant Professor in the Department of Civil Engineering, National Institute of Technology Kurukshetra Haryana, India. Dr. Prabhakar has seven years of teaching experience at the National Institute of Technology, Kurukshetra, and research experience in the field of surface water hydrology, planning and management of water resources, drought analysis, and climate change. He has published research papers in peer-reviewed and SCI journals in water resources, groundwater, rainfall trend, climate change. He has supervised MTech and BTech students in civil engineering. He is an active member of National Institute of Technology Kurukshetra Library and faculty in charge of the Technical Society of Civil Engineering Department of NIT Kurukshetra. Dr. Prabhakar is also a member of the Association of Agrometeorologists and American Society of Civil Engineering.

Dr. Ajay Krishna Prabhakar acquired his BTech (Civil Engineering) from Bundelkhand Institute of Engineering and Technology, Jhansi, Uttar Pradesh, India; MTech (Water Resources Development and Management) from Indian Institute of Technology, Roorkee; and PhD (Water Resources Engineering) from National Institute of Technology, Kurukshetra, Haryana, India.

Contents

Contributors

M. L. Angurala
Central Ground Water Board, North Western Region, Chandigarh, India

Anjali
National Institute of Hydrology, Roorkee 247667, Uttarakhand, India

Manohar Arora
National Institute of Hydrology, Roorkee, Uttarakhand, India

Rafig Azzam
Department of Engineering Geology and Hydrogeology, RWTH Aachen University, Lochnestr 52064, Aachen, Germany

Triambak Baghel
Water Engineering and Management, Asian Institute of Technology, Bangkok, Thailand

Klaus Baier
Department of Engineering Geology and Hydrogeology, RWTH Aachen University, Lochnestr 52064, Aachen, Germany

Ankit Chakravarti
Faculty of Hydraulics and Water Resources Engineering, School of Civil and Environmental Engineering, Ambo University, Technology Campus, Ambo, Ethiopia. E-mail: ankitchakravarti@gmail.com

Surendra Kumar Chandniha
Department of Soil and Water Engineering, BRSM CAET&RS, IGKV, Mungeli, Chhattisgarh, India; E-mail: chandniha.surendra@gmail.com

G. K. Das
Department of Agrometeorology, IGKV, Raipur, India

Pragya Das
Department of Soil and Water Engineering, SV CAET & RS, IGKV, Mungeli, Chhattisgarh, India

Swagatam Das
Water Resources Systems Division, National Institute of Hydrology, Roorkee, Uttarakhand, India

Abhay Guleria
Civil Engineering Department, Indian Institute of Technology, Delhi 110016, India

Priyanka Gunjan
Department of Water Resources Development & Management, Indian Institute of Technology, Roorkee, India; E-mail: pri.gunjan@gmail.com

Chanchal Gupta
Research Associate, Natural Resource Department, TERI University (SAS), New Delhi 110070, India; E-mail: chanchal.iisc@gmail.com

J. B. Gurav
Department of Civil Engineering, Amrutvahini College of Engineering, Sangamner 422608, Maharashtra (Savtiribai Phule Pune University, Pune), India

Sanjay K. Jain
Water Resource System Division, National Institute of Hydrology, Roorkee 247667, India

Sukant Jain
Surface Water Hydrology Division, National Institute of Hydrology, Roorkee 247667, Uttarakhand, India

Rahul Kumar Jaiswal
Central India Hydrology, Regional Centre, WALMI Campus, Near Kaliasot Dam, Post Ravishankar Nagar, Bhopal, India

Nitin Joshi
Department of Civil Engineering, Indian Institute of Technology, Jammu 181221, India

R. U. Kamodkar
Department of Civil Engineering, Government Polytechnic, Jalgaon 425001, Maharashtra, India; E-mail: ravikamodkar@gmail.com

M. L. Kansal
Department of Water Resources Development & Management, Indian Institute of Technology, Roorkee 247667, Uttarakhand, India; E-mail: mlk@wr.iitr.ac.in

Gopal Krishan
National Institute of Hydrology, Roorkee, Uttarakhand, India; E-mail: drgopal.krishan@gmail.com

Aman Kumar
Water Engineering and Management, Central University of Jharkhand, India

Amit Kumar
School of Geography, University of Nottingham, United Kingdom

Dheeraj Kumar
Innovante Water Solutions Pvt. Ltd., Noida 201304, India; E-mail: ceo@inno-vante.com

Neeraj Kumar
Innovante Water Solutions Pvt. Ltd., Noida 201304, India

Priti Kumar
Department of Water Engineering and Management Central University of Jharkhand, India

Sumant Kumar
Scientist-D, Groundwater Hydrology Division, National Institute of Hydrology, Roorkee, India; E-mail: sumantks@gmail.com

Vinod Kumar
Groundwater Hydrology Division, National Institute of Hydrology, Roorkee, India

Anil Kumar Lohani
Surface Water Hydrology Division, National Institute of Hydrology, Roorkee, Uttarakhand, India; E-mail: aklnih@gmail.com

Prabhash Kumar Mishra
Surface Water Hydrology Division, National Institute of Hydrology, Roorkee, Uttarakhand, India; E-mail: erprabhash@gmail.com

S. K. Mishra
Department of Water Resources Development & Management, Indian Institute of Technology, Roorkee, India

Anoop Nagar
National Institute of Hydrology, Roorkee, Uttarakhand, India

Gaurav Kant Nigam
Krishi Vigyan Kendra–Korba, IGKV, Raipur 492012, India; E-mail: er.nigamgk@gmail.com

Osheen
Department of Water Resources Development & Management, Indian Institute of Technology, Roorkee 247667, Uttarakhand, India; E-mail: osheen@wr.iitr.ac.in

Lalit Pal
Department of Civil Engineering, Indian Institute of Technology, Roorkee, India

Manish Pandey
Department of Civil Engineering, NIT Warangal 506004, Telangana, India

Himanshu Panjiar
Faculty of Material Science and Engineering, Jimma University, Jimma, P.O. Box 378, Ethiopia; E-mail: hpanjiar92@gmail.com

Nitesh Patidar
National Institute of Hydrology, Roorkee 247667, Uttarakhand, India; E-mail: npatidar.nihr@gov.in

Ajay Krishna Prabhakar
Department of Civil Engineering, National Institute of Technology, Kurukshetra, India; E-mail: ajayprabhakar2@gmail.com

Yempali Priyanka
Water Engineering and Management, Central University of Jharkhand, India

Ravi Raj
Water Engineering and Management, Central University of Jharkhand, India

Preeti Rajput
Department of Civil Engineering, Government Engineering College Bilaspur, Koni 495009, India

Rakesh Rana
Central Ground Water Board, North Western Region, Chandigarh, India

D. G. Regulwar
Department of Civil Engineering, Government College of Engineering, Aurangabad 431005, India

Ravi K. Saini
Groundwater Hydrology Division, National Institute of Hydrology, Roorkee, India

Arnab Sarma
Department of Civil Engineering, The Assam Royal Global University, Betkuchi, Guwahati 781035, Assam, India; E-mail: arnab.sarma@rgi.edu.in

Monika Sharma
Department of Environmental & Water Resources Engineering, University Teaching Department, Chhattisgarh Swami Vivekanand Technical University Bhilai, Bhilai 491107, Chhattisgarh, India

Nayan Sharma
Innovante Water Solutions Pvt. Ltd., Noida 201304, India

Himanshu Sharma
Department of Civil Engineering, Institute of Infrastructure Technology Research and Management, Ahmedabad 380026, Gujarat, India; E-mail: himfce@gmail.com

Vikash Shivhare
Natural Resource Department, TERI University (SAS), New Delhi 110070, India

Hemant Singh
Water Resources Systems Division, National Institute of Hydrology, Roorkee, Uttarakhand, India

K. K. Singh
Department of Civil Engineering, National Institute of Technology, Kurukshetra, India

Manish Kumar Sinha
Department of Engineering Geology and Hydrogeology, RWTH Aachen University,
Lochnestr, 4-20, 52064, Aachen, Germany; E-mail: manishsinha200389@gmail.com

Karan Solanki
Department of Civil Engineering, Institute of Infrastructure Technology Research and Management,
Ahmedabad, Gujarat 380026, India

N. Sudarsan
National Institute of Hydrology, Roorkee, Uttarakhand, India

Saurabh Suman
Department of Water Engineering and Management Central University of Jharkhand, India

Shakti Suryavanshi
Departmentof Civil Engineering, Sam Higginbottom Institute of Agriculture, Technology & Sciences,
Allahabad 211007, Uttar Pradesh, India

Deepak Swami
School of Engineering, Indian Institute of Technology, Mandi 175001, India;
E-mail: deepak@iitmandi.ac.in

Chandni Thakur
School of Engineering, Indian Institute of Technology, Mandi 175001, India

Nity Tirkey
Department of Water Engineering and Management, Central University of Jharkhand, India

M. P. Tripathi
Department of Soil and Water Engineering, S.V. College of Agricultural Engineering and Technology,
IGKV–Indira Gandhi Agricultural University, Raipur, Chhattisgarh, India

Mukesh Kumar Verma
Chhattisgarh Swami Vivekanand Technical University Bhilai, Bhilai 491107, Chhattisgarh, India

Vivek Verma
Water & Environmental Engineering, Florida International University, Miami, USA;
E-mail: vivekverma10@gmail.com

Pratibha Warwade
Water Engineering and Management, Central University of Jharkhand, India;
E-mail: pratibhawarwade@gmail.com

Alka Yadav
Department of Water Resources Development & Management, Indian Institute of Technology,
Roorkee 247667 (Uttarakhand), India; E-mail: alkayadav2012@gmail.com

Abbreviations

ABC	artificial bee colony
ACF	autocorrelation function
ACO_RCM	Ant Colony Optimization_Rating Curve Method
ADD	annual drought duration
ADE	advection-dispersion equation
ADI	apportion disorder index
ADP	annual drought peak
ADS	annual drought severity
AE	apportionment entropy
AEM	analytical element method
AHP	analytic hierarchy process
AMO	Atlantic Multidecadal Oscillation
ANNs	artificial neural networks
ARIMA	autoregressive integrated moving average
ARX	linear autoregressive
AR5	fifth assessment report
ATDD	asymptotic time-dependent dispersion
AWSI	Arab Water Sustainability Index
BMP	best management practices
CAM4	Community Atmosphere Model
CCIS	Canadian Climate Impacts Scenarios
CCSM4	Community Climate System Model Version 4
CFD	computational fluid dynamics
CL	confidence limits
CNN	convolutional neural network
CPD	cumulative probability distribution
CVFD	control-volume finite-difference
CV	coefficient of variation
DADI	decadal apportionment disorder index
DAE	decadal apportionment entropy
DEM	digital elevation model
DO	dissolved oxygen
DSS	decision support system
DWS	domestic water supply

EC	electric conductivity
ED	ecological deficit
EFI	ecological footprint index
EKF	extended Kalman Filter
EnKF	ensemble Kalman Filter
ENSO	El Niño Southern Oscillation
ER	ecological remainder
ERT	electrical resistivity tomography
ESI	Environmental Sustainability Index
FCR	feeder canal release
FDM	finite difference method
FEM	finite element method
FFBP	feed forward back propagation
FFNN	feed forward neural network
FIS	fuzzy inference system
FLP	fuzzy linear programming
FMGP	fuzzy multi-objective goal programming
FSDP	fuzzy stochastic dynamic programming
FPP	fuzzy parametric programming
GA_RCM	Genetic Algorithm_Rating Curve Method
GCM	Global Circulation Model
HAWQS	hydrologic and water quality system
HMR	homogeneous monsoon regions
HRUs	hydrologic response units
HUCs	hydrologic unit codes
GCMs	global circulation models
GIS	geographical information system
GPS	global positioning system
GRNN	generalized regression neural network
GW	groundwater
GWS	global wavelet spectrum
HELP	Hydrology, Environment, Life, and Policy
HIS	hydrological information system
IBI	index-based built-up index
IDF	intensity frequency
IDW	inverse distance weighting
IITM	Indian Institute of Tropical Management
IMD	Indian Metrological Department
IOD	Indian Ocean Dipole Mode Index

IPCC	intergovernmental panel on climate change
ISM	Indian Summer Monsoon
ISMR	Indian summer monsoon rainfall
ISO	Implicit Stochastic Optimization
ITM	Innovative Trend Method
IWSI	Integrated Wetland Sustainability Index
LAI	leaf area index
LDWA	Lake and Waterway Development Authority
LTDD	linear time-dependent dispersion
LULC	land use/cover
LVQ	learning vector quantization
LMR	linear multiple regression
MADI	mean apportion disorder index
MAPE	mean absolute prediction error
MDADI	mean decadal apportion disorder index
MDI	marginal disorder index
ME	marginal entropy
MESMIS	management systems using indicators
MHR	monsoon homogeneous region
MIM	mobile-immobile model
MK	Mann-Kendall
MMK	modified Mann-Kendall
MLP	multilayer perceptron
MLR	multiple linear regression
MSE	mean of squared error
MSI	multi-spectral instrument
MUSLE	modified universal soil loss equation
MWP	Mann–Whitney–Pettitt
NCEP	National Center of Environmental Prediction
NDBI	normalized difference built-up index
NDISI	normalized difference impervious surface index
NDVI	normalized difference vegetation index
NID	neural interpretation diagram
NIH	National Institute of Hydrology
NIR	near-infrared
NOAA	National Oceanic and Atmospheric Administration
NSE	Nash-Sutcliffe efficiency
OECD	Organization for Economic Cooperation and Development
PBIAS	percentage bias

PCA	principle component analysis
pdf	probability density function
PDO	pacific decadal oscillation
PNN	probabilistic neural network
PRISM	Regressions on Independent Slopes Model
PSO	practical swarm optimization
PSO_RCM	Particle Swarm Optimization_Rating Curve Method
PSR	pressure-state-response
QBO	quasi-biennial oscillation
QUMP	quantifying uncertainty in model predictions
RANS	Reynolds Averaged Navier-Stokes
RBF	radial basis function
RCM	rating curve method
RCPs	Representative Concentration Pathways
RMSE	root mean of squared error
RNN	recurrent neural networks
RR	rainfall runoff
RSR	root mean square error
SAVI	soil adjusted vegetation index
SDP	stochastic dynamic programming
SDR	sediment delivery ratio
SDSM	statistical downscaling model
SDSS	spatial decision support system
SEEI	sustainable ecological evaluation indicators
SIUWM	The Sustainability Index for Integrated Urban Water Management
SNHT	Standard Normal Homogeneity Test
SPI	standardized precipitation index
SSE	sum of squared error
SSE	Sen slope estimator
SST	sea surface temperature
STS	simple time series
SVM	support vector machine
SWAP	soil-water-atmosphere-plant
SWAT	soil and water assessment tool
SWI	Sustainable Water Index
SWIR	short wave infrared
SWS	sub-watersheds
SY	sediment yield

TDNN	time delay neural network
TIR	thermal infrared
TSI	total solar irradiance
TSS	total suspended sediment
UI	urban index
ULSE	universal soil loss equation
USGS	United States Geological Survey
UTs	union territories
WDS	water distribution system
WECD	World Commission on Environment and Development
WJWSI	West Java Water Sustainability Index
WNN	wavelet neural network
WQI	water quality index
WSI	Watershed Sustainability Index
WT	wavelet transform

Preface

Advances in Hydrology and Climate Change: Historical Trends and New Approaches in Water Resources Management summarizes recent trends in the water sector using a variety of innovative approaches. The chapters address the overall scenarios involving issues and challenges in the water sector.

Further, the chapters focus on providing a better understanding of climatic uncertainty in the hydrometeorological datasets and its application in hydrological modeling. In hydrometeorological parameters, precipitation is the most pronounced parameter that majorly affects living and nonliving creatures in nature. For any analysis, historical observations are important to better understand the actual behavior system.

However, this book highlights the hydrometeorological parameters and its behavior identification using different statistical and nonparametric approaches. Uncertainty of climatic variability depends on several factors, that is, its erratic distribution, topography, seasonal variation, land use change, and anthropogenic activities, etc. and proper justification is a massive and a time-consuming task. Hence, the chapters include overall interconnection between different parameters of the hydrological cycle. Water budgeting/water balance/hydrological modeling is nothing but near-to-approximate estimation of water allocations in different components, that is, surface water contribution, groundwater contribution, storage, and other losses. For this purpose, hydrological modeling approaches are included in this book. In the water sector, mainly two conditions are most common: water surplus and water deficit. Proper estimation of water surplus (wet condition), water deficit (drought condition), and drought analysis are the basis of some chapters that are included in this book.

Simultaneously, qualitative and quantitative parameter handling is also very complicated in research activity. Hence, soft computing application is the best approach to address such problems. This book includes a chapter associated with soft computing applications in the water sector.

The chapters are very helpful for upcoming researchers and academicians to understand properly complex, nonliner solutions. In remote places, surveys are time consuming and not feasible as per an economic point of view. So, how remote sensing and GIS-based techniques can be applicable in water sector is also emphasized.

In the past, climate change has been the most pronounced topic worldwide, and its impact is also very hazardous. For a future prospective, climate change scenarios based on future projections are also included in the chapters. Climate change-based scenarios are useful for future water resources and agricultural planning and mitigation under consideration of climatic uncertainty.

This book, *Advances in Hydrology and Climate Change*, covers recent trends in water sectors toward solving existing problems in different areas. The book comprises 22 chapters and all the chapters are directly or indirectly associated with water-related problems. This book focuses on the surface and groundwater-related issues and its sustainable solution.

This book will be a standard reference work for the disciplines such as water resources, soil and water engineering, engineering hydrology, groundwater hydrology, climate change, agrometeorology, agriculture, ecology, and environmental science as well as it will be a way forward toward strategy formulation to combat hydrology and climate changes.

The editors would appreciate receiving comments from readers that may assist in the development of the future editions.

—Surendra Kumar Chandniha, PhD
Anil Kumar Lohani, PhD
Gopal Krishna, PhD
Ajay Prabhakar, PhD

CHAPTER 1

Parameter Estimation for Solute Transport through Coupled Stratified Porous Media Using Time-Dependent Dispersion Coefficient

DEEPAK SWAMI[1*], CHANDNI THAKUR[1], ABHAY GULERIA[2], and NITIN JOSHI[3]

[1]*School of Engineering, Indian Institute of Technology, Mandi 175001, India*

[2]*Civil Engineering Department, Indian Institute of Technology, Delhi 110016, India*

[3]*Department of Civil Engineering, Indian Institute of Technology, Jammu 181221, India*

Corresponding author. E-mail: deepak@iitmandi.ac.in

ABSTRACT

This study presents a formulation of parameter estimation technique to estimate solute transport parameters. Finite-difference numerical model coupled with Levenberg-Marquardt parameter estimation algorithm is developed and presented. Parameter estimation results show that for a unique estimation, the inverse procedure is limited to three unknown transport parameters. While estimating two or three unknown parameters, the presence of equilibrium sorption coefficients of mobile and immobile domains results in nonunique estimates. For the unique solution of equilibrium sorption coefficient of both

Advances in Hydrology and Climate Change: Historical Trends and New Approaches in Water Resources Management. Surendra Kumar Chandniha, Anil Kumar Lohani, Gopal Krishan, & Ajay Krishna Prabhakar (Eds.)

the domains, at least one parameter should be measured experimentally. Asymptotic time-dependent dispersion (ATDD) model results in lesser number of nonunique estimate combinations in comparison with linear time-dependent dispersion (LTDD) model. Finally, unknown transport parameters are estimated from observed data using developed model in a hydraulically coupled stratified porous media. The results show that the ATDD model fits observed data more effectively than LTDD model.

1.1 INTRODUCTION

Transport of reactive or nonreactive solute transport through heterogeneous porous medium specifically for a stratified porous media is a complex problem. The asymmetry in breakthrough curves demonstrating early arrival and long tailing have been observed for these stratified and heterogeneous media which reflects anomalous transport (Srivastava and Brusseau, 1996; Li and Brusseau, 2000; Levy and Berkowitz, 2003; Cortis and Berkowitz, 2004). Several studies have reported that utility of advection–dispersion equation (ADE) lacks for simulating anomalous transport behavior (Van Genuchten and Wierenga, 1976; Li et al., 1994; Levy and Berkowitz, 2003; Gao et al., 2009, 2010; Swami et al., 2014; Sharma et al., 2016). Therefore, more complex modeling approaches which captures anomalous transport and considers physical and chemical heterogeneities of the porous media have been discussed (Dagan, 1988; Brusseau et al., 1989, 1997; Cortis and Berkowitz, 2005; Kartha and Srivastava, 2008; Swami et al., 2013; Aquino et al., 2015; Dentz et al., 2015; Joshi et al., 2015; Mehmani and Balhoff, 2015). These approaches emphasize on the physical partitioning of the porous media considering pore distribution and rate of solute mass exchange (Van Genuchten and Wierenga, 1976, 1977; Brusseau et al., 1989; Haggerty and Gorelick, 1995; Joshi et al., 2015; Swami et al., 2013, 2014). Mobile–immobile model (MIM) consists of an ADE with a mass transfer term and partitions porous media into stagnant (immobile) and flowing (mobile) zones (Van Genuchten and Wierenga, 1976, 1977; Gao et al., 2009). MIM describes asymmetric breakthrough curves by incorporating a mass transfer coefficient which takes into account the mass exchange between flowing and stagnant zones (Nielsen et al., 1986; Brusseau et al., 1989; Brusseau, 1991). In addition to this, scale dependency of dispersion coefficient further enhances the complexity in modeling solute transport (Dagan, 1988; Gelhar et al., 1992; Zhou and Selim, 2003a). Lumping of dispersion coefficient throughout the domain is a trivial approximation and fails to portray the wider spectrum of temporal breakthrough curves through stratified aquifers (Pickens and Grisak, 1981a, 1981b; Gelhar et al., 1992; Logan, 1996; Zhou and Selim, 2003a, 2003b; Schulze-Makuch, 2005; Kumar et al., 2006, 2008; Swami et al., 2014). Some of the studies have also shown existence of scale and time-dependent dispersion (from now onwards TDD) prior to asymptotic regime (Dagan, 1988; Jaiswal et al., 2009; Gao et al., 2010; Sharma and Srivastava, 2011; Selim, 2014; Guérin and Dean, 2017). Apart from heterogeneity, resident time and mean travel distance also influence the dispersion prior to asymptotic regime (Warrick et al., 1971; Gelhar et al., 1979,1992; Sauty, 1980; Pickens and Grisak, 1981a,1981b; Sudicky, 1986; Basha and El-Habel, 1993; Gelhar, 1993; Logan, 1996; Schulze-Makuch, 2005). The breakthrough curve simulated using scale-dependent dispersion (from now onward SDD) have demonstrated a steeper concentration profile compared with the TDD coefficient (Sharma and Srivastava, 2011; Selim, 2014). Due to the formulation, a major limitation of SDD is to grow continuously without upper bound (Swami et al., 2016b). As numerical calculation updates the concentration in liquid phase after each time step, therefore, application of TDD for a heterogeneous medium is found more suitable (Natarajan, 2016). Therefore, to capture the liquid phase concentration with better approximation, ADE with a TDD coefficient was proposed (Barry and Sposito, 1989; Basha and El-Habel, 1993; Zou et al., 1996; Sander and Braddock, 2005; Kumar et al., 2006, 2008). There are a few studies which presented MIM incorporating TDD coefficient (Sharma and Srivastava, 2011). Hence, it can be said that the mobile–immobile formulation is limited to constant dispersion models when simulating experimental breakthrough curves (BTCs). Further, this is also essential to estimate the MIM model parameters with greater accuracy and precision for correct quantification of solute transport. Estimation of transport parameters remains complex because many input parameters vis-a-vis dispersivity, mass transfer coefficient, distribution coefficient, etc. are difficult to measure at a lab scale or aquifer scale (Yeh, 1986; Kool et al., 1987). The estimation of subsurface solute transport parameters have been carried out by utilizing parameter estimation algorithms (Yeh, 1986; Kool et al., 1987; Kool and Parker, 1988; Barth and Hill, 2005a, 2005b; Ratha et al., 2009;Ojha et al., 2011; Joshi et al., 2013). Levenberg algorithm was used in HYDRUS model to estimate the water flow and heat transport parameters for unsaturated soil (Nakhaei and Šimůnek, 2014; Mashayekhi et al., 2016). Levenberg-Marquardt optimization algorithm has also been used to estimate the virus transport parameters (Ratha et al., 2009; Ojha et al., 2011; Joshi et al., 2013). Coupling of numerical codes with optimization algorithms is also reported in the literature which has been found as a promising approach for the estimation of virus transport parameters (Ratha et al., 2009). Several studies have been carried out to check the robustness and efficacy of optimization algorithm (Barth and Hill, 2005a, 2005b; Ojha et al., 2011). Sensitivity toward errors and biases in data brought in by objective function and the effects these on the parameter estimation has been studied (Ratha et al., 2009; Prasad et al., 2012). The objective function tends to induce some errors and bias if more than two parameters are estimated simultaneously, and it leads to nonuniqueness of parameters (Prasad et al., 2012). It has been observed that the above-mentioned parameter estimation studies are limited to constant dispersion models only (Ratha et al., 2009; Ojha et al., 2011; Joshi et al., 2013). To fill the gap identified from the literature review, parameter estimation algorithm for stratified porous media using MIM with TDD function is presented. Findings of this study would help in identifying which set of parameters can be estimated uniquely and which set of parameters are to be obtained experimentally for modeling solute transport through stratified porous media using MIM simulations.

This study presents an extension of MIM by inclusion of arbitrary TDD coefficient and its solution further by means of finite-difference method. Second, solute transport parameters are estimated using Levenberg-Marquardt inverse optimization algorithm. The performance of the algorithm is tested by using synthetically generated solute concentration data and issues related to identifiability of parameters are also discussed. Sensitivity analysis broadly determines the range of time-dependent transport parameters within which estimation is unique and feasible. Finally, developed numerical model with inverse optimization algorithm is used to simulate solute concentration data from the column experiment.

1.2 GOVERNING EQUATIONS

The conceptualization of MIM is based on bifurcating the representative elementary volume of heterogeneous porous media into mobile (flowing) region or immobile (stagnant) zones. This portioning is based on physical movement of contaminant with groundwater flow in pores. The mobile zone comprises of well-connected pore structure, whereas immobile zone represents the water content which is irreducible. Due to the presence of two continua of porous formation which is in hydraulic coupling, transport in mobile zone is governed by advection and dispersion only whereas in immobile zone, transport is majorly governed by diffusive gradient. Considering linear sorption isotherm for the sake of simplicity, governing equations for MIM can be written as:

$$\left(\theta_m + f\rho_b K_{d_m}\right)\frac{\partial C_m}{\partial t} = \theta_m D_{(t)}\frac{\partial^2 C_m}{\partial x^2} - v_m\theta_m\frac{\partial C_m}{\partial x} - \omega\left(C_m - C_{\mathrm{im}}\right) - \left(\theta_m\mu_{lm} + f\rho_b K_{d_m}\mu_{\mathrm{sm}}\right)C_m \tag{1.1}$$

$$\left(\theta_{\mathrm{im}} + (1-f)\rho_b K_{d_{\mathrm{im}}}\right)\frac{\partial C_{\mathrm{im}}}{\partial t} = \omega\left(C_m - C_{\mathrm{im}}\right) - \left(\theta_{\mathrm{im}}\mu_{\mathrm{lim}} + (1-f)\rho_b K_{d_{\mathrm{im}}}\mu_{\mathrm{sim}}\right)C_{\mathrm{im}}. \tag{1.2}$$

where, θ is the porosity, C is the liquid phase concentration (ML^{-3}), ρ_b is the bulk density of the porous medium (ML^{-3}), ω is the mass transfer coefficient (T^{-1}), v is the pore water velocity (LT^{-1}), f is the fractions of instantaneous sorption sites, K_d is the distribution coefficient (L^3/M), μ_l is the solution phase first-order decay coefficients (T^{-1}), μ_s is the first-order decay coefficients in adsorbed phase (T^{-1}), and $D_{(t)}$ is the dispersion coefficient at time t (L2/T). For all the parameters m and im are the subscript for mobile and immobile region, respectively.

Two types of TDD have been considered:

a) Linear time-dependent dispersion function (LTDD):

$$D_{(t)} = D_0\frac{t}{K_L} + D_m. \tag{1.3}$$

b) Asymptotic time-dependent dispersion function (ATDD):

$$D_{(t)} = D_0\frac{t}{t + K_A} + D_m. \tag{1.4}$$

where D_0 is the maximum dispersion coefficient (L^2/T), t is the time (T), D_m is the diffusion coefficient (L^2/T), K_L is the LTDD coefficient (T), and K_A is the ATDD coefficient (T).

1.2.1 INITIAL AND BOUNDARY CONDITIONS

Initially, the medium is solute-free:

$$C_m(x,0) = C_{\mathrm{im}}(x,0) = 0. \tag{1.5}$$

1.2.2 BOUNDARY CONDITIONS

Dirichlet type boundary condition at inlet for a continuous source is:

$$C_m(0,t) = C_0. \tag{1.6}$$

Neumann type boundary condition at the outlet is:

$$\left(\frac{\partial C_m(x,t)}{\partial x}\right)_{(x=L,t)} = 0. \tag{1.7}$$

where C_0 is the concentration of solute at the inlet (ML^{-3}).

1.3 NUMERICAL SCHEME FOR SOLUTE TRANSPORT MODEL

The MIM model in the presented study was solved using Crank-Nicolson, one of the finite-difference scheme with arbitrary TDD coefficient. First-order upwind scheme was used to discretize advection term in the mobile transport equation to contain numerical dispersion associated with the central weighting scheme (Zheng and Bennett,2002). Second-order central difference scheme was used to discretize the dispersion terms. A first-order forward difference scheme was used to discretize the temporal terms. Following are the details of the discretization:

$$R_m\left(\frac{C_{mi}^{t+1} - C_{mi}^{t}}{\Delta t}\right) = \frac{\theta_m D_{(t+1)}}{2}\left(\frac{C_{mi+1}^{t+1} - 2C_{mi}^{t+1} + C_{mi-1}^{t+1}}{\Delta x^2}\right) + \frac{\theta_m D_{(t)}}{2}$$
$$\left(\frac{C_{mi+1}^{t} - 2C_{mi}^{t} + C_{mi-1}^{t}}{\Delta x^2}\right) - \frac{\theta_m v_m}{2}\left(\frac{C_{mi}^{t+1} - C_{mi-1}^{t+1}}{\Delta x}\right) - \frac{\theta_m v_m}{2}\left(\frac{C_{mi}^{t} - C_{mi-1}^{t}}{\Delta x}\right) \tag{1.8}$$
$$-\frac{\omega}{2}\left(C_{mi}^{t+1} - C_{imi}^{t+1}\right) - \frac{\omega}{2}\left(C_{mi}^{t} - C_{imi}^{t}\right) - \frac{A_1}{2}C_{mi}^{t+1} - \frac{A_1}{2}C_{mi}^{t}.$$

$$R_{\text{im}}\left(\frac{C_{\text{im}i}^{t+1} - C_{\text{im}i}^{t}}{\Delta t}\right) = \frac{\omega}{2}\left(C_{mi}^{t+1} - C_{\text{im}i}^{t+1}\right) + \frac{\omega}{2}\left(C_{mi}^{t} - C_{\text{im}i}^{t}\right)$$
$$-\frac{A_2}{2}C_{\text{im}i}^{t+1} - \frac{A_2}{2}C_{\text{im}i}^{t}. \tag{1.9}$$

where

$$R_m = \left(\theta_m + f\rho_b K_{dm}\right). \tag{1.10}$$

$$R_{\text{im}} = \left(\theta_{\text{im}} + (1-f)\rho_b K_{d\text{im}}\right). \tag{1.11}$$

$$A_1 = \left(\theta_m \mu_{lm} + f \rho_b K_{dm} \mu_{\text{sm}}\right). \tag{1.12}$$

$$A_2 = \left(\theta_{\text{im}} \mu_{\text{lim}+(1-f)\rho_b K_{dim} \mu_{\text{sim}}}\right). \tag{1.13}$$

Equations 1.8 and 1.9 on rearrangement can be written as:

$$A_7 C_{mi}^{t+1} - \left(A_3 D_{(t)}\right) C_{mi+1}^{t+1} - \left(A_3 D_{(t)} + A_4\right) C_{mi-1}^{t+1} - A_5 C_{\text{im}i}^{t+1} = A_{10} C_{mi}^{t} + \left(A_3 D_{(t)}\right) C_{mi+1}^{t} + \left(A_3 D_{(t)} + A_4\right) C_{mi-1}^{t} + A_5 C_{\text{im}i}^{t}. \tag{1.14}$$

$$A_8 C_{\text{im}i}^{t+1} - A_9 C_{mi}^{t+1} = A_{11} C_{imi}^{t} + A_9 C_{mi}^{t}. \tag{1.15}$$

Substituting the following:

$$A_3 = \frac{\theta_m}{\Delta x^2}\left(\frac{\Delta t}{2R_m}\right). \tag{1.16}$$

$$A_4 = \frac{v_m \theta_m}{\Delta x}\left(\frac{\Delta t}{2R_m}\right). \tag{1.17}$$

$$A_5 = \frac{\omega \Delta t}{2R_m}. \tag{1.18}$$

$$A_6 = \frac{A_1 \Delta t}{2R_m}. \tag{1.19}$$

$$A_7 = 1 + \left(2A_3 D_{(t)}\right) + A_4 + A_5 + A_6. \tag{1.20}$$

$$A_8 = 1 + \left(\frac{\omega \Delta t}{2R_{\text{im}}}\right) + \left(\frac{A_2 \Delta t}{2R_{\text{im}}}\right). \tag{1.21}$$

$$A_9 = \left(\frac{\omega \Delta t}{2R_{\text{im}}}\right). \tag{1.22}$$

$$A_{10} = 1 - \left(2A_3 D_{(t)}\right) - A_4 - A_5 - A_6. \tag{1.23}$$

$$A_{11} = 1 - \left(\frac{\omega \Delta t}{2R_{\text{im}}}\right) - \left(\frac{A_2 \Delta t}{2R_{\text{im}}}\right). \tag{1.24}$$

Here, Δt is the computational time step, Δx is grid size along the length of domain. The equations described above were solved by using the Thomas algorithm and the concentration values were evaluated for various time levels.

1.4 INVERSE PROBLEM AND SOLUTION ALGORITHM

The transport parameters $(K_{dm}, K_{dim}, D_0, K_A, or\ K_L, \omega)$ were estimated using the optimization method coupled with MIM incorporating TDD. Utilizing the concept of nonlinear optimization, inverse problem was formulated and the parameters were estimated by minimizing the objective function (Ratha et al., 2009; Joshi et al., 2013). The objective function (ϕ) is as follows:

$$\min_{b} \phi = \left(C^* - \hat{C}\right)^T \left(C^* - \hat{C}\right). \tag{1.25}$$

where C^* (x, t) is the vector of observed concentration, $\hat{C}$ (x, t) is the vector of simulated concentration obtained by solving MIM incorporating TDD coefficient for a given parameter vector b, $b = \{D_0, K_A\ or\ K_L, \omega, K_{dm}, K_{dim}\}^T$ is the parameter vector.

Nonlinear objective function was obtained for parameter vector b and iterative procedure was used for minimization. At each iteration i, parameter correction vector Δb^i was determined so that

$$b^{i+1} = b^i + \Delta b^i. \tag{1.26}$$

In this study, Levenberg-Marquardt optimization algorithm was used for parameter estimation which is given as

$$\left(J^T J + \beta D^T D\right)\Delta b = -J^T e. \tag{1.27}$$

where β is a positive scalar and D – $diag$ $(d_1, d_2, \ldots, d_p)$ which is a scaling matrix that considers differences in the magnitude of the sensitivities for the different parameters. J is the sensitivity matrix, and e is the vector of residuals ($C^* - \hat{C}$).

The columns of the Jacobian Matrix J contain the partial derivatives of the residuals e with respect to the elements of the parameter vector b. The Jacobian has dimension $n \times p$, where n is the number of observations and p is the number of unknown parameters to be estimated (Ratha et al., 2009; Joshi et al., 2013). The elements of D are updated as given by Kool and Parker (1988). Step direction and step length depend upon the parameter β. A larger value of β will result in the steepest descent direction, while $\beta = 0$ will result in the full Gauss-Newton step. Initially, a large value of β is assumed, so that the minimization will move in the right direction. Equation 1.13) can be represented by the normal equations for the linear least-squares problem (Osborne 1976; Moré 1978; Kool and Parker 1988) as

$$\min_{\Delta b} e^T e = \left(\begin{matrix} J \\ \frac{1}{\beta^2 D} \end{matrix} \right) \Delta b - \left(\begin{matrix} -e \\ 0 \end{matrix} \right)^2 \quad (1.28)$$

The minima was find out using orthogonal techniques to avoid the computation of ill-conditioned $J^T J$. It was observed that due to inclusion of constraints, algorithm converged to true value even when initial guess value differed by an order of magnitude (Cobb et al., 1982; Kool et al., 1987; Ratha et al., 2009; Ojha et al., 2011; Joshi et al., 2013). Modifications and constraints were incorporated in the Levenberg-Marquardt algorithm in this study.

1.5 VALIDATION OF THE NUMERICAL MODEL

The presented numerical solution was compared with the analytical solution of one-dimensional solute transport equation for time-dependent and constant dispersion coefficients. Numerical solution was validated by scaling down to advection–dispersion model by keeping the value of mass transfer coefficient (ω) and saturated porosity of immobile region equal to zero. In addition, the present numerical solution was also validated with the analytical solution proposed by Basha and El-Habel, 1993. For validation, following input parameters were used: $L = 100$ m, $\theta = 0.35$, $\omega = 0$, $v_m = 0.25$ m/day, total simulation time = 200 days, $R = 1$, $\rho_b = 2.11$ gm/cm^3, D_0 for ATDD function = 1 m^2/day, $K_A = 50$ days, diffusion and first-order decay were not considered. For the simulation Peclet number ($P_e = v_m \Delta x / D_{(t)}$) was less than 2 and Courant number ($Cr = v_m \Delta x / D_{(t)}$) was kept less than 1 to reduce the numerical oscillation. Evolution of solute plume for continuous source is shown in Figure 1.1. The numerical model shows good agreement with the analytical solution.

In the second case, the numerical solution of present model was validated with the analytical solution developed by Barry and Sposito (1989). In this case, MIM with ATDD was downscaled to ADE with constant dispersion using zero values for ω, θ_{im}, and K_A. For validation, the specified parameters were as follows: $L = 5$ cm, $v_m = 1$ cm/day, total simulation time = 2.5 days, $R = 1$, $D_0 = 0.05$ cm^2/day, ignoring effective molecular diffusion and first-order decay. Figure 1.2 represents that numerical model shows good agreement with the analytical solution proposed by Barry and Sposito (1989).

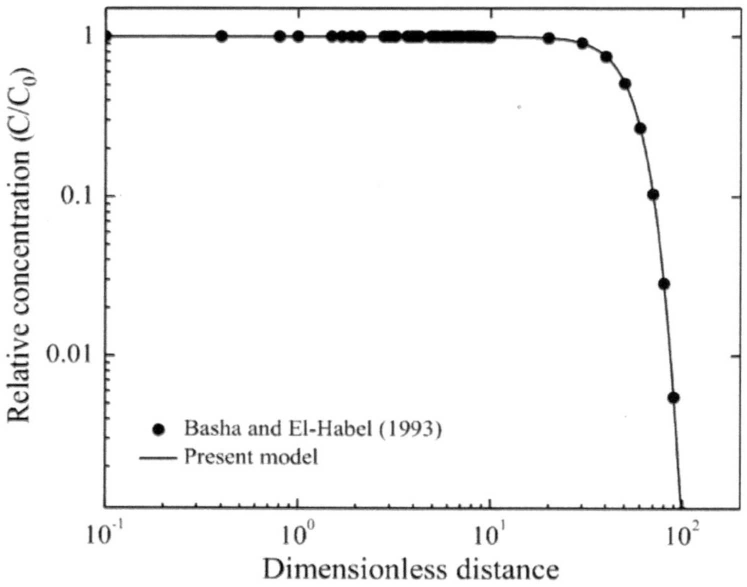

FIGURE 1.1 Concentration profile for nonreactive solute at time T = 200 for $K_A = 50$.
Source: Reprinted with permission from Guleria et al., 2019. © Springer.

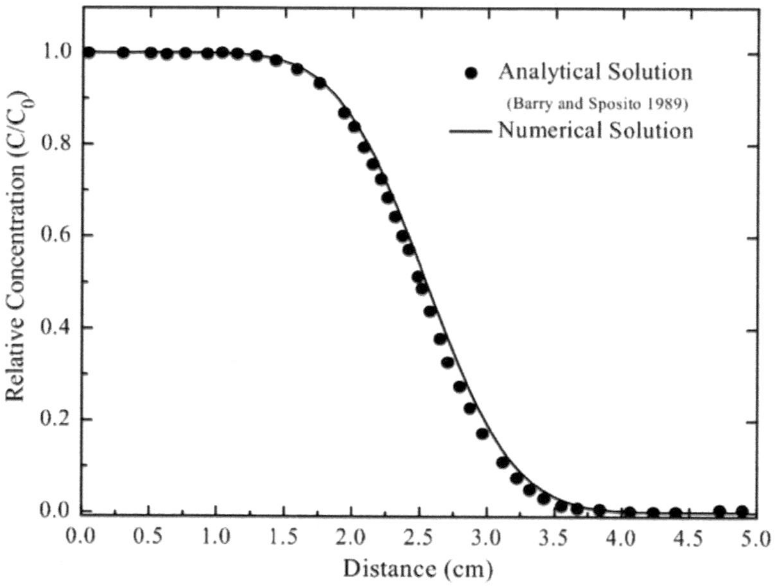

FIGURE 1.2 Spatial concentration profile for nonreactive solute at time, T = 2.5 days.
Source: Reprinted with permission from Guleria et al., 2019. © Springer.

1.6 EFFECTIVENESS OF PARAMETER ESTIMATION ALGORITHM

Inverse problem requires initial estimate of the parameters to be optimized, spatial locations of observation points and measured concentration data with corresponding time. Therefore, in this study, hypothetical concentration as a function of space and time was generated by solvingeqs 1.1 and 1.2, subjected to initial condition given by eq 1.4 and boundary conditions given by eqs 1.5 and 1.6. The transport parameters for MIM with LTDD and ATDD models used in simulation are given in Table 1.1. Solute concentrations at the discrete times of 1000, 2000, 3000, 7000, and 10,000 min and at discrete distances from the inlet boundary 40, 80, 120, 160, 200, 240, 280, 320, 360, and 400 cm were generated by solving eqs 1.1 and 1.2. These generated solute concentrations were used as observed data to estimate the unknown transport parameters.

TABLE 1.1 Input Parameters Used for Generating Hypothetical Solute Concentration Data.

Parameter	L	D_0	**Asymptotic time-dependent coefficient (K_A)**	θ_m	θ_{im}
Value	400 cm	4.51 cm^2/min	200 min	0.33	0.42
Parameter	f	Kd_m	Kd_{im}	ω	ρ_b
Value	0.44	0.00134 mL/g	0.0121 mL/g	2.12E-05 min^{-1}	1.5212 g/cm^3
Parameter	q	v_m	t_{max}	t_{pulse}	Linear time-dependent coefficient (K_L)
Value	0.04 cm/min	0.121 cm/min	10,080 min	2880 min	500 min

For testing the robustness of algorithm, the number of unknown parameters to be determined were varied from one to four. Initial guess value of individual parameters was varied in the orders of 10 with respect to observed value to check the efficiency of the optimization algorithm. Performance of algorithm for MIM with LTDD and ATDD is discussed in detail in the subsequent sections.

1.6.1 MOBILE–IMMOBILE MODEL WITH LTDD

In this section, the solute transport parameters such as D_0, K_L, ω, Kd_m, and Kd_{im} were estimated from the hypothetical solute concentration. Uniqueness of inverse solution was tested by evaluating various scenarios with a different number of unknown parameters. The parameter estimation for MIM with LTDD model is described in the following subsections.

1.6.1.1 CASE 1: ESTIMATION OF ONE UNKNOWN PARAMETER

For this case, only one unknown parameter was estimated with other parameters kept constant as their corresponding observed value used for generating hypothetical solute concentration. Furthermore, two subcases were considered. First and second subcases were represented as underestimation and overestimation, respectively. In underestimation, initial guess value was kept lesser than observed value by an order of magnitude while for overestimation, kept an order higher than observed value. This was done to investigate the influence of TDD function on parameter estimation over a large range of transport parameters. Table 1.2 presents initial guess and optimal estimated values of the transport parameters D_0, K_L, ω, Kd_m, and Kd_{im}. It was observed that the algorithm successfully converged to the observed value for three parameters (D_0, K_L, and ω) in both the subcases (underestimate and overestimate). However, while estimating Kd_m and Kd_{im} in underestimation case, a significant deviation from the observed values was noticed. Number of iterations for convergence to the optimal solution is less in the underestimation subcase as compared with overestimation. Estimation of sorption distribution coefficient in mobile and immobile regions for MIM with LTDD model is less accurate in comparison with other transport parameters as described in Table 1.2.

TABLE 1.2 Parameter Estimates for Linear Time-Dependent Dispersion Model (MIML)—One Unknown Parameter.

Parameter	True value	Underestimated			Overestimated		
		Initial guess	Estimated value	NOI	Initial guess	Estimated value	NOI
D_0(cm^2/min)	4.51	0.451	4.5086	10	45.1	4.5112	14
K_L(min)	500	50	499.9994	12	5000	500.1235	14
ω (min^{-1})	2.12E-05	2.12E-06	2.1136E-05	10	2.12E-04	2.1247E-05	14
Kd_m (mL/g)	0.00134	0.000134	0.0003011	3	0.0134	0.001404	11
Kd_{im}(mL/g)	0.0121	0.00121	0.001815	1	0.121	0.04956	4

NOI, number of iterations; MIML, mobile immobile with linear dispersivity

1.6.1.2 CASE 2: ESTIMATION OF TWO UNKNOWN PARAMETERS

Case 2 presents the estimation of two unknown parameters with other parameters kept constant as the corresponding observed values. Therefore, 10 combinations of two unknown parameters (D_0, K_L), (D_0, K_{d_m}), (D_0, $K_{d_{im}}$), (D_0, ω), (ω, K_L), (ω, K_{d_m}), (ω, $K_{d_{im}}$), (K_L, K_{d_m}), (K_L, $K_{d_{im}}$), and (K_{d_m}, $K_{d_{im}}$) were possible. Further, four subcases were considered for each of these combinations. In the first subcase (underestimation), initial guess values of both unknown parameters were kept lesser than observed value by one order of magnitude. However, for the second subcase (overestimation), both unknown parameters were kept higher than the observed value by an order of magnitude. In the third subcase, first unknown parameter was kept lesser while second unknown parameter was kept higher than the observed value by an order of magnitude. Whereas, in the fourth subcase, first unknown parameter was kept higher and second unknown parameter was kept lesser than observed value by an order of magnitude. Table 1.3 shows the initial guess and optimal estimated value for all 10 combinations. This study observed that optimization yielded unique values of transport parameters (D_0, $K_{d_{im}}$) for combination in all the four subcases. However, optimization resulted in nonunique estimates of transport parameters for some of the subcases of remaining nine combinations.

The effectiveness of inverse problem was evaluated by plotting objective function against a pair of optimized parameters. Each of the contour was obtained by solving the solute transport equation with appropriate initial and boundary conditions. Convexity of the objective function (ϕ) in the parametric space was studied for all 10 combinations. In this section, contours of the objective function in (D_0, $K_{d_{im}}$) and (D_0, K_L) parametric spaces for MIM with LTDD model are shown.

Objective function was convex in nature for (D_0, $K_{d_{im}}$) parametric space as shown in Figure 1.3. Figure 1.3 shows that there is only one global minimum at observed values in (D_0, $K_{d_{im}}$) parametric space. As only two parameters (D_0, $K_{d_{im}}$) appear within a single contour plot analysis, the behavior of objective function in various parametric planes is an indication of the uniqueness of solution in these planes only and not in the full parametric space. Local minima may exist elsewhere in the full parametric space but does not appear in the selected parametric space. Figure 1.4 shows that the objective function has many local minima in (D_0, K_L) parametric space. As a result, the optimal estimation deviates significantly from the observed value.

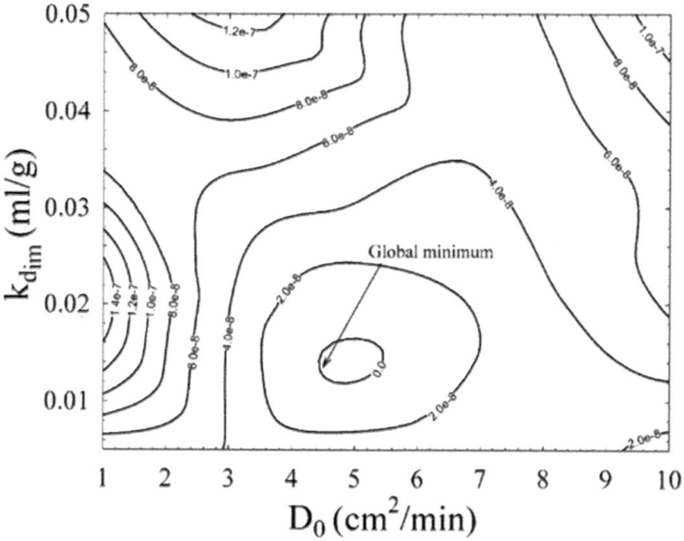

FIGURE 1.3 Contour of objective function in D_0 and K_{dim} parametric space for linear time-dependent dispersion (MIML) model.

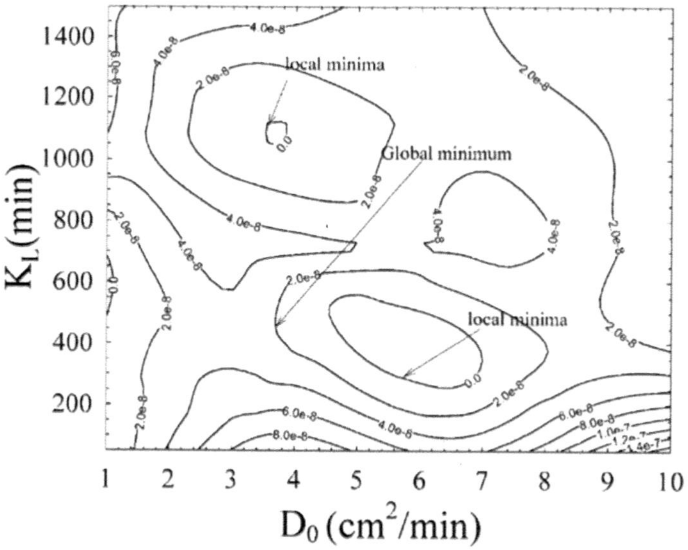

FIGURE 1.4 Contours of objective function in D_0 and K_L parametric space for MIML model.

Table 1.3 and contour plots of objective function in various parametric spaces show that incorporation of linear TDD function along with sorption

TABLE 1.3 Parameter Estimates for MIML Model–Two Unknown Parameters.

Parameter	True value	Underestimated			Overestimated			Mixed- I			Mixed- II		
		Initial guess	Estimated value	NOI	Initial guess	Estimated value	NOI	Initial guess	Estimated value	NOI	Initial guess	Estimated value	NOI
D_0(cm^2/min)	4.51	0.451	4.5125	16	45.1	4.5122	34	0.451	4.5109	20	45.1	4.5118	18
Kd_{im}(mL/g)	0.0121	0.00121	0.01415		0.121	0.01809		0.121	0.01724		0.00121	0.01064	
Unidentified parameters													
D_0 (cm^2/min)	4.51	0.451	4.5083	18	45.1	19.4763	31	0.451	4.5126	15	45.1	4.5098	18
ω (min^{-1})	2.12E-05	2.12E-06	2.1129E-05		2.12E-04	4.6606E-02		2.12E-04	2.1261E-05		2.12E-06	2.1207E-05	
D_0 (cm^2/min)	4.51	0.451	0.451	1	45.1	45.0997	1	0.451	17.6277	16	45.1	6.5382	16
K_L (min)	500	50	49.9999		5000	5000.026		5000	1954.1400		50	724.7598	
ω (min^{-1})	2.12E-05	2.12E-06	2.1271E-05	18	2.12E-04	2.1196E-05	17	2.12E-06	2.1188E-05	21	2.12E-04	6.6745E-02	33
K_L (min)	500	50	499.5565		5000	499.9392		5000	499.9104		50	115.6401	
ω (min^{-1})	2.12E-05	2.12E-06	2.1389E-05	13	2.12E-04	2.1234E-05	18	2.12E-06	2.1145E-05	16	2.12E-04	2.1546E-05	18
Kd_m (mL/g)	1.34E-03	1.34E-04	7.8623E-04		1.34E-02	1.3272E-03		1.34E-02	1.4744E-03		1.34E-04	6.7899E-04	
K_L (min)	500	50	502.7679	11	5000	499.822	15	50	496.7301	25	5000	503.557	18
Kd_m (mL/g)	1.34E-03	1.34E-04	1.1513E-04		1.34E-02	1.3710E-03		1.34E-02	2.7793E-03		1.34E-04	1.6191E-05	
Kd_m (mL/g)	1.34E-03	1.34E-04	3.0150E-04	3	1.34E-02	1.3131E-03	11	1.34E-04	3.0149E-04	3	1.34E-02	1.3256E-03	11
Kd_{im} (mL/g)	0.0121	0.00121	0.002723		0.121	0.01039		0.121	0.07744		0.00121	0.009856	
D_0 (cm^2/min)	4.51	0.451	4.4893	15	45.1	4.5138	31	0.451	4.4944	18	45.1	4.5133	11
Kd_{im} (mL/g)	1.34E-03	1.34E-04	4.6983E-04		1.34E-02	1.4378E-03		1.34E-02	5.7992E-04		1.34E-04	1.4879E-03	
ω (min^{-1})	2.12E-05	2.12E-06	2.1221E-05	12	2.12E-04	2.1257E-05	16	2.12E-06	2.1222E-05	18	2.12E-04	2.1245E-05	16
Kd_{im} (mL/g)	0.0121	0.00121	0.000292		0.121	0.01089		0.121	0.02694		0.00121	0.0004091	
K_L (min)	500	50	500.1185	10	5000	500.2505	18	50	499.8430	22	5000	500.1122	18
Kd_{im} (mL/g)	0.0121	0.00121	0.01245		0.121	0.02982		0.121	0.02069		0.00121	0.008046	

distribution coefficient in the mobile and immobile regions predominantly makes the inverse problem unidentifiable, resulting in nonunique estimates. Performance of LTDD model was also found unsatisfactory even for the estimation of two unknown parameters. Therefore, estimation of three unknown parameters was not carried out for LTDD model.

1.6.2 MIM MODEL WITH ATDD

In this section, solute transport parameters of asymptotic dispersion model (D_0, ω, K_A, K_{d_m} and $K_{d_{im}}$) were estimated from the hypothetical solute concentrations. Efficiency and robustness of optimization algorithm for ATDD dispersion model was tested by considering different scenarios which are discussed below.

1.6.2.1 CASE 1: ESTIMATION OF ONE UNKNOWN PARAMETER

In this case, only one unknown parameter was estimated using optimization algorithm. Other parameters were kept constant to corresponding observed values used for generating hypothetical solute concentrations. Two subcases were considered further, which were listed as underestimation and overestimation in the same way as described in Section 6.1.1. Table 1.4 presents the observed values of transport parameters (D_0, K_A, ω, K_{d_m}, and $K_{d_{im}}$) used to generate the hypothetical solute concentrations, initial guess for underestimation and overestimation subcases, estimated value of the parameters at convergence and corresponding number of iterations.

TABLE 1.4 Parameter Estimates for Asymptotic Time-Dependent Dispersion Model (MIMA)–One Unknown Parameter.

Parameter	True value	Underestimated			Overestimated		
		Initial guess	Estimated value	NOI	Initial guess	Estimated value	NOI
D_0(cm^2/min)	4.51	0.451	4.5071	9	45.1	4.5129	15
K_A(min)	200	20	199.3229	8	2000	200.9150	13
ω (min^{-1})	2.12E-05	2.12E-06	2.1183E-05	8	2.12E-04	2.1249E-05	14
K_{d_m}(mL/g)	0.00134	0.000134	0.000452	3	0.0134	0.001575	10
$K_{d_{im}}$ (mL/g)	0.0121	0.00121	0.00182	1	0.121	0.0203	8

MIMA, mobile immobile with asymptomatic dispersivity

Results show that the algorithm converges to observed value for three parameters (D_0, K_A, and ω) with good accuracy for both the subcases. Also, optimization algorithm resulted in a nonunique estimate of K_{d_m} and $K_{d_{im}}$ in underestimation subcase. Number of iterations for convergence are lesser for the underestimation subcase as compared with overestimation as shown in Table 1.4.

1.6.2.2 CASE 2: ESTIMATION OF TWO UNKNOWN PARAMETERS

This case presents the estimation of two unknown parameters and treats other three parameters as constant to observed values used for generating the hypothetical data. In this estimation procedure, 10 combinations of two unknown parameters (D_0, K_A), (D_0, K_{d_m}), (D_0, $K_{d_{im}}$), (D_0, ω), (ω, K_A), (ω, K_{d_m}), (ω, $K_{d_{im}}$), (K_A, K_{d_m}), (K_A, $K_{d_{im}}$), and (K_{d_m}, $K_{d_{im}}$) were considered. For each of these combinations, four subcases were considered in the same way as described in Section 6.1.2. Table 1.5 presents the initial guess, observed value, optimal estimated value at convergence, and corresponding number of iterations. However, optimization resulted in nonunique estimates of transport parameters for three combinations (D_0, K_{d_m}), (ω, $K_{d_{im}}$), and (K_A, $K_{d_{im}}$). It is also observed from Table 1.5 that the optimization algorithm converged to observed values in a fewer number of iterations when both the unknown parameters are underestimated in most of the combination cases. Contour plots showed the sensitivity of the objective functions with respect to change in two parameter values while other transport parameters were held constant at observed value. Contour plots of objective function in the parametric spaces (D_0, K_A), (D_0, K_{d_m}), (D_0, ω), (ω, K_A), (ω, $K_{d_{im}}$), (K_A, $K_{d_{im}}$) are shown below.

Contour plot of objective function in D_0-K_A parametric space is shown in Figure 1.5. Value of D_0 was varied from 3.0 to 15.0 cm^2/min while asymptotic TDD coefficient (K_A) was varied from 50 to 1200 min. Minimum value of the objective function (ϕ = 4.12E-10) was obtained at observed values of transport parameters given as: D_0 = 4.51 cm^2/min and K_A = 200 min. It is observed that there was only one global minimum at the observed values of parameters in (D_0, K_A) parametric space. Figure 1.5 shows that the objective function is more sensitive to K_A than D_0 and the movement toward higher values of K_A from the observed value results in a steep variation in the objective function. Figure 1.6 provides the contour plot of objective function in (D_0, ω) parametric space. Value of D_0 was varied from 1.0 to 10.0 cm^2/min while mass transfer coefficient (ω) was varied from 5E-06 to 3.5E-05 min^{-1}. The minima of objective function (ϕ = 4.08E-10) was obtained at observed value of parameters given as: $D_0 = 4.51$ cm^2/min and mass transfer coefficient (ω) = 2.12E-05 min^{-1} in (D_0, ω) parametric space. Only one global minimum of objective function was obtained in (D_0, ω) parametric space as shown in Figure 1.6. During optimization, it was observed that for (D_0, ω) and (D_0, K_A) combination, the optimization resulted in unique solution.

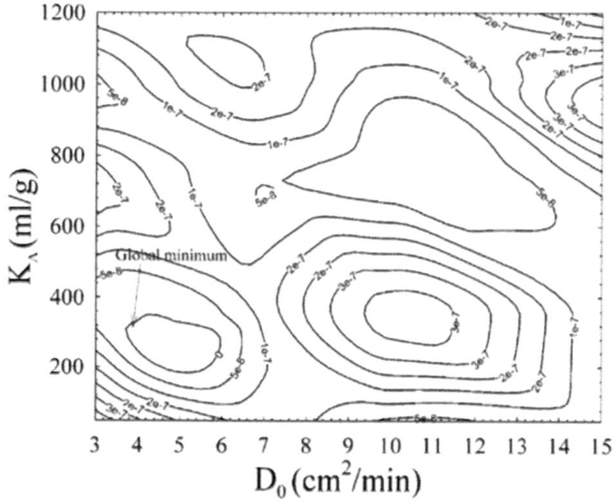

FIGURE 1.5 Contour of objective function in D_0 and K_A parametric space for asymptotic time-dependent dispersion (MIMA) model.

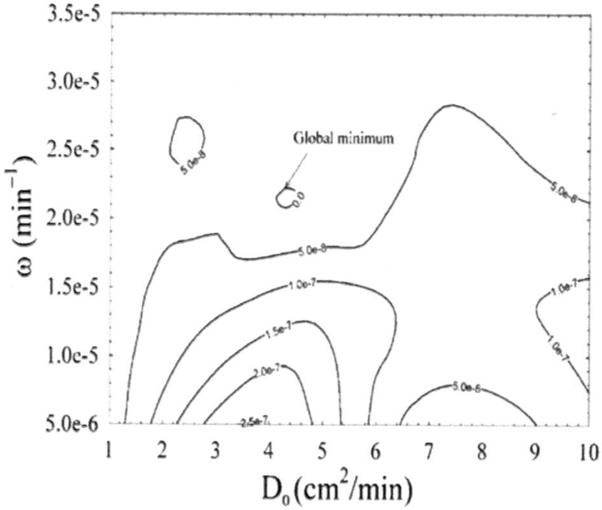

FIGURE 1.6 Contour of objective function in D_0 and ω parametric space for MIMA model.

TABLE 1.5 Parameter Estimates for MIMA Model—Two Unknown Parameters.

Parameter	True value	Underestimated			Overestimated			Mixed- I			Mixed- II		
		Initial guess	Estimated value	NOI	Initial guess	Estimated value	NOI	Initial guess	Estimated value	NOI	Initial guess	Estimated value	NOI
D_0 (cm^2/min)	4.51	0.451	4.5079	15	45.1	4.5083	18	0.451	4.5100	18	45.1	4.5092	19
ω (min^{-1})	2.12E-05	2.12E-06	2.1153E-05		2.12E-04	2.1196E-05		2.12E-04	2.1212E-05		2.12E-06	2.1208E-05	
D_0 (cm^2/min)	4.51	0.451	4.5119	17	45.1	4.5115	22	0.451	4.5208	16	45.1	4.5208	26
K_A (min)	200	20	200.3716		2000	200.6989		2000	202.1312		20	203.1350	
D_0 (cm^2/min)	4.51	0.451	4.5087	13	45.1	4.5091	23	0.451	4.5088	13	45.1	4.5118	18
$K_{d_{im}}$ (mL/g)	0.0121	0.00121	0.01176		0.121	0.01252		0.121	0.01247		0.00121	0.01216	
ω (min^{-1})	2.12E-05	2.12E-06	2.1201E-05	17	2.12E-04	2.1190E-05	15	2.12E-06	2.1187E-05	17	2.12E-04	2.1216E-05	18
K_A (min)	200	20	200.0900		2000	199.8886		2000	199.5896		20	199.9260	
ω (min^{-1})	2.12E-05	2.12E-06	2.1214E-05	13	2.12E-04	2.1192E-05	15	2.12E-06	2.1202E-05	15	2.12E-04	2.1187E-05	20
K_{d_m} (mL/g)	1.34E-03	1.34E-04	1.3022E-03		1.34E-02	1.3427E-03		1.34E-02	1.3307E-03		1.34E-04	1.3275E-03	
K_A (min)	200	20	200.13	9	2000	199.472	15	20	199.9156	16	2000	199.6299	25
K_{d_m} (mL/g)	1.34E-03	1.34E-04	1.3272E-03		1.34E-02	1.3864E-03		1.34E-02	1.3394E-03		1.34E-04	1.3872E-03	
K_{d_m} (mL/g)	1.34E-03	1.34E-04	1.3389E-03	9	1.34E-02	1.3412E-03	12	1.34E-04	1.2849E-03	11	1.34E-02	1.3456E-03	12
$K_{d_{im}}$ (mL/g)	0.0121	0.00121	0.01255		0.121	0.01224		0.121	0.01162		0.00121	0.01303	
Unidentified parameters													
D_0 (cm^2/min)	4.51	0.451	4.4778	11	45.1	4.5101	28	0.451	4.5097	14	45.1	4.5091	24
K_{d_m} (mL/g)	1.34E-03	1.34E-04	1.0436E-04		1.34E-02	1.3267E-03		1.34E-02	1.3559E-03		1.34E-04	1.2942E-03	
ω (min^{-1})	2.12E-05	2.12E-06	2.12E-05	13	2.12E-04	2.1185E-05	17	2.12E-06	2.1198E-05	13	2.12E-04	2.1188E-05	17
$K_{d_{im}}$ (mL/g)	0.0121	0.00121	0.01233		0.121	0.01393		0.121	0.01642		0.00121	0.00149	
K_A (min)	200	20	199.8479	13	2000	200.4118	16	20	200.1376	14	2000	200.6048	16
$K_{d_{im}}$ (mL/g)	0.0121	0.00121	0.01204		0.121	0.01246		0.121	0.01249		0.00121	0.00255	

Convexity of the objective function in the (D_0, K_{d_m}) parametric space is studied as shown in Figure 1.7. There are many local minima in (D_0, K_{d_m}) parametric space as shown in Figure 1.7 which indicates that the optimal estimation of algorithm depends upon the initial guess values. The response contour plot of objective function is quite flat in (D_0, K_{d_m}) parametric space in comparison to other parametric spaces, suggesting that the objective function is less sensitive to D_0.

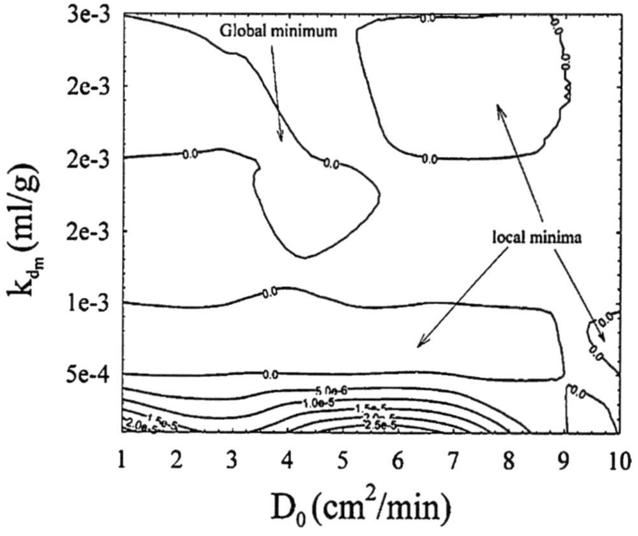

FIGURE 1.7 Contour of objective function in D_0 and K_{d_m} parametric space for MIMA model.

Figure 1.8 shows the contour of the objective function in the parametric space (ω, $K_{d_{im}}$). Value of ω was varied from 5E-06 to 4.5E-05 min^{-1}while sorption distribution coefficient of immobile region ($K_{d_{im}}$) was varied from 0.005 to 0.025 mL/g. Figure 1.8 shows more than one local minimum in contour plot of objective function in (ω, $K_{d_{im}}$) parametric space. Thus, it can be concluded that the presence of sorption distribution coefficient of immobile region predominantly results in nonunique estimates for (ω, $K_{d_{im}}$) combination. Finally, contour plot of objective function in (K_A, $K_{d_{im}}$) parametric space is shown in Figure 1.9. It clearly shows that the objective function has many local minima in the parametric space (K_A, $K_{d_{im}}$). Values of local minima are spread over a broad range around true value of parameters (K_A, $K_{d_{im}}$). Therefore, the optimal estimate deviated considerably from its observed value.

Also, systematic assessment of sensitivity of objective function toward solute transport parameters is carried out effectively using contour plots of objective function in various parametric spaces.

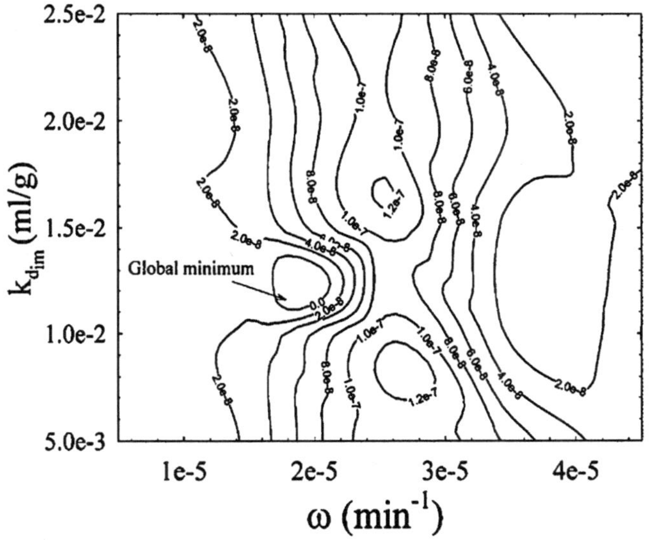

FIGURE 1.8 Contour of objective function in ω and Kd_{im} parametric space for MIMA model.

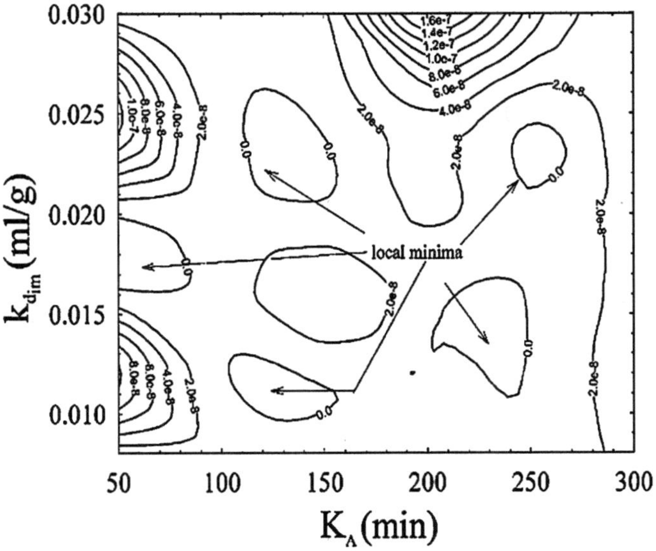

FIGURE 1.9 Contour of objective function in K_A and Kd_{im} parametric space for MIMA model.

1.6.2.3 CASE 3: ESTIMATION OF THREE UNKNOWN PARAMETERS

This case presents the simultaneous estimation of three unknown parameters with other parameters equal to observed value used to generate hypothetical solute concentrations. Ten combinations of parameters were considered (c), (D_0, K_{d_m}, $K_{d_{im}}$), (D_0, ω, K_{d_m}), (ω, K_A, K_{d_m}), (D_0, ω, $K_{d_{im}}$), (ω, K_A, $K_{d_{im}}$), (D_0, K_A, K_{d_m}), (D_0, K_A, $K_{d_{im}}$), (ω, K_{d_m}, $K_{d_{im}}$), and (K_A, K_{d_m}, $K_{d_{im}}$). Three subcases were considered further, which are listed as underestimation, overestimation, and mixed-I. In the first subcase (underestimation), initial guess was kept lesser than the observed value by one order while for the second subcase (overestimation), higher than observed value by an order. In the third subcase (mixed-I), either two unknown parameters were underestimated while the third unknown parameter was overestimated by one order or two unknown parameters were overestimated while the third unknown parameter was underestimated. Table 1.6 presents the initial guess, observed values, optimal estimated parameters at convergence and the corresponding number of iterations for all the subcases of 10 combinations.

It is observed that the optimization converges to observed value of transport parameters for six combinations. However, optimization resulted in nonunique estimates of transport parameters for four particular combinations (D_0, K_{d_m}, $K_{d_{im}}$), (D_0, ω, K_{d_m}), (D_0, ω, $K_{d_{im}}$), and (D_0, K_A, $K_{d_{im}}$). Optimization algorithm converges to true value in a lesser number of iterations when all unknown parameters are underestimated in most of the combinations. Nonunique estimate cases are observed when (D_0, K_{d_m}) is considered with any other parameter (K_A ω *or* $K_{d_{im}}$). From these observations, it can be said that the presence of equilibrium sorption coefficient of the mobile and immobile leads to nonunique estimate for MIM with ATDD model. It is observed from Tables 1.3 to 1.5 that the LTDD model gives a large number of nonunique parameter combinations as compared with ATDD model. It is observed that the contour plots of objective function in all parametric planes facilitated the sensitivity analysis and identifiability issues related to optimization algorithm. It can be concluded that the prediction of observed solute concentrations using ATDD model are more accurate than using LTDD model.

1.7 ESTIMATION OF TRANSPORT PARAMETERS USING OBSERVED EXPERIMENT DATA OF SOLUTE TRANSPORT

In this section, we demonstrate the Levenberg-Marquardt algorithm to estimate the unknown transport parameters for the experiment conducted

TABLE 1.6 Parameter Estimates for MIMA Model—Three Unknown Parameters.

Parameter	True value	Underestimated			Overestimated			Mixed-I		
		Initial guess	Estimated value	NOI	Initial guess	Estimated value	NOI	Initial guess	Estimated value	NOI
D_0 (cm^2/min)	4.51	0.451	4.4896	41	45.1	4.5177	30	0.451	4.4903	39
ω (min^{-1})	2.12E-05	2.12E-06	2.1239E-05		2.12E-04	2.1179E-05		2.12E-04	2.1277E-05	
K_A (min)	200	20	196.5495		2000	200.9367		20	196.1278	
ω (min^{-1})	2.12E-05	2.12E-06	2.1166E-05	25	2.12E-04	2.1197E-05	16	2.12E-06	2.1153E-05	26
K_A (min)	200	20	198.0171		2000	200.0121		20	197.7412	
$K_{d_{im}}$ (mL/g)	1.34E-03	1.34E-04	1.5847E-03		1.34E-02	1.3049E-03		1.34E-02	1.6612E-03	
ω (min^{-1})	2.12E-05	2.12E-06	2.1229E-05	17	2.12E-04	2.1198E-05	15	2.12E-06	2.1203E-05	17
K_A (min)	200	20	199.6613		2000	199.9592		20	200.1811	
$K_{d_{im}}$ (mL/g)	0.0121	0.00121	0.01224		0.121	0.01256		0.121	0.01234	
D_0 (cm^2/min)	4.51	0.451	4.5101	16	45.1	4.5086	27	45.1	4.5189	48
K_A (min)	200	20	199.9968		2000	199.578		20	201.9674	
K_{d_m} (mL/g)	0.0121	0.00121	0.01218		0.121	0.01173		0.121	0.01195	
ω (min^{-1})	2.12E-05	2.12E-06	2.1207E-05	16	2.12E-04	2.1179E-05	17	2.12E-06	2.1207E-05	19
K_{d_m} (mL/g)	1.34E-03	1.34E-04	1.3431E-03		1.34E-02	1.3451E-03		1.34E-02	1.3773E-03	
$K_{d_{im}}$ (mL/g)	0.0121	0.00121	0.01234		0.121	0.01289		0.121	0.01241	
K_A (min)	200	20	200.0694	12	2000	199.457	15	20	199.3502	14
K_{d_m} (mL/g)	1.34E-03	1.34E-04	1.3294E-03		1.34E-02	1.3917E-03		1.34E-02	1.4197E-03	
$K_{d_{im}}$ (mL/g)	0.0121	0.00121	0.010001		0.121	0.01285		0.121	0.01326	

TABLE 1.6 *(Continued)*

Parameter	True value	Underestimated			Overestimated			Mixed-I		
		Initial guess	Estimated value	NOI	Initial guess	Estimated value	NOI	Initial guess	Estimated value	NOI
Unidentifiable parameters										
D_0 (cm^2/min)	4.51	0.451	4.5113	25	45.1	4.5109	28	0.451	4.5079	23
K_{d_m} (mL/g)	1.34E-03	1.34E-04	1.3647E-03		1.34E-02	1.3232E-03		1.34E-04	1.2622E-03	
$K_{d_{im}}$ (mL/g)	0.0121	0.00121	0.01208		0.121	0.01296		0.121	0.01239	
D_0 (cm^2/min)	4.51	0.451	4.4914	20	45.1	4.51788	32	0.451	4.5232	22
ω (min^{-1})	2.12E-05	2.12E-06	2.1398E-05		2.12E-04	2.1068E-05		2.12E-06	2.1062E-05	
K_{d_m} (mL/g)	1.34E-03	1.34E-04	7.7181E-04		1.34E-02	1.6944E-03		1.34E-02	1.7119E-03	
D_0 (cm^2/min)	4.51	0.451	4.5096	19	45.1	4.5085	18	0.451	4.5093	16
ω (min^{-1})	2.12E-05	2.12E-06	2.1196E-05		2.12E-04	2.1199E-05		2.12E-06	2.1198E-05	
K_{d_m} (mL/g)	0.0121	0.00121	0.01209		0.121	0.007615		0.121	0.0122	
D_0 (cm^2/min)	4.51	0.451	4.5103	21	45.1	4.5159	25	0.451	4.5092	24
K_A (min)	200	20	201.0103		2000	203.1192		20	199.0811	
K_{d_m} (mL/g)	1.34E-03	1.34E-04	1.1933E-03		1.34E-02	1.0273E-03		1.34E-02	1.4307E-03	

by Swami et al. (2016b). They conducted five experiments at various flow rates in a stratified porous media with reactive and conservative solutes. Experiments were conducted in a rectangular tank of dimensions 400 × 100 cm. In this case study, chloride BTC's observed at various down-gradient distances (400, 300, 200, 150, 100 and 50 cm) for natural soil-gravel stratification are simulated separately using MIM with LTDD and ATDD models. The known input parameters used for the simulation are $\rho_b = 1.5212$ gm/cm^3, $\theta_m = 0.33$, $\theta_{im} = 0.42$, $f = 0.44$, $q = 0.2$ cm/min, $v_m = 0.606$ cm/min, $L = 400$ cm and simulation time (t) = 3060 min. This study specified continuous source condition at up-gradient end of the rectangular tank. Optimized unknown parameters D_0, K_A, K_L, and ω for various down-gradient distances are summarized in Annexure I. Chloride has been selected in this study as a conservative tracer and the values of distribution coefficients of mobile and immobile regions are kept zero to simulate conservative solute transport. For LTDD, three parameters D_0, K_L, and ω were estimated while for asymptotic dispersion model D_0, K_A, and ω were estimated. Three subcases are considered for parameter estimation at each of the down-gradient distances from source. In the first and second subcases, initial guess of all the unknown parameters are kept small and large respectively. In the third subcase, either two unknown parameters are kept smaller while the third unknown parameter is kept larger or two unknown parameters are kept larger while the third unknown parameter is kept smaller. Error indices such as RMSE, r^2, and NSE are calculated for both the models. The parameter estimates corresponding to minimum RMSE, maximum NSE, and maximum r^2 value were finally considered as optimal estimates and were used to estimate chloride concentrations. The optimal estimate of parameters at 400 cm down-gradient distance from source for MIM with LTDD model are $D_0 = 16.23$ cm^2/min, $\omega = 9.38\text{E-}05$ min^{-1}, and $K_L = 281.89$ min. While for MIM with ATDD model $D_0 = 16.85$ cm^2/min, $\omega = 7.92\text{E-}05$ min^{-1}, and $K_A = 38.15$ min which are the optimal parameter estimates at 400 cm distance from the input. Figure 1.10 shows the experimentally observed and simulated chloride concentrations at the optimal parameter values for both the MIM with LTDD and ATDD models. Asymmetric breakthrough curves are obtained as shown in Figure 1.10 which clearly demonstrates that the porous medium system is influenced by the presence of physical nonequilibrium. Figure 1.10 suggested that the parameters estimated using inverse optimization algorithm for MIM with TDD match reasonably well with the observed concentration. It is observed that the TDD function captures peak of chloride concentration reasonably well. Figure 1.10(A) and 1.10(B) show the predicted BTC's at 400 and 300 cm distance from the input respectively for both the models. TDD model is simulating chloride concentration reasonably well at late times and simulation from LTDD and ATDD converges. However, ATDD is simulating initial breakthrough concentration with slightly higher accuracy as compared with LTDD model. Similar simulation capabilities are observed for 200 and 150 cm from the input as shown in Figure 1.10(C) and 1.10(D). Figure 1.10(F) shows the BTC at 50 cm from the input. Early initial breakthrough concentration and broader peak concentration are observed which is due to proximity of source concentration for longer duration of time. Figure 1.10 and Annexure I show that the parameter estimation algorithm coupled with MIM model combined with TDD is a powerful tool to simulate solute transport for a heterogeneous and complex porous media. Also, Annexure I shows that the nonfiction behavior of mass transfer. Also, the limitation of MIM with TDD is highlighted using log–log plot that it fails to capture early initial breakthrough concentration. Results suggested that the MIM with ATDD model performs better in capturing solute transport through porous media in comparison to LTDD model.

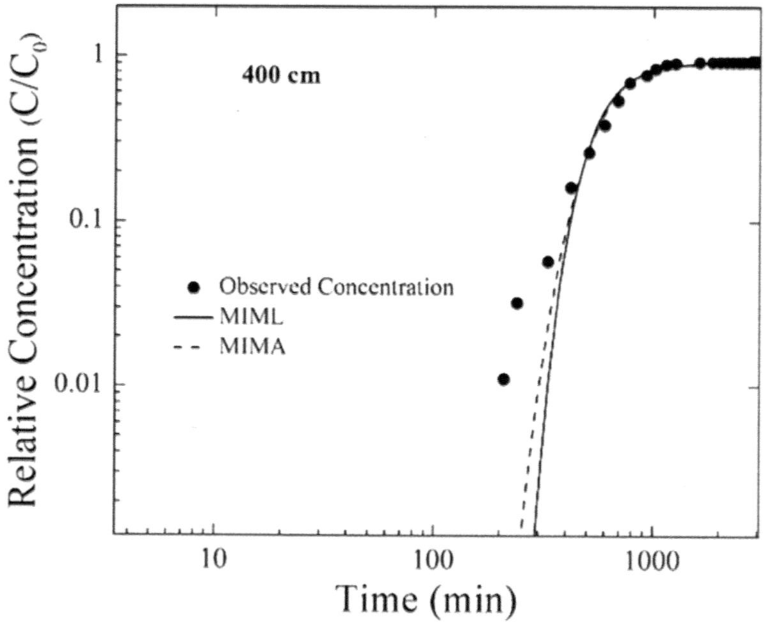

FIGURE 1.10(a) Breakthrough curve of solute transport at 400 cm down-gradient distance for $v_m = 0.606$ cm/min.

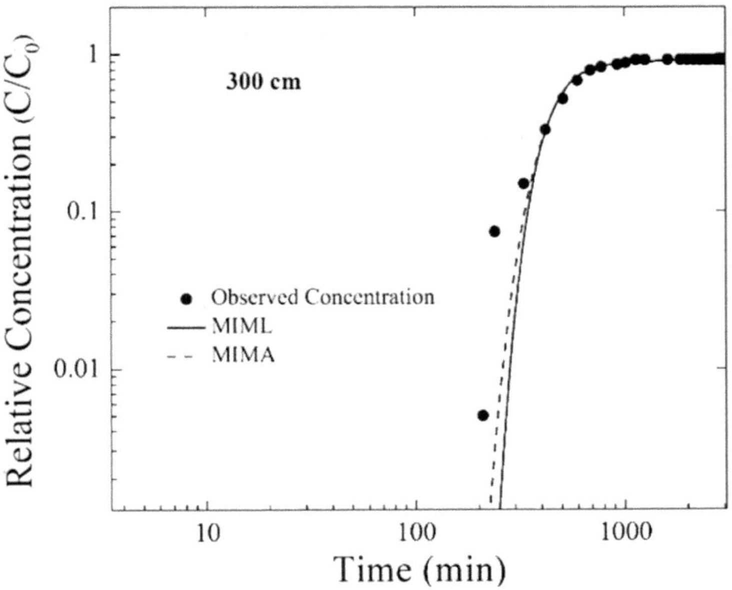

FIGURE 1.10(b) Breakthrough curve of solute transport at 400 cm down -gradient distance for v_m = 0.606 cm/min.

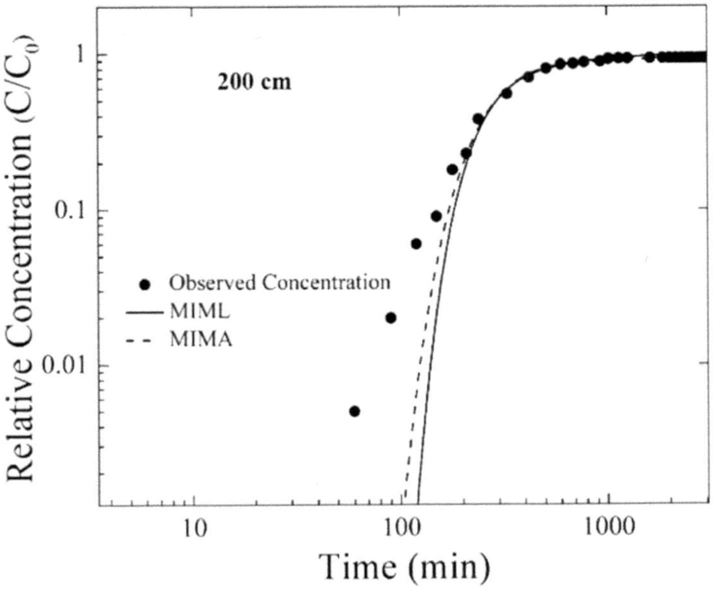

FIGURE 1.10(c) Breakthrough curve of solute transport at 200 cm down-gradient distance for v_m = 0.606 cm/min.

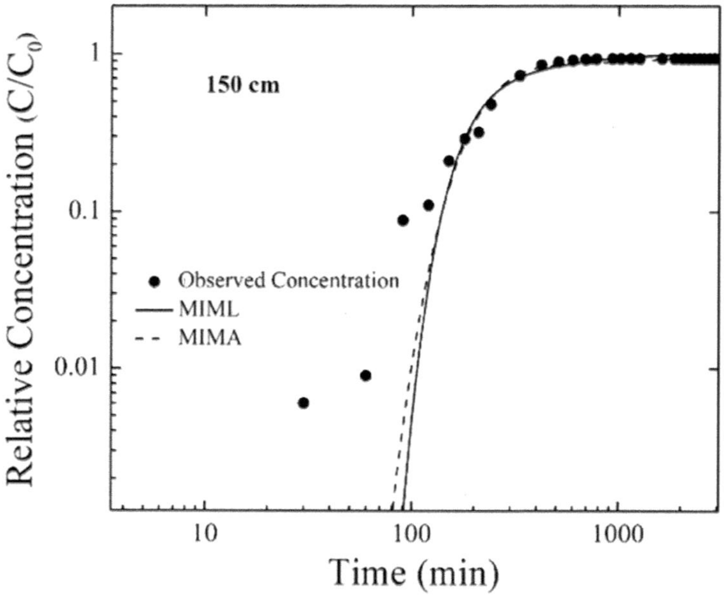

FIGURE 1.10(d) Breakthrough curve of solute transport at 150 cm down-gradient distance for $v_m = 0.606$ cm/min.

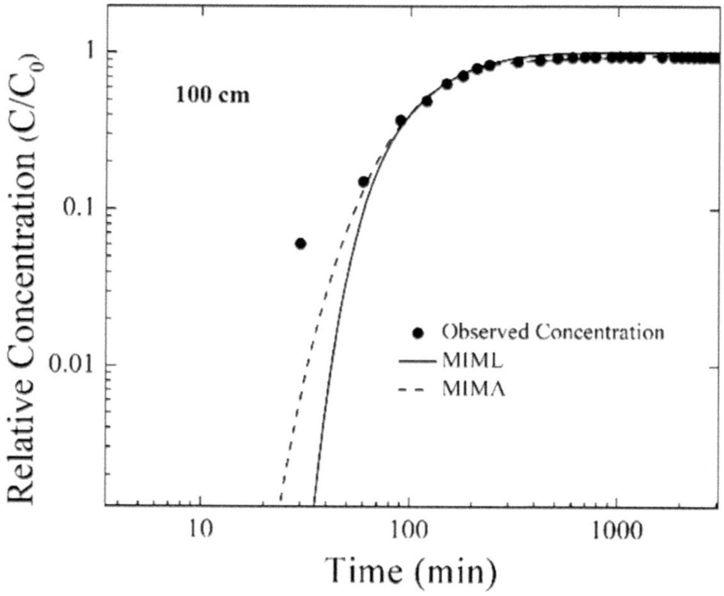

FIGURE 1.10(e) Breakthrough curve of solute transport at 100 cm down-gradient distance for $v_m = 0.606$ cm/min.

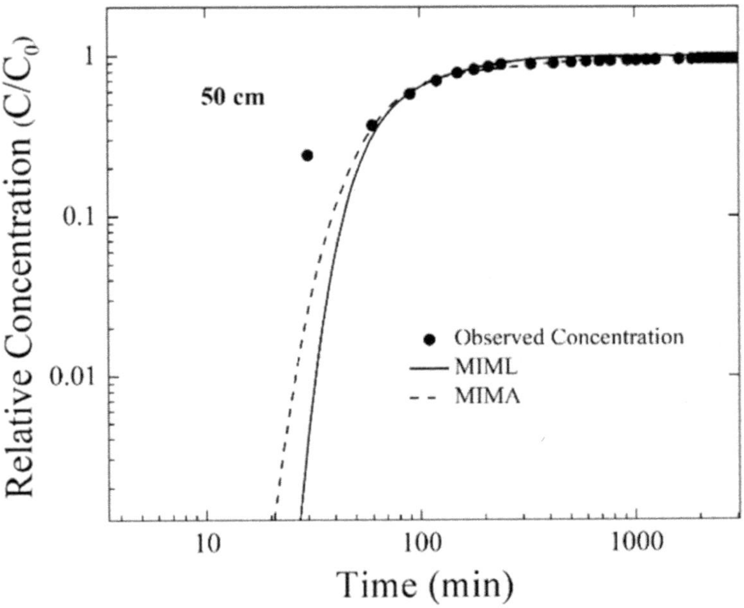

FIGURE 1.10(f) Breakthrough curve of solute transport at 50 cm down-gradient distance for v_m = 0.606 cm/min.

1.8 CONCLUSIONS

The major findings and contributions of the study are summarized as under:

- It is impossible to simultaneously estimate all the five unknown transport parameters uniquely with optimization procedure. Parameter estimation algorithm is limited to simultaneous unique estimation of three unknown parameters. For the unique solution of equilibrium sorption coefficient of mobile and immobile region, at least one parameter should be estimated experimentally.
- LTDD along with sorption distribution coefficient predominantly makes the inverse problem unidentifiable, resulting in nonunique estimates. However, in the case of ATDD model, the presence of equilibrium sorption coefficient of the mobile and immobile region makes the inverse problem unidentifiable.
- It is found that the model prediction with optimal parameter estimates matches reasonably well with experimental data. It is seen that when MIM with ATDD model is used to simulate solute concentration data,

a much better fit to the observed data is obtained in comparison to LTDD model.

KEYWORDS

- **solute transport**
- **coupled stratified porous media**
- **mobile-immobile model**
- **time-dependent dispersion**
- **parameter estimation**

REFERENCES

Aquino, T.; Aubeneau, A.; Bolster, D. Peak and Tail Scaling of Breakthrough Curves in Hydrologic Tracer Tests. *Adv. Water Resour.* **2015,** *78,* 1–8.

Barry, D. A.; Sposito, G. Analytical Solution of a Convection-Dispersion Model with Time-Dependent Transport Coefficients. *Water Resour. Res.* **1989,** *25,* 2407–2416.

Barth, G.; Hill, M. C. Numerical Methods for Improving Sensitivity Analysis and Parameter Estimation of Virus Transport Simulated Using Sorptive–Reactive Processes. *J. Contam. Hydrol.* **2005a,** *76,* 251–277.

Barth, G. R.; Hill, M. C. Parameter and Observation Importance in Modelling Virus Transport in Saturated Porous Media—Investigations in a Homogenous System. *J. Contam. Hydrol.* **2005b,** *80,* 107–129.

Basha, H. A.; El-Habel, F. S. Analytical Solution of the One-Dimensional Time-Dependent Transport Equation. *Water Resour. Res.* **1993,** *29,* 3209–3214.

Brusseau, M. L. Application of a Multiprocess Nonequilibrium Sorption Model to Solute Transport in a Stratified Porous Medium. *Water Resour. Res.* **1991,** *27,* 589–595.

Brusseau, M. L.; Hu, Q.; Srivastava, R. Using Flow Interruption to Identify Factors Causing Nonideal Contaminant Transport. *J. Contam. Hydrol.* **1997,** *24,* 205–219.

Brusseau, M. L.; Jessup, R. E.; Rao, P. S. C. Modeling the Transport of Solutes Influenced by Multiprocess Nonequilibrium. *Water Resour. Res.* **1989,** *25,* 1971–1988.

Cobb, P. M.; McElwee, C. D.; Butt, M. A. Analysis of Leaky Aquifer Pumping Test Data: An Automated Numerical Solution Using Sensitivity Analysis. *Groundwater* **1982,** *20,* 325–333.

Cortis, A.; Berkowitz, B. Anomalous Transport in 'Classical' Soil and Sand Columns. *Soil Sci. Am. J.* **2004,** *68,* 1539–1548.

Cortis, A.; Berkowitz, B. Computing 'Anomalous' Contaminant Transport in Porous Media: The CTRW MATLAB Toolbox. *Groundwater* **2005,** *43,* 947–950.

Dagan, G. Time-Dependent Macrodispersion for Solute Transport in Anisotropic Heterogeneous Aquifers. *Water Res. Res.* **1988,** *24,* 1491–1500.

Dentz, M.; Kang, P. K.; Le Borgne, T. Continuous Time Random Walks for Non-Local Radial Solute Transport. *Adv. Water Resour.* **2015,** *82*, 16–26.

Gao, G.; Feng, S.; Zhan, H.; Huang, G.; Mao, X. Evaluation of Anomalous Solute Transport in a Large Heterogeneous Soil Column with Mobile-Immobile Model. *J. Hydrol. Eng.* **2009,** *14*, 966–974.

Gao, G.; Zhan, H.; Feng, S.; Fu, B.; Ma, Y.; Huang, G. A New Mobile-Immobile Model for Reactive Solute Transport with Scale-Dependent Dispersion. *Water Resour. Res.* **2010,** *46*, 1–16.

Gelhar, L. W. *Stochastic Subsurface Hydrology*; Prentice Hall, 1993.

Gelhar, L. W.; Gutjahr, A. L.; Naff, R. L. Stochastic Analysis of Macrodispersion in a Stratified Aquifer. *Water Resour. Res.* **1979,** *15*, 1387–1397.

Gelhar, L. W.; Welty, C.; Rehfeldt, K. R. A Critical Review of Data on Field-Scale Dispersion in Aquifers. *Water Resour. Res.* **1992,** *28*, 1955–1974.

Van Genuchten, M. T.; Wierenga, P. J. Mass Transfer Studies in Sorbing Porous Media I. Analytical Solutions. *Soil Sci. Am. J.* **1976,** *40*, 473–480.

Van Genuchten, M. T.; Wierenga, P. J. Mass Transfer Studies in Sorbing Porous Media: II. Experimental Evaluation with Tritium. *Soil Sci. Am. J.* **1977,** *41*, 272–278.

Guleria, A., Swami, D., Sharma, A. et al. Non-reactive solute transport modelling with time-dependent dispersion through stratified porous media. *Sādhanā* **44,** 81 (2019). https://doi.org/10.1007/s12046-019-1056-6

Guérin, T.; Dean, D. S. Universal TDD Properties for Diffusion in a One-Dimensional Critically Tilted Potential. *Phys. Rev. E, APS* **2017,** *95*, 12109.

Haggerty, R.; Gorelick, S. M. Multiple-Rate Mass Transfer for Modeling Diffusion and Surface Reactions in Media with Pore-Scale Heterogeneity. *Water Resour. Res.* **1995,** *31*, 2383–2400.

Jaiswal, D. K.; Kumar, A.; Kumar, N.; Yadav, R. R. Analytical Solutions for Temporally and Spatially Dependent Solute Dispersion of Pulse Type Input Concentration in One-Dimensional Semi-Infinite Media. *J. Hydro-Environ. Res.* **2009,** *2*, 254–263.

Joshi, N.; Ojha, C. S. P.; Sharma, P. K.; Madramootoo, C. A. Application of Nonequilibrium Fracture Matrix Model in Simulating Reactive Contaminant Transport through Fractured Porous Media. *Water Res. Res.* **2015,** *51*, 390–408.

Joshi, N.; Ojha, C. S. P.; Sharma, P. K.; Surampalli, R. Y. Parameter Identification of Virus Transport in Porous Media. Using Equilibrium and Non-Equilibrium Models. *J. Environ. Chem. Eng.* **2013,** *1*, 1099–1107.

Kartha, S. A.; Srivastava, R. Effect of Slow and Fast Moving Liquid Zones on Solute Transport in Porous Media. *Trans. Porous Media* **2008,** *75* (2), 227–247.

Kool, J. B.; Parker, J. C. Analysis of the Inverse Problem for Transient Unsaturated Flow. *Water Resour. Res.* **1988,** *24*, 817–830.

Kool, J. B.; Parker, J. C.; Van Genuchten, M. T. Parameter Estimation for Unsaturated Flow and Transport Models—A Review. *J. Hydrol.* **1987,** *91*, 255–293.

Kumar, G. S.; Sekhar, M.; Misra, D. Time Dependent Dispersivity Behavior of Non-Reactive Solutes in a System of Parallel Fractures. *Hydrol. Earth Syst. Sci. Discuss.* **2006,** *3*, 895–923.

Kumar, G. S.; Sekhar, M.; Misra, D. Time-Dependent Dispersivity of Linearly Sorbing Solutes in a Single Fracture with Matrix Diffusion. *J. Hydrol. Eng.* **2008,** *13*, 250–257.

Levy, M.; Berkowitz, B. Measurement and Analysis of Non-Fickian Dispersion in Heterogeneous Porous Media. *J. Contam. Hydrol.* **2003,** *64*, 203–226.

Li, L.; Barry, D. A.; Cuiligan-Hensley, P. J.; Bajracharya, K. Mass Transfer in Soils with Local Stratification of Hydraulic Conductivity. *Water Resour. Res.* **1994,** *30*, 2891–2900.

Li, Z.; Brusseau, M. L. Nonideal Transport of Reactive Solutes in Heterogeneous Porous Media–6:Microscopic and Macroscopic Approaches for Incorporating Heterogeneous Rate-Limited Mass Transfer. *Water Resour. Res.* **2000,** *36*, 2853–2867.

Logan, J. D. Solute Transport in Porous Media with Scale-Dependent Dispersion and Periodic Boundary Conditions. *J. Hydrol.* **1996,** *184*, 261–276.

Mashayekhi, P.; Ghorbani-Dashtaki, S.; Mosaddeghi, M. R.; Shirani, H.; Nodoushan, A. R. M. Different Scenarios for Inverse Estimation of Soil Hydraulic Parameters from Double-Ring Infiltrometer Data Using HYDRUS-2D/3D. *Int. Agrophys.* **2016,** *30*, 203–210.

Mehmani, Y.; Balhoff, M. T. Mesoscale and Hybrid Models of Fluid Flow and Solute Transport. *Rev. Mineral. Geochem.* **2015,** *80*, 433–459.

Moré, J. J. The Levenberg-Marquardt Algorithm: Implementation and Theory. *Numer. Analy.* **1978,** 105–116.

Nakhaei, M.; Šimůnek, J. Parameter Estimation of Soil Hydraulic and Thermal Property Functions for Unsaturated Porous Media Using the HYDRUS-2D Code. *J. Hydrol.* **2014,** *62* (1), 7–15.

Natarajan, N. Effect of Distance-Dependent and TDD on Non-Linearly Sorbed Multispecies Contaminants in Porous Media. *ISH J. Hydrol. Eng.* **2016,** *22*, 16–29.

Nielsen, D. R.; Th. Van Genuchten, M.; Biggar, J. W. Water Flow and Solute Transport Processes in the Unsaturated Zone. *Water Resour. Res.* **1986,** *22*, 89S–108S.

Ojha, C. S. P.; Hari Prasad, K. S.; Ratha, D. N.; Surampalli, R. Y. Virus Transport through Unsaturated Zone: Analysis and Parameter Identification. *J. Hazard. Toxic Radioactive Waste* **2011,** *16*, 96–105.

Osborne, M. R. Nonlinear Least Squares—the Levenberg Algorithm Revisited. *ANZIAM J.* **1976,** *19*, 343–357.

Pickens, F.; Grisak, E.; Directorate, I. W.; Canada, E. Modelling of Scale-Dependent Dispersion in Hydrogeologic Systems **1981,** *17*, 1701–1711.

Pickens, J. F.; Grisak, G. E. Scale-Dependent Dispersion in a Stratified Granular Aquifer. *Water Resour. Res.* **1981a,** *17*, 1191–1211.

Pickens, J. F.; Grisak, G. E. Modeling of Scale-Dependent Dispersion in Hydrogeologic Systems. *Water Resour. Res.* **1981b,** *17*, 1701–1711.

Prasad, K. S. H.; Ratha, D. N.; Ojha, C. S. P. Identification of Virus Transport Parameters in Groundwater: Data Errors and Bias. *J. Hydro-Environ. Res.* **2012,** *6*, 41–50.

Ratha, D. N.; Hari Prasad, K. S.; Ojha, C. S. Analysis of Virus Transport in Groundwater and Identification of Transport Parameters. *Pract. Period. Hazard. ToxicRadioactive Waste Manage.* **2009,** *13*, 98–109.

Sander, G. C.; Braddock, R. D. Analytical solutions to the Transient, Unsaturated Transport of Water and Contaminants through Horizontal Porous Media. *Adv. Water Resour.* **2005,** *28*, 1102–1111.

Sauty, J. P. An Analysis of Hydrodispersive Transfer in Aquifers. *Water Resour. Res.* **1980,** *16*, 145–158.

Schulze-Makuch, D. Longitudinal Dispersivity Data and Implications for Scaling Behavior. *GroundWater* **2005,** *43*, 443–456.

Selim, H. M. *Transport & Fate of Chemicals in Soils: Principles & Applications*; CRC Press, Inc., 2014.

Sharma, P. K.; Shukla, S. K.; Choudhary, R.; Swami, D. Modeling for Solute Transport in Mobile–Immobile Soil Column Experiment. *ISH J. Hydrol. Eng.* **2016,** *22*, 204–211.

Sharma, P.; Srivastava, R. Concentration profiles and Spatial Moments for Reactive Transport through Porous Media. *J. Hazard. Toxic Radioactive Waste* **2011,** *16*, 125–133.

Srivastava, R.; Brusseau, M. L. Nonideal Transport of Reactive Solutes in Heterogeneous Porous Media: 1. Numerical Model Development and Moments Analysis. *J. Contam. Hydrol.* **1996,** *24*, 117–143.

Sudicky, E. A. A Natural Gradient Experiment on Solute Transport in a Sand Aquifer: Spatial Variability of Hydraulic Conductivity and Its Role in the Dispersion Process. *Water Resour. Res.* **1986,** *22*, 2069–2082.

Swami, D.; Sharma, P. K.; Ojha, C. S. P. Experimental Investigation of Solute Transport in Stratified Porous Media. *ISH J. Hydrol. Eng.* **2013,** *19*, 145–153.

Swami, D, *Study on Reactive Solute Transport through Porous Medium*; Indian Institute of Technology Roorkee: Roorkee, Uttarakhand, India, 2014.

Swami, D.; Sharma, P. K.; Ojha, C. S. P. Simulation of Experimental Breakthrough Curves Using Multiprocess Non-Equilibrium Model for Reactive Solute Transport in Stratified Porous Media. *Sadhana* **2014,** *39*, 1425–1446.

Swami, D.; Sharma, A.; Sharma, P. K.; Shukla, D. P. Predicting Suitability of Different Scale-Dependent Dispersivities for Reactive Solute Transport through Stratified Porous Media. *J. Rock Mech. Geotech.* **2016a,** *8*, 921–927.

Swami, D.; Sharma, P. K.; Ojha, C. S. P. Behavioral Study of the Mass Transfer Coefficient of Nonreactive Solute with Velocity, Distance, and Dispersion. *J. Environ. Eng.* **2016b,** *143*, 1–10.

Turco, M.; Kodešová, R.; Brunetti, G.; Nikodem, A.; Fér, M.; Piro, P. Unsaturated Hydraulic Behaviour of a Permeable Pavement: Laboratory Investigation and Numerical Analysis by Using the HYDRUS-2D Model. *J. Hydrol.* **2017,** *554*, 780–791.

Warrick, A. W.; Biggar, J. W.; Nielsen, D. R. Simultaneous Solute and Water Transfer for an Unsaturated Soil. *Water Resour. Res.* **1971,** *7*, 1216–1225.

Yeh, W. W. Review of Parameter Identification Procedures in Groundwater Hydrology: The Inverse Problem. *Water Resour. Res.* **1986,** *22*, 95–108.

Zhou, L.; Selim, H. M. Application of the Fractional Advection-Dispersion Equation in Porous Media. *Soil Sci. Am. J.* **2003a,** *67*, 1079–1084.

Zhou, L.; Selim, H. M. Scale-Dependent Dispersion in Soils: An Overview. *Adv. Agron.* **2003b,** *80*, 223–263.

Zou, S.; Xia, J.; Koussis, A. D. Analytical Solutions to Non-Fickian Subsurface Dispersion in Uniform Groundwater Flow. *J. Hydrol.* **1996,** *179*, 237–258.

ANNEXURE I

Down-gradient distance	Model	Parameter	Case 1				Case 2				Case 3			
			Initial guess	Estimated value	RMSE	NOI	Initial guess	Estimated value	RMSE	NOI	Initial guess	Estimated value	RMSE	NOI
400 cm	MIML	D_0 (cm^2/min)	5.0	43.65	0.03119	11	20	52.06	0.03091	10	5.0	16.23	0.03117	6
		ω (min^{-1})	1E-05	9.78E-05			4E-05	1.025E-04			5E-05	9.384E-05		
		K_L (min)	500	746.08			2000	866.34			500	281.89		
	MIMA	D_0 (cm^2/min)	5.0	22.09	0.02604	11	40.0	44.61	0.02739	7	5.0	16.85	0.02618	8
		ω (min^{-1})	1E-05	7.9245e-5			5E-05	8.0031E-05			5E-05	7.9254E-05		
		K_A (min)	50	101.59			500	452.42			50	38.15		
300 cm	MIML	D_0 (cm^2/min)	4	13.8881	0.02760	16	20	44.2594	0.02761	7	4	8.9156	0.02723	9
		ω (min^{-1})	1E-05	1.09847E-04			5E-05	1.1053E-05			5E-05	1.146E-04		
		K_L (min)	400	426.473			2000	1364.756			400	271.3806		
	MIMA	D_0 (cm^2/min)	4	17.3004	0.02360	16	30	25.1051	0.02411	7	4	13.6218	0.02324	18
		ω (min^{-1})	1E-05	1.00005E-04			5E-05	1.0228E-04			5E-05	9.991E-05		
		K_A (min)	40	230.0853			400	446.8916			40	131.9194		
200 cm	MIML	D_0 (cm^2/min)	3	11.6257	0.04256	15	15	114.9444	0.0413	19	3	15.1170	0.04098	26
		ω (min^{-1})	1E-05	3.0589E-04			5E-05	3.8395E-04			5E-05	3.9908E-04		
		K_L (min)	300	110.4207			1500	960.718			300	123.7978		
	MIMA	D_0 (cm^2/min)	3	42.4231	0.02702	28	20	105.3412	0.03038	12	3	22.5390	0.02495	23
		ω (min^{-1})	1E-05	2.0324E-04			5E-05	2.7807E-04			5E-05	1.7923E-04		
		K_A (min)	30	234.8584			300	718.9724			30	72.7586		

ANNEXURE I *(Continued)*

Down-gradient distance	Model	Parameter	Case 1				Case 2				Case 3			
			Initial guess	Estimated value	RMSE	NOI	Initial guess	Estimated value	RMSE	NOI	Initial guess	Estimated value	RMSE	NOI
150 cm	MIML	D_0 (cm^2/min)	2	17.6082	0.05257	16	10	113.4066	0.05305	14	2	22.8986	0.05275	11
		ω (min^{-1})	1E-05	5.4122E-04			5E-05	6.0641E-04			5E-05	5.6546E-04		
		K_L (min)	200	175.7630			1000	1054.112			200	224.6512		
	MIMA	D_0 (cm^2/min)	2	7.8645	0.04053	15	15	8.2291	0.04032	17	2	8.1826	0.04043	13
		ω (min^{-1})	1E-05	2.6201E-04			5E-05	2.5948E-04			5E-05	2.6113E-04		
		K_A (min)	20	9.6872			200	12.0606			20	11.7309		
100 cm	MIML	D_0 (cm^2/min)	2	310.4335	0.05357	26	10	114.5937	0.05314	31	2	76.6943	0.05384	14
		ω (min^{-1})	1E-05	2.5162E-03			5E-05	1.6560E-03			5E-05	2.8405E-03		
		K_L (min)	200	701.2153			1000	304.5646			200	166.308		
	MIMA	D_0 (cm^2/min)	2	13.4544	0.02654	30	15	13.6068	0.02686	41	2	13.5047	0.02681	34
		ω (min^{-1})	1E-05	3.4282E-04			5E-05	3.5976E-04			5E-05	3.5804E-04		
		K_A (min)	20	0.03242			200	0.07929			20	0.03478		
50 cm	MIML	D_0 (cm^2/min)	1	6.5167	0.06445	26	5	35.0768	0.06440	15	1	10.7511	0.06438	15
		ω (min^{-1})	1E-05	2.3884E-03			5E-05	2.3354E-03			5E-05	2.3035E-03		
		K_L (min)	100	55.5207			800	305.4379			100	94.0338		
	MIMA	D_0 (cm^2/min)	1	4.4776	0.05180	35	10	4.3188	0.05168	19	1	4.5711	0.05186	19
		ω (min^{-1})	1E-05	1.4603E-03			5E-05	1.4432E-03			5E-05	1.4637E-03		
		K_A (min)	10	4.4691			100	3.7831			10	4.8939		

CHAPTER 2

Wavelet Transform-Based Trend Analysis of Drought Variables over Homogenous Monsoon Regions of India

HIMANSHU SHARMA[1*], NITIN JOSHI[2], and SHAKTI SURYAVANSHI[3]

[1]*Department of Civil Engineering, Institute of Infrastructure Technology Research and Management, Ahmedabad, Gujarat 380026, India*

[2]*Department of Civil Engineering, Indian Institute of Technology Jammu, India*

[3]*Department of Civil Engineering, Sam Higginbottom Institute of Agriculture, Technology & Sciences, Allahabad, Uttar Pradesh 211007, India*

[*]*Corresponding author. E-mail: himfce@gmail.com*

ABSTRACT

In this study, trends as well as dominant and significant periodicity of metrological drought variables were analyzed for all regions of India and five homogenous monsoon regions of India. Monsoon, nonmonsoon and annual standardized precipitation index (SPI) along with annual drought duration (ADD), annual drought severity (ADS), and annual drought peak (ADP) time series were analyzed. Discrete wavelet transform-based Mann-Kendall test was used to identify trends and dominant periodic components affecting trend, whereas global wavelet spectrum obtained from continuous wavelet transform was used to identify significant periodicities in drought variables.

Advances in Hydrology and Climate Change: Historical Trends and New Approaches in Water Resources Management. Surendra Kumar Chandniha, Anil Kumar Lohani, Gopal Krishan, & Ajay Krishna Prabhakar (Eds.)

A significant decreasing trend in monsoon SPI was observed in central and north east India, whereas significant decreasing trend was also observed in annual SPI in north eastern India. No significant trends were observed in northwest monsoon drought, ADD, ADS, and ADP time series. Both high- (2–8 year periodicity) and low-frequency (16–64 year) components were observed to influence the trend in drought variables. However, for most of the regions, these dominant periodic components were not significant.

2.1 INTRODUCTION

Indian economy depends on agriculture. In 2013, agriculture and allied activities contributed to about 13.7% of GDP (gross domestic production) and involves about 50% of the work force (Sharma, 2017; Kumar et al., 2016). In India, a major portion of the agricultural sector completely depends on monsoon. In India, about 70–90% of annual mean precipitation is associated with southwest monsoon (Kumar et al., 2013) and failure of monsoon may result in droughts. Monsoon droughts have significant negative impact on the economy of the country. During 1987–1988 drought, there was a loss of 36 million tons of food production and during 2002 drought, the GDP was reduced by 1% (Gadgil et al., 2003). In India, out of 130 years (1875–2004), there were 27 drought years, therefore, on an average, a drought occurs once in 5 years. As per Indian Metrological Department (IMD), if the seasonal rainfall anomaly averaged over the country is less than 10% of its long-term average, than that year is termed as a drought year (Kumar et al., 2013).

There are a number of studies which analyze trends and variability in drought variables over India (Chowdhury et al., 1989; Sikka, 1999; SinhaRay and Shewale, 2001; Guhathakurta and Rajeevan, 2008; Naresh Kumar et al., 2009; Pai et al., 2011; Kumar et al., 2013; Suryavanshi et al., 2014; Das et al., 2015; Joshi et al., 2016). In India, monsoon drought has been found to be associated with sea surface temperature (SST) anomalies of tropical Pacific Ocean and Indian Ocean (Meehl and Arblaster, 2002). There are a number of studies which report that climate indices, such as El Niño Southern Oscillation (ENSO), Atlantic Multidecadal Oscillation (AMO), and Indian Ocean Dipole Mode Index (IOD) significantly influence the monsoon drought (Kumar Krishna et al., 1999; Kumar et al., 2013; Ganguli and Janga Reddy, 2014a, 2014b). However, Kumar Krishna et al. (1999) reported that relationship between ENSO and droughts is nonstationary and have weakened in the recent years.

Wavelet transform has been widely used for the analysis of nonstationary time series (Torrence and Compo, 1998; Maheswaran and Khosa, 2012; Karthikeyan and Nagesh Kumar, 2013; Nourani et al., 2014). Wavelet transformation in conjunction with the Mann–Kendall (MK) test have been widely used for the identification of trends and periodic components in hydrometeorological variables (Zume and Tarhule, 2006; Partal and Küçük, 2006; de Artigas et al., 2006; Almasri et al., 2008; Adamowski et al., 2009; Partal, 2012; Nalley et al., 2012, 2013; Sang, 2013; Adarsh and Janga Reddy, 2015; Araghi et al., 2015; Rashid et al., 2015; Joshi et al., 2016). There are a very few studies which apply wavelet transform to analyze the trends and periodicity associated with the drought variables over India. Joshi et al. (2016) used wavelet transform to analyze seasonal trends, dominant and significant periods of variability of drought variables for 30 rainfall subdivisions in India over 141 years. They reported that drought occurrences increased significantly in the subdivisions of northeast and central India during the second half of 20th century. The found that both short-term (2–8 years) and decadal (16–32 years) periodicities in drought variables were found to influence the trend. They found that these dominant periodic components were not significant for most of the geographical subdivisions.

This study analyzes the drought trend and periodicity for the five homogenous monsoon regions and at all India scale. In this study, the trend analysis is performed at a coarser spatial scale as compared with that of Joshi et al. (2016). This study would give insight into the trend and dominant periodic components influencing the trends in drought variables at a larger scale and the results obtained from this study are compared with that of Joshi et al. (2016).

2.2 THEORETICAL BACKGROUND

2.2.1 STANDARDIZED PRECIPITATION INDEX

In order to quantify drought onset, severity, and duration, a number of drought indices have been developed. This study used a rainfall-based drought index, SPI (McKee et al., 1993), which has been widely used to quantify meteorological droughts (Mo, 2008; Shukla and Wood, 2008; Mishra et al., 2010; Mishra and Cherkauer, 2010). For SPI computation, a probability distribution was first fitted to the monthly precipitation time series. The present study utilizes a two-parameter Gamma distribution and estimates the distribution parameters using the maximum likelihood method. A cumulative probability distribution for the time series was derived from the estimated parameters and then transformed into a standard normal random variable with a mean of zero and a unit variance to obtain the SPI value (Naresh Kumar et al., 2009). The following steps describe the procedure for computation of SPI.

1. Compute mean ($\bar{x}$), standard deviation (σ) and skewness coefficient (S_k) of precipitation data using eqs 2.1–2.3, respectively

$$\bar{x} = \frac{\sum_{i=1}^{N} x_i}{N} \tag{2.1}$$

Here, N = No. of precipitation observations

$$\sigma = \sqrt{\frac{\sum_{i=1}^{N} (x_i - \bar{x})^2}{N-1}} \tag{2.2}$$

$$S_k = \frac{N}{(N-1)(N-2)} \sum_{i=1}^{N} \frac{(x_i - \bar{x})^3}{\sigma^3} \tag{2.3}$$

2. Then the precipitation values are converted to log-normal value and the values of log mean (U), shape parameter (β), and scale parameters (α) are computed using eqs 2.4–2.7.

$$\bar{x}_{ln} = \ln \bar{x} \tag{2.4}$$

$$U = \bar{x}_{ln} - \frac{\sum_{i=1}^{N} \ln x_i}{N} \tag{2.5}$$

$$\beta = \frac{1 + \sqrt{1 + \frac{4U}{3}}}{4U} \tag{2.6}$$

$$\alpha = \frac{\bar{x}}{\beta} \tag{2.7}$$

3. The cumulative probability distribution (CPD) is computed after excluding cases of zero precipitation from the observations:

$$F(x) = \frac{\int_0^x x^{\alpha-1} e^{-\frac{x}{\beta}} dx}{\beta^{\alpha} \tilde{A}(\alpha)} \tag{2.8}$$

4. In case observations have days of no precipitation or zero precipitation, then the following formulation is used for generation of CPD:

$$G(x) = p + (1 - p)F(x) \tag{2.9}$$

5. Now, the cumulative probability value *G(x)* is converted into a standard normal variate Z, having zero mean and unit standard deviation by using following formulations:

   For $0 < G(x) < 0.5$

$$Z = \text{SPI} = -\left(t - \frac{c_o + c_1 t + c_2 t^2}{d_o + d_1 t + d_2 t^2 + d_3 t^3}\right) \tag{2.10a}$$

   For $0.5 < G(x) < 1$

$$Z = \text{SPI} = \left(t - \frac{c_o + c_1 t + c_2 t^2}{d_o + d_1 t + d_2 t^2 + d_3 t^3}\right) \tag{2.10b}$$

   Here,

   For $0 < G(x) < 0.5$

$$t = \sqrt{-\ln\left[G(x)\right]^2} \tag{2.11a}$$

   For $0.5 < G(x) < 1$

$$t = \sqrt{-\ln\left[1 - G(x)\right]^2} \tag{2.11b}$$

   Also,

$$c_o = 2.515517 \tag{2.12a}$$

$$c_1 = 0.802583 \tag{2.12b}$$

$$c_2 = 0.010328 \tag{2.12c}$$

$$d_1 = 1.432788 \tag{2.12d}$$

$$d_2 = 0.189269 \tag{2.12e}$$

$$d_3 = 0.001308 \tag{2.12f}$$

SPI values were computed at three distinct time scales, namely, monsoon period, nonmonsoon period, and yearly. In India, southwest monsoon period ranges from June to September, therefore, a 4-SPI September, which uses aggregated precipitation form June–September is considered as a monsoon drought indicator. In India, October–December period is referred as northeast monsoon. Northeast monsoon is a major period of rainfall over Southern

Peninsula particularly in the subdivisions of costal Andhra Pradesh, Rayalaseema, Tamil Nadu, Pondicherry. Therefore, a 3-SPI December is used to quantify the droughts during northeast monsoon season. In India, water year comprises from June to next year May, therefore, 12-SPI May is used to quantify the yearly droughts.

Additionally, annual drought duration (ADD), severity (ADS), and peak (ADP) were also computed using a 6-month SPI (6-SPI). 6-SPI is obtained by fitting gamma distribution to the 6-month aggregated precipitation time series. When the SPI value for a month remains below a specified threshold, then the month is termed as a drought month and ADD is the total number of drought months in a year. The threshold limit is taken as 20 percentile SPI. ADS is the cumulative value of SPI for the drought months in a given year and ADP is the minimum SPI value during the drought months of a given year.

2.2.2 WAVELET TRANSFORM

Wavelet transform has been widely used in the fields of signal processing, geophysics, and hydrology (Daubechies, 1992; Rao and Bopardikar, 1998; Zheng et al., 2013). In hydrology, wavelet transform was successfully applied for time series forecasting, cluster analysis, and the identification of trends and periodicities in hydrological variables (Maheswaran and Khosa, 2012; Karthikeyan and Nagesh Kumar, 2013; Bharath and Srinivas, 2015; Nalley et al., 2013; Adamowski et al., 2013; Araghi et al., 2015). Most of the hydrological time series are nonstationary in nature and may contain multispectral signals, having their frequencies varying from hourly to multi-decadal (Torrence and Compo, 1998; Nourani et al., 2014). Thus, in order to understand and interpret the nonstationarity existing in the time series, localization of both time, and frequency is needed. Fourier theory can estimate different frequency components present in a signal but cannot estimate when in time those frequencies occurred. Unlike Fourier transform, wavelet transform provides both time and frequency localization, and therefore, it is a more suitable method to analyze the nonstationarity in the time series (Daubechies, 1992; Karthikeyan and Nagesh Kumar, 2013).

A wavelet is a small wave of a finite length in space having a zero mean (Torrence and Compo, 1998; Lau and Weng, 1995). Wavelet transform (WT) uses a flexible window function which can be dilated and shifted in such a manner that it matches the time and frequency domain for the signal spectra (Lau and Weng, 1995; Percival and Walden, 2000). In order to match the high frequency component of the spectra, a narrow window is utilized while for low-frequency components relatively a wider window is used. Like Fourier transform, WT can be performed both at discrete and continuous domains. This is discussed in further detail in the following sections.

2.2.2.1 CONTINUOUS WAVELET TRANSFORM

Consider a time series, which has a continuous scale. For continuous wavelet transform (CWT), the mother wavelet function (Ψ), as defined by Partal and Küçük (2006):

$$\Psi(\eta) = \Psi(s,\gamma) = \frac{1}{\sqrt{s}}\Psi\left(\frac{t-\gamma}{s}\right) \tag{2.13}$$

where t is time, s is the wavelet scale factor, which ranges from 0 to $+\infty$, and γ is the translation factor of the wavelet over the time series. The parameter γ is for the shifting of the mother wavelet in the time domain, whereas the parameter, s, governs the dilation or compression of the mother wavelet. When s is less than 1, it corresponds to a high frequency function and vice versa. The mother wavelet should satisfy the conditions outlined in eqs 2.14a and 2.14 b:

$$C_\psi = \int_R \frac{|\hat{\psi}(\eta)|^2}{|\eta|} d\eta < \infty \tag{2.14a}$$

$$\int_{-\infty}^{+\infty} \Psi(\eta)\, d\eta = \hat{\Psi}(0) = 0 \tag{2.14b}$$

In CWT, the wavelet coefficients (W_ψ) are calculated by convolving xt with the scaled and translated versions of the wavelet, Ψ(η) (Partal and Küçük, 2006), as seen in equation 15:

$$W_\psi(s,\gamma) = \frac{1}{\sqrt{s}} \int_{-\infty}^{\infty} x(t)\psi^*\left(\frac{t-\gamma}{s}\right)\delta t \tag{2.15}$$

W_ψ represents wavelet power at different scales and * indicates the complex conjugate. In CWT, the wavelet power spectrum is obtained by plotting wavelet power ($|W_\psi(s)|^2$) against both time and frequency. The wavelet power spectrum provides a lot of information about the time series, which may be difficult to interpret (Westra and Sharma, 2006). Therefore, the wavelet power spectrum is often averaged over time and scale, and the result becomes easier to interpret and provides the valuable information about the time series. The time averaging of power for a given frequency is known as a global wavelet spectrum (GWS). The GWS is tested against a background red noise or white noise spectrum at a 95% confidence interval (Kestin et al., 1998; Torrence and Compo, 1998).

2.2.2.2 *DISCRETE WAVELET TRANSFORM*

Discrete wavelet transform (DWT) is commonly used for trend analysis and forecasting (Karthikeyan and Nagesh Kumar, 2013; Nalley et al., 2013), and operates on dyadic scales, thus limiting the scaling and shifting functions to discrete values (Adarsh and Janga Reddy, 2015). For a time series, the mother wavelet function (Ψ) can be written as:

$$\psi_{(a,b)}\left(\frac{t-\gamma}{s}\right)=\frac{1}{\left(s_0\right)^{a/2}}\psi\left(\frac{t-b\gamma_0 s_0^a}{s_0^a}\right) \tag{2.16}$$

where Ψ is the mother wavelet. *a* and *b* are integers that represent the amount of dilation and translation of the wavelet, respectively. So is the dilation step so that $s_o > 1$, and γ_o is the location variable whose value should be greater than zero. The wavelet coefficients for DWT (Partal and Küçük, 2006) are:

$$W_\Psi\left(a,b\right)=\frac{1}{\left(2\right)^{a/2}}\sum_{t=0}^{N-1}x_t\Psi\left(\frac{t}{2^a}-b\right) \tag{2.17}$$

The values of s_o and γ_o are 2 and 1, respectively (Mallat, 1989; Daubechies, 1992).

Inverse discrete wavelet transform can be performed to reconstruct a time series from the wavelet coefficients. The reconstruction of the signal can be written as:

$$x(t)=A_n(t)+\sum_{i=1}^{n}D_i(t) \tag{2.18}$$

Detailed components correspond to wavelet coefficients while approximations correspond to scaling function coefficients. The approximate component A_n at any level n is a subsampled output from a low-pass filter. The detailed component D_i at different levels ($i = 1,2,3 \ldots n$) is a subsampled output from a high-pass filter. The approximate components give information about the low frequency or mean behavior of the signal, whereas the detailed components give information about the different frequency components of the original signal, which range from high to low (Adarsh and Janga Reddy, 2015).

2.2.3 THE MANN-KENDALL TEST FOR TIME SERIES TRENDS

The MK test (Mann, 1945; Kendall, 1975) is a nonparametric test commonly used to evaluate the presence of trends in a time series. The test MK statistic S is defined as:

$$S = \sum_{i=1}^{n-1} \sum_{j=i+1}^{n} \operatorname{sgn}(x_j - x_i) \tag{2.19}$$

where

x_j is the sequential data value, n is the length of the data set and *sgn (t)* is defined as.

$$\operatorname{sgn}(t) = \begin{cases} 1, & \text{for } t > 0 \\ 0, & \text{for } t = 0 \\ -1, & \text{for } t < 0 \end{cases} \tag{2.20}$$

when $n \geq 8$, the test statistic S is approximately normally distributed with a mean and variance as follows:

$$E(S) = 0 \tag{2.21}$$

$$\operatorname{Var}(S) = \frac{n(n-1)(2n+5) - \sum_{i=1}^{m} t_i(t_i - 1)(2t_i + 5)}{18} \tag{2.22}$$

where m is the number of tied groups, and t_i is the size of the ith tie group. The standardized test statistic Z is computed as follows:

$$Z_{\text{MK}} = \begin{cases} \dfrac{S-1}{\sqrt{\operatorname{Var}(S)}}, & \text{for } S > 0 \\ 0 \quad , & \text{for } S = 0 \\ \dfrac{S+1}{\sqrt{\operatorname{Var}(S)}}, & \text{for } S < 0 \end{cases} \tag{2.23}$$

The standardized MK statistic Z follows the standard normal distribution with a zero mean and unit variance. If $|Z| \geq Z_{1-(\alpha/2)}$, the null hypothesis is rejected at the significance level α (5% in this study). For autocorrelated data (i.e., when lag-1 autocorrelation is significant), the original MK test underestimates the variance of the data (Suryavanshi et al., 2014). Therefore, Hamed and Rao (1998) proposed a modified MK test in which variance of the test statistic is modified to take into account the autocorrelation of the data. The modified variance is given as:

$$V^*(S) = \mathrm{Var}(S) \times \frac{n}{n_s^*} = \frac{n(n-1)(2n+5)}{18} \times \frac{n}{n_s^*} \tag{2.24}$$

where, n/n_s^* represents a correlation due to the autocorrelation in the data. The n/n_s^* is evaluated using the following equation:

$$\frac{n}{n_s^*} = 1 + \frac{2}{n(n-1)(n-2)} \sum_{i=1}^{n-1} (n-i)(n-i-1)(n-i-2)\rho_s(i) \tag{2.25}$$

where n is the actual number of observations, and $\rho_s(i)$ is the autocorrelation function of the ranks of the observations. The autocorrelation of ranks of observations $\rho_s(i)$ is related to the parent autocorrelation function and is given as (Kendall, 1975):

$$\rho(i) = 2\sin\left(\frac{\pi}{6}\rho_s(i)\right) \tag{2.26}$$

The significance of the trends is tested by comparing the standardized test statistic Z:

$$Z = \frac{S}{[V^*(S)]^{0.5}} \tag{2.27}$$

A significance level of $\alpha = 0.1$ is used for the autocorrelation of the ranks,$\rho_s(i)$, which produces the best overall empirical significance level.

2.2.4 DECOMPOSITION OF DATA USING DWT

DWT provides valuable information about the periodic structure of the time series (Araghi et al., 2015). DWT was used into set of approximate (A) to decompose the time series into approximate (A) and detailed components (D).

Primary step for DWT is the selection of appropriate mother wavelet. Haar, Symlets, Daubechies family of wavelets are commonly used for the trend analysis (Sang, 2013; Nalley et al., 2013; Adarsh and Janga Readdy, 2014; Araghi et al., 2015). In this study, Daubechies (db) family of wavelet was used. After the selection of mother wavelet, the number of decomposition needs to be selected. There are a number of recommended methods to select appropriate decomposition levels. Equation 2.12 has been most commonly used to compute the decomposition levels (de Artigas et al., 2006; Misiti et al., 2013), and therefore, it is used in this study.

$$L = \frac{\log\left(\frac{n}{2v-1}\right)}{\log(2)} \tag{2.28}$$

where L is the maximum number of decomposition levels, v is the number of vanishing moments of the db wavelet, and n is the number of data points in time series which rounded off to the next higher dyadic scale in this data length of 141 years, and therefore n is taken as 2^8=256.

As DWT uses a convolution process, therefore, a finite length signal leads to border distortion. In order to eliminate the border distortion, appropriate border extension method must be selected. Border extension technique extends the length of the signal and a smooth signal reconstruction is ensured (Su et al., 2011; Nalley et al., 2012, 2013). There are three border extension methods, namely, zero-padding, symmetrization, and periodic padding which are commonly used. Wavelet type was varied from db4 to db10, level of decomposition for each wavelet type was computed from eq 2.12 and all the three border extension methods were used. The relative error (R_e) criterion as given by eq 2.13 is used to select the most appropriate combination (Nalley et al., 2012, 2013).

$$R_E = \frac{|Z_a - Z_o|}{|Z_o|} \tag{2.29}$$

where Z_o and Z_a are MK-Z values for the original and A component of the last decomposition level, respectively. The best combination of wavelet type, decomposition level and border decomposition is selected which minimizes the relative error.

2.2.5 GLOBAL WAVELET SPECTRUM

Significant periodic components in the drought variables were estimated through CWT-based GWS. GWS provides an unbiased estimate of true power spectrum. There are various studies which use GWS for the analysis of significant periodic components (Partal and Küçük, 2006; Westra and Sharma, 2006; Partal, 2012). In order to reduce the border distortion effects, the time series is padded with zeros and a Morlet wavelet was used for the analysis. GWS is obtained by time averaging CWT power spectra at each scale and is given by

$$\bar{W}^2 = \frac{1}{N}\sum_{n=0}^{N-1}\left|W_{\psi}(s)\right|^2 \tag{2.30}$$

In the present study, the code given by Torrence and Compo (1998) was used to compute GWS (http://atoc.colorado.edu/research/wavelets/). GWS provides an unbiased estimate of true power spectrum and has been used to analyze the presence of significant periodic components in a time series. The distribution of the local wavelet spectrum is given by

$$\frac{W^2 s}{\sigma^2} \Rightarrow \frac{1}{2} P_k \chi_2^2 \tag{2.31}$$

For the red noise spectrum, P_k is given as the following (Torrence and Compo, 1998):

$$P_k = \frac{1-\alpha^2}{1+\alpha^2 - 2\alpha\cos\left(2\pi k / N\right)} \tag{2.32}$$

where P_k is the mean spectrum at the Fourier frequency k corresponding to the wavelet scale s; χ_2^2 is the chi-square value for the chosen confidence limit. Appropriate background spectrum was chosen based on lag-1 serial correlation. GWS was tested against background noise spectrum at a 95% confidence interval to obtain significant periods of variability.

2.3 STUDY AREA AND DATA

India is divided into six homogenous monsoon regions, namely, western Himalaya, northwest (NWIND), central northeast (CNEIN), northeast (NEIND), west central (WCIND), peninsular (PENIN). Location map of the study area has been shown in Figure 2.1. western Himalaya, being a hilly region, is not considered in this study. For all the homogenous monsoon regions, monthly precipitation time series for the period of 141 years (i.e., from 1871 to 2012) was obtained from the Indian Institute of Tropical Management (IITM) Pune (http://www.tropmet.res.in). The monthly precipitation time series were constructed using the data from 1476 metrological station almost uniformly spaced throughout the country. This data set has been used in a number of studies (Kumar et al., 2010; Adarsh and Janga Reddy, 2014; Mondal et al., 2014; Joshi et al., 2016). Additionally, all India rainfall time series is also considered in this study.

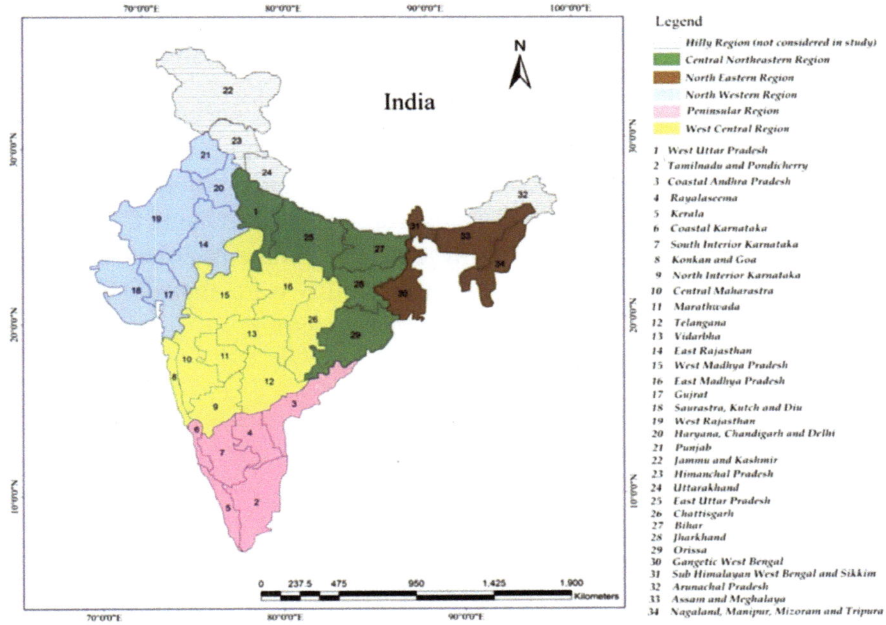

FIGURE 2.1 Location map of study area.

2.4 METHODOLOGY

For each homogenous region, DWT was used to analyze the trend and periodicity for the drought variables. The following steps are involved in the analysis:

1. Monthly time series of rainfall was obtained for the five homogenous rainfall region and all India were used to compute the 4-SPI September, 3-SPI December, and 12-SPI May indicating droughts during southwest monsoon season, northeast monsoon season, and entire water year. From the 6-SPI time series ADD, ADS, and ADP for each region were also computed.
2. For each data set, the presence of serial correlation was checked using Lag-1 autocorrelation coefficients at 5% significant level.
3. Using DWT, each data set was decomposed into detailed (D) and approximate (A) component. For the present analysis, the Daubechies (db) wavelet was considered as a mother wavelet. Most appropriate combination of wavelet function (between db4 and db10), level of decomposition, and padding was selected using minimum relative error (R_e) criterion as given by eq 2.29.

4. MK test either original or modified MK test was applied on the original time series, A and D components and combination of A and D components were used.
5. The MK-Z values for the observed time series were compared with the MK-Z values of the A+D component combinations to identify the dominant periodic components influencing the trend.
6. Significant periodic components were determined using CWT-based GWS. GWS was obtained by averaging wavelet power spectrum over time. Significant periods of variability were obtained be testing GWS against white noise background spectrum.

2.5 RESULTS

2.5.1 PRELIMINARY DATA ANALYSIS

Table 2.1 shows the statistical properties of drought variables for all India and homogenous monsoon regions. PENIN had highest number of drought months followed by NWIND and CNEIN. Presence of seasonality and autocorrelation in drought variables was checked through the autocorrelation function (ACF). No significant autocorrelations were observed in the drought variables for All India and homogenous monsoon regions (ACF not shown here).

TABLE 2.1 Statistical Properties of Rainfall and Drought Variables.

Regions	Average annual rainfall (mm)	Min 6-SPI	Max 6-SPI	Max ADS	Max ADP	Drought months
ALLIN	90.508	−3.983	3.175	19.31	3.983	344
NWIND	45.530	−4.52	3.115	22.12	4.52	346
WCIND	89.337	−4.635	2.859	22.32	4.635	339
CNEIN	99.775	−3.582	3.122	18.33	3.582	346
NEIND	170.940	−3.602	2.973	24.85	3.602	342
PENIN	96.903	−3.345	3.117	25.4	4.481	348

2.5.2 DWT-BASED TREND ANALYSIS OF 4-SPI SEPTEMBER, 3-SPI DECEMBER, AND 12-SPI MAY

Table 2.2 presents the trend analysis for the original data, A, D, A+D, and A+ D+ D combination for 4-SPI September. For the original and approximate components of 4-SPI September time series, CNEIN and NEIND regions show a significant decreasing trend. Although, 4-SPI September time series for all India and WCIND also showed a decreasing trend, but was insignificant at 5% significant level. None of the details component showed significant increasing/decreasing trend. A decreasing trend in 4-SPI September indicates drier condition during the monsoon season, which is due to decrease in the monsoon rainfall. GWS was used to identify the significant periodic components against a background white noise spectrum. Figure 2.2 shows the GWS for the 4-SPI September time series.

TABLE 2.2 MK-Z Values of the 4SPI_September Time Series for the Original Data, Approximation (A) and Detail (D) Components..

	ALLIN	NWIND	WCIND	CNEIN	NEIND	PENIN
Orig	−1.95	0.12	−1.61	−2.62*	−3.72*	0.67
Aprox	−1.87	0.03	−1.62	−2.64*	−3.72*	0.67
D1	−0.13	−0.26	−0.24	−0.12	−0.18	−0.20
D2	−0.18	−0.28	0.26	−0.10	0.17	0.40
D3	0.25	−0.07	−0.24	−0.83	−0.67	−0.40
D4	0.42	0.59	0.29	0.65	−0.18	0.11
D5	0.02	0.40	–	0.32	–	1.33
A+D1	−1.75	−0.08	−2.14*	−1.75	−4.02*	0.31
A+D2	−1.78	0.04	**−1.97***	−2.23*	−4.54*	1.19
A+D3	−2.96*	0.51	−3.68*	−5.32*	−5.81*	**0.69**
A+D4	−2.83*	0.92	−3.18*	−2.45*	−1.38	2.06
A+D5	−2.70*	0.77	–	−2.51*	–	2.33
A+D1+D2	−1.35	−0.34	−2.58*	−1.06	−2.92*	0.44
A+D1+D3	−1.33	−0.25	−1.99*	−1.63	−3.06*	0.00
A+D1+D4	−1.24	**0.08**	−2.05*	−1.60	−3.03*	2.07
A+D1+D5	**−1.71**	0.39		−2.16*		0.65
A+D2+D3	−1.58	0.04	−5.16*	**−2.55***	**−3.82***	0.74
A+D2+D4	−1.24	0.39	−2.11*	−1.72	−3.32*	1.60
A+D2+D5	−2.48*	0.44		−2.57*		1.56
A+D3+D4	−1.34	1.51	−2.09*	−3.41*	−3.09*	1.37
A+D3+D5	−3.40*	0.82	–	−4.40*	–	1.57
A+D4+D5	−2.66*	1.79	–	−1.05	–	2.44*

Note: Significant trend at 5% significant level is denoted by an asterisk and dominant periodicities are denoted by bold letters.

For CNEIN and NEIND regions, a high frequency component having a periodicity of 2–8 year found to be dominant. However, from GWS analysis, these dominant periodic components were not found to be statistically significant. Monsoon SPI in ALLIN and NWIND were found to be affected by both high (2–4 year) and low-frequency (16–32 year) components. Based on the GWS analysis, it was found that for NWIND, a 16-year period component was significant and for WCIND, a 64-year periodic component was significant. Joshi et al. (2016) also reported a significant decreasing trend in monsoon SPI for ASMEG (Assam and Meghalaya), NMMAT (Nagaland, Manipur, Mizoram and Tripura), and GNWBL (Gangetic West Bengal) subdivisions of NEIND (North-East India), EMPRA (Eastern Madhya Pradesh) and CHHAT (Chhattisgarh) subdivisions of CNEIN (Central India).

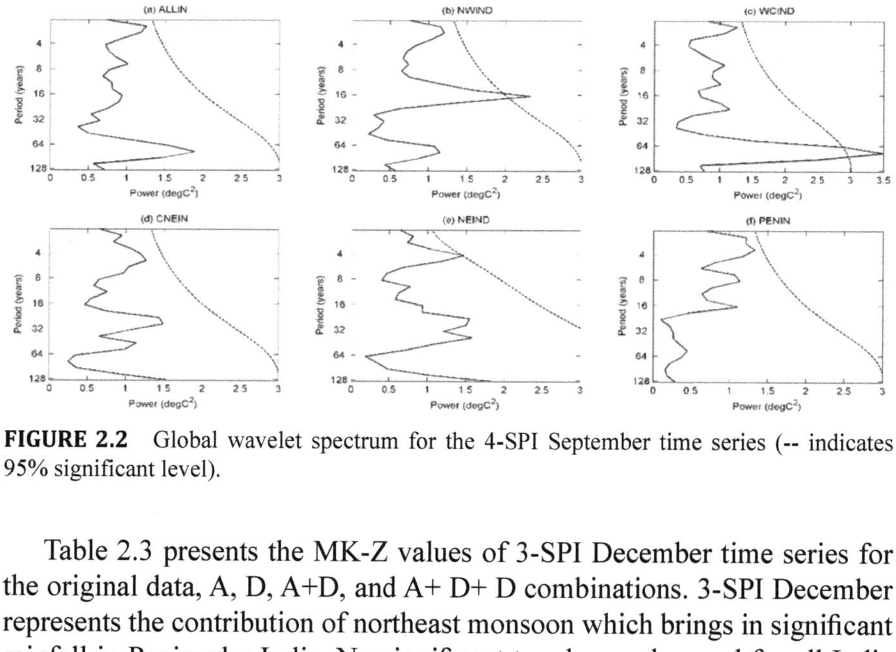

FIGURE 2.2 Global wavelet spectrum for the 4-SPI September time series (-- indicates 95% significant level).

Table 2.3 presents the MK-Z values of 3-SPI December time series for the original data, A, D, A+D, and A+ D+ D combinations. 3-SPI December represents the contribution of northeast monsoon which brings in significant rainfall in Peninsular India. No significant trend was observed for all India and homogenous monsoon regions. A nonsignificant but a decreasing trend was observed for CNEIN, whereas all the other subdivision reports a nonsignificant increasing trend in 3-SPI December time series. Global wavelet spectrum is shown in Figure 2.3. PENIN exhibits a short-term periodicity of 4–8 years. For ALLIN and NWIND both short-term (2 year) and long-term (16 year) periodic components were found to influence the drought trend; whereas, for WCIND, CNEIN, and NEIND, long-term periodic component (16–32 year) was found to influence the trend. However, none of the periodic components were found to be significant. Joshi et al. (2016) also reported that for the non-monsoon SPI at subdivisional scale, both high and low-frequency components, corresponding to 4–8 and 16 year periods influence the trend. Joshi et al. (2016) reported a significant increasing trend in nonmonsoon SPI for KERLA, COKNT subdivisions of PENIN regions, TELNG subdivision of WCIND, and WSRST subdivision of NWIND region. Although present study reports an increasing trend in WCIND, NWIND, and PENIN regions, but it was not statistically significant.

TABLE 2.3 MK-Z Values of the 3-SPI December Time Series for the Original Data, Approximation (A) and Detail (D) Components.

	ALLIN	NWIND	WCIND	CNEIN	NEIND	PENIN
Orig	0.79	0.38	0.34	−0.51	1.08	0.45
Aprox	0.94	0.72	0.35	−0.52	0.82	1.85
D1	0.35	−0.06	−0.18	0.31	−0.13	0.11
D2	0.38	0.24	0.07	0.44	−0.08	−0.40
D3	−0.23	0.30	0.55	0.12	−0.06	−0.39
D4	0.48	0.57	0.13	−2.18*	0.30	0.15
D5	–	–	−0.71	−1.49	0.67	–
D6	–	–	–	–	−0.13	–
A+D1	1.91	0.74	0.83	0.68	−0.18	0.87
A+D2	1.21	0.94	1.24	0.33	−0.58	0.49
A+D3	1.51	0.93	2.14*	0.45	0.56	1.31
A+D4	1.75	0.53	0.66	−0.81	0.65	2.62*
A+D5	–	–	−0.04	**−0.50**	2.23*	–
A+D6	–	–	–	–	1.82	–
A+D1+D2	2.08*	0.56	0.57	0.26	−0.49	0.22
A+D1+D3	1.24	0.94	0.97	0.04	−0.17	0.42
A+D1+D4	**0.94**	**0.44**	1.02	0.16	0.05	2.04*
A+D1+D5	–	–	0.51	−0.12	1.56	–
A+D1+D6	–	–	–	–	0.14	–
A+D2+D3	1.22	0.74	1.25	0.06	−0.19	**0.43**
A+D2+D4	0.95	0.44	1.39	0.00	−1.12	0.90
A+D2+D5	–	–	−0.06	−0.32	1.51	–
A+D2+D6	–	–	–		−0.29	–
A+D3+D4	0.51	0.46	1.12	−0.14	0.58	1.56
A+D3+D5	–	–	0.47	−0.14	1.81	–
A+D4+D5	–	–	**0.34**	−0.11	**0.95**	–
A+D4+D6	–	–	–	−1.09	2.48*	–
A+D5+D6	–	–	–	–	1.44	–

Note: Significant trend at 5% significant level is denoted by asterisk and dominant periodicities are denoted by bold letters.

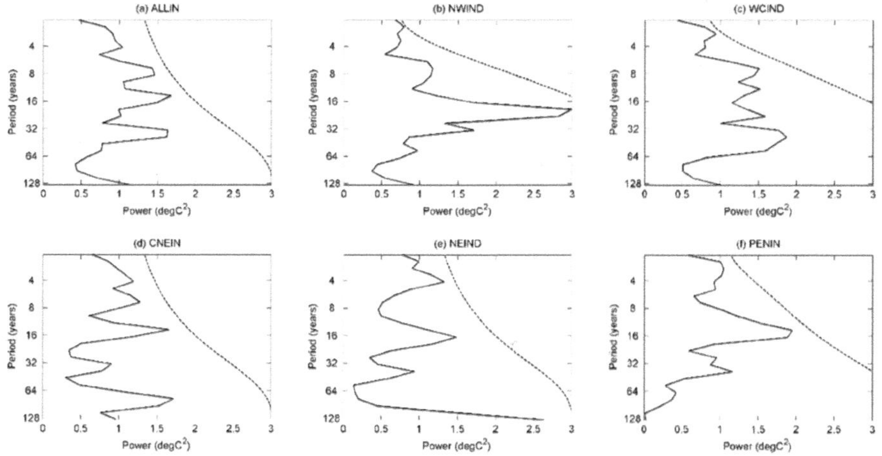

FIGURE 2.3 Global wavelet spectrum for the 3-SPI December time series (– indicates 95% significant level).

Table 2.4 presents MK-Z values for 12-SPI May time series. NEIND showed a significant decreasing trend in the original data and A component, whereas PENIN showed a significant trend in A and D5 (64 year periodic) components. Although, ALLIND, WCIND, and CNEIN showed a decreasing trend in the original data and approximate components but were not significant. The annual SPI time series were influenced by both high (2–4 year) and low-frequency (8–16 year) components. Significant periodic components were also identified as shown in Figure 2.4 and none of the periodic components were found to be significant except for NWIND where a 16-year periodicity was significant. Joshi et al. (2016) reported a deceasing trend in NMMAT and an increasing trend in GNWBL subdivision ofNEIND. They also reported a decreasing trend in EMPRA and CHHAT subdivision of CNEIN. The present study reports a significant decreasing trend in NEIND and nonsignificant but a decreasing trend in the CNEIN region.

2.5.3 DWT-BASED TREND ANALYSIS OF ADD, ADS, AND ADP

Table 2.5 presents the MK-Z values of the ADD time series, it can be seen that none of the regions showed a significant trend in ADD. Although an increasing nonsignificant trend were observed in drought duration for CNEIN and NEIND regions. Trends in ADD were found to be influenced by both high frequency (2–8 year) and low-frequency components (16–32 year). Trends in ALLIND and WCIND were influenced by 8-year periodic component, NWIND was influenced by 4-year periodic component. Trends in PENIN was influenced by 2 and 4-year periodic component and trend in NEIND was influenced by a 16-year periodic component. For CENIN both short-term period component of 2 year and long-term period component of 32 year were found to influence the trend. Significant periods of availability were identified using GWS as shown in Figure 2.5. For ALLIN a 64-year periodicity was found to be significant, for NWIND, a periodicity of 2 and 16 year, for WCIND, a periodicity of 2 year, whereas for PENIN, a periodicity of 8 year was found to be significant.

TABLE 2.4 MK-Z Values of the 12-SPI-May Time Series for the Original Data, Approximation (A), and Detail (D) Components.

	ALLIN	NWIND	WCIND	CNEIN	NEIND	PENIN
Orig	–0.20	0.69	–1.65	–0.73	–2.11*	1.26
Aprox	–0.14	0.71	–1.61	–0.75	–2.29*	3.18*
D1	–0.18	–0.09	0.01	0.19	–0.30	0.19
D2	0.09	–0.03	0.42	–0.12	0.34	–0.39
D3	–0.28	0.38	–0.46	–0.24	–0.79	0.23
D4	–1.26	0.26	0.92	–0.20	–1.50	0.67
D5	-	-	-	0.13	–0.59	5.87*
A+D1	–0.27	0.42	–1.19	–0.63	–0.53	1.45
A+D2	0.12	0.78	–2.29*	–1.31	–1.22	1.77
A+D3	–0.51	1.05	–2.13*	–0.44	–1.06	2.34*
A+D4	–0.51	0.51	–1.10	–1.62	–1.76	4.80*
A+D5	-	-	-	–1.33	–0.88	7.63*
A+D1+D2	–0.27	0.19	–1.02	–0.59	–0.52	**1.10**
A+D1+D3	–0.44	0.77	–1.07	–0.36	–0.52	1.88
A+D1+D4	–0.60	0.45	–0.67	**–0.77**	–1.45	1.79
A+D1+D5	–	–	–	–0.87	-0.62	2.20*
A+D2+D3	**–0.19**	0.87	**–1.67**	–1.24	–0.92	1.64
A+D2+D4	–0.36	**0.60**	–0.95	–0.97	–1.82	2.54*
A+D2+D5	–	–	–	–1.38	–1.37	2.20*
A+D3+D4	–0.51	1.17	–1.88	–1.16	**–2.48***	2.53*
A+D3+D5	–	–	–	–0.89	–0.94	3.53*
A+D4+D5	–	–	–	–1.25	–0.35	6.34*

Note: Significant trend at 5% significant level is denoted by asterisk and dominant periodicities are denoted by bold letters.

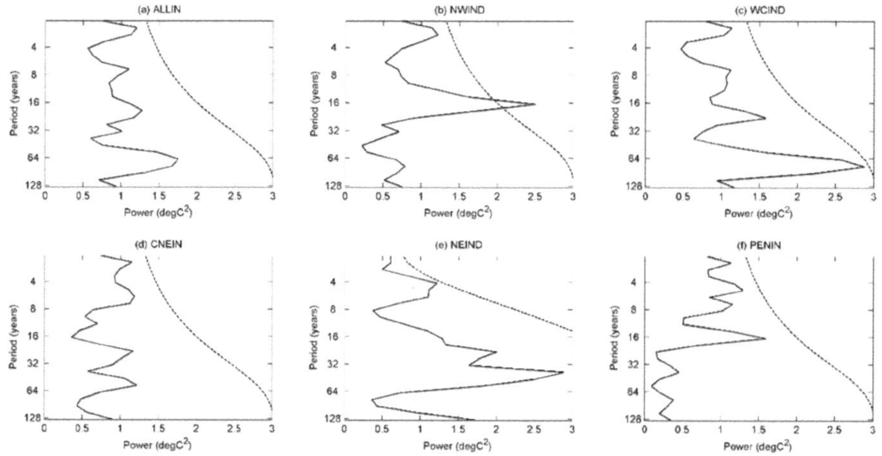

FIGURE 2.4 Global wavelet spectrum for the 12-SPI May time series (– indicates 95% significant level).

MK-Z values for ADS is presented in Table 2.6, it is seen that none of the regions shows a significant trend in ADS. For NEIND and CENIN regions, drought severity shows an increasing but nonsignificant trend. Trends in the regions were influenced by both high and low-frequency components. For CENIN and PENIN regions, a 4-year periodic component influenced the trend. For WCIND, a 16 year and for ALLIND, a 64-year periodic component influenced the trend. For NEIND, 4, 8 and 64-year periodic component influenced the trend. For NWIND, 8 and 16-year periodic component influenced the trend. Figure 2.6 shows the GWS and it is seen that for NWIND, a periodicity of 16 year, for CNEIN and PENIN, a periodicity of 8 year and for WCIND a periodicity of 2 year was found to be significant.

Table 2.7 presents the MK-Z values of the ADP time series. It can be seen that none of the regions showed a significant trend in ADP. Although for NEIND, drought peak shows an increasing trend but it was not statistically significant. ADP time series was also found to be influenced by both high and low-frequency components. For ALLIN and NWIND, 8 and 16-year periodic component influenced the trend. For WCIND, 2 and 16-year periodic component influenced the trend. For CNEIN, 4 and 8-year periodic component influenced the trend. For NEIND, an 8-year periodic component and for PENIN, a 2 and 4-year periodic component influenced the trend. Figure 2.7 shows the GWS for ADP and it can be seen that for ALLIND and WCIND, a periodicity 2 year, for NWIND, a periodic component of 4 year was found to be significant.

TABLE 2.5 MK-Z Values of the ADD Time Series for the Original Data, Approximation (A), and Detail (D) Components.

	ALLIN	NWIND	WCIND	CNEIN	NEIND	PENIN
Orig	−0.53	−0.01	−0.08	0.67	1.36	−0.42
Aprox	−0.52	−0.01	−0.09	0.72	1.44	−0.18
D1	−0.20	0.04	−0.28	0.08	−0.17	−0.03
D2	0.00	**−0.07**	0.04	−0.08	0.04	0.34
D3	0.31	0.15	**−0.08**	0.23	0.11	−0.07
D4	0.25	−0.19	−0.34	−0.41	−0.02	0.11
D5	–	–	−0.13	0.07	–	–
D6	–	–	–	4.07	–	–
A+D1	−1.11	1.10	0.12	0.45	2.40*	−1.11
A+D2	−0.86	0.48	−0.32	0.13	3.06*	−0.52
A+D3	**−0.49**	0.97	0.22	0.51	3.03*	−1.27
A+D4	0.50	0.92	0.43	1.58	**1.17**	−2.82*
A+D5	–	–	0.37	0.24	–	–
A+D6	–	–	–	3.94*	–	–
A+D1+D2	−1.33	0.77	−0.40	0.44	2.68*	**−0.44**
A+D1+D3	−0.86	0.28	0.12	0.24	2.53*	−1.07
A+D1+D4	−0.27	0.82	0.05	0.20	2.50*	−1.14
A+D1+D5	–	–	0.04	**0.67**	–	–
A+D1+D6	–	–	–	1.90	–	–
A+D2+D3	−0.89	0.28	**−0.07**	−0.06	3.35*	−0.57
A+D2+D4	−0.32	0.001	−0.47	0.13	1.99*	−1.11
A+D2+D5	–	–	−0.08	0.34	–	–
A+D2+D6	–	–	–	1.70	–	–
A+D3+D4	0.05	0.43	0.21	0.27	1.89	−2.04*
A+D3+D5	–	–	0.61	−0.14	–	–
A+D3+D6	–	–	–	2.50*	–	–
A+D4+D5	–	–	0.14	**0.63**	–	–
A+D4+D6	–	–	–	4.32*	–	–
A+D5+D6	–	–	–	2.33*	–	–

Note: Significant trend at 5% significant level is denoted by an asterisk and dominant periodicities are denoted by bold letters.

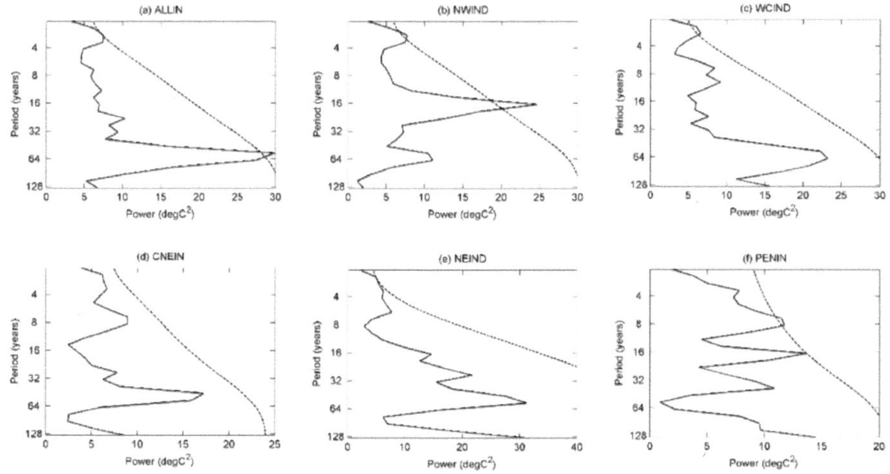

FIGURE 2.5 Global wavelet spectrum for the ADD time series (– indicates 95% significant level).

2.6 DISCUSSION

The present study considers of metrological drought which occurs due to deficit in rainfall. There are numerous studies which report teleconnections of Indian summer monsoon rainfall (ISMR) with various climate indices, such as ENSO, QBO, AMO, PDO, etc. (Gadgil et al., 2007; Kumar et al., 2013; Adarsh and Janga Reddy, 2015; Krishnamurthy and Krishnamurthy, 2015). Several studies reported a negative correlation between ISMR and ENSO (Kumar et al., 2013). Similarly, Krishnamurthy and Krishnamurthy (2015) reported a positive correlation between AMO and monsoon rainfall. Malik et al. (2017) reported that wet and dry decades of ISMR are influenced by AMO, PDO, and total solar irradiance (TSI). Therefore, climate indices, such as ENSO, AMO, PDO play a predominant role in causing droughts over India.

For the monsoon drought, both high- and low-frequency components were found to influence the trend, however, GWS analysis revealed that their effect was not significant. For NWIND, a 16-year periodic component, for NEIND, a 4-year periodic component, and for WCIND, a 64-year periodic component was significant. Lower frequency components having periodicities of 16–32 and 64 years may be related to the Pacific Decadal Oscillation (PDO) and the AMO, respectively. High frequency components with a periodicity of 2–4 years and 4–8 years have been related to the Quasi-Biennial Oscillation (QBO) and the ENSO phenomenon (Kumar et al., 2013).

TABLE 2.6 MK-Z Values of the ADS Time Series for the Original Data, Approximation (A), and Detail (D) Components.

	ALLIN	NWIND	WCIND	CNEIN	NEIND	PENIN
Orig	−0.89	−0.19	−0.38	0.45	0.99	−0.74
Aprox	−0.90	−0.26	−0.28	0.77	0.85	−1.22
D1	0.001	0.38	0.04	−0.34	0.11	−0.18
D2	0.06	−0.33	−0.22	−0.39	−0.11	0.28
D3	0.001	0.60	0.75	0.91	0.87	0.30
D4	−0.07	0.07	−0.28	0.04	**0.93**	0.1
D5	0.63	–	–	–	*2.45**	–
D6	−0.58	–	–	–	1.59	–
A+D1	−1.16	−0.07	−0.64	0.49	1.59	−1.66
A+D2	−0.33	−0.99	−1.01	**0.46**	2.40*	**−0.69**
A+D3	−1.05	−0.39	−0.06	3.51*	8.18*	−2.45*
A+D4	0.14	−0.88	**−0.46**	1.85	*2.32**	−2.68*
A+D5	−1.04	–	–	–	1.42	–
A+D6	**−0.99**	–	–	–	1.77	–
A+D1+D2	−0.64	−0.59	−1.61	−0.39	1.19	−0.49
A+D1+D3	−0.27	0.004	−0.29	0.96	1.97*	−2.01*
A+D1+D4	−0.42	−0.54	−0.23	0.40	1.96*	−1.90
A+D1+D5	−0.46	–	–	–	1.83	–
A+D1+D6	−0.27	–	–	–	4.81*	–
A+D2+D3	−0.77	−0.59	−0.89	0.87	1.13	−1.38
A+D2+D4	0.58	−1.14	−1.476	0.68	2.91*	−1.51
A+D2+D5	0.12	–	–	–	2.29*	–
A+D2+D6	−0.89	–	–	–	**0.99**	–
A+D3+D4	−0.89	**−0.09**	−0.003	2.42*	0.85	−3.99*
A+D3+D5	0.001	–	–	–	0.11	–
A+D3+D6	0.06	–	–	–	−0.10	–
A+D4+D5	0.002	–	–	–	0.87	–
A+D4+D6	−0.07	–	–	–	**0.93**	–
A+D5+D6	0.63	–	–	–	*2.45**	–

Note: Significant trend at 5% significant level is denoted by an asterisk and dominant periodicities are denoted by bold letters.

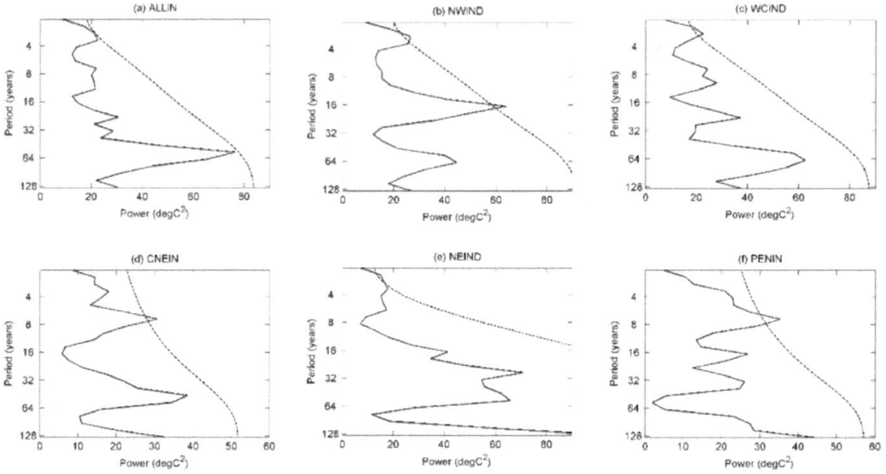

FIGURE 2.6 Global wavelet spectrum for the ADS time series (– indicates 95% significant level).

There are very few studies which related northeast (NE) monsoon with climate indices (Bhanu Kumar, 2003; Kripalani and Kumar, 2004). Khole and De (2003) reported that NE monsoon in southeastern peninsula is higher during El-nino years and vice versa. This study reports that for NE monsoon, although PENIN exhibits a dominant periodicity of 4–8 years, it was not found to be significant. For Annual SPI, only NWIND exhibited a significant 16-year periodic component. In most of the regions, ADD, ADS, and ADP time series exhibited a significant short-term periodic component of 2–8 year. Therefore, drought severity, duration and peak could be related to QBO and ENSO phenomenon.

2.7 CONCLUSION

Wavelet transform in conjunction with MK test was used to identify presence of trend and significant periods of variability in the drought variables for all India and homogenous rainfall regions of India over a period of 141 years. A significant decreasing trend in monsoon SPI was observed in CNEIN and NEIND regions. It indicates the presence of dry condition resulting from decrease in the monsoon rainfall. Furthermore, NEIND also exhibited a decreasing trend in annual SPI. No significant trends were observed in ADD, ADS, and ADP time series. Dominant periodic components influencing trends were identified using DWT, however, for most of the regions, these dominant periodic components were not found to be significant at a 95% significance level as tested using GWS. Drought variables exhibit both high (2–8-year periodicity) and low-frequency (16–64 year) components. Therefore, periodicities corresponding to QBO (2–4 years), ENSO (4–8 years) and PDO (16–32 years) influenced the trends in drought variables over India. The findings of this study were in agreement with that of Joshi et al. (2016), although spatial averaging has reduced the magnitude of trends but still significant trends were observed in some regions especially in central and northeast India.

TABLE 2.7 MK-Z Values of the ADP Time Series for the Original Data, Approximation (A), and Detail (D) Components.

	ALLIN	NWIND	WCIND	CNEIN	NEIND	PENIN
Orig	−1.38	−0.46	−1.17	−0.63	0.81	−0.67
Aprox	−1.45	−1.74	−1.19	−0.53	0.75	−1.15
D1	−0.57	0.13	−0.24	−0.05	0.22	0.13
D2	−0.55	−0.21	0.15	−0.32	0.04	0.22
D3	−0.01	0.58	0.18	0.75	0.07	0.13
D4	−0.14	0.27	1.10	−0.42	−0.28	−2.07*
D5	–	–	–	–	–	−1.02
A+D1	−3.86*	−1.76	−1.83	−0.89	2.33	−1.28
A+D2	−4.44*	−2.23*	−2.34*	−1.11	3.25	−1.39
A+D3	−2.57*	−4.22*	−2.55*	−0.93	1.97*	−3.87*
A+D4	−1.66	−1.29	−0.88	−2.13	**0.51**	−4.14*
A+D5	–	–	–	–	–	−2.49*
A+D1+D2	−3.29*	−1.63	−1.91	−0.92	1.39	**−0.54**
A+D1+D3	−3.05*	−1.22	−1.89	−0.31	1.41	−1.03
A+D1+D4	−2.69*	−1.38	**−1.11**	−1.10	1.64	−1.56
A+D1+D5	–	–	–	–	–	−1.53
A+D2+D3	−3.59*	−2.27*	−1.95	**−0.88**	2.76*	−1.35
A+D2+D4	−3.54*	−1.53	−2.50*	−1.09	2.09*	−3.15*
A+D2+D5	–	–	–	–	–	−1.22
A+D3+D4	**−1.34**	**−0.94**	−2.16*	−1.50	1.23	−3.74*
A+D3+D5	–	–	–	–	–	−3.63*
A+D4+D5	–	–	–	–	–	−2.12*

Note: Significant trend at 5% significant level is denoted by an asterisk and dominant periodicities are denoted by bold letters.

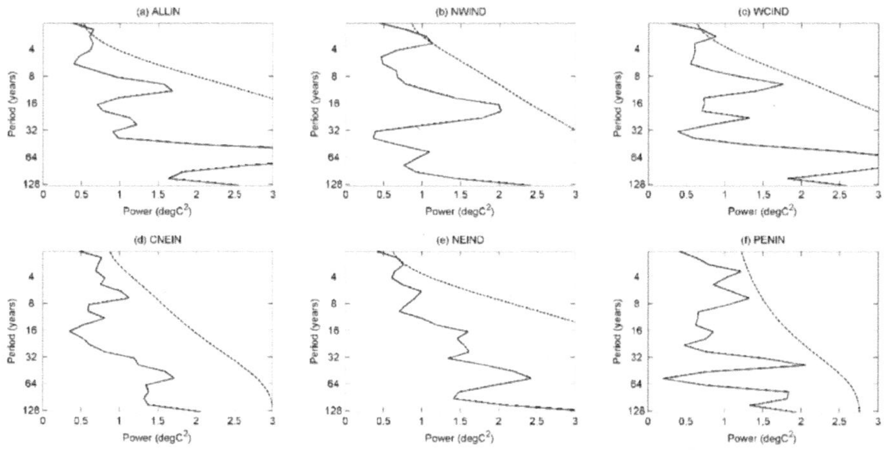

FIGURE 2.7 Global wavelet spectrum for the ADP time series (– indicates 95% significant level).

KEYWORDS

- **trend analysis**
- **drought**
- **wavelet transform**
- **standardized precipitation index**
- **Mann–Kendall test**

REFERENCES

Adamowski, K.; Prokoph, A.; Adamowski, J. Development of a New Method of Wavelet Aided Trend Detection and Estimation. *Hydrol. Process* **2009,** *23* (18), 2686–2696.

Almasri, A.; Locking, H.; Shukar, G. Testing for Climate Warming in Sweden during 1850–1999 Using Wavelet Analysis. *J. Appl. Stat.* **2008,** 35 (4), 431–443.

Araghi, A.; Baygi, M. M.; Adamowski, J.; Malard, J.; Nalley, D.; Hasheminia, S. M. Using Wavelet Transforms to Estimate Surface Temperature Trends and Dominant Periodicities in Iran Based on Gridded Reanalysis Data. *Atmos. Res.* **2015,** 155, 52–72.

Adarsh, S.; Janga Readdy, M. Trend Analysis of Rainfall in Four Meteorological Subdivisions of Southern India Using Nonparametric Methods and Discrete Wavelet Transforms. *Int. J. Climatol.* doi: 10.1002/joc.4042.

Bhanu Kumar, O. S. R. U. Influence of the ENSO and the IOD on Indian Winter Monsoon Rainfall. *14th Global Warming Conference*; Boston, May 27–31, 2003.

Bharath, R.; Srinivas, V. V. Delineation of Homogeneous Hydrometeorological Regions Using Wavelet Based Global Fuzzy Cluster Analysis. *Int. J. Climatol.* doi:10.1002/joc.4318.

Chowdhury, A.; Dandekar, M. M.; Raut, P. S.Variability in Drought Incidence over India–a Statistical Approach. *Mausam* **1989,** *40*, 207–214.

Daubechies, I. *Ten Lectures on Wavelets, 61, Society for Industrial and Applied Mathematics*; Philadelphia, PA, 1992.

de Artigas, M. Z.; Elias, A. G.; de Campra, P. F. Discrete Wavelet Analysis to Assess Long-Term Trends in Geomagnetic Activity. *Phys. Chem. Earth* **2006,** *31*, 77–80.

Das, P. K.; Datta, D.; Sharma, J. R.; Dadhwal, V. K. Trends and Behaviour of Meteorological Drought (1901–2008) over Indian Region Using Standardized Precipitation–Evapotranspiration Index. *Int. J. Climatol.* doi: 10.1002/joc.4392.

Gadgil, S.; Vinayachandran, P. N.; Francis, P. A. Droughts of the Indian Summer Monsoon: Role of Clouds over the Indian Ocean. *Curr. Sci.* **2003,** *85* (12), 1713–1719.

Ganguli, P.; Janga Reddy, M. Evaluation of Trends and Multivariate Frequency Analysis of Droughts in Three Meteorological Subdivisions of Western India. *Int. J. Climatol.* **2014a,** *34*, 911–928. http://dx.doi.org/10.1002/joc.3742.

Ganguli, P.; Janga Reddy, M. Ensemble Prediction of Regional Droughts Using Climate Inputs and the SVM–Copula Approach. *Hydrol. Process.* **2014a,** *28*, 4989–5009.

Guhathakurta, P.; Rajeevan, M. Trends in the Rainfall Pattern over India. *Int. J. Climatol.* **2008,** *28*, 1453–**1469**. http://dx.doi.org/10.1002/ joc.

Hamed, K. H.; Ramachandra Rao, A. A Modified Mann–Kendall Trend Test for Auto Correlated Data. *J. Hydrol.* **1998,** *204*, 182–196.

Joshi, N.; Gupta, D.; Suryavanshi, S.; Adamowski, J.; Madramootoo, C. A. Analysis of Trends and Dominant Periodicities in Drought Variables in India: A Wavelet Transform Based Approach. *Atmos. Res.***2016,** *182*, 200–220.

Kestin, T. S.; Karoly, D. J., Yano, J.; Rayner, N. A. Time– Frequency Variability of ENSO and Stochastic Simulations. *J. Clim.* **1998,** *11*, 2258–2272.

Khole, M.; De, U. S. A Study on North East Monsoon Rainfall over India. *Mausam* **2003,** *54* (2), 419–426.

Kumar, V.; Jain, S. K.; Singh, Y. Analysis of Long-Term Rainfall Trends in India. *Hydrol. Sci. J.* **2010,** *55* (4), 484–496.

Karthikeyan, L.; Nagesh Kumar, D Predictability of Nonstationary Time Series Using Wavelet and EMD Based ARMA Models. *J. Hydrol.* **2013,** *502*, 103–119.

Kendall, M. G. *Rank Correlation Methods*; Charles Griffin: London.

Kripalani, R. H.; Kumar, P. Northeast Monsoon Rainfall Variability over South Pe Indian Ocean Dipole Mode. *Int. J. Climatol.* **2004,** *24* (10), 1267–1282. http://dx.doi.org/10.1002/joc.1071

Krishnamurthy, L.; Krishnamurthy, V. Teleconnections of Indian Monsoon Rainfall with AMO and Atlantic Tripole. *Clim. Dyn.* **2015,** *46* (7–8), 2269–2285.

Kumar, H.; Suryavanshi, S.; Joshi, N. Study of Climatic Variability in India- A Review. *Int. J. Sci. Technol. Eng.* **2016,** *3* (1), 348–354.

Kumar Krishna, K.; Rajagopalan, B.; Cane, M. A. On the Weakening Relationship between the Indian Monsoon and ENSO. *Science* **1999,** *284*, 2156–2159.

Kumar. N. K.; Rajeevan, M.; Pai, D. S.; Srivastava, A. K.; Preethi, B. On the Observed Variability of Monsoon Droughts over India. *Weather Clim. Extremes* **2013,** *1*, 42–50.

Lau, K. M.; Weng, H. Climate Signal Detection Using Wavelet Transform: How to Make a Time Series Sing. *Bull. Am. Meteorol. Soc.* **1995,** *76,* 2391–2402.

Malik, A.; Brönnimann, S.; Stickler, A.; Raible, C. C.; Muthers, S.; Anet, J.; Rozanov, E.; Schmutz, W. Decadal to Multi-Decadal Scale Variability of Indian Summer Monsoon Rainfall in the Coupled Ocean-Atmosphere-Chemistry Climate Model SOCOL-MPIOM. *Clim. Dyn.* **2017,** 49, 3551–3572.

Mann, H. B. Nonparametric Tests against Trend. *Econometrica* **1945,** *13* (3), 245–259.

Mallat, S. G. A Theory for Multiresolution Signal Decomposition: The Wavelet Representation. *IEEE Trans. Pattern Anal. Mach. Intel.* **1989,** *11* (7), 674–693.

McKee, T. B.; Doesken, N. J.; Kliest, J. The Relationship of Drought Frequency and Duration to Time Scales. *Proc. 8th Conf. Appl. Climatol.*, 17–22 January; American Meteorological Society: Boston, MA, Anaheim, CA, 1993; pp 179–184.

Meehl, G. A.; Arblaster, J. The Tropospheric Biennial Oscillation and Asian–Australian Monsoon Rainfall. *J. Clim.* **2002,** *15* (7), 722–744.

Maheswaran, R.; Khosa, R. Comparative Study of Different Wavelets for Hydrologic Forecasting. *Comput. Geosci.* **2012,** *46,* 284–295.

Mishra, V.; Cherkauer, K. A. Retrospective Droughts in the Crop Growing Season: Implications to Corn and Soybean Yield in the Midwestern United States. *Agric. For. Meteorol.* **2010,** *150* (7), 1030–1045.

Mishra, V.; Cherkauer, K. A.; Shukla, S. Assessment of Drought Due to Historic Climate Variability and Projected Future Climate Change in the Midwestern United States. *J. Hydrometeor.* **2010,** *11,* 46–68.

Misiti, M.; Misiti, Y.; Oppenheim, G.; Poggi, J. M. *Wavelet Toolbox, User's Guide*; MathWorks Inc, USA, 2013, 593 pp.

Mondal, A.; Khare, D.; Kundu, S Spatial and Temporal Analysis of Rainfall and Temperature Trend of India. *Theor. Appl. Climatol.* **2014,** 1–16. doi: 10.1007/s00704-014-1283-z.

Mo, K. C. Model-Based Drought Indices over the United States. *J. Hydrometeorol.* **2008,** *9* (6), 1212–1230.

Naresh Kumar, M.; Murthy, C. S.; Sesha Sai, M. V. R.; Roy, P. S. On the Use of Standardized Precipitation Index (SPI) for Drought Intensity Assessment. *Meteorol. Appl.* **2009,** *16* (3), 381–389.

Nalley, D.; Adamowski, J.; Khalil, B. Using Discrete Wavelet Transforms to Analyze Trends in Streamflow and Precipitation in Quebec and Ontario (1954–2008). *J. Hydrol.* **2012,** *475,* 204–228.

Nalley, D.; Adamowski, J.; Khalil, B.; Ozga-Zielinski, B. Trend Detection in Surface Air Temperature in Ontario and Quebec, Canada during 1967–2006 Using the Discrete Wavelet Transform. *Atmos. Res.* **2013,** 132–133, 375–398.

Nourani, V.; Baghanam, A.; Adamowski, J.; Kisi, O. Applications of Hybrid Wavelet-Artificial Intelligence Models in Hydrology: A Review. *J. Hydrol.* **2014,** *514,* 358–377.

Pai, D. S.; Latha Sridhar, P.; Guhathakurta, H. H. R. District-Wide Drought Climatology of the South West Monsoon Season over India Based on Standardized Precipitation Index (SPI). *Nat. Hazards* **2011,** *59,* 1797–1813.

Partal, T. Wavelet Analysis and Multi-Scale Characteristics of the Runoff and Precipitation Series of the Aegean Region (Turkey). *Int. J. Climatol.* **2012,** *32* (1), 108–120.

Partal, T.; Küçük, M. Long-Term Trend Analysis Using Discrete Wavelet Components of Annual Precipitations Measurements in Marmara Region (Turkey). *Phys. Chem. Earth* **2006,** *31,* 1189–1200.

Percival, D. B.; Walden, A. T. *Wavelet Methods for Time Series Analysis*; Cambridge University Press: Cambridge, 2000.

Rao, R. M.; Bopardikar, A. S. *Wavelet Transforms, Introduction to Theory and Applications*; Addison-Wesley, 1998; 310 pp.

Rashid, M. M.; Beecham, M.; Chowdhury, R. K. Assessment of Trends in Point Rainfall Using Continuous Wavelet Transforms. *Adv. Water Resour.* **2015,** *82*, 1–15. http://dx.doi.org/ 10.1016/j.advwatres.2015.04.006.

Sang, Y. F. Improved Wavelet Modeling Framework for Hydrologic Time Series Forecasting. *Water Resour. Manag.* **2013,** *27*, 2807–2821.

Sikka, D. R. Monsoon Drought in India. No.2. Center for Ocean–Land–Atmosphere Studies, Center for the Application of Research on the Environment, 1999.

Sharma, V. P. *Oilseed Production in India: The Problems and Prospects*; Springer: New Delhi, India, 2017.

Shukla, S.; Wood, A. W. Use of a Standardized Runoff Index for Characterizing Hydrologic Drought. *Geophys. Res. Lett.* **2008,** *35*, L02405.

SinhaRay, K. C.; Shewale, M. P. Probability of Occurrence of Drought in Various Subdivisions of India. *Mausam* **2001,** *52*, 541–546.

Su, H.; Liu, Q.; Li, J. Alleviating Border Effects in Wavelet Transforms for Nonlinear Time-Varying Signal Analysis. *Adv. Electr. Comput. Eng.* **2011,** *11*, 55–60.

Suryavanshi, S.; Pandey, A.; Chaube, U. C.; Joshi, N. Long Term Historic Changes in Climatic Variables of Betwa Basin, India. *Theor. Appl. Climatol.* **2014,** *117* (3–4), 403–418.

Torrence, C.; Compo, G. P. A Practical Guide to Wavelet Analysis. *Bull. Am. Meteorol. Soc.* **1998,** *79*, 61–78.

Westra, S.; Sharma, A. Dominant Modes of Interannual Variability in Australian Rainfall Analyzed Using Wavelets. *J. Geophys. Res.* **2006,** *111*, D0**5102**. doi:10.1029/2005JD005996.

Zume, J.; Tarhule, A. Precipitation and Stream Flow Variability in Northwestern Oklahoma, 1894–2003. *Phys. Geogr.* **2006,** *27* (3), 189–205.

Zheng, J.; Peng, S. P.; Liu, M. C.; Liang, Z A Novel Seismic Wavelet Estimation Method. *J. Appl. Geophys.* **2013,** *90*, 92–95.

CHAPTER 3

Quantitative Assessment of Glacier Changes between 1998 and 2011 in the Parvati River Basin, Western Himalaya, India

CHANCHAL GUPTA[1*], VIKASH SHIVHARE[1], and SANJAY K. JAIN[2]

[1]*Natural Resource Department, TERI University (SAS), New Delhi 110070, India*

[2]*Water Resource System Division, National Institute of Hydrology, Roorkee 247667, India*

[*]*Corresponding author. E-mail: chanchal.iisc@gmail.com*

ABSTRACT

Remote sensing techniques, since its inception, have provided an efficient way to acquire data about glaciers in data void regions. The quantitative assessment of glaciers and their temporal dynamism require more attention as the glaciers are continuously shrinking and getting altered in the Himalayan region due to climatic variability. The Parvati river basin, a tributary of the Beas River in the western Himalayan region, was selected for the glacier change investigation using remotely sensed data in time and space. Various multisource data sets were used, that is, topographic maps from a survey of India and Landsat TM and ETM+ for 1998, 2000, 2010, and 2011 were used to study the glacier changes in the basin. Manual digitization of glaciers has been performed for the investigation of glacier changes. Glacier areas were digitized from the Landsat imageries on a different time frame and computed

Advances in Hydrology and Climate Change: Historical Trends and New Approaches in Water Resources Management. Surendra Kumar Chandniha, Anil Kumar Lohani, Gopal Krishan, & Ajay Krishna Prabhakar (Eds.)

as 16.80% in 1998, 16.08% in 2000, 14.23% in 2010, and in 2011. It was observed from the study that 2.83% of glacier area is lost from 1998 to 2011. Changes in the glaciated area were observed as 10.68 km^2 from 1998 to 2000 whereas 3.89 km^2 from 2010 to 2011 in the basin. Further, the alteration of glaciers area was estimated with respect to slope and aspect. As a result, the recession is observed in glacier cover areas in all slope ranges whereas the glacier changes in different directions indicate that the south or southwest facing glacier decreases at a higher rate compared with other direction.

3.1 INTRODUCTION

The cryosphere has several parts, among them, the glacier is the most important, which plays a significant role in the study of climate change and its adverse effects (Wessels et al., 2002; Oerlemans et al., 1998; Rai et al., 2016). Numerous studies have reported that a total number of 9575 glaciers were identified in the Indian Himalayan region (Raina and Srivastava, 2008). The Himalaya, with a total glacier covering of around 33,000 km^2, has the most extensive collection of glaciers outside the polar region (Bhambari et al., 2011, Mir et al., 2014). The primary perennial rivers of Asia, such as Indus, Ganga, and the Brahmaputra, are mainly fed by these glaciers and are the only source of fresh water in the region (Immerzeel et al., 2010). Most of the Himalayan glaciers are debris-covered and produce a large number of sediments. Apart from this, water discharge from these glaciers significantly contributes to the Himalayan Rivers' runoff (Singh and Jain, 2002, Immerzeel et al., 2010). The discharge water from these glaciers is utilized for various purposes in the downstream areas, such as irrigation and hydropower generation (Mir et al., 2014). From the climate change point of view, these glaciers are now well recognized as the most reliable indicators (IPCC, 2007). These glaciers in the Himalayan regions are continuously retreating, melting, and shrinking due to climatic instability, which causes the formation of Glacier Lake and its outburst flood (Fujita et al., 2001). These fragile environmental and climatic instability indicators produce much concern for glacier changes and alterations due to global warming. Many studies reveal that the impact of climate change on glaciers is occurring in the Himalayan region, and the glaciers have been retreating since the end of the little ice age, and it has increased during the past few decades (Kulkarni et al. 2007; Bolch et al. 2008; Bhambri et al.2011, Mir et al. 2014; Murtaza and Romshoo, 2017). The immediate impact of this rise of temperature affects the glacier recession rate in the Himalayas. Changes in river flow in the Himalayan origin rivers is highly influenced by precipitation (Seasonal and monsoon) and snow and glacier (Immerzeel et al. 2010; Mir et al., 2014; Murtaza and Romshoo, 2017). The recession or loss of the Himalayan glacier has adversely affected the water resources. The lifecycle of the glacier has been affected due to climatic variability and changes in recent decades. It is observed that the northwest Himalaya region has warmed significantly at a higher rate than the global average. A significant rise of 1.6°C from 1998 to 2002 has been reported in the northwest Himalaya (Bhutiyani et al., 2007). Due to rising temperatures in the Himalayas, glaciers are fading more rapidly than the other regions (Shekhar et al., 2010). Due to rapid melting, most glaciers have retreated remarkably in the past two decades (Fujita et al., 1997).

Therefore, mapping and monitoring of these glaciers are critical in observing the changes in the glaciated area of the Himalayas due to climate change. Various methods were developed and employed to monitor and study the glacier changes and recession rate in the Himalayas based on field investigation (Wagnon et al. 2007; Dobhal et al. 2008; Mehta et al. 2011) and remote sensing satellite imagery and topographical maps (Kulkarni, 1992; Pankaj et al. 2012; Mir et al. 2014; Murtaza and Romshoo, 2017, Shukla et al. 2016). The field-based approach for studying the glacier behavioral dynamics has several constraints for measurement. Complex terrain, remoteness, cold climatic condition, etc., of the Himalayan glacier, create many regular monitoring and data collection obstacles through field measurements and direct measurement difficulties. Thus, the remote sensing technique provides alternatives for the measurement of glacier dynamics and changes. Remote sensing provides multitemporal and multispectral and sensor satellite images to investigate the glacier health and behavior with the advantage of the synoptic view over a large area. These remotely sensed data and methods allow automated and semi-automated approaches to measuring glacier extent and changes in the rugged Himalayan terrain along with field-based glaciological measurement (Bolch et al. 2010; Mir et al. 2014; Murtaza 2015; Shukla et al. 2016).

The remote sensing technique is widely used to predict and quantify climatic variability in many forms. It provides real-time spatial and temporal information of natural resources and their physical properties, such as glacier area, length, surface elevation, albedo, equilibrium line altitude, terminus position, volume, accumulation/ablation rates, and accumulation area ratio from which glacier mass balance can be inferred. These techniques are beneficial for inaccessible areas (Lama et al., 2015). Multispectral and temporal data has abundant potential for mapping and monitoring sizeable spatial coverage of glaciers. Most of the glacier change studies from the Indian Himalayas are based on a comparison of satellite data with topographic maps (Bhambri et al., 2011). GIS is an essential and effective technique for assessing the temporal glacier area changes and glacial modeling, as remote sensing derived information can be well integrated with the database for evaluating the temporal glacier changes and snow cover areas (Racoviteanu et al., 2008). Some landmark studies using remote sensing data have been done to estimate the glacier retreat for 466 glaciers in Chenab, Parbati, and Baspa basins (Kulkarni et al., 2007), showing overall deglaciation of 21% during 1962–2001.

The glacier change/recession in the Parvati basin is a significant concern among the scientific community as the retreat rate is remarkable. Many studies suggest that a significant number of glaciers are shrinking and retreating due to climatic instability and the Parvati basin is one of them (Kulkarni et al., 2002; Ewertowski and Rzeszewski, 2007; Bolch et al., 2010). Nowadays, this valley records a significant change in glaciers retreating and shrinking down to the basin as we know that global climate is regularly changing in many ways. However, the quantification and monitoring of these changes are essential for determining future scenarios and trends. This study aims to analyze the glacier changes in the Parvati river basin, which includes the extraction of the glacier cover area from the satellite images at different times and to show the overall glacier cover changes/recession in a defined time frame.

3.2 STUDY AREA

The Parvati River, one of Beas River's major tributaries, rises from Man-Talai Glacier below the Pin-Parvati-Pass in the greater Himalaya's western slope at an altitude of 5200 m above MSL. It passes through the north direction and flows down with a gradual curve in a slender valley. Before joining the Beas River, it turns to the west after passing 7 km from its origin and meets the Beas River at Bhuntar, located 10 km southwards from the Kullu district of Himachal Pradesh, an elevation of 1096 m above MSL. Parvati river basin is located in hilly and mountain ranges, and several rocks types, such as phyllite, slate, quartzite, limestone, schist, and granite can be seen in the basin with varying altitudinal ranging from 1096 to 6250 m above MSL and are rich in floral and faunal diversity (Biswas, 2016). Parvati glacier is one the largest glacier found in the Parvati valley and is fed by almost 36 glaciers (Kulkarni et al., 2005). The Parvati basin is selected as a basic unit of study for assessing the trends and scenario of glacier area change between 1998 and

2011. The geographical area of Parvati river basin is covering approximately 1485 km^2 and extends from 31°45′ to 32°13′ North latitude 77°80′ to 77°52′ East longitude. A location map of the study area is depicted in Figure 3.1.

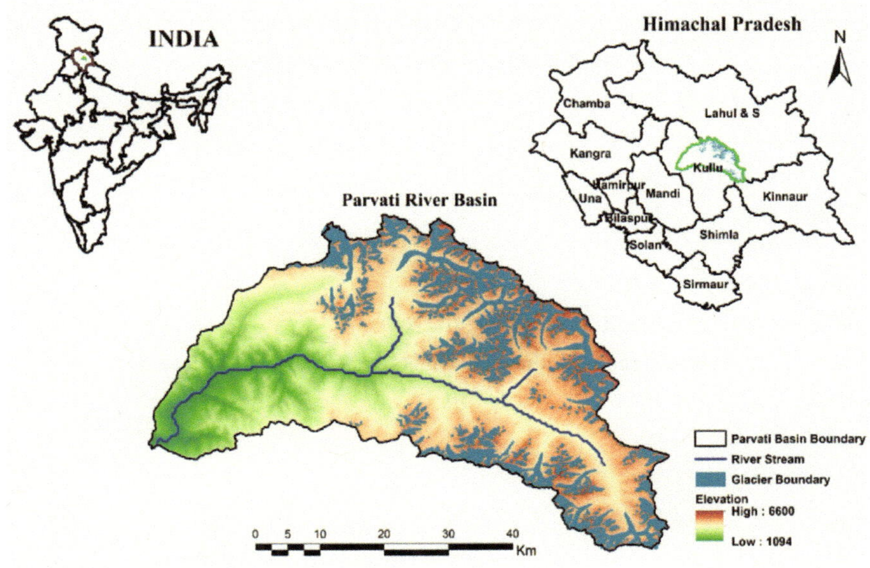

FIGURE 3.1 Location map of study area.

3.3 METHODOLOGY

Various multisource data sets were used, that is, topographic maps from a survey of India and satellite images from Landsat TM and ETM+ were used to monitor the glacier changes in the basin and obtained from the USGS website, and Digital Glacier Atlas was used as a base map for the identification of glacier cover extent. The basin was spread in several tiles, which have been mosaicked using ERDAS Imagine software. The Parvati river basin was delineated from the Terra Aster DEM. Landsat TM and ETM+ satellite images for the years 1998, 2000, 2010, and 2011 were used in this study. The specifications of its sensors are given here in Table 3.1.

Topographical maps and Glacier Atlas of India have been used and downloaded from the web (http://www.lib.utexas.edu) at a scale of 1:250000. Toposheets number NH 43-04 (Shimla) and NI 43-16 (Palampur) cover the study's whole basin. These toposheets were spatially georeferenced using boundary coordinates with projection UTM zone 43 WGS 84 and mosaic to mask out the area of interest (Parvati river basin).

TABLE 3.1 Details of Satellite Data Used in This Study.

Spectral band (μm)	Spectral channel	Spatial resolution (m)	Radiometric resolution
	Details of Landsat Thematic Mapper Sensor		
0.45–0.52 (B1)	Blue	30	8-bit
0.52–0.60 (B2)	Green	30	8-bit
0.63–0.69 (B3)	Red	30	8-bit
0.79–0.90 (B4)	NIR	30	8-bit
1.55–1.75(B5)	SWIR	30	8-bit
10.40–12.50 (B6)	TIR	120	8-bit
2.08–2.35 (B7)	SWIR	30	8-bit
	Details of Landsat Enhanced Thematic Mapper (ETM+) sensor		
0.45–0.515 (B1)	Blue	30	8-bit
0.525–0.605 (B2)	Green	30	8-bit
0.63–0.69 (B3)	Red	30	8-bit
0.75–0.90 (B4)	NIR	30	8-bit
1.55–1.75(B5)	SWIR	30	8-bit
10.40–12.5 (B6)	TIR	60	8-bit
2.09–2.35 (B7)	SWIR	30	8-bit
0.52–0.90(B8)	PAN	15	8-bit

3.4 DATA PROCESSING

ERDAS Imagine software and Arc-GIS software were used to generate the digital database and preprocessing of the satellite data. The entire databases required for the glacier change study for Parvati river basin have been developed to achieve all current study objectives. In the present study, glacier distribution and glacier change of Parvati river basin for the years 1998, 2000, 2010, and 2011 were mapped using remotely sensed data. ASTER DEM was processed to delineate stream network and basin boundary through the hydrological process, and outlet was defined at the Bhuntar where Parvati river joins with Beas river. The main procedures and the various steps followed for the study's main concern are explained sequentially. A brief flow chart of the entire methodology for glacier mapping and glacier change study in the Parvati basin is depicted below in Figure 3.2.

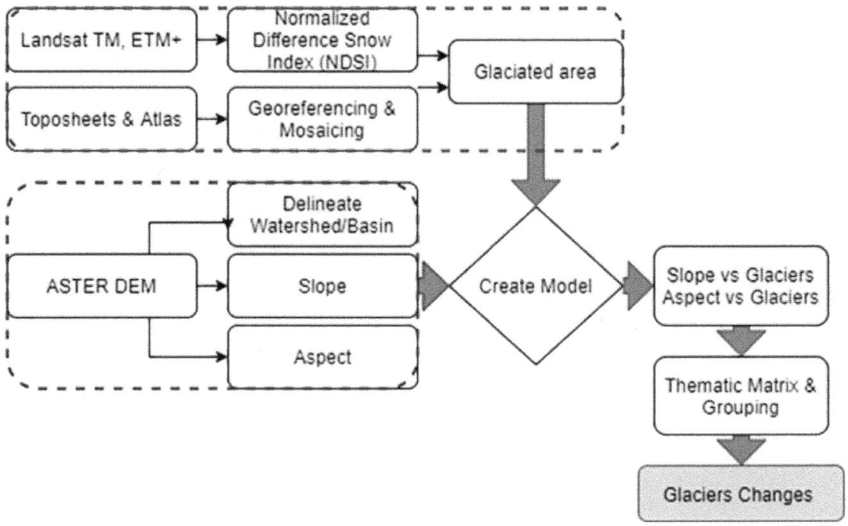

FIGURE 3.2 Flow chart for glacier mapping and glacier change study.

The basin boundary of the study area was delineated through the hydrologic process of DEM. The multispectral satellite data were used to extract/trace the glacier extent for the years 1998, 2000, 2010, and 2011. In this study, a slope map is prepared with the help of DEM at 10°C intervals. The slope in the study area varies from 0°C to 80°C in order to find the glacier over all the eight slope classes, and the slope map was intersected with extracted glacier map. Figure 3.5 represents the slope map of Parvati basin with slope ranges of 0–10°, 10–20°, 20–30°, 30–40°, 40–50°, 50–60°, 60–70°, and 70–80° were intersected with the extracted glacier extent.

Aspect, which is defined as the orientation of slope face, that is, north, east, west, south, northeast, southeast, northwest, and southwest, is a crucial constituent of topography and can have a substantial impact on snow and glacier distribution processes (Dexter, 1986; Meiman, 1970). Baumgartner and Apfl (1997), suggested that a north–south difference in snow melt is expected due to solar illumination and shadowing. The aspect map of the study area was prepared using DEM. The glaciers of the Parvati basin are manually digitized from the satellite images for the years 1998, 2000, 2010, and 2011, where topographic maps and glacier atlas from a survey of India were considered as a reference or a base map to trace the glacier coverage and classified based on stream network, such as gl_1, gl_2, etc. The area of glacier cover was calculated using Arc-GIS software. The map of a digitized glacier with satellite images is shown in Figure 3.3.

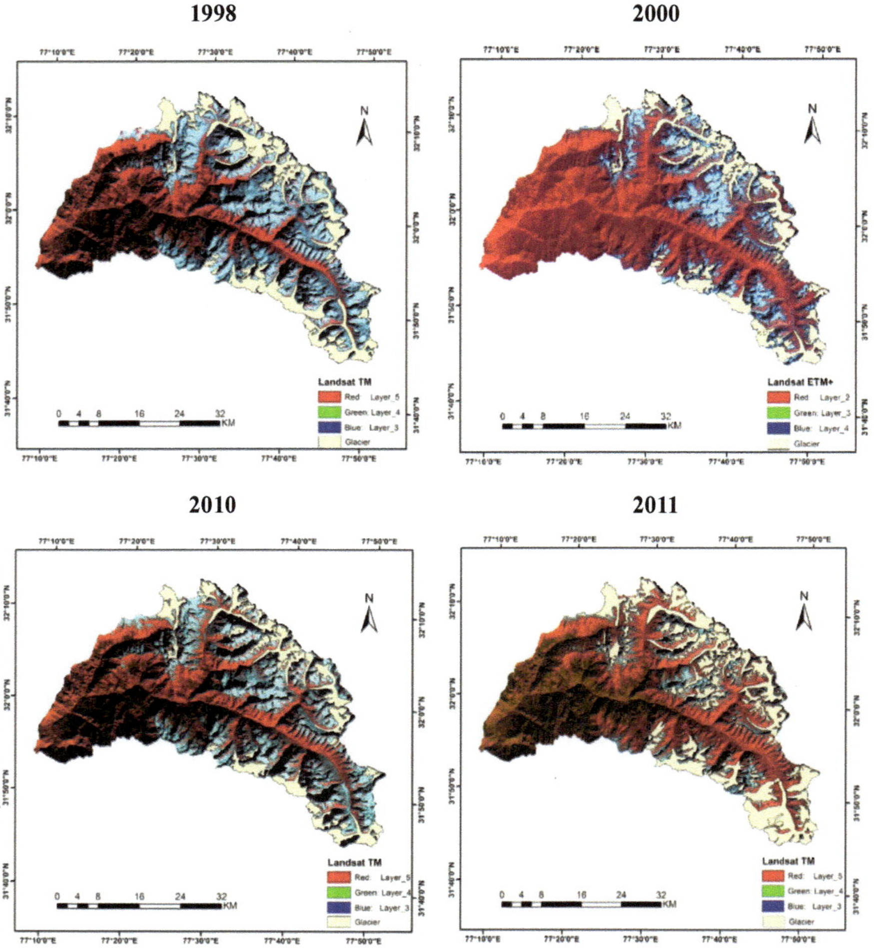

FIGURE 3.3 Delineation of glacier boundary for different years in Parvati basin.

3.5 RELATIONSHIP BETWEEN GLACIER AREA VERSUS ASPECT AND SLOPE

Aspect and slope maps have been generated from the ASTER DEM. An aspect map was generated in eight classes which shows the direction of the slopes. The DEM was taken as input, and conditional functions have been defined in the model like "(map ge 0 and map le 23)1," "(map gt 23 and map le 68)2," "(map ge 68 and map le 112)3" same, as eight classes were generated and converted it into a thematic map. The model has been built successfully, and the histogram is generated from the output image, and the area is calculated using the formula given below:

$$(\text{"Histogram"} * 0.09 * 0.01)$$

Similar procedures were being followed for the slope as well. Glacier images were digitized from the satellite imagery for respective years and were converted into a raster format. This raster file of the glacier was taken as an input to discriminate the glacier and nonglacier areas. The model's generated outputs are classified into three categories based on the function, where one is classified as glacier cover area and the other one as nonglacier cover area. The matrix was generated for the glacier and aspect through the model.

3.6 RESULTS AND DISCUSSIONS

3.6.1 CHANGE IN GLACIER AREA

All the data were processed to find spatial changes in the overall glacier area and glacier area with respect to slope and aspect for the years 1998, 2000, 2010, and 2011. In this study, the entire glacier area was mapped through Landsat imageries for the respective years. The total area in the Parvati basin is calculated and found to be about 249.36 km^2 in 1998, 238.68 km^2 in 2000, 211.31 km^2 in 2010, and 207.41 km^2 in 2011, respectively (Fig. 3.4). This is 16.79%, 16.07%, 14.23%, and 13.97% area of the total basin area (1484.50 km^2). The significant changes in the glacier area are notified and they show a decreasing trend. Change in glacier area from 1998 to 2000 is 10.68 km^2 (decrease) and from 2010 to 2011, it is 3.89 km^2 (decrease). These changes specify the effect of climate change on the glacier that glacier is continuously shrinking and retreating in the basin.

3.6.2 GLACIER AREA VERSUS SLOPE

The distribution of glaciers in various slope categories was analyzed for all the years. The maximum slope area was observed in 30.0°–40.0° slope class while the maximum glacier area comes under the slope range of 40°–60°. The slope appears to be less critical than elevation in glacier formation. The effects of slope on glacier area distribution for all the years are shown in Figure 3.5. In all the slope classes, the Glacier area shows a decreasing trend from 1998 to 2011, with the most significant reduction between 1998 and 2000.

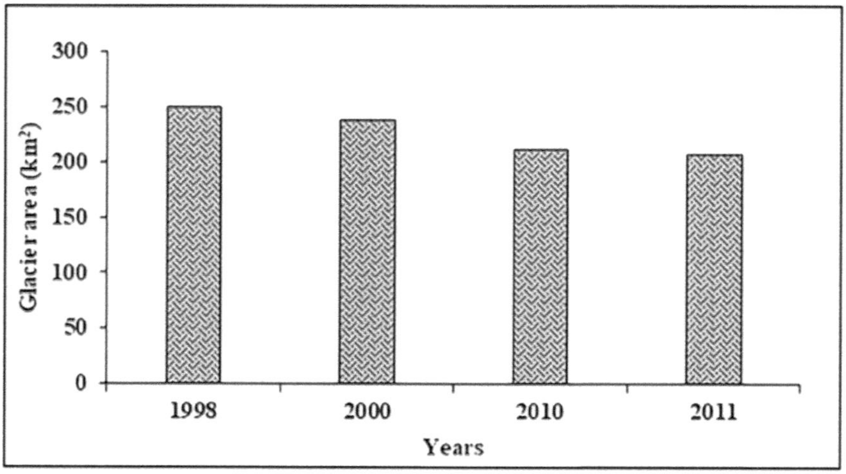

FIGURE 3.4 Area of all glaciers in km^2 for the years 1998, 2000, 2010, and 2011.

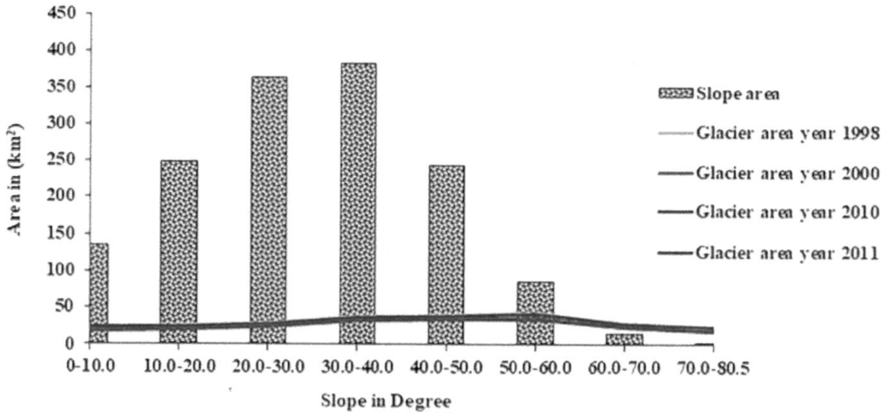

FIGURE 3.5 Distribution of glacier area (km^2) concerning the slope ranges.

As shown in Figure 3.5, the glacier area is higher for mainly 50°–60° slope ranges in 1998 and is calculated as 42.81 km^2 than to take it forward, the glacier area in 2011 is calculated as 35.21 km^2. The area of the glacier is decreasing with time variation. It shows that the glacier area is declining in all slope ranges with time. A clear picture of the change is given below in Table 3.2. As the findings show, the glacier area from 1998 to 2011 is continuously decreasing with slope and time, indicating the shrinking and retreat of the glacier from the higher degree of slope to lower ranges in the geographical region.

TABLE 3.2 Distribution of Glaciers Area Concerning Slope Ranges for the Years 1998, 2000, 2010, and 2011.

Slope (degree)	Slope area (km^2)	Glacier area km^2 (1998)	Glacier area km^2 (2000)	Glacier area km^2 (2010)	Glacier area km^2 (2011)
0–10	134.49	23.48	23.06	19.2	19.19
10–20	247.96	24.77	23.81	20.72	20.54
20–30	363.56	28.43	26.84	24.77	23.92
30–40	383.12	37.47	35.65	33.26	32.5
40–50	242.98	40.11	37.84	35.07	34.08
50–60	85.24	42.81	40.75	35.56	35.21
60–70	15.56	28.83	27.59	23.68	23.07
70–80.5	0.97	23.42	23.1	19.01	18.85
Total	**1473.91**	**249.36**	**238.68**	**211.31**	**207.41**

3.6.3 GLACIER AREA VERSUS ASPECT

The effect of aspect on glacier dynamics is shown in Figure 3.6. Since the watershed is located on the southern slope of the western Himalayan region, the north-west (NW) aspect accounts for the smallest area, while the southwest (SW) aspect has the highest (Table 3.3). Figure 3.6 shows that glacier extent was highest in the SW direction and it was minimum in the NW directions. This is due to the fact that south-facing aspects receive more solar radiation, which often enhances glacier melting between storms, resulting in less accumulation on these aspects over the years (She et al., 2015). Several temperature index-based snowmelt runoff modeling studies have comprised the effect of solar radiation on snowmelting, and there is a strong theoretical basis that direct solar radiation may cause the reduction of snow and glaciers (Shukla et al. 2016). The glacier area is consequently retreating in the southwestern and southern direction from 1998 to 2011, and the minor changes are notified on the northern side. In 1998, the southwestern side's glacier area was 42.81 km^2, 40.75 km^2 in 2000, 35.56 km^2 in 2010, and 35.21 km^2 in 2011. The calculated statistics specified that the glacier shrank mostly in southwestern and southern directions at a higher rate than the other sides of the region. The change statistics of the basin are given in Table 3.3.

Hence, all the results are successfully analyzed, and it can be believed based on the current study that the glacier area is changing due to climate variability in the geographical region, and the glacier area of the Parvati river basin has a decreasing trend.

TABLE 3.3 Changes of Glacier Area Concerning Aspect/Direction.

Direction	Glacier area km² (1998)	Glacier area km² (2000)	Glacier area km² (2010)	Glacier area km² (2011)
N	23.48	23.06	19.2	19.19
NE	24.77	23.81	20.72	20.54
E	28.43	26.84	24.77	23.92
SE	37.47	35.65	33.26	32.5
S	40.11	37.84	35.07	34.08
SW	42.81	40.75	35.56	35.21
W	28.83	27.59	23.68	23.076
NW	23.42	23.1	19.01	18.85
Total	**249.36**	**238.68**	**211.31**	**207.41**

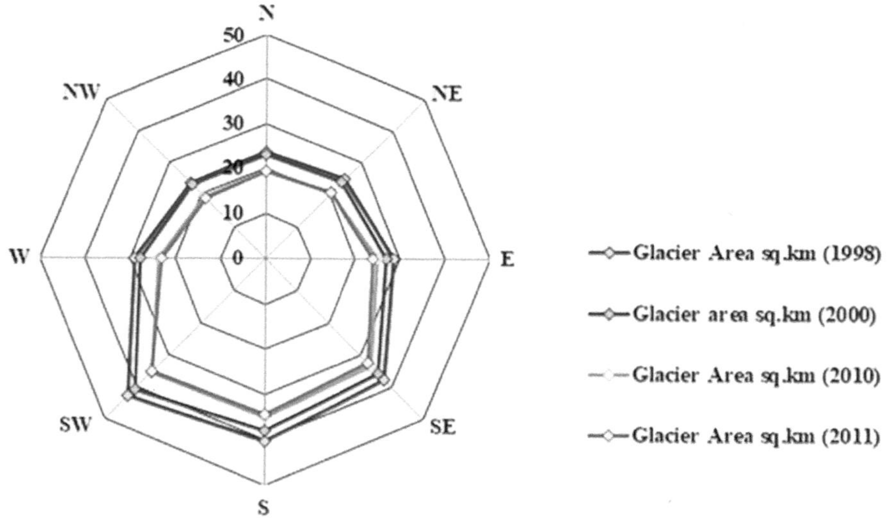

FIGURE 3.6 Glacier area changes concerning the aspect.

3.7 CONCLUSIONS

Remote sensing and GIS offer an excellent agreement for glacier monitoring and mapping for data void regions and lacking traditional glaciological methods. This paper shows a well agreement for glacier mapping and change assessment for void data regions like the Western Himalayas. Various multi-source data sets were used in the study, that is, Landsat images and ASTER

DEM are efficient for mapping and monitoring glacier areas. The basin's overall glacier area is calculated and found to be 249.36 km^2 in 1998, 238.68 km^2 in 2000, 211.31 km^2 in 2010, and 207.41 km^2 2011, respectively. This is 16.79%, 16.07%, 14.23%, and 13.97% of the entire catchment area (1484.50 km^2). Further, the Parvati basin's glaciers have shown a recession with an area loss of about 10.6 km^2 during the years 1998–2000 and 3.89 km^2 from 2010 to 2011. The study exhibits the glacier area change concerning slope range and aspect for the years 1998, 2000, 2010, and 2011.

Recession of glaciers indicates that there has been a warming trend in the Western Himalaya. The glacier change concerning slope is determined, and it was found that the glacier area is more significant in 1998 and 2011. The glacier area is mainly decreasing in all slope ranges. The glacier's calculated area in 1998 was 42.81 km^2, 40.75 km^2 in 2000, 35.56 km^2 in 2010, and 35.21 km^2 in 2011 on the 50°C–60°C slope range. This indicates that the glacier area gradually decreases with time at all slope ranges (higher to the lower range). The cross-relationship between aspect and glacier area was determined for the given year, and it was found that the glacier area was continuously changing from 1998 to 2011. These changes are mainly noted more on the southwestern side and least on the basin's northern side. This is because south-facing aspects receive more solar radiation, which often enhances the melting rate of the glacier (Shukla et al., 2016). In 1998, the southwestern facing glacier area was 42.81 km^2, 40.75 km^2 in 2000, 35.56 km^2 in 2010 and 35.21 km^2 in 2011. It shows that the glacier shrank at a higher rate with time in the basin. Therefore, it is concluded that glacier area is continuously decreasing with time due to climatic variability and requires more focus on continuous monitoring of glacier through various approaches. Additionally, a systematic, continuous, and detailed monitoring should be carried forwarded to assess the pattern of fluctuations in the coming decades and its linkage with water resources availability.

KEYWORDS

- **glacier change**
- **Parvati River basin**
- **Himalaya**
- **remote sensing**
- **GIS**

REFERENCES

Baumgartner, M. F.; Apfl, G. M. Remote Sensing, Geographic Information Systems and Snowmelt Runoff Models-an Integrated Approach; IAHS Publication, 1997; pp 73–82.

Bhambri, R.; Bolch, T.; Chaujar, R. K.; Kulshreshtha, S. C. Glacier Changes in the Garhwal Himalaya, India, from 1968 to 2006 Based on Remote Sensing. *J. Glaciol.* **2011,** *57* (203), 543–556.

Bhutiyani, M. R.; Kale, V. S.; Pawar, N. J. Long-Term Trends in Maximum, Minimum and Mean Annual Air Temperatures across the Northwestern Himalaya during the Twentieth Century. *Climatic Change* **2007,** *85* (1–2), 159–177.

Biswas, S. S. Analysis of GIS Based Morphometric Parameters and Hydrological Changes in Parbati River Basin, Himachal Pradesh, India. *J. Geog. Nat. Disasters* **2016,** *6* (2), 1–8.

Bolch, T.; Buchroithner, M.; Pieczonka, T.; Kunert, A. Planimetric and Volumetric Glacier Changes in the Khumbu Himal, Nepal, since 1962 Using Corona, Landsat TM and ASTER Data. *J. Glaciol.* **2008,** *54* (187), 592–600.

Bolch, T.; Menounos, B.; Wheate, R. Landsat-Based Inventory of Glaciers in Western Canada, 1985–2005. *Remote Sens. Environ.* **2010,** *114* (1), 127–137.

Dexter, L. R. Aspect and Elevation Effects on the Structure of the Seasonal Snow Cover in Colorado; Doctoral Dissertation, University of Colorado, 1986.

Dobhal, D. P.; Gergan, J. T.; Thayyen, R. J. Mass Balance Studies of the Dokriani Glacier from 1992 to 2000, Garhwal Himalaya, India. *Bull Glaciol Res.* **2008,** *25*, 9–17.

Ewertowski, M.; Rzeszewski, M.; *Changes of Glacier Extent: Using GIS to Integration Maps, Remote Sensing Material and Field Surveying*; Central Spitsbergen Case Study: Petuaniabukta, 2007.

Fujita, K.; Kadota, T.; Rana, B.; Kayastha, R. B.; Ageta, Y. Shrinkage of Glacier AX010 in Shorong Region, Nepal Himalayas in the 1990s. *Bull. Glaciol. Res.* **2001,** *18*, 51–54.

Fujita, K.; Nakawo, M.; Fujii, Y. and Paudyal, P. Changes in Glaciers in Hidden Valley, Mukut Himal, Nepal Himalayas, from 1974 to 1994. *J. Glaciol.* **1997,** *43* (145), 583–588.

Hasnain, S. I. Impact of Climate Change on Himalayan Glaciers and Glacier Lakes. In *Proceedings of the Taal 2007: The 12th World Lake Conference*, 2008; pp 1088–1091.

Immerzeel, W. W.; Van Beek, L. P.; Bierkens, M. F. Climate Change Will Affect the Asian Water Towers. *Science* **2010,** *328* (5984), 1382–1385.

IPCC. *Climate Change 2007: Synthesis Report. Contribution of Working Groups I, II and III to the Fourth Assessment Report of the Intergovernmental Panel on Climate Change* [Core Writing Team; Pachauri, R. K., Reisinger, A., Eds.]; IPCC: Geneva, Switzerland, 2007; p 104.

Knight, P. G. *Glaciers*; Psychology Press, 1999.

Kulkarni, A. V. Mass Balance of Himalayan Glaciers Using AAR and ELA Methods. *J. Glaciol.* **1992,** *38*, 101–104.

Kulkarni, A. V.; Bahuguna, I. M.; Rathore, B. P.; Singh, S. K.; Randhawa, S. S.; Sood, R. K.; Dhar, S. Glacial Retreat in Himalaya Using Indian Remote Sensing Satellite Data. *Curr. Sci.* **2007,** *92* (1), 69–74.

Kulkarni, A. V.; Mathur, P.; Rathore, B. P.; Alex, S.; Thakur, N.; Kumar, M. Effect of Global Warming on Snow Ablation Pattern in the Himalaya. *Curr. Sci.* **2002,** *83* (2), 120–123.

Kulkarni, A. V.; Rathore, B. P.; Mahajan, S.; Mathur, P. Alarming Retreat of Parbati Glacier, Beas basin, Himachal Pradesh. *Curr. Sci.* **2005,** *88*, 1844–1850.

Lama, L.; Kayastha, R. B.; Maharjan, S. B.; Bajracharya, S. R.; Chand, M. B.; Mool, P. K. Glacier area and volume changes of Hidden Valley, Mustang, Nepal from~ 1980s to 2010 Based on Remote Sensing. *Proc. Int. Assoc. Hydrol. Sci.* **2015,** *368*, 57–62.

Mehta, M.; Dobhal, D. P.; Bhist, M. P. S. Change of Tipra Glacier in the Garhwal Himalaya, India, between 1962 and 2008. *Progress Phys. Geogr.* **2011,** *35*, 721–738.

Meiman, J. R. Snow Accumulation Related to Elevation, Aspect, and Forest Canopy. In Snow Hydrology: Proceedings. Workshop Seminar on Snow Hydrology, Fredericton, Feb 28–29, 1969, 1970; pp 35–47.

Mir, R. A.; Jain, S. K.; Saraf, A. K.; Goswami, A. Glacier Changes Using Satellite Data and Effect of Climate in Tirungkhad Basin Located in Western Himalaya. *Geocarto Int.* **2014,** *29* (3), 293–313.

Murtaza, K. O.; Romshoo, S. A. Recent Glacier Changes in the Kashmir Alpine Himalayas, India. *Geocarto Int.* **2017,** *32* (2), 188–205.

Oerlemans, J.; Anderson, B.; Hubbard, A.; Huybrechts, P.; Johannesson, T.; Knap, W. H.; Schmeits, M.; Stroeven, A. P.; Van de Wal, R. S. W.; Wallinga, J.; Zuo, Z. Modelling the Response of Glaciers to Climate Warming. *Clim. Dyn.* **1998,** *14* (4), 267–274.

Pankaj, A.; Kumar, P.; Mishra, A. Extraction of Glacio-Geomorphological Units of Tons River Watershed Based on Remote Sensing Techniques. *J. Indian Soc. Remote Sens.* **2012,** *40*, 725–734. doi:10.1007/s12524-011-0151-x.

Racoviteanu, A. E.; Arnaud, Y.; Williams, M. W.; Ordonez, J. Decadal Changes in Glacier Parameters in the Cordillera Blanca, Peru, Derived from Remote Sensing. *J. Glaciol.* **2008,** *54* (186), 499–510.

Rai, P. K.; Mohan, K.; Mishra, V. N.; Biswari, M.; Sharma, A.; Rai, S. Changing Regimes of Gangotri and Surrounding Glaciers: A Case Study of Garhwal Himalaya, India. *Remote Sens. App.: Soc. Environ.* **2016,** *3*, 53–72.

Raina, V. K. *A State-of-art Review of Glacial Studies, Glacial Retreat and Climate Change*; Discussion Paper. MOEF, New Delhi, 2009.

She, J.; Zhang, Y.; Li, X.; Feng, X. Spatial and Temporal Characteristics of Snow Cover in the Tizinafu Watershed of the Western Kunlun Mountains. *Remote Sens.* **2015,** *7* (4), 3426–3445.

Shekhar, M. S.; Chand, H.; Kumar, S.; Srinivasan, K.; Ganju, A. Climate-Change Studies in the Western Himalaya. *Ann. Glaciol.* **2010,** *51* (54), 105–112.

Shukla, S.; Kansal, M. L.; Jain, S. K. Snow Cover Area Variability Assessment in the Upper Part of the Satluj River Basin in India. *Geocarto Int.* **2016,** 1–22.

Singh, P.; Jain, S. K. Snow and Glacier Melt in the Satluj River at Bhakra Dam in the Western Himalayan Region. *Hydrol. Sci. J.* **2002,** *47* (1), 93–106.

Wagnon, P.; Linda, A.; Arnaud, Y.; Kumar, R.; Sharma, P.; Vincent, C.; Pottakkal, J. G.; Berthier, E.; Ramanathan, A.; Hasnain, S. I.; Chevallier, P. Four Years of Mass Balance on Chhota Shigri Glacier, Himachal Pradesh, India, a New Benchmark Glacier in the Western Himalaya. *J Glaciol.* **2007,** *53*, 603–611.

Wessels, R. L.; Kargel, J. S.; Kieffer, H. H. ASTER Measurement of Supraglacial Lakes in the Mount Everest Region of the Himalaya. *Ann. Glaciol.* **2002,** *34* (1), 399–408.

CHAPTER 4

Soft Computing Technique in the Water Sector: Artificial Neural Network Approach

HIMANSHU PANJIAR[1*] and ANKIT CHAKRAVARTI[2]

[1]*Faculty of Matersial Science and Engineering, Jimma University, Jimma, Ethiopia*

[2]*Faculty of Hydraulics & Water Resources Engineering, School of Civil and Environmental Engineering, Ambo University, Technology Campus, Ambo, Ethiopia*

[*]*Corresponding author. E-mail: hpanjiar92@gmail.com*

ABSTRACT

Soft computing techniques are general statistical tools used in various ways to fit the data-based process and satisfy the dire need of a specific application. Artificial neural network (ANN) is one of the soft computing techniques and could be applied effectively in water sector management. This study provides an overall view about a type of soft computing techniques that is an ANN, which covers its history, few of their kinds with training algorithm, model testing, shortcoming with training, and finally its effective application in one of the water sector managements. There are numerous ANNs existing today based on various neural network structures and training algorithms and it is not feasible to cover all, so in this study only three types of ANN which are generally used, namely, feed-forward backpropagation, generalized regression, and radial basis function neural networks are discussed in detail with their

Advances in Hydrology and Climate Change: Historical Trends and New Approaches in Water Resources Management. Surendra Kumar Chandniha, Anil Kumar Lohani, Gopal Krishan, & Ajay Krishna Prabhakar (Eds.)

training algorithms. And finally, the development, application of rainfall-runoff model and its implementation using ANN in water sector are discussed.

4.1 INTRODUCTION

In this modern time, all the statistical things are worked out using various soft computing techniques as per the requirement. And the field of various water sector management experiencing huge data-based problems due to different water resources utility expansion globally and that has to be addressed from time to time, otherwise, water resources imbalance in supply and demand will be raised and can go out of control. In the case of statistical-based study, the soft computing techniques are very useful and their application in water sector management was found to be very promising (Yang, 1996; Cigizoglu and Alp, 2004; Cigizoglu and Alp, 2006; Tayfur, 2012; Oyebode and Stretch, 2018).

Soft computing is a term first coined by Professor Lotfi Zadeh (Zadeh, 1994), also known for the concept of fuzzy logic. The soft computing is a series of techniques relying on computer and software-based platform instead of hard computing methods and that deals with the practical data in the same way as humans deal with them. The foundation of soft computing is actually based on practical input, common sense, knowledge about relation between input and output, and reasoning on artificial as well as natural ideas.

The overall classification of soft computing techniques used in the field of water sector is shown in Figure 4.1. The major coverage of soft computing in the field of water sector is addressed by neural network (mostly used term is artificial neural network (ANN)), fuzzy logic, support vector machine and genetic algorithm. All these disciplines of soft computing can be employed on the problem individually as required or as hybrids of techniques (Sivanandam and Deepa, 2011; Tayfur, 2012).

So, it is better to understand little about soft computing properties and further each methods of soft computing in instructive and informative manner, and finally their direct application in the field of water sector, while the present study is limited to one of the soft computing techniques, that is, ANN.

4.1.1 SOFT COMPUTING PROPERTIES

One of the main properties of soft computing is to have reasoning and thinking characteristics of human mind as closely as possible, by which its primary goal is to follow the way as human mind works. Further, a specific property of soft computing which makes it attractive for statistical analysis is learning from experimental data, and derives their internal relation from interpolating or approximating to produce outputs from previously unobserved inputs by using trained model which captured the input–output relation from previous learning. So, soft computing based on approximate models rather than exact model as in the case of hard computing.

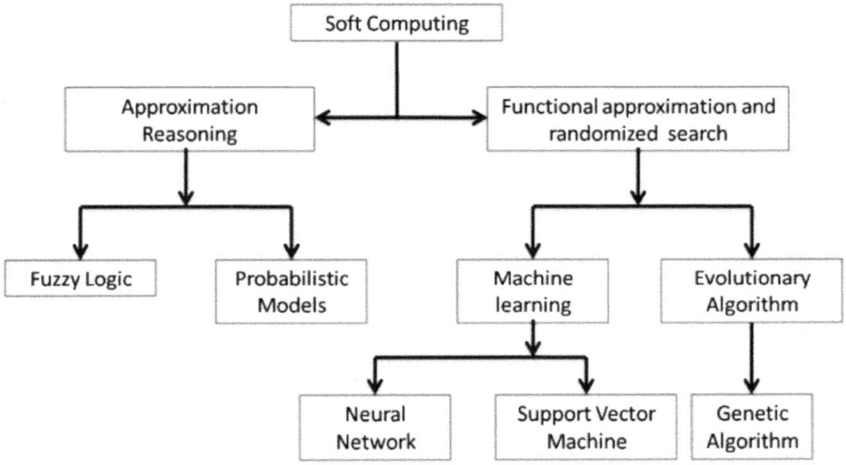

FIGURE 4.1 Classification of soft computing.

Based on the above properties, the real-world problems which are pervasively imprecise and unsure and can be considered under the premises of soft computing, as the rationale behind soft computing is to exploit the approximation, tolerance for inaccuracy, and uncertainty to achieve robustness and tractability with low solution cost. Out of all classified methods under soft computing, the field of probabilistic reasoning is also sometimes included due to its better control of randomness and uncertainty, while not explored much for its application in water sector. So, with this brief introduction of soft computing, we further focus on the ANN, as this soft computing method is considered for the study and its effective application in water source management is also discussed.

4.2 ARTIFICIAL NEURAL NETWORKS

ANNs are just a small imitation capability in mathematical form which was inspired by the information-processing capability of a brain. ANNs are intemperately parallel computing system comprising of a huge number of processors with numerous interconnections. The main ANN model effort is to apply some organizational principles which are believed to exist in the human brain (Mohiuddin and Mao, 1994). ANN is made based on various functions that a human brain does, apart from intelligence. ANN represents the concept of human brain function (i.e., biological neural networks) and has many desirable characteristics such as:

a. Massive parallelism
b. Ability to learn
c. Ability to generalize
d. Inherent contextual information processing
e. Very low-energy consumption during ANN model development.

ANN research has experienced its initialization and further development in three periods of extensive research activity. In the beginning, the first peak came in the 1940s was due to a pioneer research work (McCulloh and Pitts, 1943). The second peak occurred in the 1960s with another famous work "Rosenblatt's perceptron convergence theorem" (Rosenblatt, 1962), and thirdly Minsky and Papert's research work with limitations of a simple perceptron (Minsky and Papert, 1996).

4.2.1 BIOLOGICAL NEURAL NETWORK OVERVIEW

A nerve cell (or neuron) is the only unit responsible for information processing in human brain, which is also called as special biological cell (see Figure 4.2). It comprises of soma (a cell body), the axon and the dendrites (axon and dendrites are tree-like branches). The soma has a nucleus, and that nucleus contains hereditary traits information and plasma. There is the molecular equipment held by plasma and that produces material required by the neuron during information processing.

A neuron receives impulses (signals) from other neighboring neurons through its dendrites (also termed as receivers) and transmits signal along the axon (transmitter). At the terminal of these strands are the synapses, a synapse is the elementary construction and functional unit between two connected neurons (one neuron axon strand and a dendrite of another neuron). When the impulse reaches the synapses terminal, certain chemical substances called neurotransmitters are released as shown in Figure 4.3 (Jain and Mao, 1994).

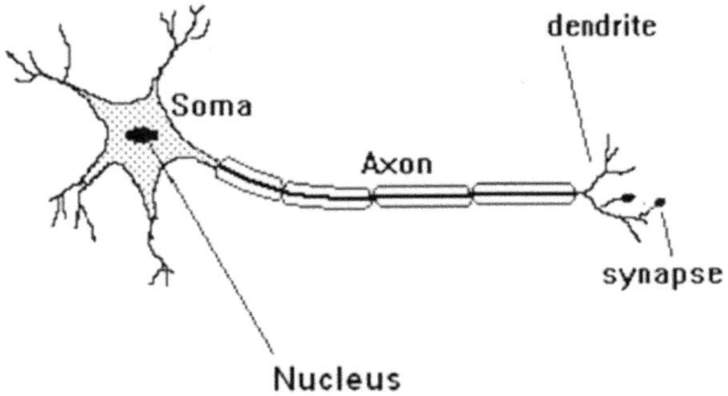

FIGURE 4.2 A sketch of a biological neuron.

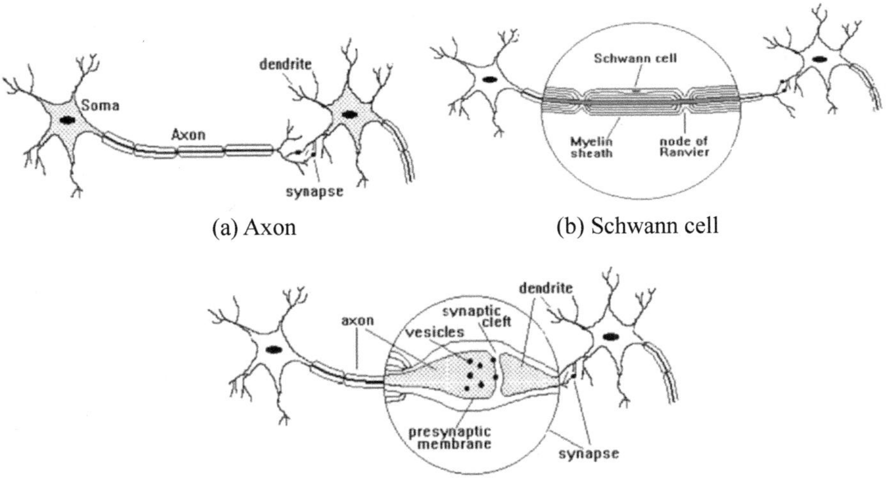

(a) Axon (b) Schwann cell

(c) Synapse

FIGURE 4.3 Schematic of neuron connection.

The neurotransmitters have the ability to diffuse across the synaptic gap, and depending on the type of the synapse, it enhances or inhibits the signals. This passing of signals through synapse adjusting the synapse's effectiveness in such a manner so that the synapse learning can take place from the participating activities. This type of learning dependence on history acts as a memory, which is also responsible for human memory. The human memory built of brain cells named as cerebral cortex, which is about 2–3 mm thick and having a large flat sheet of neurons with a surface area of about 2200 cm^2 approximately. The cerebral cortex contains approximately 10^{11} neurons, and these neurons are massively connected. Every neuron in cerebral cortex is connected to 10^{13}–10^{14} other neurons. In total, the human brain contain approximately 10^{14}–10^{15} interconnections.

4.2.2 BASICS OF COMPUTATIONAL MODEL OF ARTIFICIAL NEURONS

As ANN is completely influenced by biological neuron network, so to create such working network artificially, there are few things to be known first, and those are the critical components of ANN that make sense of neural network in such a way that it can be used for some real-world practical applications. Actually, ANN consists of four main components which include input vector, synaptic weight, bias, and intermediate nodes or output which depends on network type, and further, few operations such as activation function operation to adjust the synaptic weight of the ANN such that the error in network is reduced and network output computation operation is generally a summation of node values. All the above items are working together to make the ANN to mimic the actual science behind particular process and to behave like a well-trained function for that particular process. To understand the ANN in better way, it is good to see the things with historical work reported in the literature, and additionally explanations of the subtopic provided wherever necessary which will develop better understanding of ANN characteristics. The artificial neuron of ANN was made analogous to the biological neuron in the following terms as given in Table 4.1.

TABLE 4.1 Analogy between Artificial Neuron and Biological Neuron.

Artificial neuron	Biological neuron
Output	Axon
Input	Dendrites
Node	Cell Nucleus (Soma)
Weights or interconnections	Synapse

In ANN, there is a network of multiple nodes. The ANN is represented by three layers, such as input layer, hidden layer (single or multiple), and an output layer. Each point of network except the input nodes is called as artificial neuron or perceptron, and sometimes that the perceptron consists of a neuron with adjustable weights also known as synaptic weight, w_j, $j=1, 2, 3\ldots, n$ and threshold (u), as shown in Figure 4.4 (McCulloh and Pitts, 1943). There are two types of perceptrons: single-layer perceptron and multilayer perceptron. Multi-layer perceptrons (MLPs) are also known as feed-forward neural networks with two or more hidden layers and have the greater processing power, while single-layer perceptrons are capable of learning only linearly separable patterns.

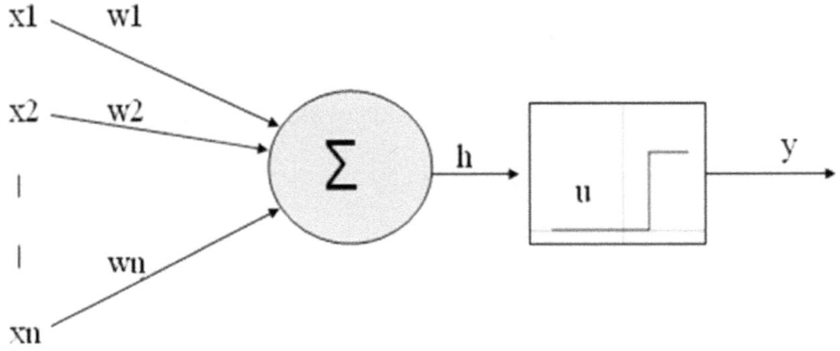

FIGURE 4.4 McCulloch-Pitts model of a neuron.

McCulloh and Pitts (1943) first proposed a computational model with binary threshold (activation function) unit for an artificial neuron in 1943. McCulloh and Pitts model contains some simplifying assumptions that actually do not reflect the true behavior of biological neurons because of the use of threshold activation function. There are various activation functions as shown in Figure 4.5 that can be used to model an artificial neuron which will reflect approximately the true behavior of a biological neuron. As the basic ANN has few things in it which has to be clear first, the further description about threshold or activation function which is represented by "u" in McCulloch-Pitts model of a neuron is described first and followed by the description of basic ANN connection pattern, and further, ANN learning methods are also summarized which are also called learning paradigms.

There are several activation functions, out of which, the sigmoid function is the most preferable due to its nonlinear characteristics, and even the ANN is generally used to capture the nonlinear relationship between data. So, in ANN, the sigmoid activation function is by far the most often used function with range from 0 to 1, and is a strictly increasing function which also shows smoothness with desired asymptotic properties. There is also a hyperbolic tangent (*tanh*) function similar to sigmoid function, which is just a rescaled logistic sigmoid function between the ranges −1 to +1.

Deciding between the *tanh* or sigmoid depends on the requirement of gradient strength as there is a point to mention which is that the gradient is weaker for sigmoid than *tanh* due to range property. The general mathematical representations of activation functions are shown in Table 4.2, while for further readings on activation functions, few literatures can be followed (Sivanandam and Deepa, 2011; Tayfur, 2012; Haykin, 1994). Threshold is a term used in ANN associated with activation function, which restricts or limits the activation function. In neural network, based on the set value of threshold, the activation functions are defined and the network output is calculated. For example, in the case of step activation function mentioned in Table 4.2, thes *f(x)* value restriction is applied at certain value of *x* and that is called as fixed-threshold value.

(a) Threshold (b) Piecewise linear (c) Sigmoid (d) Gaussian

FIGURE 4.5 Different types of activation functions.

TABLE 4.2 Mathematical Representations of Activation Functions.

Activation functions	Mathematical representations (general)	Remarks
Step	$f(x)=\begin{cases}1x>0\\0x<0\end{cases}$	Also called as unit step function if output is 1 at threshold value
Piecewise linear (also called as Rampage function)	$f(x)=\begin{cases}1x>1\\ax-1<x<1\\1x<-1\end{cases}$	In case of linear only $f(x)=a.\,x$
Sigmoid	$f(x)=\dfrac{1}{1+e^{-x}}$	In case of hyperbolic tangent $f(x)=\dfrac{2}{1+e^{-x}}-1$
Gaussian	$f(x)=\exp\left(\dfrac{-x^2}{v}\right)$ where v is variance of the distribution	Gaussian activation function maps have normal bell-shaped distribution for the input–output relation

ANN can be viewed as a weighted directed graph, in which nodes (artificial neurons) and directed edges (with weights) are present. The nodes are connected by directed edges. The weights of directed edges if positive, then that correspond to excitatory synapses, while vice versa for inhibitory once.

On the basis of ANN connection pattern, it is grouped into two following categories:

a) Feed-forward networks, graphs with no loops.
b) Recurrent (or feedback) networks, graphs with loops occurrence because of feedback connections.

In most common family of feed-forward ANN, called MLP, nodes are arranged in layers that have unidirectional connections as shown in Figure 4.6, where all the arrows shown carry a synaptic weight themselves. Different types of connectivity yields different network behaviors. In general, the feed-forward ANNs are static.

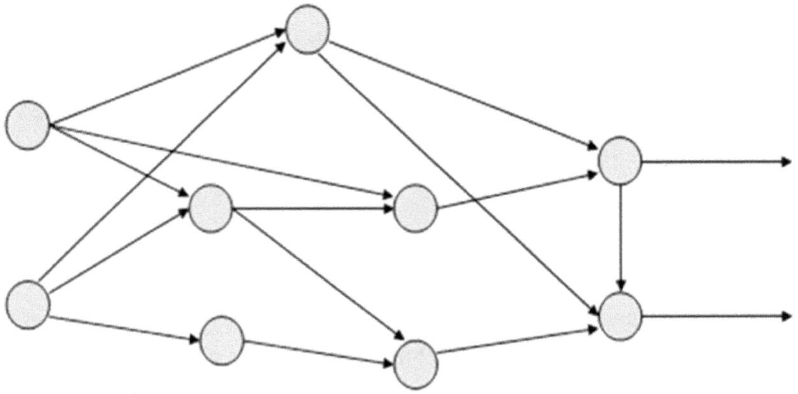

FIGURE 4.6 Feed-forward network.

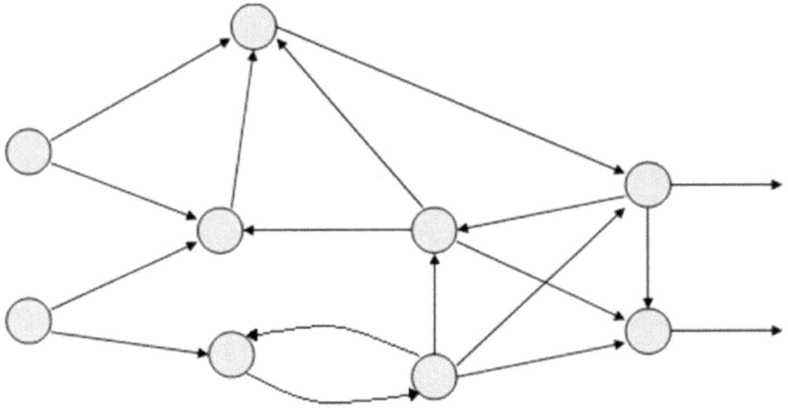

FIGURE 4.7 Recurrent or feedback network.

Recurrent or feedback networks are dynamic systems as shown in Figure 4.7, whenever a new input set is fed into network, the node outputs are computed. Due to feedback paths, the inputs to each node are then modified, which finally leads the network to enter a new state.

The networks during training usually learn and adjust the connection weights from the available training patterns. Performance is improved over time by this iteratively updating of the weights in the ANN (Mohiuddin and Mao, 1994). This ability of ANN to automatically learn from examples builds their attractive and exciting features. To design or understand an ANN learning process, at first, we must have the environment information in which a neural network operates, and that result in a model. This model is referred as a learning paradigm (Haykin, 1994). While at second phase, we must realize how network weights are getting updated, that is where network learning rules govern the updating or adjusting process. The weights of the network adjusting by some learning rules are referred to as learning algorithm. The three main types of learning paradigms which are described here under and also shown in Figure 4.8:

a. Supervised learning
b. Unsupervised learning
c. Reinforcement learning

In the case of supervised learning process (learning with a teacher), the network is supplied with a correct output against every input pattern. The weights in network are determined and adjusted during ANN training to allow the ANN to produce outputs near possible value to the known observed outputs. The schematic of supervised learning is shown in Figure 4.8a.

In the case of unsupervised learning process (learning without a teacher), the network does not require a correct output related with each input vector in the training dataset. During this ANN training, it explores the underlying structure information in the data, or correlations between patterns in the data, and further organizes these patterns into categories from founded correlations. The schematic of unsupervised learning is shown in Figure 4.8b.

Reinforcement learning is like supervised learning in nature but differs in error correction method. In the case of supervisor learning, the output value is known for each input pattern and its difference goes back to network for the adjustment of network weight while in the case of reinforcement learning, the actual output value is not used for learning purpose. Instead, the network response acts as critic information such as the observed output is only 60% correct or so. So here, the exact information is not available, only critic information is available. This "critic information"-based learning is known as reinforcement learning and the feedback sent back to network is called as reinforcement signal as shown in Figure 4.8c. So, we can say that the reinforcement learning is similar to the supervised learning, as the network gets some feedback regarding error in the final output from its environment.

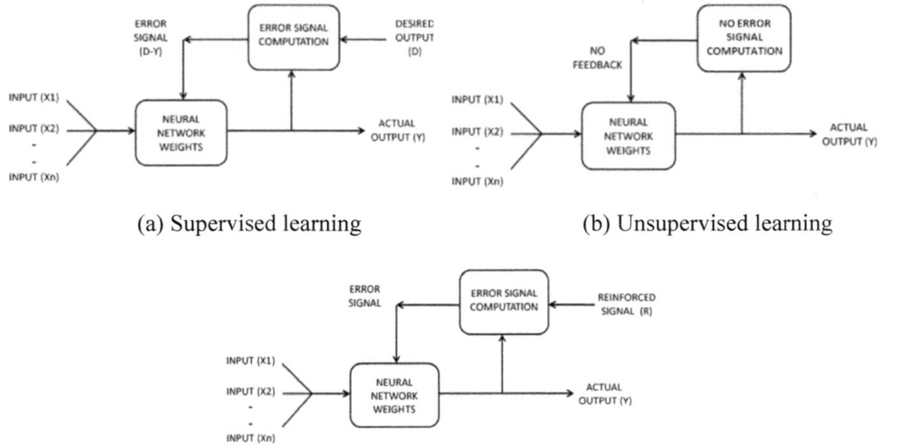

(a) Supervised learning (b) Unsupervised learning

(c) Reinforcement learning

FIGURE 4.8 Network learning paradigms.

Learning hypothesis necessity is to address the three practical and fundamental issues related to learning from samples data: network capacity, the computational complexity, and sample complexity. Each one of them plays a major role in the development of a good ANN model. The network capacity concerns about how many network patterns can be stored, and further what decision boundaries and functions a network can form. The sample complexity actually decides the number of network training patterns required to train the network to assure a valid generalization. There may be chances that too few patterns may cause "overfitting." The computational complexity refers to the actual time required for learning algorithm/rules to decide a solution from training patterns. There are several learning rules, out of which one has to be implemented in the process of ANN learning, as it is known that basically learning process are categorized into two groups, one is supervised learning and another is unsupervised learning. In the case of supervised learning, there is a trainer to train the ANN which influences the decision on size of testing and training data sets, type of the training (such as backpropagation, radial basis, perceptron training, etc.), network structure (such as single or multiple hidden layers and number of neurons in it), learning rate (α), and the number of iterations (epochs). In the case of the unsupervised learning, the network searching for trends in the input signals (say x_i) and further adaptation takes place accordingly. In all the learning rules, initially, connection weights (say w_{ij}) are initialized with random numbers, and further with training, these weights are adjusted from old (say w_{ij}^{old}) to new (say w_{ij}^{new}) to meet the requirement of less error in network or actual output y with respect to the desired or target output y_t.

In each learning process, there are various learning rules, out of which any rule can be implemented as per requirement. In the case of supervised learning, some of the commonly employed learning rules are as follows:

a. Perceptron learning or error correction learning
b. Delta learning
c. Widrow–Hoff learning
d. Boltzmann learning

(a) Perceptron learning: The perceptron learning rule is based on the synaptic weight error correction principle only during supervised learning, as the neural network training starts by random weight initialization and further the error reduces with number of iterations (epochs) for any ANN. In the supervised learning paradigm, the network is fed with a desired output for each input vector. And during learning, the ANN output, say y generated by the network may not be equal to the observed/target output y_t. The base principle of perceptron learning rule is to use the error signal $(y_t - y)$ to adjust the network connection weights to scale down the error (Anderson and Rosenfeld, 1988). In perceptron learning rule, the connection weight adjustment is formulated as (Tayfur, 2012):

$$w_{ij}^{\text{new}} = w_{ij}^{\text{old}} - \alpha \left[y_t - f\left(w_{ij} x_i \right) \right] x_i \tag{4.1}$$

Where $f(w_{ij} x_i)$ is the term that represents the summation operation on the product of respective input and connection weights, the eq 4.1 is applicable to MLPs as this is actually used for ANN model development for any application process, while the same function is applicable to single-layer perceptron with required modification according to network.

(b) Delta learning: The delta learning rule is used during supervised learning. In this learning, the weights between the connections update so as to minimize the difference between target value and the network output value. The main aim is to reduce the error over all training patterns. Each pattern error is reducing by this learning, one at a time. In perceptron learning rule the connection weight adjustment is formulated as (Tayfur, 2012):

$$w_{ij}^{\text{new}} = w_{ij}^{\text{old}} - \alpha\left(\frac{\partial E(w_{ij})}{\partial w_{ij}}\right) \tag{4.2}$$

In eq 4.2, the term in the right-hand side, inside the bracket $\left(\frac{\partial E(w_{ij})}{\partial w_{ij}}\right)$, represents the negative gradient for weight w_{ij} as per the generalized delta rule, which is multiplied by the learning rate. The generalized delta rule is actually based on the steepest gradient descent method, where the gradient vector being set to the negative direction vector. Therefore, the solution always follows a zig-zag path as trying to achieve a minimum error position. Sometimes, it is also possible to get trapped in a local minimum, but the actual desire is to achieve the global minimum position where one achieves connection weights optimal values.

(c) **Widrow–Hoff learning:** This learning rule is used during supervised learning, and it is very similar to the perceptron learning rule. The basic characteristic of this learning is that this learning is independent of an activation function and is formulated as (Tayfur, 2012):

$$w_{ij}^{\text{new}} = w_{ij}^{\text{old}} - \alpha\left[y_t - \left(w_{ij}x_i\right)\right]x_i \tag{4.3}$$

As seen in this learning rule, there is no need for the data standardization since eq 4.3 does not involve any activation function.

(d) **Boltzmann learning:** This learning rule is similar to perceptron learning method and is used during supervised learning. In this learning algorithm, each individual neuron, in addition to the network output, is taken into account, and the difference between the system probability distributions is considered here for synaptic weight adjustment instead of a direct difference between the network result value and the observed value as in the case of perceptron learning. Boltzmann learning rule is substantially slower than the perceptron learning rule.

In the case of unsupervised learning, some of the commonly employed learning rules are as follows: (1) Hebbian learning, (2) Competitive learning, and (3) Learning vector quantization.

1. **Hebbian learning:** This learning is actually instigated by the weight adjustment mechanism of biological neural. This method converts a neuron inability to learn and enables it to develop cognition with response to external stimuli whenever some input is given in to the network. This learning rule is not used in the supervised learning, as it can be seen from its formulation (eq 4.4) that there is no involvement of any target output value (Tayfur, 2012):

$$w_{ij}^{\text{new}} = w_{ij}^{\text{old}} - \alpha\left[f\left(w_{ij}x_i\right)\right]x_i \tag{4.4}$$

2. **Competitive learning:** This learning rule is a form of unsupervised learning, in which each hidden nodes compete for the right to react to a subset of the input vector data. This learning process is usually applied to ANN that contains a hidden layer, which is also known as competitive layer. Each competitive neuron is described by a weight vector w_{ij} and computes the similarity measure between the weight vector and the input data x_n. For every input vector data, each competitive neuron is computing its value separately and "compete" with each other to check similarity with particular input vector. The winner competitive neuron sets its output as "1" and all other competitive neurons set their output as '0". The competitive neuron which wins in the competition is called a "winner-take-all" neuron. This learning rule is highly suitable to find clusters within data.
3. **Learning vector quantization (LVQ):** This learning rule is a form of unsupervised learning, in which pattern, classification is taking place. LVQ is a special class of competitive learning and uses supervised learning methodology, in which the set of training patterns with known classifications (i.e., input vector and its respective output) is given to the network with an initial distribution of the reference weight vectors. LVQ classify an input vector by assigning it to the same class as that of the output unit, which has its weight vector very close to the input vector, so LVQ adjusts the boundaries between categories to minimize existing misclassification (Sivanandam and Deepa, 2011).

4.2.3 ANN STRUCTURES AND TRAINING ALGORITHMS

ANNs are parameterized nonlinear models built by training algorithm and used for empirical regression and classification modeling. Their flexibility makes them attractive due to easy adaptability of process data and able to discover more complex relationships in data than traditional linear statistical model. ANN models are usually trained on a set of practical process observation and are examples of input and output data. A set of coefficients (called weights) generated as an outcome of this training and a specification of the functions, and further, these in combination relates the input to the output. The training process involves a search for the optimum nonlinear relationship between the inputs and the outputs.

4.2.3.1 BASIC ANN STRUCTURE

The basic structure property of any neural network is that each node in a layer is connected with all the nodes in another layer, and so on as per the network structure. As ANN is similar to linear regression in terms of output node value computation, linear functions of the input nodes are operated on by an activation function in such a manner that each input contributes to every hidden unit as shown in Figure 4.9 for all type of ANN structures. After knowing all the basics about the connection between artificial neuron and biological neuron, this is the time when one can easily pick up the ANN structure and related notations to be used in ANN.

For developing initial understanding about ANNs, the basic ANN structures are shown in Figure 4.9 and the terminologies were defined which are to be used in that while computing network output. In ANN structure, there are multiple layers, such as input layer, hidden layer (single or multiple), and an output layer. Each layer is having one or more nodes, and each node of network except the input nodes is called as artificial nodes or perceptron, as input layer nodes are constant data fed to the network by user. There are many existing ANN structures but few basic sketches of ANN are shown in Figure 4.9a–c. The following notations are used to define the network parameters, where *'w'* and *'b'* representing the weights and biases.

Even though as shown in Figure 4.9 the networks are looking very clear to understand, and generally represented everywhere in literature in same way, but still many things are not shown in the network and even showing those things in the network will make network very complex to understand such as the connections to show updates of synaptic weights and the incorporation of activation function, which are missing in all those networks of Figure 4.9.

For example, let us suppose if the computation for a hidden layer node h_j has to be done in Figure 4.9a, then we can write the equation for general calculation of the hidden layer nodes (eq 4.5), equation for specific node of the hidden layer for h_1 (eq 4.6).

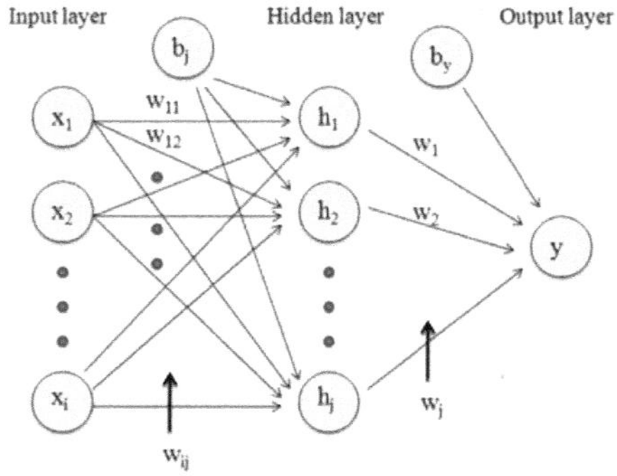

(a) ANN structure with three layers and an output

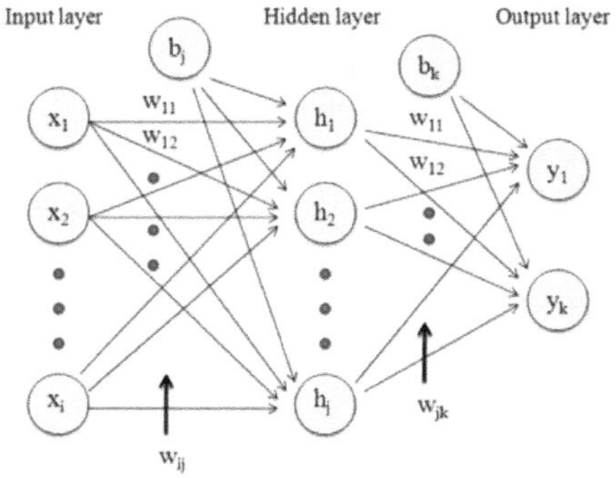

(b) ANN structure with three layers and multiple outputs

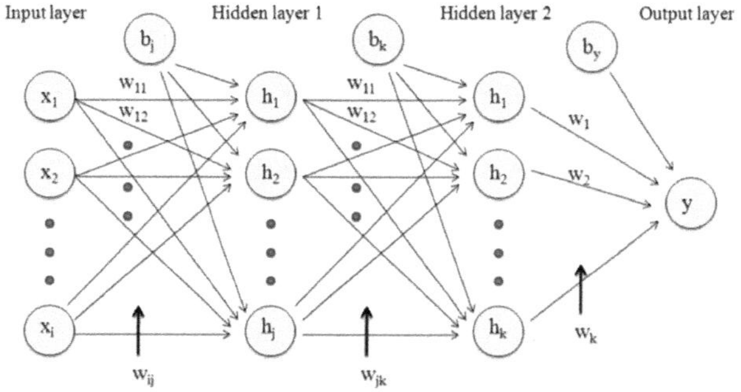

(c) ANN structure with four layers and an output

FIGURE 4.9 Basic ANN structures.

General for h_j

$$h_j = \varphi\left(\sum_{i=1}^{n}\left(x_i w_{ij} + b_j\right)\right) \tag{4.5}$$

Specific for h_1 (here i = 1,2,..n)

$$h_1 = \varphi\left(\sum_{i=1}^{n}\left(x_i w_{i1} + b_1\right)\right) \tag{4.6}$$

Specific for h_1 without activation function φ (here i = 1,2,..n)

$$h_1 = \sum_{i=1}^{n}\left(x_i w_{i1} + b_1\right) \tag{4.7}$$

Equation 4.7 shows the calculation of h_1 node of hidden layer without activation function as this makes the nodal calculation to be linear regression based with limited learning power. If the activation function is not applied on the node, then node output signal becomes a simple linear function and that is not suitable for ANN, as here the activation function work is to introduce nonlinear real-world properties to ANN.

There are many ANNs existing today based on various neural network structures and training algorithms, such as feed-forward backpropagation (FFBP) neural network (sometimes also called as feed-forward neural network—FFNN), radial basis function (RBF) neural network, generalized regression neural network (GRNN), convolutional neural network (CNN), recurrent neural networks (RNN), time delay neural network (TDNN), wavelet neural network (WNN), etc. Sometimes, even a term MLP is also used in ANN, which is a class of feed-forward neural network. It is difficult to cover all the ANNs at any point of time, here, only three types of ANN which are generally used, namely, FFBP neural network, RBF neural network, and GRNN are discussed in detail with their training algorithms.

4.2.3.2 ANN TRAINING ALGORITHMS

Before ANN application to any real-world problem, the network has to be trained first. And before training of ANN, the data must be normalized in the range of threshold value for selected activation functions such as 0 to 1 for sigmoid and ±1 for hyperbolic tangent activation function. The normalization can be done by using the following equations with respect to selected activation function, for example, purpose of sigmoid and hyperbolic tangent function:

Activation function: Sigmoid and its normalization equation are given as:

$$p_n = \frac{p_0 - p_{\min}}{p_{\max} - p_{\min}} \tag{4.8}$$

Activation function: Hyperbolic tangent and its normalization equation are given as:

$$p_n = 2\left(\frac{p_0 - p_{\min}}{p_{\max} - p_{\min}}\right) - 1 \tag{4.9}$$

Where, p_0 is the observed data, p_n is the normalized scaled data, and p_{max}, p_{min} are the maximum and minimum observed data points, respectively. The above eqs 4.8 and 4.9 were used to normalize the inputs and output data in order to provided consistency for the analysis. An important stage in developing an ANN model is the computation and adjustment of its weight matrix through training. And before starting of the training procedure, the connection weights (synaptic weights) and biases must be assigned random values. The allocated random values must have to be small since the inputs and outputs are compressed to (0, 1) or (−1, 1) as per the selected activation function. It was also found in literature that the range for the connection weights (synaptic weights) and biases random values is in general set between −3 and +3, while as per the practical rule random values range is to be set between −3/n and +3/n, where *n* is the number of neurons in the input layer (Tayfur, 2012).

Network training: The network after training is expected to yield the target output based upon the input set, and to perform that the network training (learning) algorithms are to be employed. These network learning algorithms in general divided into two group's unsupervised learning and supervised learning. Now when the whole database is normalized, further the database is randomized before training in order to group the network input information into training and validation data sets in a fair manner. As the complete database has to undergo in two different network learning process, so the whole database grouped into two data sets, and further one of the data sets is used for testing with unsupervised learning while the other for training with supervised learning.

In unsupervised or adaptive training algorithm, only input vector data from the grouped data sets (data sets contains both input and output data) is provided to the network. In iterative manner, the network automatically adapts its connection weights during network training to cluster input patterns into similar properties classes. This type of network training often referred as self-adaptation or organization. A grade system is used to rate the network training performance over a series of training patterns. This type of network does not involve any external tutor and shows how the trained models behave with unseen data, as the network relies only on the input data set.

In supervised training algorithm, there is an external tutor involved in the training of the ANN, here user decides the network learning rate, training type, and number of iterations. In this type of network training, whole grouped data sets (data sets contains both input and output data) is provided to the network. After network fed with data sets, the network undergoes in learning process to find the nonlinear relationship between network input and output data with the help of supervised learning algorithms. There are many supervised learning algorithms, but here the discussion of learning algorithms is restricted for three ANN structures FFBP, RBF, and GR neural networks as mentioned earlier. FFBP neural network uses back-propagation algorithm, RBF neural network uses RBF algorithm, and GR neural network uses generalized regression algorithm.

(A) Backpropagation Algorithm

This algorithm is used in feed-forward backpropagation neural network, where only three layers are there input, hidden and output layer.

The network is less complex due to limitation in layers, but as discussed earlier the shown network is not describing the complete function of the network as the schematic of FFBP neural network is shown in Figure 4.10 (left) while the internal computation of a hidden layer node h_1 is shown in right which is similarly applicable to all the nodes of network except input layer nodes.

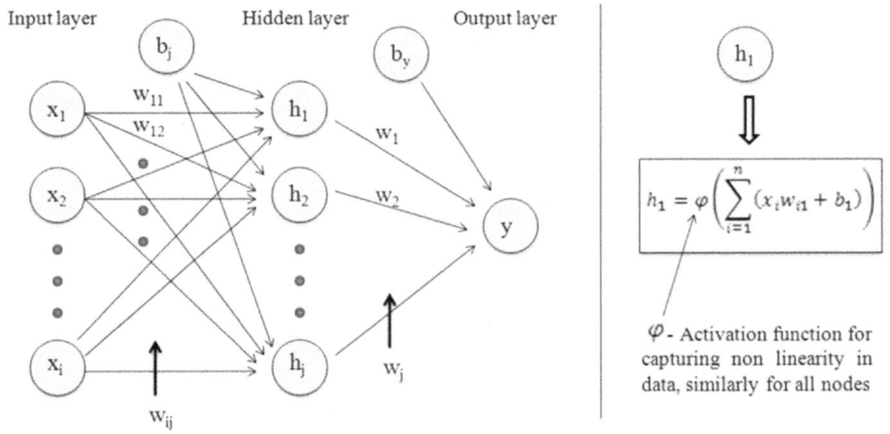

FIGURE 4.10 Schematic of FFBP neural network.

The backpropagation algorithm consists of mainly two activities, namely, forward pass and backward pass. In forward pass, the network activity is propagated from the input layer to hidden layer nodes for computation of output. While in backward pass, the network activity is propagated from the output layer to hidden layers and from hidden layers to input layer.

The activities of forward pass can simply summarized as: (1) Training pairs from input data sets are selected randomly for supervised and unsupervised learning and fed into the input neurons, (2) The fed inputs are multiplied by random initialized weights first time and then summed to form the complete net information at each node except input layer nodes, (3) Further, the net information at each node is applied with an activation function to produce the actual node output, (4) The generated node output is passed to each neuron in a successive layer, and (5) The process is continued until the network output is produced. While the activities of backward pass can simply summarized as: (1) Based on the overall network error, the error is computed for each output unit, (2) The error for each hidden node is computed by propagating errors layer by layer, (3) All the connection weights are updated by the generalized delta rule method.

$$h_j = \varphi\left(\sum_{i=1}^{n} w_{ij} x_i + b_j\right) \tag{4.10}$$

$$y = \varphi\left(\sum_{j=1}^{m} w_j h_j + b_y\right) \tag{4.11}$$

For FFBP the neural network is shown in Figure 4.10, the combination of eqs 4.10 and 4.11 with number of input variables (n), value of input variables, a set of weights, biases, number of hidden nodes (m) defines the network completely (Bhadeshia, 1999; Panjiar et al., 2015).

Where φ is the activation function, and in the case of FFBP neural network, mostly sigmoid or hyperbolic tangent activation function is used. The bias is designated b with suffix as per the network layer and that is analogous to the constant in linear regression equation. The strength of the activation function in each case is determined by the weight w_{ij}. The final output of the network is defined as a linear function of hidden nodes and bias. Above set of equation is valid for forward pass only, but in the case of backward pass, the adjustment of all network connection weights takes place as per generalized delta rule, where the each connection weights are adjusted according to overall training error. The simplified way of representing the weight adjustment is given in eq 4.12 while for explanation refer generalized delta rule or delta learning from the literature (Sivanandam and Deepa, 2011; Tayfur, 2012).

$$w_{ij}^{\text{new}} = w_{ij}^{\text{old}} + \Delta w_{ij} \tag{4.12}$$

This process of forward pass and backward pass activities continues till the network error reduces to acceptable limit.

(B) Radial Basis Function Algorithm

M. J. D. Powell developed the RBF, which is a classification and functional approximation neural network. The theoretical basis of the RBF approach is the field of multivariate functions interpolation. This algorithm is used in RBF neural network, which is composed of three layers, such as input, hidden, and output layer as shown in Figure 4.11. The output layer can have single or multiple output nodes, and that can be designed as per the requirement.

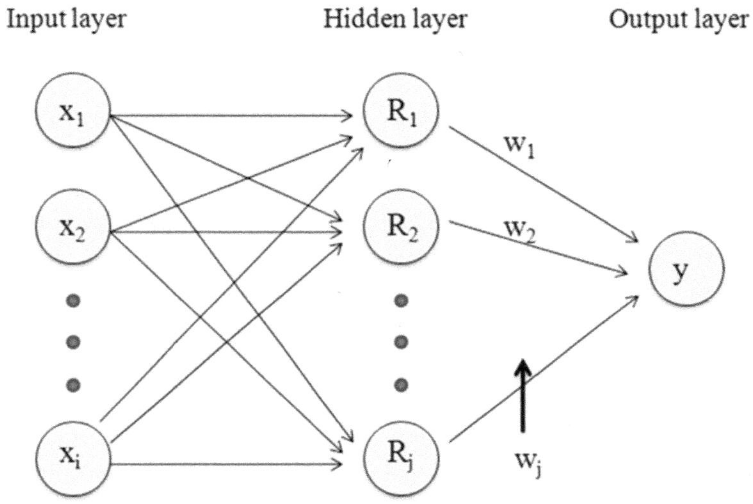

FIGURE 4.11 Schematic representation of RBF neural network.

Here, in RBF neural network, the hidden layer nodes are represented by RBFs (R_1, R_2... R_j). A RBF is a function $R(x)$ with real-value following the norm represents the Euclidean distance (which in simple words is the ordinary distance between two points), and that the value only depends on the distance from another point called a center c, then $R(x) = R(\|x-c\|)$ or from the origin where $c=0$, then $R(x) = R(\|x\|)$. The characteristic feature of function $R(x)$ is that their response increases or decreases monotonically with distance from a central point. So the RBF approach able to construct a linear function space, which depends on the positions of the earlier known data points according to an arbitrary distance measure, and typically used to develop an approximating function in the following form for any arbitrary function:

$$y = \sum_{j=1}^{m} w_j R\left(\|x-c_i\|\right) + b \tag{4.13}$$

Where $\|x-c_i\|$ denotes Euclidean norm, y is the approximating function, m is the number of RBFs, R is the real-valued function, w_j is the connection weight coefficient, and $b \in \Pi_n^k$ is a polynomial of degree at most n (fixed in advance) in k variables (Taurino et al., 2003). The interpolation problem is to determine the real coefficients w_j and the polynomial term b (and b_1, b_2... for multiple outputs). In some cases, the polynomial term in eq 4.13 can be omitted and the final network output y is represented as a sum of the product of appropriate coefficient w_j and m RBFs, each associated with a different center c_i as shown in eq 4.14.

$$y = \sum_{j=1}^{m} w_j R\left(\left\|x - c_i\right\|\right) \tag{4.14}$$

The hidden layer in RBF neural network performs a determined nonlinear transformation using RBFs with no adjustable parameters. Generally, RBFs used an activation function in hidden layer neurons, and that activation function (φ) may be sigmoidal function or Gaussian function or multiquadric function (Sivanandam and Deepa, 2011; Broomhead and Lowe, 1988). In general, the Gaussian function is used for RBFs (Sivanandam and Deepa, 2011; Tayfur, 2012; Broomhead and Lowe, 1988) and is expressed as:

$$R_j = \varphi(\| x_i - c_i \|) = \exp\left[-\sum_{i=1}^{n} \frac{\| x_i - c_i \|^2}{2\sigma^2} \right] \tag{4.15}$$

In eq 4.15, x_i represents input node value, c_i is the center of the clusters, σ is the standard deviation of the clusters, and R_j is RBF. The nonlinearity is enforced in RBF through network training with provided set of data. Once all the RBFs in the hidden layer is finalized, the RBF neural network needs only to train the output layer in a linear summation of the product of RBFs and connection weights between hidden layer and output layer, respectively.

The most important task in RBF neural network is to determine c_i (c_1, c_2... c_n) centers of clusters. It is well known that clustering is the classification of things into different groups. Here, it means the division of a data set into many subsets which are called clusters. In each of the subset, a common trait is supposed to be shared by the data. The trait sighted here is often understood as proximity according to some defined distance measure. The computational work of classifying whole input data set into k clusters is always referred to as *k*-clustering. Hence, in RBF neural network *k*-means clustering is often utilized to form the groups of training set and select the center for each group. So each point of input layer assigned to the cluster whose center (also called centroid) is the nearest by the *k*-means clustering. That selected center of the cluster is the average of all the points in that cluster, which means the coordinate of the center, is the arithmetic mean for each dimension property separately over all the points in the cluster.

For this a set of m arbitrary (generally nonlinear) "basis" functions *RBFs* $= \varphi$ ($\|x_i - c_j\|$) is introduced, where $\{x_i; i = 1,2,..,m \, and \, x_i \in \mathbb{R}^n\}$, and the vectors $c_j \in \mathbb{R}^n$ $\{c_j; j = 1,2,\ldots, m\}$ are the centers of the basis functions and taken to be sample data points. The function $\{s : \mathbb{R}^n \rightarrow \mathbb{R}\}$ which satisfies the interpolation conditions are considered through random generation by *k*-means clustering and finally termed as c_i $\{i = 1, 2,\ldots, k\}$, the function initially defined as:

$$s(x_i) = f_i; i = 1,2,..,m \tag{4.16}$$

Now to compute the connection weights between hidden layer nodes and output layer node a term (A_{ij}) assigned for each RBF, so we can write $A_{ij} = \varphi(\| x_i - c_j \|)$ $\{with i, j = 1,2\ldots m\}$ for better understanding and from the RBF neural network and eq 4.14 with $y = s(x_i)$, we can write:

$$s(x_i) = \sum_{j=1}^{m} w_j \varphi(\| x_i - c_j \|) = \sum_{j=1}^{m} w_j A_{ij}\, x_i \in \mathbb{R}^n \tag{4.17}$$

Inserting the interpolation conditions mentioned in eqs 4.16 into 4.17, gives the following set of linear equations for the coefficients $\{w_j\}$.

$$\begin{pmatrix} f_1 \\ \vdots \\ \vdots \end{pmatrix} = \begin{pmatrix} A_{11} & \cdots & A_{1m} \\ \vdots & \ddots & \vdots \\ A_{m1} & \cdots & A_{\mathrm{mm}} \end{pmatrix} \begin{pmatrix} w_1 \\ \vdots \\ \vdots \end{pmatrix} \tag{4.18}$$

Given the existence of the inverse to matrix A with elements A_{ij}, eq 4.18 uniquely defines the coefficients w_j through $\{w = A^{-1}f\}$. Since RBFs is equal to the number of data points, calculating the inverse of m x m matrix R_i becomes intractable in practice.

The RBF algorithm steps can be summarized as: (1) Select the number of *k* clusters, (2) Randomized generation of *k* clusters and calculate the cluster centers, or straight way generate *k* random points as cluster centers, (3) Allot each input data point to the nearest cluster center, (4) Again compute the new cluster centers, and (5) Repeat the previous two steps until some convergence criterion is satisfied.

The major advantages of RBFs algorithm are its simplicity and speed that allows it to run on large data sets. This kind of RBF function approximation is generally used in the control of nonlinear systems and time-series predictions exhibiting chaotic behavior. The fact by Broomhead and Lowe work (Broomhead and Lowe, 1988) that the RBF plays the important role of activation functions in the RBF neural network, eq 4.14 with sufficient count of RBFs, can closely approximate any continuous function in a bounded interval.

(C) Generalized Regression Algorithm

Generalized regression algorithm is used in generalized regression neural network (GRNN). Generalized regression algorithm does not need iterative training as it approximates any arbitrary function between output and input vectors. This algorithm is directly drawing the function estimate from the training data. It is based on a standard statistical technique called kernel regression and used for the estimation of probability density function of continuous variables. GRNN was initially suggested by Specht in 1991 (Specht, 1991). GRNN consists of four layers, namely, input layer $\{x_i; i = 1,2\ldots\}$ as first layer, pattern layer $\{p_j; j = 1,2\ldots\}$ as second layer, summation layer $\{s_k; k = 1,2\ldots\}$ as third layer, and an output layer (y) is the final layer as shown in Figure 4.12. The output layer can have single or multiple output nodes and that can be designed as per the requirement. One of the important points to be noted here is that each output node should be connected to two summation layer nodes (Specht, 1991).

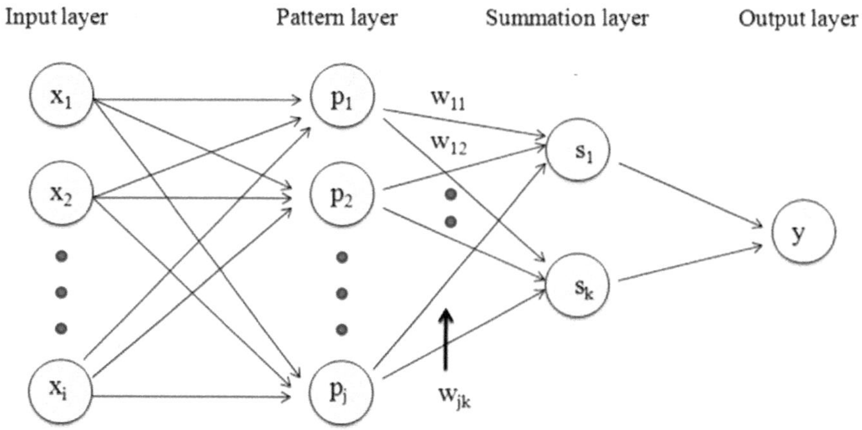

FIGURE 4.12 Schematic representation of GRNN.

In GRNN, the input layer nodes are simply distribution units, which supply all of the measurement variables X to all of the nodes on the second layer, the pattern units. The pattern unit is dedicated to one exemplar, similarly each pattern layer node dedicated to unique exemplar, and when a new vector X^i is entered into the network, it is subtracted from the stored vector representing each cluster center. Either the absolute values or the squares of the differences are summed and fed into a nonlinear activation function (Specht, 1991). The activation function normally used is the Gaussian function, although there are many alternate activation functions present and also could be used here. Each of the pattern layer nodes output are passed on to the each summation layer nodes. The summation layer nodes perform a dot product between a weight vector and a vector composed of the signals from the pattern layer nodes. The summation layer node that generates a probability estimate of $\hat{f}(X,Y)$ sums the outputs of the pattern nodes weighted by the number of observations each cluster center represents. The output unit simply divides $w_{jk} \cdot f(X, y)$ by $f(X, y)$ to yield the desired estimate of $\hat{Y}(X)$.

The whole network computation of GRNN is designed on the basis of probability estimator and probability density function approach. Generalized regression algorithm is a method for estimating a joint probability density function (pdf) of x and y given only a training set. The regression of a dependent variable y on an independent x calculates the nearly probable value for y, given x and a training set. The regression method generates the calculated value of y which minimizes the mean squared error (MSE) of the network training (Cigizoglu and Alp, 2006). The following notations are used here to establish the network relations:

x	–	Input random variables
X	–	A measured value set of input random variables, supplied to each pattern node
$\hat{Y}(X)$ *or* y	–	Desired estimated output or Network output
Y	–	Average of output data
X^i	–	A sample value set of input random variables
Y^i	–	Sample output data for X^i
$E[y \vert X]$	–	Conditional mean
$f(X, y)$	–	Probability estimator specific
p	–	dimension of the vector variable x
n	–	number of sample observations
$\hat{f}(X, Y)$	–	Probability estimator general
σ	–	Width of the estimating kernel, can be selected by using holdout method
w_{jk}	–	Connection weight between one layer *jth* node and another layer *kth* node

Let suppose, for a vector random variable x and a scalar random variable y, there is a function $f(x,y)$ represents a known joint continuous pdf. Then let X^i be a particular measured value of the random variable x. The conditional mean of network output of y given X^i (also called as the regression of y on X^i) is given by

$$E[y|X] = \frac{\int_{-\infty}^{\infty} yf(X,y)\,dy}{\int_{-\infty}^{\infty} f(X,y)\,dy} \tag{4.19}$$

When the probability density function $f(x,y)$ is not known, it should usually be calculated from a sample of observations of x and y. For pdf, the probability estimator (eq 4.20) is used which is based upon sample values X^i and Y^i of the random variables x and y:

$$\hat{f}(X,Y) = \frac{1}{2\pi^{(p+1)/2}\sigma^{(p+1)}}\frac{1}{n}\sum_{i=1}^{n}\exp\left[-\frac{\left(X-X^i\right)^T(X-X^i)}{2\sigma^2}\right] \exp\left[-\frac{(Y-Y^i)^2}{2\sigma^2}\right] \tag{4.20}$$

A physical interpretation of the probability estimate $\hat{f}(X, Y)$ is that it assigns sample probability of width σ for each sample X^i and Y^i, and the probability estimate is the sum of those sample probabilities. Substituting the joint probability estimate (eq 4.20) into the conditional mean (eq 4.19), gives the desired conditional mean of y given X. Particularly, combining eqs 4.19 and 4.20 and swapping the order of integration and summation yields the desired conditional mean, designated as $\hat{Y}(X)$ (Specht, 1991):

$$\hat{Y}(X) = \frac{\int_{-\infty}^{\infty} y\sum_{i=1}^{n}\exp\left[-\frac{\left(X-X^i\right)^T(X-X^i)}{2\sigma^2}\right]\exp\left[-\frac{(Y-Y^i)^2}{2\sigma^2}\right]dy}{\int_{-\infty}^{\infty}\sum_{i=1}^{n}\exp\left[-\frac{\left(X-X^i\right)^T(X-X^i)}{2\sigma^2}\right]\exp\left[-\frac{(Y-Y^i)^2}{2\sigma^2}\right]dy} \tag{4.21}$$

Allocating a scalar function D_i^2 to reduce the above eq 4.21 is given as

$$D_i^2 = (X-X^i)^T(X-X^i) \tag{4.22}$$

Further applying declared scalar function in eq 4.22 with computation of integrations will yield the following solution:

$$\hat{Y}(X) = \frac{\sum_{i=1}^{n} w_{jk}\exp\left(-\frac{D_i^2}{2\sigma^2}\right)}{\sum_{i=1}^{n}\exp\left(-\frac{D_i^2}{2\sigma^2}\right)} \tag{4.23}$$

where in eq 4.23

$\hat{Y}(X)$ is the prediction value of input X, w_{ij} is the activation weight for the pattern layer neuron at k, and $\exp(-D_i^2/2\sigma^2)$ is the RBF kernel (Gaussian kernel).

The resulting regression (eq 4.23), which involves summations over the observations, can be directly relevant to numerical data problems. So, once the GRNN is trained with user input data, the model is ready to predict output with the new sample input data.

4.2.4 ERRORS IN ANN TRAINING

Generally, the error (*E*) is the difference between the network output and the desired output after the completion of training, so the error can be defined as follows (eq 4.24):

$$E = y_t - y \tag{4.24}$$

There are many types of error considerations are present, such as sum of squared error (SSE), mean of squared error (MSE), root MSE (RMSE), and so on. But mostly, SSE or MSE is used to find the overall error of ANN training. For example, suppose for FFBP, the overall error (E_D) for trained model is the SSE between the desired output y_t and computed network output y as given here in eq 4.25.

$$E_D = \sum_{p=1}^{t} (y_t - y)^2 \tag{4.25}$$

where, the subscript p refers to the patterns (observations) with a total of t patterns. To minimize the overall error during the ANN training, any learning rule can be employed, where the idea is to find a minimum of E by repeatedly computing minima of error function. There is also an effect of learning rate parameter on the overall error, as it can be seen from Figure 4.13 that for the case of bad training, the learning rate parameter may be chosen very less or more, while at the same time, if learning rate parameter for training is optimal, then the overall error will be reducing at every iteration as shown by good training curve.

Even some times during training, the overall error may be selected as local minima value instead of global optima as shown in the bad training case in Figure 4.13 if there is no change in synaptic weights of the ANN for few number of consecutive iterations. While in the case of good training, it can be seen that there is very less or no deviation in the overall error after sometime of the training, that means there is no significant change in synaptic weights of the ANN even if iterations continued. For that reason, it is better to fix the number of iterations or epochs for the ANN training in such a way that the overall error is almost constant over final iterations.

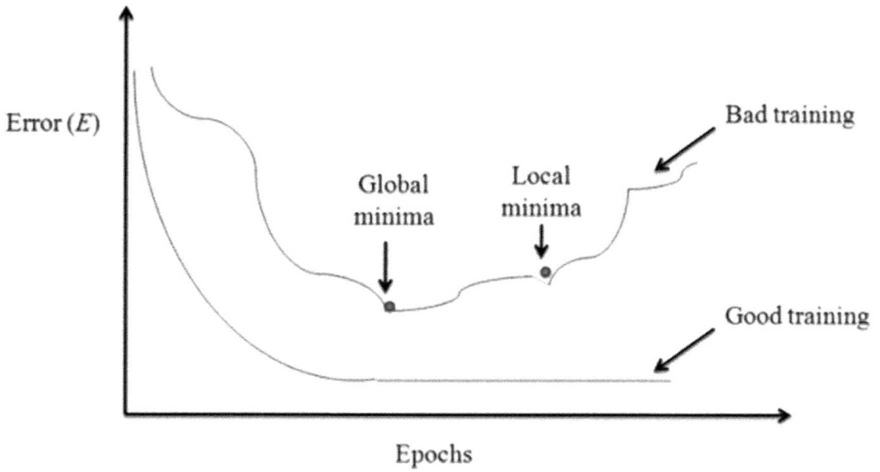

FIGURE 4.13 Change in overall error with respect to number of iterations during training.

4.3 ANN MODEL TESTING

The first step in model testing comes after successful training of a network, and in model testing, one can check the network output by entering any input vector into the network without providing the output value for entered input vector. Just recall, that ANN model can produce the outputs within the activation function range which employed during the training procedure.

So to obtain the desired outputs during model testing and simulation, the produced model outputs need to be denormalized first according to the related employed standardization method (as mentioned earlier in section *ANN training algorithms*). Denormalization is a scheme used on a previously normalized database to increase the computation performance, as this process is trying to improve the read performance of a database. The denormalization process can be performed as follows. For instance, if one employs hyperbolic tangential tanh(x) activation function, so eq 4.9 for normalization, the denormalization can be performed as:

$$y_i = p_0 = \frac{\left[2p_{\min} + (p_n + 1)(p_{\max} - p_{\min})\right]}{2} \tag{4.26}$$

In eq 4.26, y_i stands for denormalized value and p_n is the network model output. And similarly, denormalization can be performed for other activation function as employed.

4.3.1 MODEL PERFORMANCE EVALUATION

To check whether the ANN training is successfully accomplished or not, it is required to feed the trained network with the same training input pairs (same used in the ANN training) to obtain ANN model outputs which is sometimes also called as validation of ANN training. Note that, during this procedure, target outputs are not fed into the network. Once the ANN model outputs are obtained, then that can be plotted against observed actual data (target outputs used in the ANN training) in order to see the performance of the ANN model (e.g., see bottom left figure in Figure 4.14).

Once the ANN training is successfully accomplished, then anyone can confidently apply the ANN model to new test data. During this, the user have to enter only single input vector test data set into the ANN model network to produce the model output. ANN model performance can be assessed by several ways, out of which, some are given as follows:

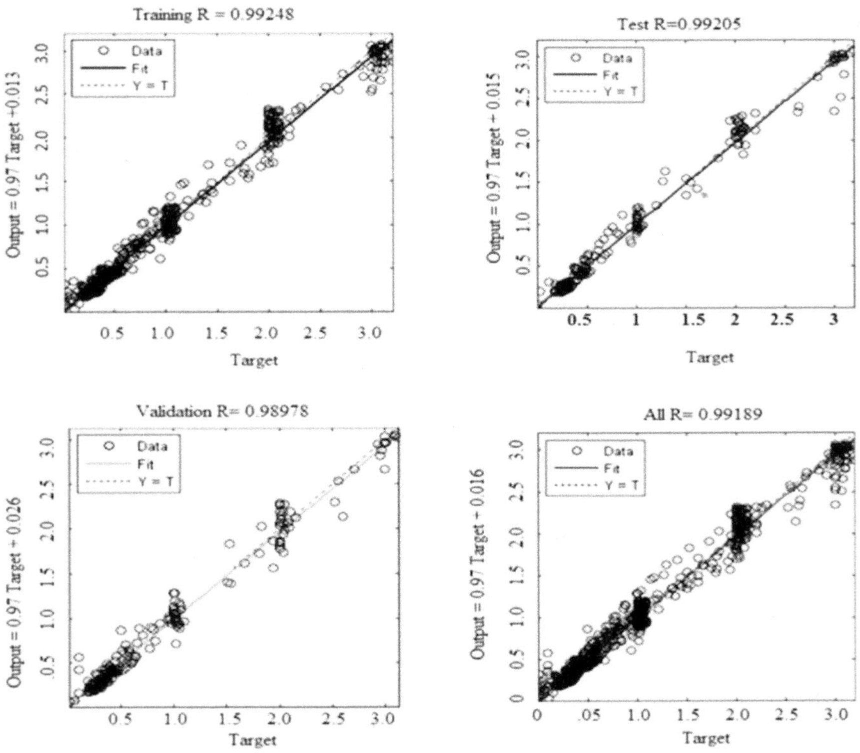

FIGURE 4.14 ANN model behaviors during training, testing and validation.

Source: Reprinted from Chakravarti et al., 2015. https://creativecommons.org/licenses/by/4.0/

A. Plotting of observed data versus ANN model output data using a scatter plot at the end of the ANN training procedure. The ANN training procedure check ANN model capability for supervised, unsupervised, validation testing and finally overall model testing, which are represented as training data plot (top left in Figure 4.14), testing data plot (top right in Figure 4.14), validation data plot (bottom left in Figure 4.14), and over all data plot (bottom right in Figure 4.14).

In all the figures of Figure 4.14, the correlation coefficient (R) value with equation for fit of a regression line is shown, where R value (near to 1 is better and vice versa) is highly significant in deciding ANN model performance in terms of its reliability (in general, the correlation coefficient is represented as R^2). Further, the fit of a regression line with equation of the fitting line to the scattered ANN model output data can provide an overall impression about the performance of the ANN model. For example, the value of correlation coefficient for over all data plot is mentioned as 0.99189 implying satisfactory training (as this value near to 1 shows a satisfactory performance) and good reliability of the ANN model.

B. Plotting of observed data versus ANN model output data using a line chart to visually check the trend. To show the same, Figure 4.15 is considered as an example from the literature (Chakravarti et al., 2015) where the prediction of runoff rate using laboratory base rainfall simulator data is done. In the study, the observed data were collected from the simulator and computed data are from ANN model. The model is 99.18% reliable as per the training outcome, and it can be seen clearly that the computed data match with the observed experimental data as in Figure 4.15.

C. Various error measurements, such as MSE, SSE, and RMSE (as discussed in the Errors in ANN training section) and even mean absolute prediction error (MAPE) is used sometimes, and it usually expresses accuracy as a percentage, and is defined as (Najah et al 2013):

$$\text{MAPE}(\%) = 100 \times \left(\frac{1}{n} \sum_{i=1}^{n} \left(\frac{|y_m - y_p|}{y_m} \right) \right) \qquad (4.27)$$

In eq 4.27, y_p is predicted data from ANN model, y_m is measured data, and n is the number of observations.

4.3.2 SHORTCOMING WITH TRAINING: OVERFITTING

Overfitting occurs often during ANN training when the database is overtrained, and this is the shortcoming of ANN training which appears when network parameters are excessively fine-tuned to training data set, such that the ANN starts attempting to fit noise component of the data set. An attempt has been made with the help of figure (Fig. 4.16) to show the difference in normal training and overtraining of ANN.

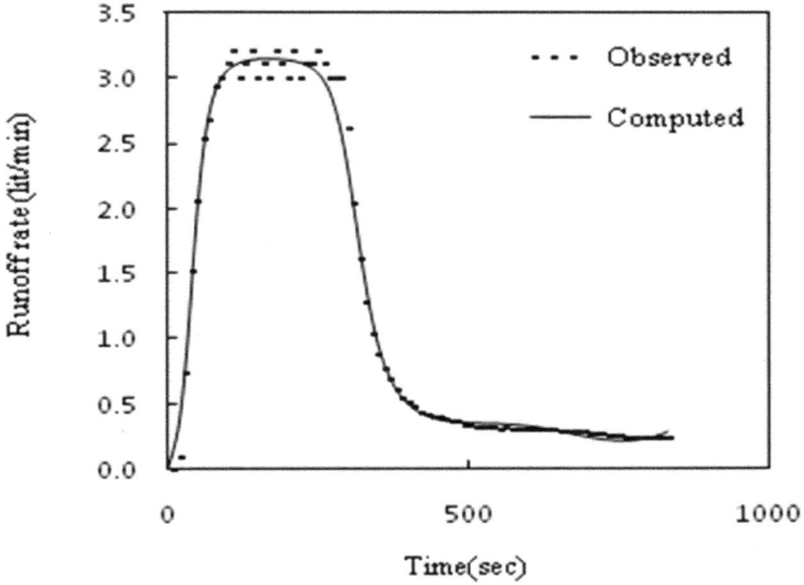

FIGURE 4.15 Comparison of ANN model predicted data and observed data.

Source: Reprinted from Chakravarti et al., 2015. https://creativecommons.org/licenses/by/4.0/

The overtraining of database with input and output vectors results in a network that memorizes individual cases rather than trends in the data set as a whole and generally occurs with less number of data set. And when this occurs, the ANN performs substantially well over the training data set, while it shows very poor performance for a testing data set. In the case of overtraining, the ANN training process should be terminated and a trial check should be performed for database and other selected parameters for training. Fresh training is generally employed to prevent overtraining after re-verification of database and other ANN training parameters.

4.4 DISCUSSION

Apart from the discussed ANNs, these can also be made with more hidden layers, sometimes also called as multilayer neural networks to capture more complexity in data. The term deep learning also came from a neural net with multiple hidden layers, and wide variety of architectures encapsulated in it. Further, in deep learning, CNN is a kind of network which is used to recognize object as an image, which is having effective application in computer vision. It is not an easy task to keep up with all the developments in the field of deep learning, as the industry and research communities have risen multiple times on their deep learning efforts and development, spawning whole new methodologies every day. With these qualities of capturing nonlinear relationships in data, the ANN is used in multiple disciplines where the role of data is therefore to understand the science behind respective experimental/field process data, whereas the scientific work showing the effective use of this approach can be found in multidisciplinary literature (Sivanandam and Deepa, 2011; Tayfur, 2012; Bhadeshia, 1999; Panjiar et al., 2015; Huber et al., 2020; Oyebode and Stretch, 2018) just to list a few. As per ANN multidisciplinary utility, this method also has found a major role in the field of water sector for modeling and simulation of water-related issues (Yang, 1996; Cigizoglu and Alp, 2004; Ulke et al., 2009; Oyebode and Stretch, 2018; Cigizoglu and Alp, 2006). The application of the ANNs in the field of water sector are discussed in next chapter, where the ANN model outputs are discussed which are associated with real-world water resources problem.

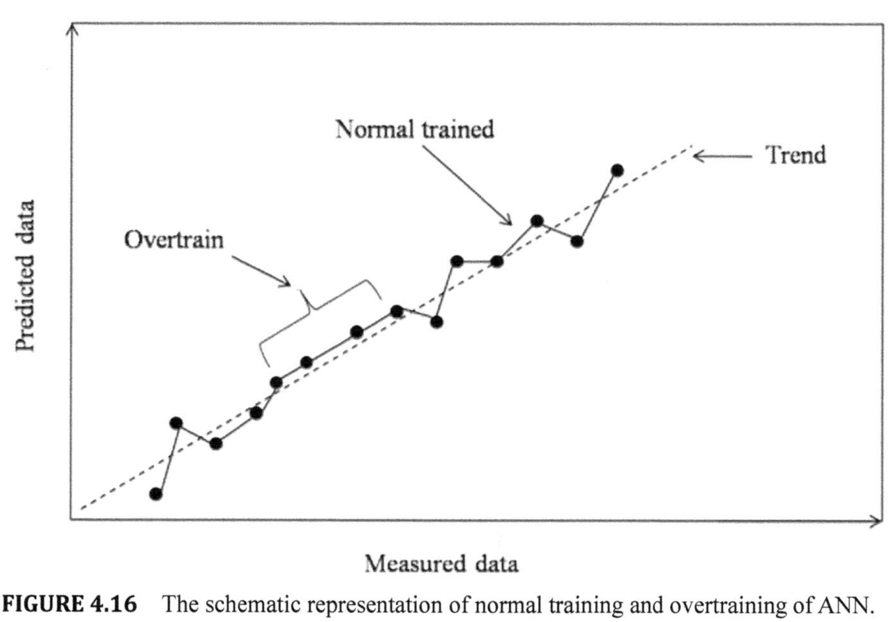

FIGURE 4.16 The schematic representation of normal training and overtraining of ANN.

4.5 CONCLUSION

A brief discussion on the soft computing techniques presented with elaborated explanation on one of the technique "ANN" is done. The present study showcases the three types of ANN, namely, feed-forward backpropagation, generalized regression, and RBF neural networks with their training algorithms. The network error and shortcoming with training were also emphasized with their consequences. A real-time data-based rainfall-runoff model development and its implementation using neural network is discussed where the model was able to capture the process and matching with experimental data pattern with overall 99.18% of reliability, which is actually intended to show the efficacy of ANN modeling in water sector application.

KEYWORDS

- **artificial neural network**
- **model development**
- **generalized regression neural network**
- **radial basis function neural network**
- **feed-forward backpropagation neural network**
- **rainfall runoff**

REFERENCES

Anderson, J. A.; Rosenfeld, E. *Neurocomputing: Foundations of Research*; MIT Press: Cambridge, 1988; pp 45–56.

Bhadeshia, H. K. D. H. Neural Networks in Materials Science. *ISIJ Int.* **1999,** *39*, 966–979.

Broomhead, D.; Lowe, D. Multivariable Functional Interpolation and Adaptive Networks. *Complex Syst.* **1988,** *2*, 321–355.

Chakravarti, A.; Joshi, N.; Panjiar, H. Rainfall Runoff Analysis Using Artificial Neural Network. *Indian J. Sci. Technol.* **2015,** *8* (14), 1–7.

Cigizoglu, H. K.; Alp, M. Generalized Regression Neural Network in Modelling River Sediment Yield. *Adv. Eng. Softw.* **2006,** *37*, 63–68.

Cigizoglu, H. K.; Alp, M. Rainfall-Runoff Modelling Using Three Neural Network Methods; Springer, ICAISC 2004, LNAI 3070, 2004; pp 166–171.

Haykin, S. *Nueral Networks: A Comprehensive Foundation.* MacMillan College Publishing Co.: New-York, 1994; pp 115–133.

Huber, N.; Kalidindi, S. R.; Klusemann, B.; Cyron, C. J. Machine Learning and Data Mining in Materials Science. *Frontiers in Material* **2020.** DOI: 10.3389/978-2-88963-651-8
Jain, A. K.; Mao, J. Nueral Networks and Pattern Recognition. In *Computational Intelligence: Imitating Life*; IEEE Press: Piscataway, 1994; pp 194–212.
McCulloh, W. S.; Pitts, W. A Logical Calculus of Ideas Immanent in Nervous Activity. *Bull. Math. Bio-phys.* **1943,** *5*, 115–133.
Minsky, M.; Papert, S. *Perceptrons: An Introduction to Computational Geometry*; MIT Press: Cambridge, MA, 1996; pp 43–80.
Mohiuddin, K.; Mao, J. A Comparative Study of Different Classifiers for Handprinted Character Recognition. In *Pattern Recognition in Practice IV*; Elsevier Science: The Netherlands, 1994; pp 437–448.
Najah, A.; El-Shafie, A.; Karim, O. A.; El-Shafie, A. H. Application of Artificial Neural Networks for Water Quality Prediction. *Neural Computing and Applications* **2013,** *22* (Suppl 1), S187–S201.
Oyebode, O.; Stretch, D. Neural Network Modeling of Hydrological Systems: A Review of Implementation Techniques. *Nat. Resour. Model.* **2018,** 1–14.
Panjiar, H.; Gakkhar, R. P.; Daniel, B. S. S. Strain-Free Graphite Nanoparticle Synthesis by Mechanical Milling. *Powder Technol.* **2015,** 25–29.
Rosenblatt, R. *Principle of Nuerodynamics*; Spartan Books: New York, 1962; pp 23–35.
Sivanandam, S. N.; Deepa, S. N. *Principles of Soft Computing*, 2nd ed.; Published by Wiley India: New Delhi, 2011.
Specht, D. F. A General Regression Neural Network. *IEEE Trans. Neural Netw.* **1991,** *2* (6), 568–576.
Taurino, A. M.; Distante, C.; Siciliano, P.; Vasanelli, L. Quantitative and Qualitative Analysis of VOCs Mixtures by Means of a Microsensors Array and Different Evaluation Methods. *Sens. Actuat.* **2003,** *93*, pp 117–125.
Tayfur, G. *Soft Computing in Water Resources Engineering: Artificial Neural Networks, Fuzzy Logic and Genetic Algorithms*; WIT Press, 2012; pp 13–51.
Ulke, A.; Tayfur, G.; Ozkul, S. Predicting Suspended Sediment Loads and Missing Data for Gediz River, Turkey. *J. Hydrol. Eng.* **2009,** *14* (9), 954–965.
Yang, C. T. *Sediment Transport Theory and Practice*; McGraw-Hill: Singapore, 1996.
Zadeh, L. A. Fuzzy Logic, Neural Networks and Soft Computing. *Communication of the ACM* **1994,** *37* (3), 77–84.

CHAPTER 5

Computation of Different Parameters in Water Resources Engineering Using Artificial Neural Network

ANKIT CHAKRAVARTI[1*], HIMANSHU PANJIAR[2], and MANISH PANDEY[3]

[1]*Faculty of Hydraulics & Water Resources Engineering, School of Civil and Environmental Engineering, Ambo University, Technology Campus, Ambo, Ethiopia*

[2]*Faculty of Material Science and Engineering, Jimma University, Jimma, Ethiopia*

[3]*Department of Civil Engineering, NIT Warangal 506004, Telangana, India*

[*]*Corresponding author. E-mail: ankitchakravarti@gmail.com*

ABSTRACT

Artificial neural network (ANN) is one of the most important soft computing techniques, which can be applied effectively for various sectors in water resources, their planning, and management. The ANN model gained interest in multidisciplinary applications and also was found to be efficient in solving complex problems related to water resources. The effective water resources management applications using ANN techniques considered in this chapter are rainfall runoff process, suspended sediment assessment, water quality assessment, seepage analysis, and flood hydrograph. The modeling of related applications parameters was analyzed and discussed using ANN, which is useful in real-time examples of water resources. It can be inferred from this work that the ANN method is one of the useful tools and is very effective to

Advances in Hydrology and Climate Change: Historical Trends and New Approaches in Water Resources Management. Surendra Kumar Chandniha, Anil Kumar Lohani, Gopal Krishan, & Ajay Krishna Prabhakar (Eds.)

solve complex water management-related modeling and prediction problems and the results are comparative to other techniques.

5.1 INTRODUCTION

In the present times, the water resources sector involves huge statistical databased problems due to the expansion in human population on the globe and has to be addressed in short period of time, otherwise imbalances in the supply and demand of water resources can create soon. Human life is directly dependent on water resources for various activities and these things are to be balanced properly if mankind has to exist. These imbalances can create major manmade disaster for mankind itself. So, the research in this area has to be updated always to cope up with their balance, and soft-computing techniques are very useful tool in that aspect to handle the complexity of the present water resource problems (Tayfur, 2012; Sivanandam and Deepa, 2011). All the soft-computing methods applications are equally important in the various fields of water resources such as modeling and simulation of rainfall runoff (RR), floods, water-logging, soil erosion, drought salty groundwater, region wise water table, etc. These are some of the major problems of water resources planning and its management for agriculture and other needs, which are discussed in detail from next section.

5.2 RAINFALL RUNOFF—THEORETICAL BACKGROUND

5.2.1 CONVERSION OF RAINFALL INTO RUNOFF

The basic understanding of RR includes the conversion of rainfall to runoff, which becomes essential if it has to be designed for creating RR model. This section gives details about the dynamics of RR model. The hydrological cycle showing RR event is presented in Figure 5.1. The available water on the earth is mostly found in oceans. Once it starts to evaporate, it goes to the atmosphere resulting in different conditions; then vapor condenses to form clouds and further become rainfall. The rainfall that falls directly on either ocean or rivers further transports it to the oceans only.

Different possibilities exist for the water that falls on the earth surface, which can be intercepted by vegetation, infiltration into the soil, and multiple site evaporation. The infiltration process starts from the unsaturated point absorbed through vegetation, in which the water returns to the atmosphere by the transpiration processes. The saturated zone water responsible for basin runoff is a part of subsurface water flow. The phenomenon of subsurface water flow back to some water body known as seepage. To know the rainfall and runoff variations, one can refer to the literature for better understanding (Rientjes, 2003; Chakravarti and Jain, 2014). A hydrograph showing major kinds of RR processes with the variation of rainfall event includes three stages (1) a rising limb, (2) a falling limb, and (3) a recession limb (De Vos, 2003). In addition, this can be seen in Figure 5.2, where days 5–7 represent a rising limb, days 8–13 show a falling limb, while further days show a recession limb.

FIGURE 5.1 Schematic presentation of rainfall into runoff.

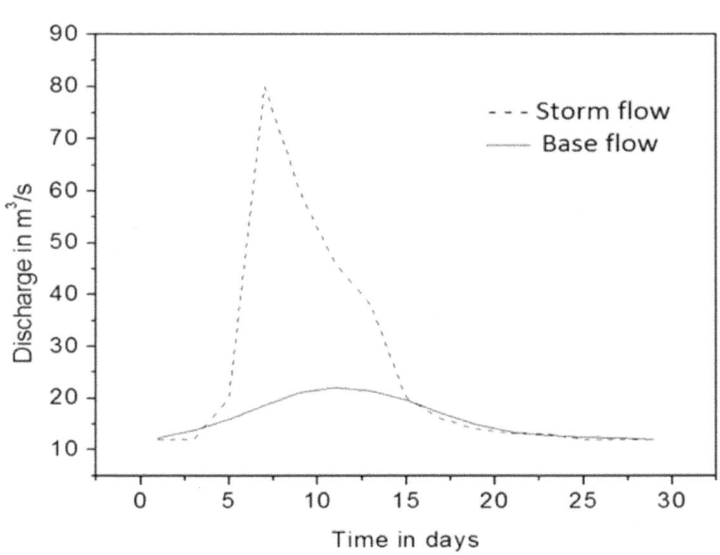

FIGURE 5.2 Hydrograph showing RR process.

5.2.2 RAINFALL-RUNOFF PROCESSES

As per Chow et al. (1988), there are three different types of overflow, that is, surface, subsurface, and groundwater. In this part, focus is given in describing the various types of overland and other flows as shown in Figure 5.3 and represented by numbers. These overland and other flows divided into many parts such as Horton overland flow (represented by (1), saturation overland flow (2), unsaturated subsurface flow (3), perched subsurface flow (4), macropore flow (5), rapid ground water flow (6), delayed ground water flow (7), and channel flow (8). Figure 5.3 will help to better understand how rainfall is getting divided into many streams after touching the earth's surface.

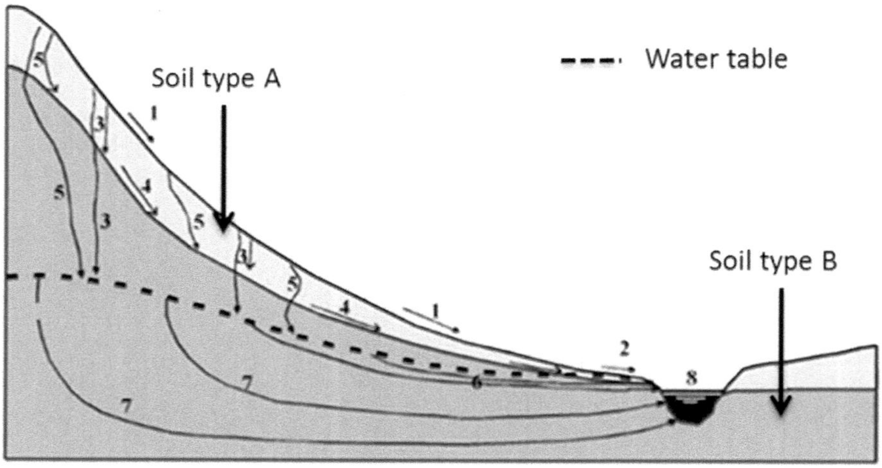

FIGURE 5.3 Various flow processes.

5.2.3 RAINFALL RUNOFF MODELING APPROACHES

RR models can predict the variation in rainfall and runoff of a basin with the given required inputs. The process of rainfall into runoff is used to forecast the stream flow useful for water planning, which includes infrastructural works, extreme flood, drought variation and control, water power generation, irrigation, drainage systems design etc. For modeling, one can approach in various ways, like the few mentioned in literature, namely, distributed, semi-distributed, and lumped model as shown in Figure 5.4 (Rientjes and Boekelman, 2001).

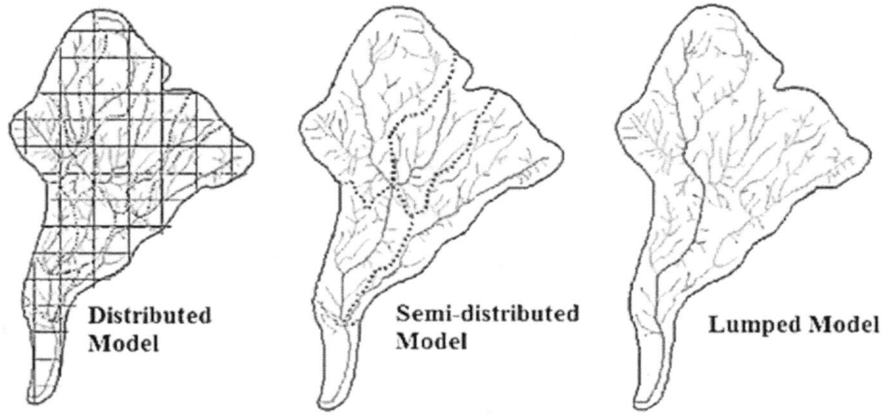

FIGURE 5.4 Distributed, semidistributed, and lumped model approach.

The conversion of rainfall into runoff becomes a highly complex event due to space and time variation of basin features (Chakravarti and Jain, 2014). As per Figure 5.4, the distributed model will have more complexity as compared to others, while the lump model has the least complexity.

5.3 RAINFALL RUNOFF MODELING USING ARTIFICIAL NEURAL NETWORKS

Various models such as artificial neural network (ANN), black box, conceptual, and physically based models were applied to solve complex hydrological events such as overland flow and other hydrological events and found promising (Abrahart et al., 1999; Beven, 2001; Cigizoglu and Alp, 2004; Dawson and Wilby, 2001; Rajurkar et al., 2002; Riad et al., 2004; De Vos and Rientjes, 2005; Jain and Srinivasulu, 2006; Chen and Zhang, 2009; Wu et al., 2010; Chakravarti et al., 2012, 2013, 2015).

From the past decades, ANNs show its ability to become one of the promising tools for hydrological modeling in case of RR event. ANNs' performance also shows the reliable output result as compared to other techniques (Chakravarti et al., 2015; Antar et al., 2006). Its capability to identify underlying connection between input and output variables without brief knowledge about the process is under study. The ANNs once trained, provide the best set of network variables and that is getting used further for output prediction with a given required set of input to the network. The ANNs used to model the RR event, and surface out the relation between input and output effectively with network weight adjustment process of RR modeling, so in general it can be inferred that ANNs can be considered as one of the efficient technique to compute complex events like RR and other hydrological events (ASCE Task Committee, 2000a 2000b; Kumar et al., 2005; Kalteh, 2008; Chakravarti et al., 2015). This section presents the dynamics of RR conversion of a catchment or basin scale, and different flow event related to this system.

Chakravarti et al. (2015) in their study conducted the laboratory experiments to know the RR process in sand bed having median size of 1 mm layer on impermeable plane area slope varies from 1% to 2%, rainfall intensity fixed at 90 mm/h. The RR model was developed using the ANN approach for the RR process and subsequently validation of experimental data. The experimental and computed runoff data shows that the ANNs predicts the RR data very well in observed hydrograph as compared with Nash–Sutcliffe efficiency, which is important for decision and forecasting in the areas of water resources planning and management and flood forecasting, etc. The ANN technique applied with the help of MATLAB platform for RR process using 586 collected data points for experimental RR process. The data used for ANN model development have various input variables (slope, rainfall intensity, and rainfall time) and an output variable runoff discharge. For ANN training, the feed-forward back propagation algorithm was used.

The variation in experimental and ANN model computed RR hydrograph presented in Figures 5.5 and 5.6, where *X*-axis shows the time of RR process and *Y*-axis represents the runoff rate. From Figures 5.5 and 5.6, it can be clearly indicated that the ANN model have the ability to match the pattern and do prediction of the laboratory overland flow process properly during rising and equilibrium stage in all events with reliability index above 95% value against Nash Sutcliffe efficiency of 90%; the reliability index is represented by coefficient of determination value between experimental and predicted data.

Further Kumar et al. (2016) also used the ANN technique to evaluate overland flow in Sarada River Basin, India with ten years data related to rainfall, out of which 7 years (2001–2007) data was used to develop the ANN model (termed as calibration), while further three years (2008–2010) data was used to validate the model (termed as validation). In their study, they employed the Feed Forward Back Propagation (FFBP) algorithm for training with optimization technique named Levenberg–Marquardt. The important input data during training of ANN was found as rainfall, rainfall lag, and lag discharge for generating the best ANN model. Even they reported that

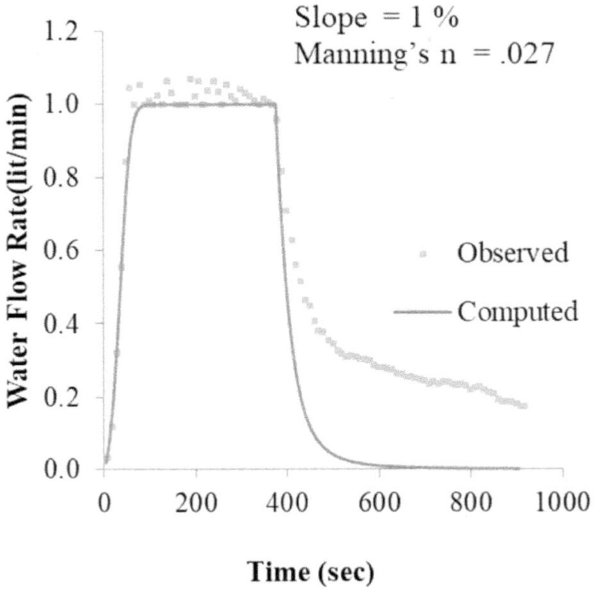

FIGURE 5.5 Variation of laboratory and calculated hydrograph (rainfall intensity 90 mm/h at 1% slope) (Chakravarti et al., 2015).

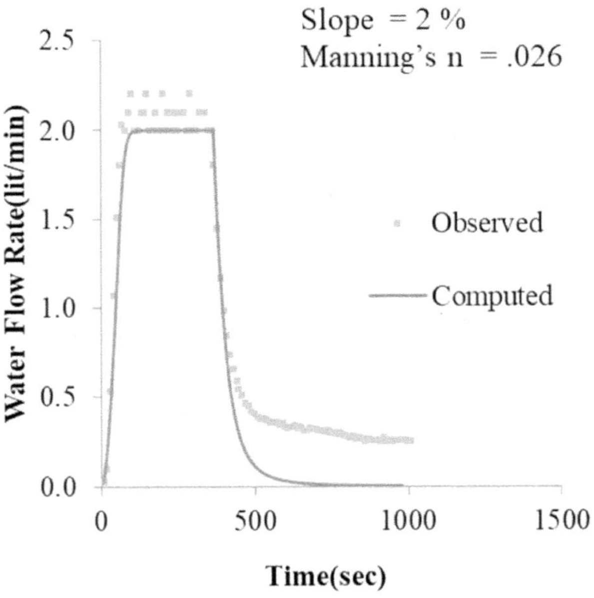

FIGURE 5.6 Variation of laboratory and calculated hydrograph (rainfall intensity 90 mm/hr at 2% slope) (Chakravarti et al., 2015)

adding lag precipitation data in input with the day precipitation increases model efficiency. The ANN model performance was determined by the term coefficient of determination (R^2) value, which was found to be 0.81 during calibration and 0.78 during validation. So, as per the training performance, the ANN model during validation run fairly provided satisfactory results as compared to observed runoff data trend. So, as per their study, it can be seen clearly that there is only three percent variation in the model output in case of validation, and that level of uncertainty is well accepted in the case of practical application of such a complex system.

In one study, Kalteh applied ANN model to forecast RR processes and explained different approaches which includes Neural Interpretation Diagram (NID), Garson's algorithm, and randomization technique to know the reaction obtained by ANN (Kalteh, 2008). The performance using ANNs shows good results not only in modeling but also showing insight through learned variables can assist modeler to describe the process. The NID is a schematic of ANN used to describe the applied network where in same way all nodes are joined by weights and getting trained based on some training algorithm which is represented here through node and link magnitude with link direction visuals, the better explanation can also be referred from literature (Ozesmi and Ozesmi1999). The ANN applied by Kalteh for identification of hydrological processes (monthly runoff in a basin in Iran) has three-layer feed-forward multi-layer perceptron (MLP) model. To do the ANN modeling, nine input variables were given to the network with one output variable. In the phase of model development during training, it was shown that the computation of root mean squared error (RMSE) and coefficient of determination (R^2) of MLP model performed and found the value of 1.27 and 0.96 for RMSE and R^2, respectively (Kalteh, 2008). This amount of R^2 value indicates that model can be reliable for the selected process, but at the same time RMSE value should also be lower enough which actually indicates the amount of uncertainty. The same MLP model was validated with the help of another set of 30 days experimental runoff data and found the value of 1.73 and 0.93 for RMSE and R^2, respectively. There is little variation in RMSE, which is around 36% increase, which contributes to the rise of uncertainty and reduction in R^2 value by around 3%, even though still the model is above 90% reliable, so it can be well accepted. So, from the study by Kalteh, also it is clear that ANN model can be used as an effective tool to model the RR process.

In one more study, Antar et al. (2006) used the ANN model to estimate the RR event in Blue Nile river basin, Ethiopia, which was sub divided into seven sub-basins. They used daily data available for 8 years (1992–1999) for ANN modeling, out of which the first 5 years data was used for model training and testing (also termed as calibration), while last 3 years data was used to validate the model. In their network of ANN, three layers were there, namely, input, hidden, and an output. The numbers of hidden nodes were taken equal to the number of input nodes. A feed-forward back propagation algorithm is applied during training. After successful training of model or calibration, the comparison of observed and computed daily hydrographs at Diem station shows that the model could simulate most of the flood peaks and the outcome of it matches with physically varied RR event. In addition, a physical distributed model that executes hydrologic basic formulas was also employed in grid base. Further, the outcome of comparative analysis between ANN and physical distributed model showed that ANN has more ability to model RR event with heavy reduction of total error. Out of all model errors, the error variance was found to be a major contributor to total model errors. Finally, their study concluded that the ANN model can be applied and if trained with suitable sets of input have a great potential to compute the RR process and also ANN model not require an intensive data set as compared to distributed physically based models.

In another study by Asadi et al., (2019) feed-forward back propagation ANN technique employed to estimate monthly runoff and focus was on the study of the effect of additional inputs hydro-geomorphic and biophysical time series in addition to regular rainfall inputs to the ANN. In their consideration of additional inputs of hydro-geomorphic and biophysical time series data of the Calliope and the Haughton river catchment, Australia, for their ANN models development separately for each river catchment, they found that ANN model with only regular rainfall inputs are lacking in their efficiency as compared with ANN model fed with additional inputs of hydro-geomorphic and biophysical time series data. The improvement in model efficiency reported were 9.77% with reduction of 24.43% RMSE value for Haughton river case, while 11.25% model efficiency improvement with 37.89% RMSE value reduction observed for Calliope river case.

Minns and Hall (1996) used ANN method to compute RR event from synthetic storm by conceptual hydrological model having a single nonlinear reservoir. Total six alternate rainfall events were conducted by changing time having various degrees of nonlinearity to make the alternate rainfall outputs. Total three conceptual models were applied; the start trials were used with a three- and four-layer ANN networks. The networks were feed with data of the considered 14 corresponding rainfall depths and three antecedent flow ordinates. Finally, with rainfall flow input varying from linear to extremely nonlinear cases, the ANN model got trained and concluded that the ANNs was capable of identifying usable relationships between discharges and antecedent rainfalls (Minns and Hall, 1996).

Mallikarjuna et al. (2012) also used ANN model to know RR variation in a Krishna basin, India. In this research work, the RR events were mapped by coupling simple time-series (STS) and linear autoregressive (ARX) models with ANN. The performance of ANN technique for modeling and forecasting of 30 day runoff at monsoon time found better by taking the residuals derived from STS and ARX models as extra inputs both with base rainfall data inputs.

5.4 PREDICTION OF SUSPENDED SEDIMENT USING ARTIFICIAL NEURAL NETWORKS

For economical design of water structures such as dam, reservoir, spillways, gates, and other hydraulic structures, it is important to know the sediment load, which is consisting of suspended sediment and bed load. The sediment consists of 75–95% of the total load is in the form of suspension (Yang, 1996; Chakravarti et al., 2013). Various research studies related to mechanics of sediment transport conducted previously, mathematically, as well as physically, and based on that various equation have been proposed to predict sediment transport phenomenon either numerically or based on developed empirical equation, but every proposed equation not applies in all conditions (Tayfur, 2003). The research based on natural event mostly used physical based and black box models, which are able to provide space and time variations of different variables, and can be described by partial differential equations (Tayfur, 2001, 2002). The developed model performance can varies as per the data availability such as topography, special and temporal variation of rainfall, soil type, vegetation type, rainfall patterns, and other hydrological data. However, all the data is very difficult to collect, also this kind of formula at the end will not give more advantages as compared to lumped models.

Due to difficulty in creating multiple input-based suspended sediment model through human calculation, soft computing technique came into picture. Initially, soft-computing techniques were also known by a name as black box model, and since the trial was started to prove these methods, so the main concentration to apply black box model in predicting the basin responses to inputs were initiated instead of different explanation about the event in the basin. The black box model was used to identify the responses of the catchment or basin based on specific set of inputs to model. Various types of black box models research can be found in survey as reported by (Bruce, 1975; Williams, 1978).

In the case of suspended sediment, different variant of models can be found, which may show linearity or nonlinearity with concentration or excess precipitation. For nonlinear model, the functional series can be formulated by nonlinear black box model as per the 2D unit sediment graph theory in case of total suspended sediment (TSS). The model is capable to show the all sediment movement nature in the basin with no availability of excess precipitation or runoff. Whereas this model has limitation, that is, the basin is time invariant, rainfall has an equal spatial variation.

The main causes of this problem in all these aforementioned modeling effort in sediment transport are as follows:

i. Difficulty to explain best correlation with concentration of sediment and other responsible variables such as stress slope, etc.
ii. In mechanics of sediment process, it is very difficult to know when the sediment particle starts to move because of its stochastic nature.
iii. The boundary is changing with sediment flow in the form of suspension or bed load and resistance to water movement is changing because of less standard and correct method to know the best equation in different roughness condition.
iv. Various unpredicted problems were found during calculating sediment load at various flow or sediment condition due to variation in hydrology, geology, and climatic variations.

Since the ending of twentieth century, an ANN a soft computing technique is getting used to predicting TSS in streams (Cigizoglu, 2004; Tayfur, 2001, 2002, 2003). The development of ANN model to calculate TSS in rivers can be done by taking discharge data as input vector, and one of these types of work was executed by some worker (Tayfur and Guldal, 2006).

The estimation of TSS by taking rainfall data has high implication. The main problem for researcher at present is to estimate the hydrologic data in ungauged catchment. Most of the areas in the earth surface come under ungauged or roughly gauged zone (Sivapalan et al., 2003). Because of that the hydrological data with rainfall data is best suitable for designer or researcher to start with. To estimate TSS in rivers by applying rainfall data was also suggested earlier, and to fix the various input data, relationship value with TSS and rainfall can firstly be calculated using eqs 5.1 and 5.2 (Tayfur and Guldal, 2006). The Cross correlation is the best-suited technique to predict the degree in which two series are matched. Assume two series data $x(i)$ and $y(i)$, where $i = 0,1,2....N-1$, the cross correlation "r" at delay t can be described as in eq 5.1:

$$r(t) = \frac{\sum_i [(x_i - \bar{x})((y_{i-1} - \bar{y})]}{\sqrt{\sum_i (x_i - \bar{x})^2} \sqrt{\sum_i (y_{i-t} - \bar{y})^2}}, \tag{5.1}$$

Where, $\bar{x}$ and $\bar{y}$ represents alternate series. Here, lagged type of itself is known as autocorrelation. For correlation coefficient, lag k of series x_0, x_1, ...x_{N-1} is described as in eq 5.2:

$$r(t) = \frac{\sum_i [(x_i - \bar{x})((x_{i+k} - \bar{x})]}{\sum_i (x_i - \bar{x})^2}, \tag{5.2}$$

At the time of correlation coefficient estimation in every lags $k=0, 1,..., N-1$, this series is known as correlogram. From their study, the cross correlations results of rainfall and total suspended sediment obtained and showed that after time lag of 4 days the correlation value near to zero, which means after lag time of 4 days no effect on TSS by rainfall. So, up to 4 days data of rainfall can be fed into the ANN to develop a model of TSS. Because of that ANN could have input neurons of rainfall in terms of days (d) for present-day and 4 days old d-1, d-2, d-3, and d-4 for 1 day old, 2-day old, and so on. By following the same way, they found that with ¾ training data and ¼ validating data of whole data yielded a very reliable ANN model with R^2 value of 0.92 and mean absolute error =34.2 mg/L, which is a very appreciable benchmark as there is no alternative model is present for the case of TSS. Here it is clear that rainfall data are very useful and multipurpose data, as present and past few-days rainfall precipitation data can be effectively used for making of ANN model for TSS.

5.5 WATER QUALITY ASSESSMENT USING ARTIFICIAL NEURAL NETWORKS

For water quality assessment, it is important to know the water distribution system (WDS) through which the availability of water to the end user are coming with acceptable pressure, flow, continuity, and water quality. The quality of water could be identifying by its chlorine concentration, odor, taste, and appearance also with others parameters. To maintain the water quality with WDS, one should know the water consumption, which is a difficult problem practically and faced by water supply department all the time with considering the part of WDS like, pipe materials, tanks, valves, etc. and other risks related to water distribution (Clement et al., 2004).

Cuesta et al. (2014) applied ANN technique and program EPANET 2.0 to know the chlorine concentration in WDS, which is also a very important issue in general to model chlorine level for drinking water process at present. This was performed by taking water quality variables, like flow, pH, temperature, etc., which permit the prediction of chlorine concentration at specific nodes or station in water supply system. With available data, Cuesta et al. created seven ANN models with different set of inputs. The inputs considered for ANN model taken from three different nodes in water supply system (pH, temperature, and turbidity from each node) with relevant supply tank data (which includes initial chlorine, free chlorine, pH, temperature, flow, and turbidity), but not even a single model had all the inputs. Cuesta et al. tried to design their models in such a manner, so that a good ANN model can find the chlorine concentration by taking fewer points inside WDS. All the ANN models were MLP-based network comprised three layers. While different models had different set of input numbers and hidden layer nodes, so their network structure varied accordingly. After training and validation, it was found that ANN with MLP 5-5-1 yielded a very good efficiency (above 90%) for the determination of chlorine level in considered water supply system, which showed that the ANN models are capable to trace out nonlinear relationship between different variables of water quality system.

In the area of water quality assessment, Gupta et al. also applied ANN technique to generate the water quality index (WQI) using various water quality parameters like dissolved oxygen (DO), pH, turbidity, *E. coli*, and electric conductivity (EC) (Gupta et al., 2019). The overall procedure was outlined with a point of acceptance "if regression value greater than 0.8 for trained ANN model," which represented the condition for the use of ANN to estimate the WQI. The dataset formulated for WQI model development using ANN includes five regular variables stated earlier, whose data also including their safe limit range. As the WQI depends on all these factors mainly, so it is better to have little knowledge about all. Turbidity shows the clarity of water, if water clarity is more, which will indicate low turbidity and having good taste. The pH value provides the acidity in the sample but it is a measurement of the potential activity of hydrogen ions (H^+). DO give the water taste, it's more content is better. If *E. coli* is available enough in water, which is not a better sign as it is responsible to create digestive problem in humans. EC indicates the availability of different chemical ions, if the value of EC is more, the percentage of ions in water will also be more. In their work,

Gupta et al. applied Fast and Cascade forward network-based ANN, all these ANNs are like FFBP neural network, only little change in networking is there for Cascade forward network where the output node is getting additional inputs from input layer directly bypassing hidden layer(s). The obtained performance of Fast and Cascade forward network was analyzed and found that Cascade forward network is the optimal network architecture in the case of WQI modeling. There is found to be selection difference in gradient and learning rate for both the cases, for Cascade forward network model, 1.1003 and 0.101 were the value of gradient and learning rate resulted overall model correlation coefficient value of 0.89. While for fast-forward network model, 0.2764 and 0.348 were the value of gradient and learning rate resulted overall model correlation coefficient value of 0.85. With that selection, it was found that at all stages of ANN model development such as training, testing, and validation the cascade-forward network model performed better. So, with the model correlation coefficient value of 0.89, the cascade-forward network model shown its ability to be employed in the field of water sector, where when the five parameters described by WHO (considered in the study) are used to determine WQI. The changes in network parameters can alter the outcomes but indicated that the higher the number of member in training dataset, the value of regression will be higher with specific selection of learning rate and gradient. With that, we can infer that the ANN models have the capability to simplify the complexity of interpretation of WQI.

Further Chen and Zhang also applied ANN technique to identify the water quality in Yellow river, China. Total 63 observations were taken in 2003–2005 in Yellow river including high, normal, and low water (Chen and Zhang, 2009). Total 10 different chemical variables were identified and considered as inputs to ANN namely NH_2, COD, CN, Pb, hydroxy benzene, DO, Cr^{+6}, Hg, Cd, and As. Different factors affect the water quality out of that selected ten factors are investigated in this study as per the water quality standard. For estimation of these collected samples, firstly, the basic statistical approach was applied to know the variables as per the water-quality standard; second, MLP neural network with back propagation and probabilistic neural networks (PNN) were used to characterize the water quality in various sampling site in different period. The correlation of water quality types and chemical measurements was sought as per their study. So, the model with water quality types and chemical measurements is built, and these models could quickly, completely, and accurately classify the water quality of the Yellow river. The water quality parameters are divided into five main classes in which class 1 indicates unpolluted class 2 indicates proper to drink; class 3 indicates able to drink; class 4 indicates only for industry; and class 5 indicates only for agriculture. Every data are randomly split in two categories: training set and crossvalidation set for training and optimizing the neural networks. After validation, it was found that ANNs gives better and acceptable results to identify various water quality parameters, which was fixed in the study from 1 to 5 classes.

In other study by Seo et al. (2016), they considered the water quality variables like temperature, DO, pH, EC, TN, TP, turbidity, and Chlorophyll-a at the downstream of Cheongpyeong dam on the North Han River to create an ANN ensemble model to determine these water-quality parameters. To decrease the influence of trend, periodicity, and stochastic component and to enhance ANN model performance during training, testing, and validation, Seo et al. implied stratified sampling method according to the distribution ratio of each parameter, and planned to forecast all these selected water quality parameters one step ahead. It was found that the seven out of eight water-quality parameters performed better with higher R^2 (0.85) except turbidity having R^2 (0.63) and five variables, that is, temperature, DO, pH, TN, and TP show good performance less than 1.0 RMSE value. In addition, they also reported that despite the effort to decrease the error with stratified sampling approach, the biased data set affected to result, and the ANN ensemble model created is not able to predict the rare event.

5.6 SEEPAGE ANALYSIS USING ARTIFICIAL NEURAL NETWORKS

This part includes the use of ANN model in exploring a locus of seepage path in an embankment dam. The embankment dam made under steady state seepage condition and because of that more seepage poses threat to integrity of a dam. This kind of dam if having excessive and improper seepage will cause unexpected failure to the dam. The study of space and time variation of seepage and infiltration, their paths in dam structure is important for safe and economic design for dam stability. The path of seepage through the dam body can be forecasted by applying mathematical model and physical theory. The electrical resistivity theory could be apply to delineate zone concerning to seepage. To decrease the seepage paths self-potential technique could be applied.

Panthulu et al. (2001) applied both the methods for Saddle dame of Som Kamla Amba Project in Rajasthan, India. In addition, they found that a dye trace test is the best test to know the path of seepage. Turkmen et al. (2002) applied these tests in Kalecik dam in Turkey, and for seepage analysis in embankment dam, mathematical models are also applied. For seepage study in saturated and unsaturated zone, finite element technique is also found as a useful method (Honjo et al., 1995). The numerical method is more useful when data varied with respect to space of real model parameters at all element of numerical mesh. However, this kind of specific data availability throughout the dam are rarely available because of the primarily time and budget limit also the numerical outcome of the highly nonlinear flow formula is prone to instability and lack of convergence sometimes. So, some soft computing technique can be used instead such as ANN. Tayfur et al. applied the ANN technique for estimation of seepage with respect to space and time and the locus of seepage path using water depth at higher and lower pools in the dam (Tayfur et al., 2005). It should be noted that horizontal screen sealing of downstream and upstream slopes of dam had important effect on level of water. If that screen is tight and there was leakage on upstream slope, that results in seepage. Tayfur et al. in their work collected the levels of water distribution measured by piezometers, lower, and upper reservoir from October 1995 to May 2002 for Jeziorsko earthfill dam in Poland for creating the two models one by using finite element method (FEM) and another by using ANN. The input data were taken as water level in upstream dam, downstream reservoir, and piezometers location, while the output data were piezometric water level. The ANN model had three layers first input, output, and hidden layers, the model was based on FFBP neural network with Levenberg–Marquardt algorithm used for training. The input, hidden, and output layers contained three neurons, five neurons, and one neuron, respectively, and some numbers taken to show the locations of the piezometers. Both FEM and ANN models were developed and applied to predict spatial and temporal variation of flow through the dam for monitoring seepage, and both models performance found satisfactory. It was found that ANN model can estimate seepage path locus in a dam body with comparative performance and, in some cases, better than the FEM model, even ANN model can detect any anomalies during seepage, which require remedial stability measures. From this study of Tayfur et al. it was also found that the ANN model creation is far easier and accurate than FEM model creation, as in this case of seepage estimation model by ANN, developed using only the lower and upper pools water levels, which are less critical parameters and easily measurable field data.

5.7 FLOOD HYDROGRAPH PREDICTIONS USING ARTIFICIAL NEURAL NETWORKS

The flood hydrograph of a river at any downstream section can be modeled and predicted based on the upstream activities in the river. The flood is a natural phenomenon occurring worldwide time to time and the flood hydrograph prediction can be an effective alarming method for downstream locations. It is well known that flood is an all-round loss incurring phenomena for any locality or regions, so this can be alleviated by some degree by some alarming model based work and soft computing methods are found to be a good option. In this context, several soft computing methods are used for flood hydrograph predictions, such as the Fuzzy logic methods (Tayfur, 2012), the ANN (Tayfur et al., 2007; Tayfur, 2012), and the GA (Tayfur and Moramarco, 2008; Tayfur et al., 2009; Tayfur, 2012). In addition, even some researcher reported the multi-step-ahead flood forecasting method in this field of hydrology (Chang et al., 2014, 2015). The movement of a flood wave along a channel length is an important parameter for flood routing in a river and also for calculating the flood hydrograph at any downstream section. Generally, two most common approaches for flood routing, the first is hydraulic approach based on numerical methods and second is hydrologic approaches, which are based on conservation of mass principle. All the methods require substantial field data; they are costly and time consuming. While the soft computing methods such as ANN is very helpful in modeling this hydrological process (Barbetta et al., 2017, Tayfur et al., 2018).

Further, to show the effective use of ANN in this field of flood hydrograph predictions, the work presented in literature by some scientist is discussed here (Tayfur et al., 2018), where they applied feed forward back propagation neural network for flood hydrograph prediction. As our discussion is limited to ANN, so the effect of ANN will be highlighted only in present section. Further for ANN modeling, the data for runoff hydrographs was observed along an equipped branch of Tiber River, central Italy, with significant lateral flow contribution, which was further used for approach testing. In that river, Ponte Felcino station is the downstream station and Santa Lucia station are the upstream station, the data collection done at every half an hour at Ponte Felcino (downstream station) and Santa Lucia (upstream station), which further constituted input vector of ANN network, whereas the target output for ANN was flow discharge measured at Ponte Felcino station. The ANN model was trained using total data point taken in testing phase was 740 by considering some month/year events (May 1995, January 1997, June 1997 and February 2004) with learning rate (α) of 0.02 and 3000 iterations. The ANN model was developed, validated using other set of data (December 1990) flood data of Ponte Felcino station and also compared with four other methods in their study. All the models including ANN-based model used there to forecast true event-based runoff hydrographs, and it was found that the efficiency of ANN in predicting these kind of flood hydrograph is appreciable and comparative with all other methods, which were used in the study namely rating curve method (RCM), genetic algorithm_rating curve method (GA_RCM), particle swarm optimization_rating curve method (PSO_RCM), and ant colony optimization_rating curve method (ACO_RCM). Even the errors were also comparable to other tested methods, the reported errors were around 7 m^3/s mean absolute error and 9 m^3/s, RMSE values on average for ANN.

5.8 CONCLUSION

Different streams of water resources sector such as RR, suspended sediment assessment in river system, water quality assessment, seepage analysis and flood hydrograph modeling, and simulation using a soft computing technique—ANN approach were presented in this study. It is found that the ANN technique is an effective tool to model and forecast specific parameters of water resources sector. In all the aforementioned cases of water sector, ANN modeling employed and compared with other theoretical and physical models, ANN always found to be comparative or better than other theoretical or physical models. As all the water resources based problems are databased problem now a days, so ANN is also a data-driven model, and due to that ANN can be accepted as one of the valuable soft computing tool for water resources planning and management.

KEYWORDS

- **artificial neural network**
- **rainfall runoff process**
- **suspended sediment assessment**
- **water quality assessment**
- **seepage analysis**
- **flood hydrograph predictions**

REFERENCES

Abrahart, R. J.; See, L. M.; Kneale, P. E. Using Pruning Algorithms and Genetic Algorithms to Optimise Network Architectures and Forecasting Inputs in a Neural Network Rainfall–Runoff Model. *J. Hydroinfo.* **1999,** *1*, 103–114.

Antar, M. A.; Elassiouti, I.; Allam, M. N. Rainfall-Runoff Modelling Using Artificial Neural Networks Technique: A Blue Nile Catchment Case Study. *Hydrological Processes* **2006,** *20,* 1201–1216.

Asadi, H.; Shahedi, K.; Jarihani, B.; Sidle, R. C. Rainfall-Runoff Modelling Using Hydrological Connectivity Index and Artificial Neural Network Approach. *Water* **2019,** *11,* 212. DOI:10.3390/w11020212

ASCE Task Committee on Application of Artificial Neural Networks in Hydrology. Artificial Neural Networks in Hydrology, I: Preliminary Concepts. *J. Hydrol. Eng., ASCE* **2000a,** *5* (2), 115–123.

ASCE Task Committee on Application of Artificial Neural Networks in Hydrology. Artificial Neural Networks in Hydrology, II: Hydrologic Applications. *J. Hydrol. Eng., ASCE* **2000b,** *5* (2), 124–137.

Barbetta, S.; Moramarco, T.; Perumal, M. A Muskingam-Based Methodology for River Discharge Estimation and Rating Curve Development under Significant Lateral Inflow Conditions. *J. Hydrol.* **2017,** *554,* 216–232.

Beven, K. J. *Rainfall-Runoff Modelling: The Primer*; Wiley Report and Publication, 2001; pp 5–13.

Bruce, R. R. Model for Runoff of Pesticides in Small Upland Watershed. *J. Environ. Qual.* **1975,** *4* (4), 541–548.

Chakravarti, A.; Jain, M. K. Experimental Investigation and Modeling of Rainfall Runoff Process. *Indian J. Sci. Technol.* **2014,** *7* (12), 2096–2106.

Chakravarti, A.; Jain, M. K.; Rohila, K. Experimental Investigation of Rainfall Runoff Process. *SWAT International Conference in Department of Civil Engineering*, IIT Delhi, India, 18–20 July, **2012,** p 91.

Chakravarti, A.; Jain, R. K.; Kothyari, U.C. Scour under Submerged Circular Vertical Jets in Cohesionless Sediments. *ISH J. Hydraul. Eng.* **2013,** *20* (1), 32–37.

Chakravarti, A.; Joshi, N.; Panjiar, H. Rainfall Runoff Analysis Using Artificial Neural Network. *Indian J. Sci. Technol.* **2015,** *8* (14), 1–7.

Chang, F. J.; Chen, P. A.; Lu, Y. R.; Huang, E.; Chang, K. Y. Real-Time Multi-Step-Ahead Water Level Forecasting by Recurrent Neural Networks for Urban Flood Control. *J. Hydrol.* **2014,** *517,* 836–846.

Chang, F. J.; Chiang, Y. M.; Ho, Y. H. Multi-Step-Ahead Flood Forecasts by Neuro-Fuzzy Networks with Effective Rainfall-Runoff Patterns. *J. Flood Risk Manage.* **2015,** *8,* 224–236.

Chen, L.; Zhang, X. Application of Artificial Neural Networks to Classify Water Quality of the Yellow River. *Fuzzy Info. Eng., ASC* **2009,** *54,* 15–23.

Chow, V. T.; Maidment, D. R.; Mays, L. W. *Applied Hydrology*; McGraw Hill Publication, 1988.

Cigizoglu, H. K.; Alp, M. *Rainfall-Runoff Modelling Using Three Neural Network Methods*; Springer, ICAISC **2004,** LNAI **3070,** 2004; pp 166–171.

Cigizoglu, H. K. Estimating and Forecasting of Daily Suspended Sediment Data by Multi Layerperceptrons. *Adv. Water Resour.* **2004,** 27, 185–195.

Clement, J.; Powell, J.; Brandt, M. *Predective Models for Water Quality in Distribution Systems*; IWA Publishing, 2004; p 106.

Cuesta, C. G. A.; Tuhovcaka, L.; Taus, M. Using artificial Neural Network Models to Assess Water Quality in Water Distribution Networks. *12th Int. Conf. Comput. Control Water Indust., CCWI***2013,** *Procedia Eng.* **2014,** 70, 399–408.

Dawson, C. W.; Wilby, R. L. Hydrological Modelling Using Artificial Neural Networks. *Progress Phys. Geog.* **2001,** *25* (1), 80–108.

De Vos, N. J. Rainfall-Runoff Modelling Using Artificial Neural Networks. Civil Engineering Informatics Group and Section of Hydrology & Ecology, TU Delft Netherlands, M. Sc. Thesis Report, 2003; pp 31–40.

De Vos, N. J.; Rientjes, T. H. M. Constraints of Artificial Neural Networks for Rainfall–Runoff Modeling: Trade-Offs and Model Evaluation. *Hydrol. Earth Syst. Sci.* **2005,** *9*, 111–126.

Gupta, R.; Singh, A. N.; Singhal, A. Application of ANN for Water Quality Index. *Int. J. Mach. Learn. Comput.* **2019,** *9* (5), 688–693.

Honjo, Y.; Giao, P. H.; Naushahi, P.A. Seepage Analysis of Tarbela Dam (Pakistan) Using Finite Element Method. *Int. J. Rock Mech. Min. Sci.* **1995,** *32* (3), 131 (A).

Jain, A.; Srinivasulu, S. Integrated Approach to Model Decomposed Flow Hydrograph Using Artificial Neural Network and Conceptual Techniques. *J. Hydrol.* **2006,** *317*, 291–306.

Kalteh, A. M. Rainfall-Runoff Modelling Using Artificial Neural Networks (ANNs): Modelling and Understanding. *Caspian J. Environ. Sci.* **2008,** *6* (1), 53–58.

Kumar, A. R. S.; Sudheer, K. P.; Jain, S. K.; Agarwal, P. K. Rainfall–Runoff Modeling Using Artificial Neural Networks: Comparison of Network Types. *Hydrol. Processes* **2005,** *19* (6), 1277–1291.

Kumar, P. S.; Praveen, T. V.; Prasad, M. A. Artificial Neural Network Model for Rainfall-Runoff. *Int. J. Hybrid Info. Technol.* **2016,** *9* (3), 263–272.

Mallikarjuna, P.; Suresh Babu, C. H.; Jagan Mohan Reddy, A. Rainfall Runoff Modeling Using Artificial Neural Networks. *ISH J. Hydraul. Eng.* **2012,** *15* (1), 24–33.

Minns, A. W.; Hall, M. J. Artificial Neural Networks as Rainfall Runoff Models. *Hydrol. Sci. J.* **1996,** *41* (3), 399–417.

Ozesmi. S. L.; Ozesmi, U. An artificial Neural Network Approach to Spatial Habitat Modelling with Interspecific Interaction. *Ecol. Model.* **1999,** *116*, 15–31.

Panthulu, T. V.; Krrishnaiah, C.; Shirke, J. M. Detection of Seepage Paths in Earth Dams Using Self Potential and Electrical Resistivity Methods. *Eng. Geol.* **2001,** *59* (3), 281–295.

Rajurkar, M. P.; Kothyari, U.C.; Chaube, U. C. Artificial Neural Networks for Daily Rainfall Runoff Modeling. *Hydrol. Sci. J.* **2002,** *47* (6), 865–877.

Riad, S.; Mania, J.; Bouchaou, L.; Najjar, Y. Rainfall Runoff Model Using an Artificial Neural Network Approach. *Math. Comput. Model.* **2004,** *40*, 839–846.

Rientjes, T. H. M.; Bokelman, R. H. Hydrological Models. Lecture Notes CThe4431 Faculty of Civil Engineering and Geosciences—Section of Hydrology and Ecology, 2001; pp 12–18.

Rientjes, T. H. M. Physically Based Rainfall-Runoff Modeling. Civil Engineering Informatics Group and Section of Hydrology & Ecology; TU Delft Netherlands, PhD Thesis Report, 2003; pp 21–30.

Seo, W.; Yun, S. H.; Choi, S. Y. Forecasting Water Quality Parameters by ANN Model Using Preprocessing Technique at the Downstream of Cheongpyeong Dam. *12th Int. Conf. Hydroinfo., HIC* **2016,** *Procedia Eng.* **2016,** *154*, 1110–1115.

Sivanandam, S. N.; Deepa, S. N. *Principles of Soft Computing*, 2nd ed.; Wiley: New Delhi, India, 2011.

Sivapalan, M.; Takeuchi, K.; Franks, S. W.; Gupta, V. K. IAHS Decade on Prediction in Ungauged Basins, Shaping an Exciting Future for the Hydrological Sciences. *Hydrol. Sci. J.* **2003,** *48* (6), 857–880.

Tayfur, G.; Moramarco, T. Predicting Hourly-Based Flow Discharge Hydrographs from Level Data Using Genetic Algorithms. *J. Hydrol.* **2008,** *352*, 77–93.

Tayfur, G. Applicability of Sediment Transport Capacity Model for Non Steady State Erosion from Steep Slopes. *J. Hydrol. Eng.* **2002,** *128* (9), 839–847.

Tayfur, G. Modeling Two Dimensional Erosion Process over Infiltrating Surface. *J. Hydrol. Eng.* **2001,** *6* (3), 259–262.

Tayfur, G. Modelling Sediment Transport. International Conference on Water and Envirnment; In *Watershed Hydrology*; Singh, V. P.; Yadava, R. N., Eds.; Allied Publication Pvt Ltd: New Delhi, India, 2003; pp 353–375.

Tayfur, G. *Soft Computing in Water Resources Engineering: Artificial Neural Networks, Fuzzy Logic and Genetic Algorithms*; WIT Press, 2012; pp 13–51.

Tayfur, G.; Guldal, V. Artificial Neural Network for Estimating Daily Total Suspended Sediment in Natural Streams. *Nordic Hydrol.* **2006,** *37* (1), 69–79.

Tayfur, G.; Barbetta, S.; Moramarco, T. Genetic Algorithm-Based Discharge Estimation at Sites Receiving Lateral Inflows. *J. Hydrol. Eng.* **2009,** *14*, 463–474.

Tayfur, G.; Moramarco, T.; Singh, V. P. Predicting and Forecasting Flow Discharge at Sites Receiving Significant Lateral Inflow. *Hydrol. Process.* **2007,** *21*, 1848–1859.

Tayfur, G.; Singh, V. P.; Moramarco, T.; Barbetta, S. Flood Hydrograph Prediction Using Machine Learning Methods. *Water* **2018,** *10* (968), 1–13. DOI: 10.3390/w10080968

Tayfur, G.; Swiatek, D.; Wita, A.; Singh, V. P. Case Study: Finite Element Method and Artificial Neural Network Models for Flow through Jeziorkoerthfill Dam in Poland. *J. Hydraul. Eng.* **2005,** *131* (6), 431–440.

Turkmen, S.; Ozguler, E.; Taga, H.; Karaogullarindan, T. Seepage Problems in the Karstic Limestone Foundation of the Kalecik Dam (South Turkey). *Eng. Geol.* **2002,** *63* (3), 247–257.

Williams, J. R. A Sediment Graph Model Based on an Instantaneous Unit Sediment Graph. *Water Resour. Res.* **1978,** *14* (4), 659–664.

Wu, C. L.; Chau, K.W.; Fan, C. Prediction of Rainfall Time Series Using Modular Artificial Neural Networks Coupled with Data-Pre-Processing Techniques. *J. Hydrol.* **2010,** *389* (1–2), 146–167.

Yang, C. T. *Sediment Transport Theory and Practice*; McGraw-Hill: Singapore, 1996.

CHAPTER 6

Morphometric Assessment of Drainage Basin for Subwatershed Prioritization Based on Land Use Land Cover Using GIS

GAURAV KANT NIGAM[1*], PREETI RAJPUT[2], and M. P. TRIPATHI[3]

[1]*Krishi Vigyan Kendra—Korba, IGKV–Indira Gandhi Agricultural University, Raipur 492012, India*

[2]*Department of Civil Engineering, Government Engineering College Bilaspur, Koni 495009, India*

[3]*Department of Soil and Water Engineering, S.V. College of Agricultural Engineering and Technology, IGKV–Indira Gandhi Agricultural University, Raipur, Chhattisgarh, India*

[*]*Corresponding author. E-mail: er.nigamgk@gmail.com*

ABSTRACT

Inappropriate use of valuable natural resource, soil, and water without any aegis is one of the most vital factors for the deterioration of watershed. Geographical information system and remote sensing technique are used for the identification and prioritization of critical watersheds through morphometric assessment and land use/cover (LULC) analysis. This investigation is an effort to assess and understand the morphometric, LULC change, and terrain characteristics, as well as to prioritize the watersheds of the upper Mahanadi basin. Database has been prepared in ArcGIS 10.4 software for morphometric parameters analysis; and DEM and drainage networks were prepared using the SRTM 30 m resolution images. LULC change analysis has been

Advances in Hydrology and Climate Change: Historical Trends and New Approaches in Water Resources Management. Surendra Kumar Chandniha, Anil Kumar Lohani, Gopal Krishan, & Ajay Krishna Prabhakar (Eds.)

carried out through LANDSAT 8 image in ERDAS-2016 image processing software. The prioritization ranks of each subwatershed were assigned after the estimation of the compound factor. A detailed prioritization map for the critical watersheds based on morphometric and LULC was prepared. These studies are highly appropriate to use for soil erosion prevention and water conservation practices for sustainable development of erosion susceptible watershed.

6.1 INTRODUCTION

Watershed is a natural hydrological boundary, which allows surface runoff to a defined channel, drain, stream, or river at a particular point (Chopra et al., 2005) and watershed management is the process of formulation and implementation of a course of action carrying out that involves modification in the natural system of watershed to achieve specified objectives (Johnson et al., 2002). The importance of the morphometric analysis of drainage basin is pivotal in understanding the geological and geomorphic behavior of drainage basin, which influences the soil erosion characteristics (Pawar-Patil and Mali, 2013). The consequence of morphometric parameters analysis is important in understanding the geomorphic and geological behavior of the watershed, which impact on the soil erosion characteristics (Strahler, 1964). Morphometric analysis involves quantifying morphometric variables, which is very useful in studies such as regional flood frequency analysis, hydrologic modeling, watershed prioritization, natural resources conservation, and management, etc. (Panhalkar et al., 2012). Many researchers across the globe and particularly from India have attempted studies on morphological analysis using RS and GIS techniques. These are basically efficient and less time consuming (Gupta and Srivastava, 2010; Horton, 1945; Brooks et al., 2006).

Remote sensing and GIS application for the morphometric and land use/cover (LULC) analysis plays vital role. The remote sensing technique is a convenient method for morphometric analysis as the satellite images provide a synoptic view of a large area (Javed et al., 2009). Prioritization of watershed on the basis of quantitative analysis of morphometric parameters and land use land cover are crucial in order to decide sustainable watershed management strategy. Morphological and LULC studies are very important to understand the behavior of a river, their drainage pattern, and erosion of river bank and also to plan for erosion control measures. In watershed management study, LULC change analysis is considered to be an important parameter and it has been given emphasis as reported in many recent research studies (Javed et al., 2009). LULC is an important component to drive various developmental indexes for land and water resource (Strahler, 1957). This study is focused on the prioritization of Upper Mahanadi basin, based on GIS concept through morphometric and LULC analysis. Soil and water are the most valuable natural resources and we need to conserve it so that the water harvesting structures prove extremely important in this regard. Thus, this study basically focusses on prioritization of upper Mahanadi basin to identify a suitable site for installation of soil erosion control structure in the subwatershed, which stands with high soil erosion status.

6.2 STUDY AREA

The upper Mahanadi basin is the major part of the Mahanadi basin, which has catchment area of 30,911.10 km^2. It joins the Mahanadi river after running a length of about 383 km upper Mahanadi basin is located between Latitude 20°16′ N to 22°41′ N and Longitude 80°25′ E to 82°35′ E (Fig. 6.1). The maximum and minimum elevations of the study area are 1140 and 201 m above MSL.

6.3 MATERIALS AND METHODS

The working methodology for this study is displayed in Figure 6.2). The topographical maps were georeferenced in Arc GIS 10.4 software. The Shuttle Radar Topography Mission (SRTM 30 m) data were used to prepare the digital elevation model (DEM) (Fig. 6.3). LULC change analysis of the upper Mahanadi basin has been carried out through LANDSAT 8 image of September 28, 2015 and October 5, 2015 by using supervised classification process in ERDAS-2016 image processing software.

The upper Mahanadi basin boundaries were demarcated and divided into 30th watersheds namely WS1 to WS30 (Fig. 6.4) in GIS environment. Hydrology tool under spatial analyst tools in ArcGIS10.4 software was used to extract the drainage network. This process involves a series of steps like fill, flow accumulation, stream order, and drainage network. The different morphometric parameters of basin such as stream length, stream order, drainage density, length of overland flow, stream frequency, bifurcation ratio, texture ratio, relief ratio, shape factor, form factor, elongation ratio, circularity ratio, and compactness constant were estimated with the GIS using standard formula and methods (Choudhari et al., 2018; Nigam et al., 2017; Smith, 1950). The definition of the various morphometric parameters used in the study and empirical formulas has been presented in Table 6.1.

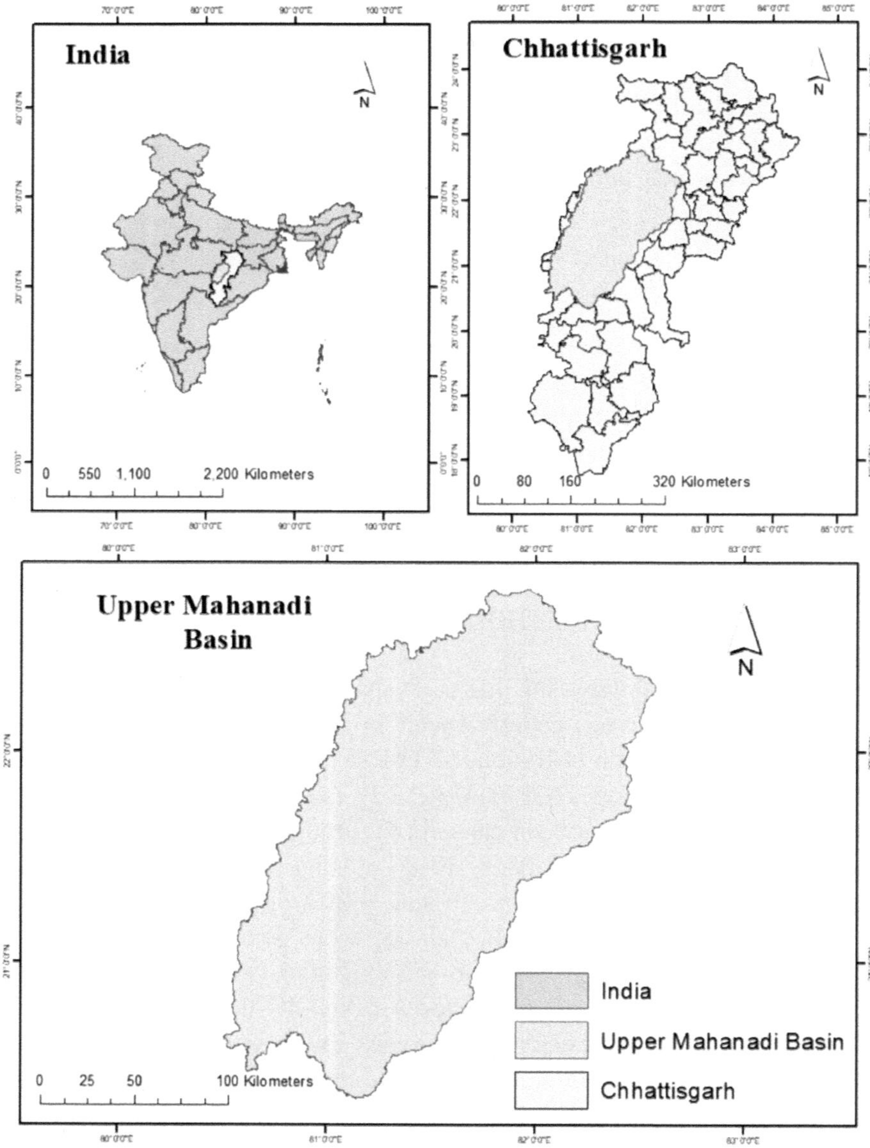

FIGURE 6.1 Study area.

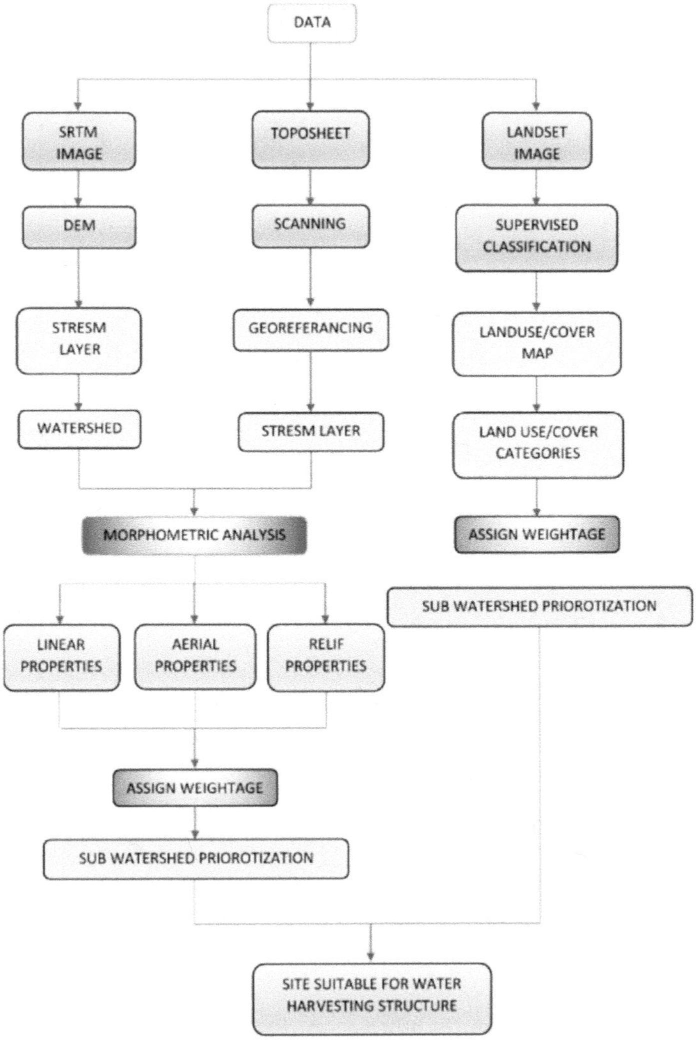

FIGURE 6.2 Flow chart of working methodology.

The LULC classification was done by supervised classification processes in different classes like lowland paddy, upland paddy, water body, forest, current fellow land, barren land, and settlement as per the land use adopted in this study area but for the prioritization only that classes were used, which strongly affect soil erosion in study area (Fig. 6.6). Classified LULC image was also performed by accuracy assessment.

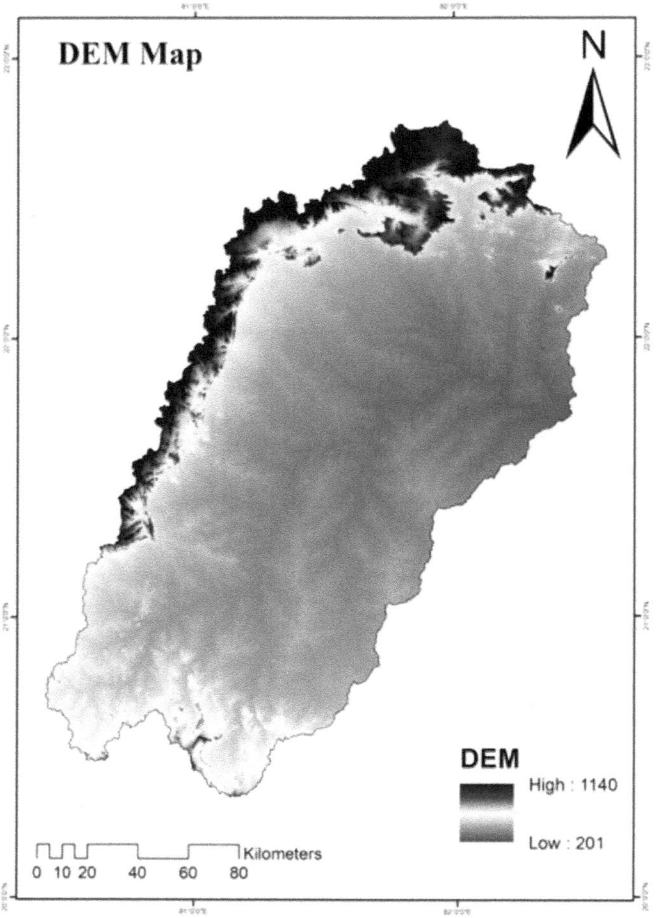

FIGURE 6.3 DEM map of upper Mahanadi basin.

Based on the morphometric and LULC information, the prioritization of the watershed is carried out to indicate the degree of soil erosion susceptibility (Javed et al., 2009). In the method of prioritization of watershed, assign the rank of each watershed based on the division of basin in to different number of watersheds. Since there are total 30 number of watersheds demarcated in the upper Mahanadi basin and rank of each watershed were assigned according to the available number of watersheds, that is, from 1 to 30.

The rank for the watershed prioritization assigned in way that the parameter, which indicate the maximum share to the erosion, assigned the rank 1 for the both in morphometry and LULC classes and the parameter, which indicate the minimum share to the erosion, assigned rank 30. Compound value of each watershed calculated by the average of the ranks of all the parameters for individual watershed, which indicate the erosion susceptibility condition of watershed. The compound value is denoted as C_p and is calculated from the following formula.

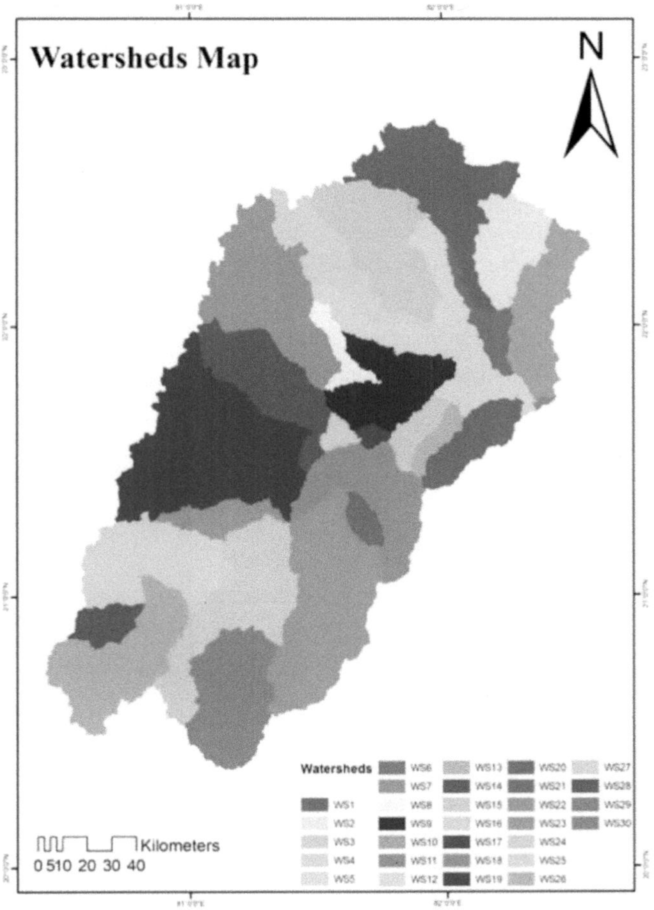

FIGURE 6.4 Watershed map of upper Mahanadi basin.

$$C_p = \frac{1}{n}\sum_{i=1}^{n} R_i$$

where C_p is compound value of individual watershed.

R_i is rank of an individual watershed for a parameter and, n is the number of parameters.

TABLE 6.1 Formulae Adopted for the Computation of Morphometric Parameters.

Morphometric parameters	Empirical formula	Reference
Stream order	Hierarchal rank	Strahler (1964)
Stream length	Length of stream (km)	Horton (1945)
Stream length ratio	$L_{ur} = L_u/L_{u-1}$ where, L_{ur} = stream length ratio L_u = mean of stream length of order 'u' (km) L_{u-1} = mean of stream length of its next lower order (km)	Horton (1945)
Bifurcation ratio	$R_b = N_u/N_u + 1$ R_b = Bifurcation ratio N_u = Total no. of stream segments of order 'u' $N_u + 1$ = No. of stream segments of the next higher order	Schumm (1956)
Length of overland flow	$L_o = 1/2D_d$ where, L_o = Length of overland flow (km) D_d = Drainage density (km/km^2)	Horton (1945)
Basin length	$L_b = 1.321A^{0.568}$ where, A = Area of the basin (km^2)	Ratnam et al. (2005)
Basin perimeter	Outer boundary of drainage basin measured in km	Schumm (1956)
Drainage density	$D_d = L_u/A$ where, D_d = Drainage density (km/km^2) L_u = Total stream length of all orders (km) A = Area of basin (km^2)	Horton (1945)
Basin area	Area from which water drains to a common stream (km^2)	Strahler (1964)
Stream frequency	$F_u = N_u/A$ where F_u = Stream frequency N_u = Total no. of streams of all orders A = Area of basin (km^2)	Horton (1932)
Texture ratio	$T = N_u/p$ where, R_t = Texture ratio N_u = Total no. of streams of all orders P = Perimeter (km)	Horton (1945)
Form factor	$R_f = A/L_b^2$ where, A = Area of basin (km^2) L_b^2 = Square of basin length (km)	Horton (1932)
Shape factor	$B_s = L_b^2/A$ where, L_b = Square of basin length (km) A= Area of basin (km^2)	Ratnam et al. (2005)

TABLE 6.1 *(Continued)*

Morphometric parameters	Empirical formula	Reference
Elongation ratio	$R_e = (4 \times A/\pi)^{0.5}/L_b$ where, R_e = Elongation Ratio A = Area of basin (km^2) L_b = Basin length	Schumm (1956)
Circulatory ratio	$R_c = 4 \times \pi \times A/P^2$ where, R_c = circulatory ratio A = Area of basin (km^2) P = Square of the perimeter (km)	Miller (1953)
Compactness constant	$C_c = 0.2821P/A^{0.5}$ where, A = Area of basin (km^2) P = Perimeter (km)	Horton (1932)

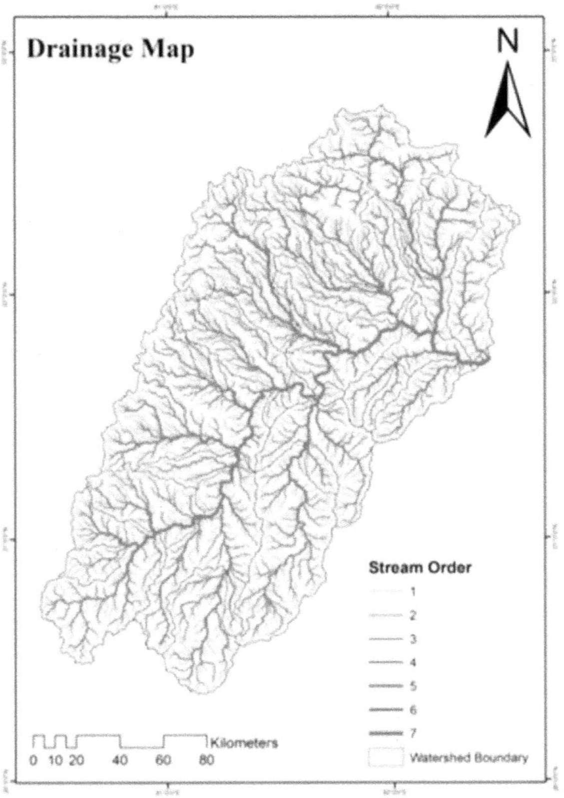

FIGURE 6.5 Drainage map of upper Mahanadi basin.

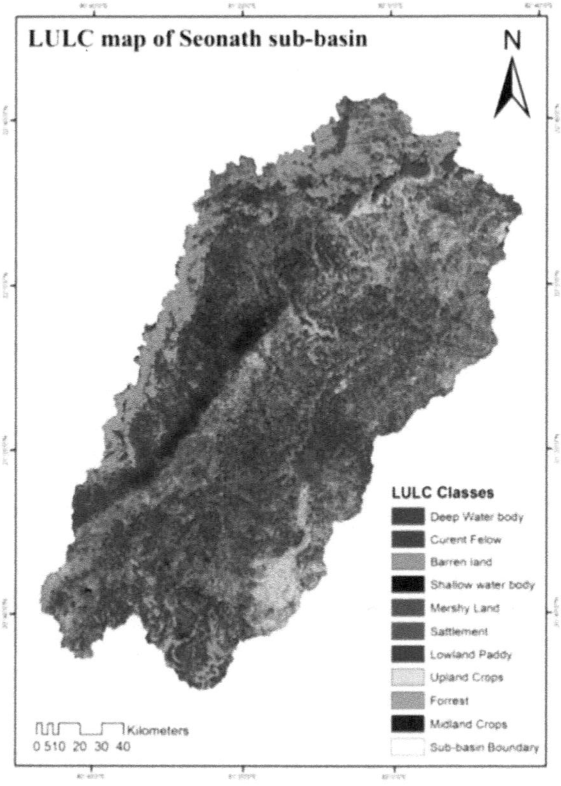

FIGURE 6.6 LULC map of upper Mahanadi basin.

6.4 RESULTS AND DISCUSSION

The linear morphological parameters of watersheds indicate the direct relationship with erosion susceptibility, that is, erosion severity increase in the watershed as a linear parameter of watershed increases and vice versa. The shape parameter of watersheds indicates the inverse relationship with erosion susceptibility, that is, erosion severity increase in the watershed as a shape parameter of watershed decreases and vice versa (Ratnam et al., 2005). Hence, for carrying out an effective prioritization of watersheds, rank 1 was assigned as the maximum value of the linear parameter, second highest value was given the rank of 2, and so forth. Shape parameters exhibit an inversely proportional relationship with soil erodibility. Thus, the minimum value of shape parameters was assigned ranked 1 and next higher value was assigned ranked as 2 and so forth whereas the maximum value was assigned ranked last in the hierarchy. In the group of watersheds, the topmost priority rank was given to that watershed, which showed the minimum value of compound factor and vice versa (Patel et al., 2012).

Area: The results of analysis showed that WS19 covers the maximum area of 3361.14 km^2, whereas WS11 has minimum area of 28.73 km^2.

Stream order: In this work, ranking of streams has been carried out based on the method proposed by Strahler (1964). On the basis of Strahler's stream ordering system, the watershed was designated as a VIIth order river basin (Fig. 6.5). The result indicated that the maximum frequency is in the case of first-order streams and it was also learnt as stream order increases the stream number decreases.

Stream number: Total number of 4199 streams were identified in the study area, out of these figure, maximum 3274 streams were found to be in first order, 701 streams in second order, 154 streams in third order, 40 streams in fourth order, 15 streams in fifth order, 12 streams in sixth order and minimum 3 streams were found in seventh order, respectively (Table 6.1). The higher amount of stream order gives an indication of lesser amount of permeability and infiltration (Strahler, 1964).

Stream length: In any particular stream order, the total measured length (in m or km), of the given stream is known as its stream length. The stream length of study area varies from 13.9 km (WS11) to 1761.8 km (WS19) (Table 6.2). The result indicated that the length of streams gradually decreased with a corresponding increase in the stream order.

Bifurcation ratio: It is the ratio between the total numbers of stream segments of a given order to that of the next higher order in a basin (Horton, 1945; Maxwell, 1955). The bifurcation ratio of the study area varies from 1.75 (WS17) to 14.21 (WS6) (Table 6.2). The result indicates that the watershed (WS17) having less structural control.

Length of overland flow: It indicates that the amount of total runoff rain water on the ground just before it gets channelized into well-defined streams network (Horton, 1945). The values of length of overland flow varies from 0.22 (WS13) to 0.28 (WS14) (Table 6.2). Higher the values of length of overland flow lower will be the relief and vice versa.

Drainage density: In the upper Mahanadi basin drainage density varied from 0.43 km/km^2 (WS13) to 0.56 km/km^2 in (WS14), respectively. The results indicated that the high drainage density indicated the presence of highly resistant permeable material with good vegetative cover and low relief.

Stream frequency: The stream frequency of the study area varied from 0.10 (WS28) to 0.17 (WS11). The results showed that the low values of stream frequency indicate low runoff and therefore low flood may occur in the watershed and vice versa.

Texture ratio: The texture ratio of the study area was found to be varied from 0.10 (WS11) to 0.79 (WS19).

Circularity ratio: The value of circularity ratio for study area was found to be varied from 0.06 (WS5) to 0.21 (WS17), respectively. The circularity ratio of the upper Mahanadi basin describes the drainage network as one possessing an almost elongated shape with medium to low relief.

Form factor ratio: The form factor ratio varied from 0.19 (WS9) to 0.36 (WS11). The results indicated that lover values show the elongated shape of watershed with low pick flow whereas higher value reflects the higher peak flows of short duration.

Elongation ratio: The minimum and maximum value of elongation ratio varied from 0.49 (WS19) to 0.68 (WS11). The results given in Table 6.2 indicate that basin is elongated (Strahler, 1964).

Compactness constant: The value of compactness constant of the upper Mahanadi basin varied from 2.20 (WS17) to 4.14 (WS5).

6.4.1 PRIORITIZATION OF WATERSHED BASED ON MORPHOMETRIC PROPERTIES

Compound values of each watershed was used to categorized the whole watershed into four priority groups very high priority (<8.5), high priority (8.6–10.5), medium priority (10.6–11.8), and low priority (>11.8). Out of 30 watersheds, WS2 and WS3 fall in very high priority; WS4, WS19, WS21, WS25, and WS29 fall in high priority; WS1, WS5, WS6, WS7, WS10, WS14, WS16, WS20, WS23, WS24, and WS27 fall in medium priority; whereas, WS8, WS9, WS11, WS12, WS13, WS15, WS17, WS18, WS22, WS26, WS28, and WS30 comes in category of low priority (Table 6.2) (Fig. 6.7).

6.4.2 PRIORITIZATION OF WATERSHED ON THE BASIS OF USING LAND USE/COVER PARAMETERS

Agriculture is most dominant category in the study area and covers more than 50% area of the basin. Higher percentage of crop covers is highly imposing in reducing the erosion rate. The overall accuracy of the LULC was 90.81% and calculated kappa coefficient of the LULC was 0.87. Prioritization of

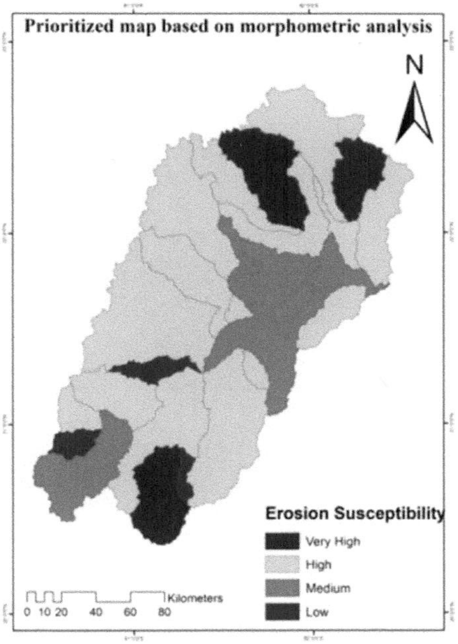

(a) Prioritization map based on morphometric analysis.

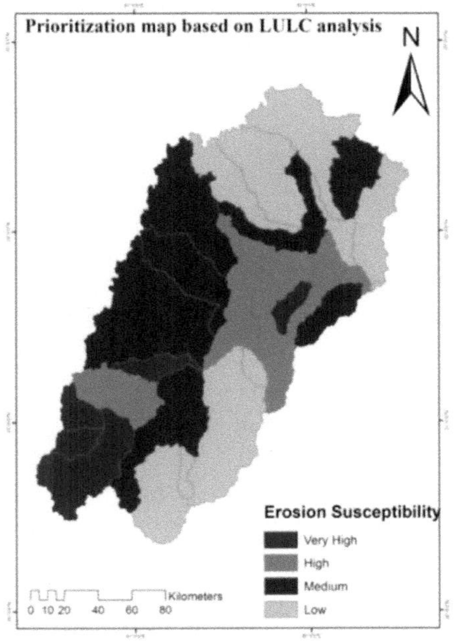

(b) Prioritization map based on LULC analysis.

FIGURE 6.7(a–d) *(Continued)*

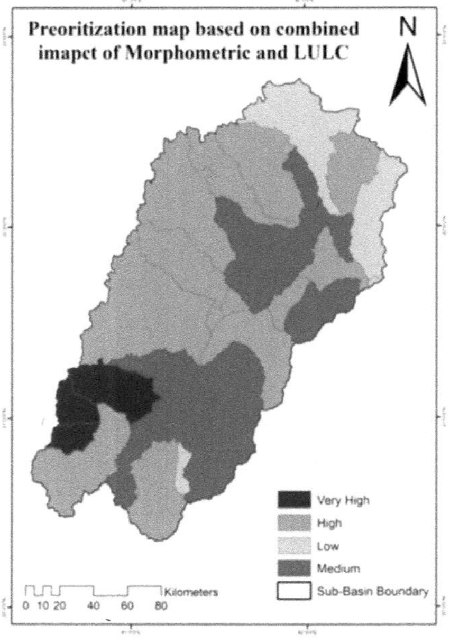

(c) Prioritization map based on combined impact of Morphometric and LULC analysis.

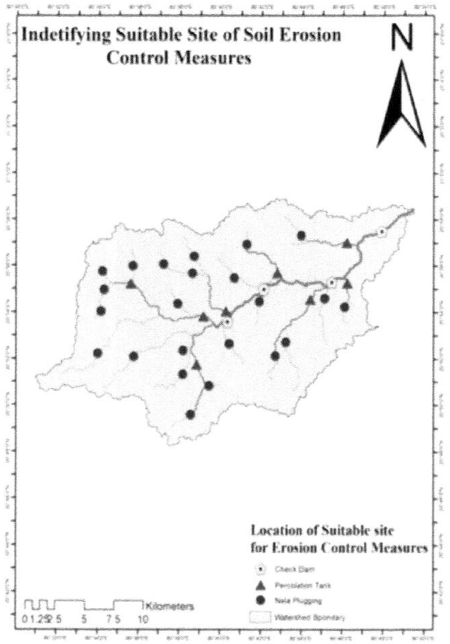

(d) Identified suitable site of soil erosion control measures.

FIGURE 6.7(a–d) Prioritization of watershed and suitable site for soil erosion control measures in upper Mahanadi basin.

TABLE 6.2 Morphometric Analysis of Watershed and Their Prioritization.

Watershed	Morphometric parameters										No of stream	Total stream length (km)	Watershed prioritization for erosion susceptibility											Erosion
	Linear parameters					Shape parameters																		
	R_b	D_d	F_u	L_o	T	R_c	R_f	B_s	R_e	C_c			R_b	D_d	F_u	L_o	T	R_c	R_f	B_s	R_e	C_c	C_p	
WS1	3.93	0.50	0.14	0.25	0.50	0.09	0.20	4.89	0.50	3.40	276	969.2	19	8	10	4	7	3	2	24	3	26	11	High
WS2	3.35	0.53	0.15	0.27	0.57	0.18	0.22	4.47	0.53	2.36	151	534.0	25	4	5	2	5	11	4	15	5	8	8.4	Very high
WS3	3.76	0.54	0.14	0.27	0.68	0.16	0.21	4.81	0.51	2.47	247	941.9	20	3	10	2	3	10	3	22	3	9	8.5	Very high
WS4	4.71	0.55	0.12	0.28	0.30	0.07	0.22	4.48	0.53	3.70	126	569.5	9	2	18	1	13	2	4	16	5	27	9.7	High
WS5	5.21	0.49	0.12	0.25	0.29	0.06	0.22	4.55	0.53	4.14	143	564.4	7	10	18	4	14	1	4	17	5	30	11	High
WS6	14.20	0.48	0.14	0.24	0.21	0.07	0.25	3.95	0.57	3.85	57	195.6	1	15	10	10	20	2	6	10	8	28	11	High
WS7	3.31	0.56	0.13	0.28	0.60	0.12	0.20	4.96	0.51	2.86	280	1200.5	26	1	15	1	4	6	2	25	3	18	10	High
WS8	11.80	0.48	0.12	0.24	0.15	0.07	0.26	3.80	0.58	3.87	36	145.4	2	15	18	10	23	2	7	7	9	29	12	Medium
WS9	4.07	0.47	0.14	0.23	0.44	0.11	0.22	4.58	0.53	3.06	166	560.3	16	18	10	15	9	5	4	18	5	22	12	Medium
WS10	5.43	0.47	0.14	0.23	0.46	0.12	0.22	4.60	0.53	2.92	168	583.3	6	18	10	15	8	6	4	19	5	20	11	High
WS11	4.00	0.48	0.17	0.24	0.10	0.16	0.36	2.76	0.68	2.54	5	13.9	18	15	1	10	24	10	18	1	14	10	12	Medium
WS12	3.42	0.47	0.16	13.00	0.39	0.11	0.24	4.21	0.55	2.98	105	303.0	24	18	2	15	11	5	5	13	6	21	12	Medium
WS13	4.13	0.43	0.17	12.00	0.27	0.13	0.27	3.67	0.59	2.79	41	102.4	13	25	1	20	15	7	10	6	10	17	12	Medium
WS14	4.01	0.56	0.11	18.00	0.41	0.12	0.21	4.67	0.52	2.90	158	775.4	17	1	20	1	10	6	3	20	4	19	10	High
WS15	3.63	0.45	0.15	13.00	0.30	0.15	0.26	3.84	0.58	2.58	50	150.4	21	23	5	16	13	9	7	8	9	12	12	Medium
WS16	5.00	0.47	0.14	0.24	0.21	0.18	0.29	3.45	0.61	2.33	21	71.2	8	18	10	10	19	11	12	3	12	7	11	High
WS17	1.75	0.49	0.14	0.25	0.17	0.21	0.31	3.25	0.63	2.20	13	47.4	29	10	10	4	22	20	15	2	13	1	13	Medium
WS18	4.25	0.46	0.13	0.23	0.46	0.10	0.21	4.67	0.52	3.09	187	642.5	11	21	15	15	8	4	3	20	4	23	12	Medium
WS19	3.56	0.52	0.13	0.26	0.79	0.13	0.19	5.27	0.49	2.75	446	1761.8	22	5	15	3	1	7	0	27	1	15	9.6	High

TABLE 6.2 *(Continued)*

Watershed	Morphometric parameters: Linear parameters					Shape parameters					No of stream	Total stream length (km)	Watershed prioritization for erosion susceptibility											Erosion
	R_b	D_d	F_u	L_o	T	R_c	R_f	B_s	R_e	C_c			R_b	D_d	F_u	L_o	T	R_c	R_f	B_s	R_e	C_c	C_p	
WS20	4.43	0.46	0.14	0.23	0.41	0.15	0.24	4.25	0.55	2.55	98	320.8	10	21	10	15	10	9	5	14	6	11	11	High
WS21	5.00	0.49	0.13	0.25	0.24	0.19	0.28	3.64	0.59	2.27	29	108.9	8	10	15	4	17	15	10	5	10	3	9.7	High
WS22	4.13	0.44	0.14	0.22	0.25	0.09	0.25	4.01	0.56	3.26	62	198.2	14	24	10	20	16	3	6	12	7	24	14	low
WS23	4.16	0.48	0.14	0.24	0.75	0.15	0.20	5.07	0.50	2.62	348	1207.1	12	15	10	10	2	9	2	26	2	13	10	High
WS24	4.08	0.47	0.15	0.23	0.54	0.09	0.20	4.88	0.51	3.32	280	899.3	15	18	5	15	6	3	2	23	3	25	12	High
WS25	3.27	0.49	0.15	0.25	0.60	0.19	0.22	4.55	0.53	2.31	168	570.0	27	10	5	4	4	15	4	17	5	5	9.6	High
WS26	3.53	0.47	0.14	0.23	0.57	0.13	0.21	4.80	0.51	2.76	232	802.5	23	18	10	15	5	7	3	21	3	16	12	Medium
WS27	2.62	0.48	0.15	0.24	0.38	0.19	0.25	4.00	0.56	2.29	65	216.3	28	15	5	10	12	15	6	11	7	4	11	High
WS28	5.57	0.46	0.10	0.23	0.23	0.19	0.26	3.90	0.57	2.32	37	169.8	5	21	25	15	18	15	7	9	8	6	13	Low
WS29	8.40	0.49	0.13	0.24	0.60	0.20	0.21	4.67	0.52	2.26	180	684.3	3	10	15	10	4	18	3	20	4	2	8.9	Very high
WS30	6.25	0.45	0.13	0.22	0.18	0.14	0.28	3.56	0.60	2.67	24	85.1	4	23	15	20	21	8	10	4	11	14	13	Low

watershed performed based on LULC response toward erosion rate. Based on the percentage of every classes of LULC of the watersheds were assigned rank. The higher percentage of classes of LULC, which indicates direct relationship with erosion, was assigned rank 1 and the same classes having minimum percentage area assigned the last rank. In the same way, the minimum percentage of the LULC classes indicated to restricts the erosion rate, assigned rank 1, and the same class having maximum percentage to assigned rank 30. At the end single value of compound factor was calculated by averaging the ranks for individual watershed.

Hence, based on LULC analysis, watersheds of upper Mahanadi basin were classified into four different priority classes, that is, very high priority (5.8–9.5), high priority (9.6–13), medium priority (13.1–15.5), and low priority (15.6–18.6). Out of 30 watersheds, WS15, WS22, WS26, WS27, and WS28 comes under very high priority; WS4, WS19, WS21, and WS25 comes under high priority; WS2, WS5, WS7, WS14, WS19, WS20, and WS24 comes under medium priority and remaining all watersheds comes under low priority class (Table 6.3).

6.4.3 PRIORITIZATION OF WATERSHED BASED ON THE COMBINED INFLUENCE OF MORPHOMETRIC AND LAND USE/COVER

The prioritization of upper Mahanadi basin was assessed based on the combined influence of Morphometric and LULC change analysis which identified the critical watershed, that is, susceptible to erosion. The results of the combined effect on watersheds indicated that the watershed (WS28) showed high-erosion susceptible than the other watershed. Watershed (WS28) indicated that sever erosion susceptible area and comes under very high categories based on morphometric parameters and LULC, while comes in high categories based on the morphometric analysis only (Table 6.4). The results indicated that the suitable soil erosion control measures should be incorporated in this critical watershed, which could keep the land from any further erosion.

6.5 CONCLUSION

Remote sensing and geographical information system approaches are found to be vital useful for the watershed prioritization and their management. The prioritization was performed based on the various morphometric parameters,

TABLE 6.3 Land Use/Cover-Based Watershed Prioritization for Erosion Susceptibility of Upper Mahanadi.

WS	LULC categories						C_p value	Erosion susceptibility
	Area	Current fellow (C_F)	Barren land (B_L)	Forrest (F)	Upland crops (U_C)	Settlement (S_L)		
WS1	Area (km^2)	9.4	103.3	587.3	352.0	48.9	18.6	Low
	Area (%)	0.5	5.3	30.2	18.1	2.5		
	Rank	20	18	24	21	10		
WS2	Area (km^2)	9.8	83.9	94.7	272.7	17.5	15.4	Medium
	Area (%)	1.0	8.4	9.5	27.3	1.7		
	Rank	16	15	13	29	4		
WS3	Area (km^2)	6.8	86.2	589.2	279.5	18.8	17.6	Low
	Area (%)	0.4	5.0	34.1	16.2	1.1		
	Rank	25	19	26	17	1		
WS4	Area (km^2)	4.7	31.0	298.1	72.1	17.9	15.6	Low
	Area (%)	0.5	3.0	28.8	7.0	1.7		
	Rank	20	23	22	9	4		
WS5	Area (km^2)	12.7	113.0	52.9	253.7	24.9	14.2	Medium
	Area (%)	1.1	9.9	4.6	22.2	2.2		
	Rank	15	13	7	27	9		
WS6	Area (km^2)	4.8	46.2	34.1	90.2	14.2	15.8	Low
	Area (%)	1.2	11.3	8.3	22.0	3.5		
	Rank	14	10	11	26	18		
WS7	Area (km^2)	12.5	75.7	772.6	68.1	27.3	14.4	Medium
	Area (%)	0.6	3.5	35.8	3.2	1.3		
	Rank	19	21	27	3	2		
WS8	Area (km^2)	6.3	35.4	19.0	52.4	6.0	11	High
	Area (%)	2.1	11.7	6.3	17.3	2.0		
	Rank	10	9	9	20	7		

TABLE 6.3 (Continued)

WS	LULC categories						C_p value	Erosion susceptibility
	Area	Current fellow (C_F)	Barren land (B_l)	Forrest (F)	Upland crops (U_C)	Settlement (S_l)		
WS9	Area (km^2)	16.2	122.4	50.0	202.9	22.3	11	High
	Area (%)	1.3	10.2	4.2	16.9	1.9		
	Rank	13	12	6	18	6		
WS10	Area (km^2)	16.3	109.9	246.2	254.5	33.5	16.6	Low
	Area (%)	1.3	8.8	19.8	20.5	2.7		
	Rank	13	15	19	24	12		
WS11	Area (km^2)	0.2	3.4	5.5	5.6	1.0	16.8	Low
	Area (%)	0.8	11.9	19.2	19.5	3.4		
	Rank	18	9	18	22	17		
WS12	Area (km^2)	15.1	83.7	68.9	71.3	13.6	11	High
	Area (%)	2.3	12.8	10.6	10.9	2.1		
	Rank	9	8	14	16	8		
WS13	Area (km^2)	6.4	36.7	21.4	24.6	9.6	11.8	High
	Area (%)	2.7	15.5	9.0	10.4	4.1		
	Rank	7	5	12	14	21		
WS14	Area (km^2)	21.0	66.1	254.3	86.8	36.2	13.6	Medium
	Area (%)	1.5	4.8	18.3	6.3	2.6		
	Rank	12	20	17	8	11		
WS15	Area (km^2)	21.6	54.8	16.7	35.2	9.9	8.6	Very high
	Area (%)	6.5	16.5	5.0	10.6	3.0		
	Rank	1	4	8	15	15		
WS16	Area (km^2)	3.0	16.4	22.0	12.9	3.9	12.2	High
	Area (%)	2.0	10.9	14.6	8.6	2.6		
	Rank	11	11	16	12	11		

TABLE 6.3 *(Continued)*

WS	LULC categories						C_p value	Erosion susceptibility
	Area	Current fellow (C_F)	Barren land (B_l)	Forrest (F)	Upland crops (U_c)	Settlement (S_l)		
WS17	Area (km²)	1.9	12.9	3.1	16.3	2.1	10.2	High
	Area (%)	2.0	13.5	3.3	17.0	2.2		
	Rank	11	7	5	19	9		
WS18	Area (km²)	39.8	185.4	39.4	106.6	68.9	10.4	High
	Area (%)	2.9	13.3	2.8	7.6	4.9		
	Rank	6	7	4	10	25		
WS19	Area (km²)	71.4	234.7	854.6	181.0	121.7	14	Medium
	Area (%)	2.1	7.0	25.4	5.4	3.6		
	Rank	10	16	20	5	19		
WS20	Area (km²)	14.1	67.7	79.1	137.7	12.7	13.4	Medium
	Area (%)	2.0	9.7	11.4	19.8	1.8		
	Rank	11	13	15	23	5		
WS21	Area (km²)	0.2	20.3	0.0	18.2	68.4	17.2	Low
	Area (%)	0.1	9.2	0.0	8.3	31.0		
	Rank	30	14	1	11	30		
WS22	Area (km²)	28.0	79.7	5.2	26.8	61.2	8.4	Very high
	Area (%)	6.2	17.5	1.1	5.9	13.5		
	Rank	2	3	2	7	28		
WS23	Area (km²)	32.8	287.2	42.3	817.4	115.3	15.8	Low
	Area (%)	1.3	11.3	1.7	32.3	4.6		
	Rank	13	10	3	30	23		
WS24	Area (km²)	44.9	130.0	372.1	48.2	80.6	13.6	Medium
	Area (%)	2.3	6.8	19.4	2.5	4.2		
	Rank	9	17	18	2	22		

TABLE 6.3 *(Continued)*

WS	**LULC categories**						**C_p value**	**Erosion susceptibility**
	Area	**Current fellow (C_F)**	**Barren land (B_l)**	**Forrest (*F*)**	**Upland crops (U_C)**	**Settlement (S_l)**		
WS25	Area (km²)	36.1	199.1	84.6	62.9	59.0	10	High
	Area (%)	3.1	17.2	7.3	5.4	5.1		
	Rank	5	3	10	5	27		
WS26	Area (km²)	58.1	305.0	153.8	148.8	35.4	8	Very high
	Area (%)	3.4	17.8	9.0	8.7	2.1		
	Rank	4	3	12	13	8		
WS27	Area (km²)	11.2	99.4	14.5	19.2	14.1	7	Very high
	Area (%)	2.5	22.3	3.2	4.3	3.1		
	Rank	8	2	5	4	16		
WS28	Area (km²)	17.3	129.0	0.0	21.6	12.8	5.8	Very high
	Area (%)	4.7	34.8	0.0	5.8	3.5		
	Rank	3	1	1	6	18		
WS29	Area (km²)	6.5	15.2	535.2	10.5	19.7	16.4	Low
	Area (%)	0.5	1.1	38.2	0.8	1.4		
	Rank	20	28	30	1	3		
WS30	Area (km²)	1.7	4.2	69.9	12.0	2.4	16	Low
	Area (%)	0.9	2.2	36.7	6.3	1.3		
	Rank	17	25	28	8	2		

TABLE 6.4 Watershed Prioritization for Erosion Susceptibility of Upper Mahanadi Basin Based on Combined Influence of Morphometric and LULC.

WS	R_b	D_d	F_u	L_o	T	R_c	R_f	B_s	R_e	C_c	C_F	B_L	S_L	U_C	F	C_p	Erosion susceptibility
WS1	19	8	10	4	7	3	2	24	3	26	20	18	24	21	10	12.4	Low
WS2	25	4	5	2	5	11	4	15	5	8	16	15	13	29	4	10.1	High
WS3	20	3	10	2	3	10	3	22	3	9	25	19	26	17	1	10.8	High
WS4	9	2	18	1	13	2	4	16	5	27	20	23	22	9	4	10.9	High
WS5	7	10	18	4	14	1	4	17	5	30	15	13	7	27	9	11.3	Medium
WS6	1	15	10	10	20	2	6	10	8	28	14	10	11	26	18	11.8	Medium
WS7	26	1	15	1	4	6	2	25	3	18	19	21	27	3	2	10.8	High
WS8	2	15	18	10	23	2	7	7	9	29	10	9	9	20	7	11.1	Medium
WS9	16	18	10	15	9	5	4	18	5	22	13	12	6	18	6	11.1	Medium
WS10	6	18	10	15	8	6	4	19	5	20	13	15	19	24	12	12.1	Low
WS11	18	15	1	10	24	10	18	1	14	10	18	9	18	22	17	12.8	Low
WS12	24	18	2	15	11	5	5	13	6	21	9	8	14	16	8	10.9	High
WS13	13	25	1	20	15	7	10	6	10	17	7	5	12	14	21	11.4	Medium
WS14	17	1	20	1	10	6	3	20	4	19	12	20	17	8	11	10.6	High
WS15	21	23	5	16	13	9	7	8	9	12	1	4	8	15	15	10.4	High
WS16	8	18	10	10	19	11	12	3	12	7	11	11	16	12	11	10.7	High
WS17	29	10	10	4	22	20	15	2	13	1	11	7	5	19	9	11.1	Medium
WS18	11	21	15	15	8	4	3	20	4	23	6	7	4	10	25	11.0	High
WS19	22	5	15	3	1	7	1	27	1	15	10	16	20	5	19	10.4	High
WS20	10	21	10	15	10	9	5	14	6	11	11	13	15	23	5	11.1	Medium

TABLE 6.4 *(Continued)*

WS	R_b	D_d	F_u	L_o	T	R_c	R_f	B_s	R_e	C_c	C_F	B_L	S_L	U_C	F	C_p	Erosion susceptibility
WS21	8	10	15	4	17	15	10	5	10	3	30	14	1	11	30	11.4	Medium
WS22	14	24	10	20	16	3	6	12	7	24	2	3	2	7	28	11.1	Medium
WS23	12	15	10	10	2	9	2	26	2	13	13	10	3	30	23	11.3	Medium
WS24	15	18	5	15	6	3	2	23	3	25	9	17	18	2	22	11.4	Medium
WS25	27	10	5	4	4	15	4	17	5	5	5	3	10	5	27	9.1	Very high
WS26	23	18	10	15	5	7	3	21	3	16	4	3	12	13	8	10.1	High
WS27	28	15	5	10	12	15	6	11	7	4	8	2	5	4	16	9.3	Very high
WS28	5	21	25	15	18	15	7	9	8	6	3	1	1	6	18	9.9	Very high
WS29	3	10	15	10	4	18	3	20	4	2	20	28	30	1	3	10.7	High
WS30	4	23	15	20	21	8	10	4	11	14	17	25	28	8	2	13.1	Low

LULC, and impact of both LULC and morphometric. The integrated remote sensing and GIS-based approaches have been incorporated, thus, proving to be of utmost appropriateness and utility as compared to the conventional methods. Further, it is understood that the remote sensing techniques are of profound suitability in the creation of an updated different thematic maps in a minimum time and very cost-effective manner and, hence, this was given greater preference in soil erosion studies for deriving input data. The watershed SW28 was identified as generally falling under very high priority. It can be concluded that the suitable sites for soil erosion control measures in critical watershed in GIS environment has been identified and they should be incorporated in that watershed, which could keep the land from any further erosion.

KEYWORDS

- **morphometric**
- **GIS**
- **watershed**
- **prioritization**
- **land use/cover**

REFERENCES

Brooks, R. P.; Wardrop, D. H.; Cole, C. A. Inventorying and Monitoring Wetland Condition and Restoration Potential on a Watershed Basis with Examples from Spring Creek Watershed, Pennsylvania, USA. *Environ. Manage.* **2006,** *38*, 673–687.

Chopra, R.; Dhiman, R. D.; Sharma, P. Morphometric Analysis of Sub-Watersheds in Gurdaspur District, Punjab Using Remote Sensing and GIS Techniques. *J. Indian Soc. Remote Sens.* **2005,** *33*, 531–539.

Choudhari, P.; Nigam, G. K.; Singh, S. K.; Thakur, S. Morphometric Based Prioritization of Watershed for Groundwater Potential of Mula River Basin, Maharashtra, India. *Geol. Ecol. Landsc.* **2018,** *2*, 256–267.

Gupta, M.; Srivastava, P. K. Integrating GIS and Remote Sensing for Identification of Groundwater Potential Zones in the Hilly Terrain of Pavagarh, Gujarat, India. *Water Int.* **2010,** *35*, 233–245.

Horton, R. E. Drainage-Basin Characteristics. *EOS Trans. Am. Geophys. Union* **1932,** *13*, 350–361.

Horton, R. E. Erosional Development of Streams and Their Drainage Basins; Hydrophysical Approach to Quantitative Morphology. *Geol. Soc. Am. Bull.* **1945,** *56*, 275–370.

Javed, A.; Khanday, M. Y.; Ahmed, R. Prioritization of Sub-Watersheds Based on Morphometric and Land Use Analysis Using Remote Sensing and GIS Techniques. *J. Indian Soc. Remote Sens.* **2009,** *37*, 261–274.

Johnson, N.; Ravnborg, H. M.; Westermann, O.; Probst, K. User Participation in Watershed Management and Research. *Water Policy* **2002,** *3*, 507–520.

Maxwell, J. The Bifurcation Ratio in Horton's Law of Stream Numbers. *Trans. Am. Geophys. Union* **1955,** *36*, 520.

Miller, V. C. *A Quantitative Geomorphic Study of Drainage Basin Characteristics in the Clinch Mountain Area Virginia and Tennessee*; Columbia Univiversity: New York, 1953.

Nigam, G.; Tripathi, M.; Ambast, S.; Kumar, L.; Khalkho, D. Morphometric Analysis of Drainage Basin Using Aerial Photographs: A Case of Karun Watershed of Seonath Sub-Basin of Chhattisgarh. *IJABR* **2017,** *7*, 623–629.

Panhalkar, S.; Mali, S.; Pawar, C. Morphometric Analysis and Watershed Development Prioritization of Hiranyakeshi Basin in Maharashtra, India. *Int. J. Environ. Sci.* **2012,** *3*, 525–534.

Patel, D. P.; Dholakia, M. B.; Naresh, N.; Srivastava, P. K. Water Harvesting Structure Positioning by Using Geo-Visualization Concept and Prioritization of Mini-Watersheds through Morphometric Analysis in the Lower Tapi Basin. *J. Indian Soc. Remote Sens.* **2012,** *40*, 299–312.

Pawar-Patil, V.; Mali, S. P. Watershed Characterization and Prioritization of Tulasi Subwatershed: A Geospatial Approach. *Int. J. Innov. Res. Sci. Eng. Technol.* **2013,** *2*, 2182–2189.

Ratnam, K. N.; Srivastava, Y.; Rao, V. V.; Amminedu, E.; Murthy, K. Check Dam Positioning by Prioritization of Micro-Watersheds Using SYI Model and Morphometric Analysis—Remote Sensing and GIS Perspective. *J. Indian Soc. Remote Sens.* **2005,** *33*, 25–38.

Schumm, S. A. Evolution of Drainage Systems and Slopes in Badlands at Perth Amboy, New Jersey. *Geol. Soc. Am. Bull.* **1956,** *67*, 597–646.

Smith, K. G. Standards for Grading Texture of Erosional Topography. *Am. J. Sci.* **1950,** 248, 655–668.

Strahler, A. N. Quantitative Analysis of Watershed Geomorphology. *EOS Trans. Am. Geophys. Union* **1957,** *38*, 913–920.

Strahler, A. N. Part II. Quantitative Geomorphology of Drainage Basins and Channel Networks. In *Handbook of Applied Hydrology*; McGraw-Hill: New York, 1964; pp 4–39.

CHAPTER 7

Prioritization of Charwa Watershed Based on Morphometric Analysis Using Remote Sensing and Geospatial Techniques

RAVI RAJ, AMAN KUMAR, YEMPALI PRIYANKA, and PRATIBHA WARWADE*

Water Engineering and Management, Central University of Jharkhand, India

Corresponding author. E-mail: pratibhawarwade@gmail.com

ABSTRACT

The behavior of watershed and hydrological analysis depends upon geo-morphometric catchment. For the development of an ungauged watershed, morphometric analysis is performed. Watersheds prioritization on the basis of stream distinctiveness of any watershed is necessary to execute sustainable water development programs primarily to counter erosion and sedimentation losses. In this study, geometrics technique had been applied on the Charwa watershed to evaluate the morphometric properties by evaluating linear, aerial, and relief aspects. Morphometric analysis had been attempted for prioritization of six mini watersheds of Charwa watershed in the river basin, which is a tributary of the Damodar had been carried out. Mini watersheds were delineated by ArcGIS 10.3 software by using digital elevation model. All the morphometric parameters like bifurcation ratio, stream frequency, stream order, stream length, circular ratio, elongation ratio, drainage density, texture ratio, form factor, and compactness ratio had been estimated

Advances in Hydrology and Climate Change: Historical Trends and New Approaches in Water Resources Management. Surendra Kumar Chandniha, Anil Kumar Lohani, Gopal Krishan, & Ajay Krishna Prabhakar (Eds.)

separately for each mini watershed using geomatics techniques. The final prioritized composite score based on morphometric parameters of every mini watershed was assigned and then consolidated scores had been calculated to find out the sensitivity of parameters. The examination revealed that stream order varied from 1 to 4. Total number of stream segment of all order was 54 in the watershed. The study emphasized the prioritization of the mini watersheds based on morphometric analysis. The total composite score of all the six mini watersheds was assigned as per erosion risk. The highest priority was given to the mini watershed with the least compound parameter value. However, the mini watersheds had been categorized into three classes as high (2.6–3.1), medium (3.1–3.5), and low (3.5–3.9) priority based on their maximum (3.9) and minimum (2.6) prioritized score.

7.1 INTRODUCTION

Morphometric properties of watershed are significant aspect for categorization of it. For sustainable development of living life, it is very necessary to implement the watershed management programs properly. Morphometric analysis of a basin and the arrangement of its stream channel can be well explained with the help of the drainage network, aerial and relief aspects, and ground slopes (Nag and Chakraborty, 2003). Geo-morphological analysis is generally used for modeling of local hydrological models of a basin to counter different hydrological limitations of the ungauged basins (Gajbhiye et al., 2014). The modern approach of quantitative analysis was given by Horton (1945) who incorporates various components such as stream segments, basin perimeter, basin area, elevation difference, slope, and role of land which is ultimately accountable for the natural development of watershed (Horton, 1945).

However, various geo-morphologist afterward modified the Horton's laws which is extremely considerable most markedly by Strahler (1952), Leopold and Miller (1956), Schumm (1956), Melton (1957), Morisawa (1957), Strahler (1957, 1958), Scheidegger (1965), Shreve (1967), and Gregory and Walling (1973). Morphometric parameters primarily depend upon bedrocks, lithology, and geological structures. Hence, the information regarding hydrology, geomorphology, geology, and land use pattern is very revealing for the dependable study of the morphometric properties of the watershed (Astras and Soulankellis, 1992). GIS and RS techniques based on the satellite imagery are used as a convenient tool for the morphometric analysis. The specific objective of the present investigation is the prioritization of mini watersheds based on the morphometric analysis of the Charwa watershed which is the part of Damodar river basin, Jharkhand, India.

7.2 DESCRIPTION OF STUDY AREA

The Charwa watershed originated from the insane river catchment and a micro watershed of Damodar basin at Jharkhand state in India. It varies from latitude 23°58′40N to 24°2′35 and longitude 85°16′40E to 85°19′E with area of the basin is about 9.459 km^2 (Fig. 7.1). The watershed covers the five villages (Govindpur, Nawadih, Mayatu, Pakrar, and Charwa) of Hazaribagh district in Jharkhand. The uppermost elevation of the watershed is 597 m and the lowest is 552 m. The Hazaribagh district received 1062 mm rainfall yearly. Basically, four seasons are there in Hazaribagh, which are per-monsoon from March to May, monsoon from June to September, post monsoon from October to November, and winter from December to February. Most of the people living in the study area come from the tribal community, their livelihood is agricultural dependent. The outlet of the watershed is situated in Hazaribagh district, the total population of the area is about 7262 (Registrar General, India, 2011) as per population density. About more than 50% of the total area of the watershed are covered with paddy, forest land, and submergence area.

7.3 MATERIALS AND METHODS

The drainage network was delineated with the help of Arcmap 10.3 using Digital Elevation Model (DEM) from Bhuvan, an Indian Geo- Platform of ISRO of each sub-watershed. In Arcmap 10.3 first of all, delineation of watershed boundary is done with a well-defined arc-GIS technique and simultaneously formation of mini watershed. The values of the morphometric parameters were estimated using the formula for each mini watershed. Composite scores were calculated and ultimately computation of basic parameters for the prioritization of mini watershed was done.

7.3.1 DELINEATION OF THE WATERSHED BOUNDARY

Most DEMs consist of local depression and sinks, that is, a cell surrounded by greater elevation value cells. This type of DEMs can provide inaccurate results and interrupt the mechanism of off-map flow directions. Hence, depression less or sink-free DEM of the study area is required for the extraction of the drainage network. For this purpose, get the Shuttle Radar Topography Mission (SRTM) image(s) of the study area. Check that downloaded DEMs are sink/void free or not, if not than using fill tool in ArcGIS software make it depression less or say sink free. Use SINK tool of the software or subtract, the original DEM forms the filled DEM to get the sinks in the original DEM. As SRTM images are already georeferenced; so there is no need of georeferencing. If the whole area comes under in one image then it is okay otherwise mosaicing is required. After this, mosaic delineates the watershed using the ArcGIS software. In the delineation of the watershed, outlet point and DEM are the two inputs and use Arc Hydro tool.

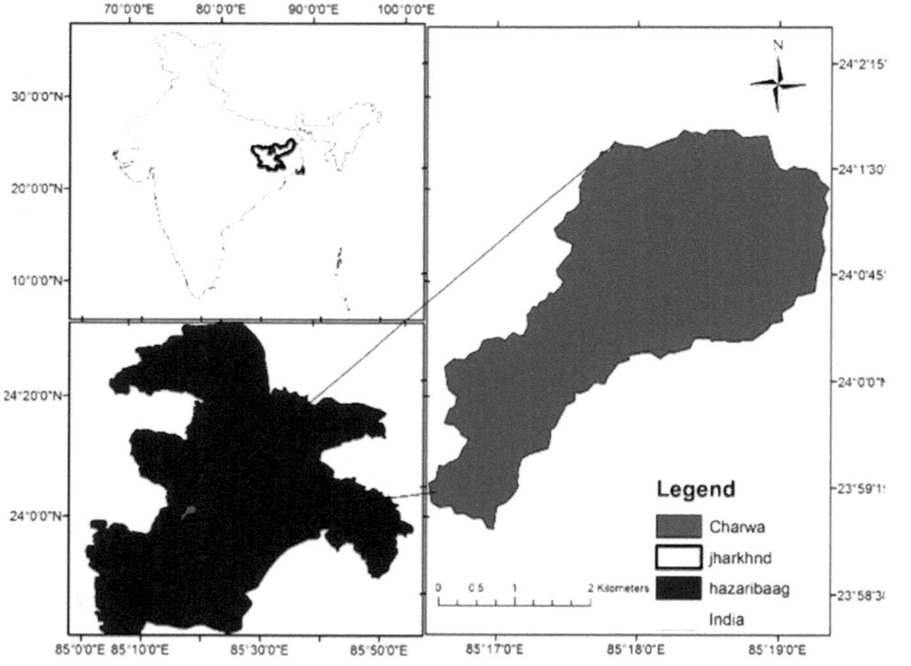

FIGURE 7.1 Location map of study area.

7.3.2 EXTRACTION OF STREAM NETWORK

After delineating sink-free DEM of the study area, stream network extraction was done the second task. And for this purpose, the flow direction tool was used on the depression less DEM to get the direction of the flow for each cell. In a (sink-free) DEM, the flow direction operation was determined into how neighboring pixel would behave if any water in a central pixel will flow naturally. Flow direction was calculated for every central pixel of a standard block of 3 by 3 pixels, each time comparing the value of the central pixel with its 8 neighbors. The output map contains flow directions as S (to the South), NE (to the North East), etc. The algorithm behind the running of the flow accumulation and flow direction operation is the D8 algorithm.

7.3.3 EXTRACTION OF THE DRAINAGE NETWORK

STREAMORDER tool of Arcmap was used on the stream raster map with the flow direction map to evaluate the order of each segment in the stream drainage network. For providing order to each stream, Strahler method of stream number was followed. It says if we have two tributaries of same order n, the linked stream would be of order $n + 1$; otherwise if the two tributaries have orders m and n with $n > m$, the link would be also of order n. Using the FLOWLENGTH tool on the flow direction map, the total flow distance from each cell to the outlet point of the basin was determined.

7.3.4 FORMATION OF MINI WATERSHEDS

For the formation of mini watersheds, STREAMLINK tool was used on the stream raster map with the flow direction map to assign a unique ID value to each stream segment. The stream link map was used as input pour point targets for the WATERSHED tool. WATERSHED tool used the flow direction map and pour point targets to determine the mini watersheds within the given micro watershed that flow to all linked stream. The delineated mini watersheds could be converted to polygon features if desired. A lower flow accumulation threshold was used to extract a more detailed drainage network, create a new stream link map (with higher number of links), and return to the WATERSHED tool.

7.3.5 COMPUTATION OF BASIC PARAMETERS

In this section, basic parameters (such as area, perimeter, stream length, stream order, stream number, and elevation) were computed for each mini watershed.

This analysis was performed extensively using the GIS and RS approach. Ultimately, drainage density (Dd), form factor (Ff), bifurcation ratio (Rb), stream frequency (Fs), texture ratio, elongation ratio (Re), circulatory ratio (Rc), and compactness ratio had been quantified with the help of standard formulae.

7.3.6 MORPHOMETRIC ANALYSIS OF WATERSHED

The morphometric analysis primarily alludes to the quantitative evaluation of hydrological parameter. This is the most usual approach to analyses watershed, with morphometry associated with interpretation and fluvially originated landforms analysis. Development of the watershed suggests that development is associated with its components such as basin area, basin length, stream segments, volume, slope, altitude, and profiles of land. The arrangement of the stream system of a drainage basin has been expressed quantitatively with its components and parameters only (Horton, 1945). Horton's laws were subsequently modified and developed by several geomorphologist, which is highly appreciable and widely accepted. The law of basin areas exhibits that the mean basin area of successive ordered streams forms a linear relationship.

Stream Order

It is very difficult to identify the behavior of the flow in any watershed. So, it is very necessary to sub-divide watersheds into smaller parts which are called as sub or mini watersheds can be defined on the basis of their area, size, and stream orders. Stream order is defined as the number of orders like first-order channels which are non-branching finger tip channel segments. Those channels which receive water from only first-ordered channels are second-order channels and likewise third-order channels are those where two second-ordered channels meet (Strahler, 1964) and so on shown in Figure 7.2. New higher order of a stream is formed if two streams of the same order are joined. Order of stream always ascending from upstream to downstream side based on watershed geomorphology.

Streams Number (Nu)

A number of streams may be defined as total counts of stream segments of different order separately. It is inversely proportional to the stream order. Stream number is denoted by Nu.

Total Stream Length (Lu)

Total stream length is evaluated as the measurement of the length of all ordered perennial streams within the catchment area of the watershed and is denoted by Lu. In general, the total stream length is measured by the 1:100,000 topographical maps.

Mean Stream Length (Lu)

Mean stream length is defined as the ratio of total stream length of a particular order to the total number of same ordered stream and is denoted by Lu.

Watershed Perimeter (Pr)

Watershed perimeter may be defined as the total length of outer boundary of the watershed. It is manually calculated by the planimeter.

Maximum Length of the Watershed (Lb)

It is the distance between the remotest point to the outlet of the watershed.

Stream Length Ratio (RL)

The stream length ratio (RL) is defined as the ratio of the mean length of the stream of a given order ($L\bar{u}$ 1) to the mean length of the streams of the next smaller order ($L\bar{u}$-1).

$$\mathrm{RL} = (L\bar{u})/2a$$

Bifurcation Ratio (Rb)

The Rb is the ratio of the number of streams in lower order (Nu) the next order (Nu + 1). It is seen that the Rb is lower in alluvial region as compared to the Himalayan zone.

$$\mathrm{Rb} = \mathrm{Nu}/(\mathrm{Nu} + 1)$$

Form Factor (Rf)

Form factor might be defined as the ratio of basin area (A) to square of the maximum length of the basin (Lb). The smaller the value of Ff, the more elongated will be the watershed. The watershed with high Ff has high peak flows of shorter duration (Horton, 1932).

$$R_f = \frac{A}{(L_b)^2}$$

Elongation Ratio (Re)

The Re is defined as the ratio of equal diameter of a circle of the same area as that of the watershed to the maximum basin length. It is demoted by Re.

$$R_e = \frac{2}{R_b}\sqrt{\frac{A}{\pi}}$$

Circulatory Ratio (Rc)

The Rc originally introduced by Miller in 1954. It is mainly influenced by the frequency of stream and its length. It is defined as the ratio of the basin area to the area of a circle having a circumference equal to the perimeter of Basin. It is denoted by the (Rc).

$$R_c = \frac{12.57A}{(P_r)^2}$$

Drainage Density

Drainage density is determined by the ratio of the total length of all the streams and rivers in all orders decided by the total area of watershed or drainage basin. It is a linear parameter of the morphometric analysis and depends upon both climate and physical characteristics of the drainage basin. It is a very sensitive indicator for the erosion calculation by the effect of topographic characteristics and the stream to the outlet. The unit of drainage density is generally expressed in terms of miles of channel per square mile or km/km^2 which indicates the proximity of the channel spacing, thus providing an idea about the quantity measure of the watershed (Strahler, 1964). It is denoted by (Dd).

$$Dd = Lu/A$$

Drainage Frequency

Drainage frequency is the number of streams per unit area of the watershed. It is mainly associated with lithology of catchment, degree of slope, stages of fluvial cycle, and amount of surface run off. It indicates the texture of the drainage network. It is expressed as

$$Fs = Nu/A$$

Texture Ratio (T)

The texture ratio is expressed as the ratio of the total number of first-or-order stream segment to the perimeter of the watershed or basin. It depends on the underlying lithology, relief aspects of the terrain, and infiltration capacity. It is expressed as

$$T = \mathrm{N1/Pr}$$

Maximum Watershed Relief (H)

Maximum watershed relief (H) or basin relief is defined as the difference between maximum and minimum elevation in watershed.

Compactness Coefficient (Cc)

According to Gravelius (1914), compactness coefficient (Cc) of watershed is calculated as the ratio of perimeter of watershed to circumference of equivalent circular area of the watershed. It is a shape parameter of a watershed which is only dependent on the slope and independent of the size of the watershed. It is expressed as Cc

$$C_c = 0.2827\left(\frac{P^{0.5}}{A}\right)$$

7.4 RESULTS AND DISCUSSION

Morphometric parameter is the principle factor to identify various characteristics of drainage basin, which are permeability of bed rock, stage of stream, shape, patterns, etc., which is very helpful to correlate with lithological characteristics (Yadav et al., 2014). In the latest study, morphometric analysis of the parameters, such as stream length, stream order, Dd, Fs, Rb, total relief, basin length, elevation difference, perimeter, length of overland flow and area, compactness co-efficient, texture ratio, Re, Ff, circulatory ratio, number of stream and total stream length of the six mini watersheds has been follow through using the mathematical equation and their results are shown in Tables 7.1–7.3.

Stream Order

The primary step of morphometric analysis is "Stream Order" which was introduced by Horton in 1932. There are six mini watersheds out of which one mini watershed has fourth ordered (SWS-5) and the remaining five, three has third ordered (MWS+1, MWS-4, MWS-5) and the last one has second ordered (MWS-2, MWS-3, MWS-6) mini watersheds which are shown in Table 7.1. Normally, when the stream order increases, the number of streams moderately decreases. And the variation in size and order of tributary basins mostly depends upon the region. The total number of all the order of streams is 54.

TABLE 7.1 No. of streams and stream length of different mini watersheds of Champua.

S. no	Mini basin name	Area (km^2)	No of streams				Stream length (km)			
			I	II	III	IV	I	II	III	IV
1	MWS1	2.22	4	2	1	–	2.055	2.099	0.080	–
2	MWS2	1.80	7	6	–	–	2.458	1.534	–	–
3	MWS3	0.79	4	3	–	–	1.332	0.818	–	–
4	MWS4	2.29	5	4	1	–	2.555	1.696	0.585	–
5	MWS5	1.37	6	2	3	1	1.289	1.103	0.953	0.211
6	MWS6	1.00	1	3	2	–	1.354	1.082	–	–

TABLE 7.2 Stream Characteristics of Mini Watershed.

Min-basin name	Area (km^2)	Perimeter (km)	Max elv.	Min elv.	Basin length (km)	Total relief (km)	No. of streams	Total stream length (km)
MWS-1	2.22	7.44	597	567	2.62	30	7	4.14
MWS-2	1.80	7.04	583	560	1.53	23	13	3.99
MWS-3	0.79	4.79	577	555	0.82	22	7	2.15
MWS-4	2.29	7.10	577	552	1.70	25	10	4.84
MWS-5	1.37	6.51	581	552	1.79	29	12	3.56
MWS-6	1.00	6.72	580	553	2.09	27	5	2.91

Stream Length

As stated by Horton's second law (1945), the characteristics of the stream length of the mini-basins certify the "laws of the stream length," which affirm that the average length of streams of each of the different orders in a drainage basin tends firmly to estimate a direct geometric ratio (Horon, 1995). Generally, logarithms of the given order of the number of streams, when plotted against the order, the obtained points lie on a straight line (Horton, 1945). Most of the drainage networks show a linear relationship with a small deviation from a straight line (Chow, 1964). In general, the total

TABLE 7.3 Result of Morphometric Analysis for Mini Watersheds.

Mini basin name	Bifurcation ratio	Drainage density	Stream frequency	Circulatory ratio	Form factor	Elongation ratio	Texture ratio	Compactness coeff.	Relief ratio	Length of overland flow
MWS1	2.000	1.860	3.153	0.503	0.323	0.840	0.537	0.340	0.011	0.268
MWS2	1.160	2.220	7.230	0.456	0.764	0.980	0.995	0.416	0.015	0.225
MWS3	1.330	2.735	8.900	0.430	1.175	1.220	0.835	0.785	0.027	0.183
MWS4	2.625	2.110	4.370	0.570	0.795	1.000	0.702	0.328	0.015	0.236
MWS5	2.222	2.600	8.780	0.405	0.426	0.737	0.922	0.526	0.016	0.192
MWS6	1.500	2.427	5.000	0.270	0.229	0.537	0.450	0.730	0.013	0.206

TABLE 7.4 Prioritized Rank of Sub-Watersheds Using the Morphometric Parameter.

Sub-basin name	Bifurcation ratio	Drainage density	Stream frequency	Circulatory ratio	Form factor	Elongation ratio	Texture ratio	Compactness coefficient	Relief ratio	Length of overland flow	Compound parameter	Final priority
SWS1	III	VI	IV	V	II	III	V	II	VI	I	3.9	V
SWS2	VI	IV	III	IV	IV	IV	I	III	III	III	3.5	III
SWS3	V	I	I	III	VI	VI	III	VI	I	VI	3.8	IV
SWS4	I	V	V	VI	V	V	IV	I	IV	II	3.8	IV
SWS5	II	II	II	II	III	II	II	IV	II	V	2.6	I
SWS6	IV	III	IV	I	I	I	VI	V	V	IV	3.4	II

length of the stream segments decreases with the stream order. Deviation from its common behavior by varying lithology and expected uplift across the Basin (Singh and Singh, 1997).

Bifurcation Ratio

The geological as well as tectonic characteristics of watershed reflect by the Rb (Gajbhaiya et al., 2014). It is a dimensionless property. Lower the value of Rb shows the partially disordered watershed without any deformation in drainage pattern and also shows poor structural disturbance (Nag, 1998). Higher the value of Rb indicates the low recharge and severe overland flow for the mini watershed. In this study, the value of the mean bifurcation ratio varies between 1.16 and 2.625 which is shown in Table 7.3.

Drainage Density

Drainage density depends upon both physical as well as climate characteristics of basins. It is a basic length scale in a landscape, which is conceded to be the transition point between scales where unstable channel farming processes yield to stables diffusive processes (Tarbotonb et al., 1992). The Dd is affected by various factors, among which infiltration capacity of the land, resistance to erosion of rocks, and climate conditions rank high (Verstappen, 1983). In Charwa watershed, the Dd of six mini watersheds varies from 1.86 to 2.735 and is shown in Table 7.3.

Stream Frequency

Drainage frequency is directly associated with the stream population per unit area of the watershed (Horton, 1932). The prioritization-wise ranking of the Fs of all mini watersheds is summarized in Table 7.3. Higher the value of drainage frequency reflects the high runoff. In this study, MWS-3 had a highest run-off value as comparison to other mini watersheds; however, ranges vary from 3.153 (MWS-1) to 8.90 (MWS-3) as shown in Table 7.3.

Circulatory Ratio (Rc)

The dendrite pattern of a watershed is indicated by the *Rc* (Miller, 1953). This is ratio (Rc) is influenced by frequency of the streams, relief and slope, frequency and length of the stream, land use/land cover (Lu/Lc), geological structures, and climatic variability of the mini-watershed (Patel et al., 2013). In the present study, the value of Rc ranges from 0.27(MWS-6) to 057(MWS-4).

Form Factor

Many researchers recommended that the value of Ff should be less than 0.7854 for a perfectly circular basin (Rekha et al., 2011; Gajbhiye et al., 2014). In the present study, it comes out with the value of Ff ranges from 0.323 to 1.35, which implies that the SWS-1 is more elongated as compared to SWS-4. However, the SWS-4 will give rise to high peak in a short interval of time.

Elongation Ratio

Elongation ratio which quantifies the shape of the catchment and it depends on the climatic and physiographic types. In the present study, the values of Re generally vary between 0.53 and 1.22. Generally, the values which are close to 1.0 represent the regions of very low relief, whereas the value which lies between 0.6 and 0.98 indicates the high relief and steep slope (Strahler, 1964). These values categorized into three namely: (1) circular (>0.9), (2) oval (0.9–0.8), and (3) elongated (<0.7).

Texture Ratio

The texture ratio depends upon the physical properties of the catchment, relief aspect of the terrain and infiltration of the soil (Vijith and Satheesh, 2006). In this investigation, its value for the mini-watersheds varies between 0.45 (MWS-6) and 0.995 (MWS-2) and classified as medium in nature. Its lesser value indicates the plain basin with a gentle gradient. The texture ratio values and assigned priority of each mini watershed as given below in Table 7.3.

Compactness Coefficient

In the present study of a circular basin the highest value was found in MWS-4 (0.73), while the lowest value was for MWS-6 (0.328). However, values of compactness coefficient of mini watersheds are compiled in Table 7.3, the feature of watershed lies oval cum elongated in nature. These types of watersheds are most hazardous from a drainage standpoint of view because it has shortest time of concentration before peak flow occurs in the basin (Ratnam et al., 2005; Javed et al., 2009).

Relief Ratio

In the present study, the relief ratio (Rh) results presented in Table 7.3 values varies from 0.010 (MWS-1) to 0.0268 (MWS-3) which indicates that large segment of the micro-watershed moderately steep slopes. High values of Rh indicate steepness of the basin as well as the intensity of erosion process operating the slopes of the watershed.

Length of Overland Fow

The value of length of overland flow in mini-watersheds varies from 0.183 (MWS-3) to 0.268 (MWS-1). These characteristics are inversely proportional to the average slope of the channel and directly relate with the length of sheet flow. The shorter length of overland flow will reach the stream quickly while the longer length indicates the mild slopes. 0.212 is the overland flow length of present study, which indicates that there is less structural disruption, low runoff conditions and having moderate overland flow.

Prioritization of Mini Watersheds

The erosion-risk assessment parameters which have been used for prioritization of watershed based on morphometric analysis are Rb, Cc, Fs, Ff, Dd, circularity ratio (Rc), drainage texture (Dt), and Re. The linear parameters such as Dd, Fs, Rb, Dt have a direct relationship with erodibility, the parameters with higher value, indicate more erodibility. As a result, the prioritization of watersheds with the highest value of linear parameters computed as rank 1, second highest value computed as rank 2, and so on, thus, the least value computed as last rank. Shape parameters such as Re, Rc, compactness coefficient, and Ff have an inverse relationship with erodibility, the parameters with the lower value indicates more erodibility. Thus, the lowest value of shape parameters evaluated as rank 1, next lower value was evaluating as rank 2 and so on and the highest value computes last rank. Therefore, the ranking of the watersheds has been determined by assigning the highest priority based on highest value assigning for linear parameters whereas lowest value assigns for shape parameters as shown in Table 7.4. Succeeding the ranks has been given to linear and aerial parameters. Ultimately, by calculating the averaging of all the parameters of each mini watershed assigned in compound parameters. However, the final priority assigned to least rating value was assigned as highest priority; next higher value was assigned second priority, and so on. Finally, the highest value of compound parameter assigned as a last rank (Fig. 7.3). The mini-watersheds are classified into three classes as high (2.4–3.2), medium (3.2–3.6), and low (>3.7) priority based on the span of C_p value. So to archive the results on the basis of morphometric analysis, MWS-5 computes as high priority, MWS-6 and MWS-2 fall in medium priority whereas MWS-4, MWS-3, and MWS-1 assign in the low priority category as given below in Table 7.3.

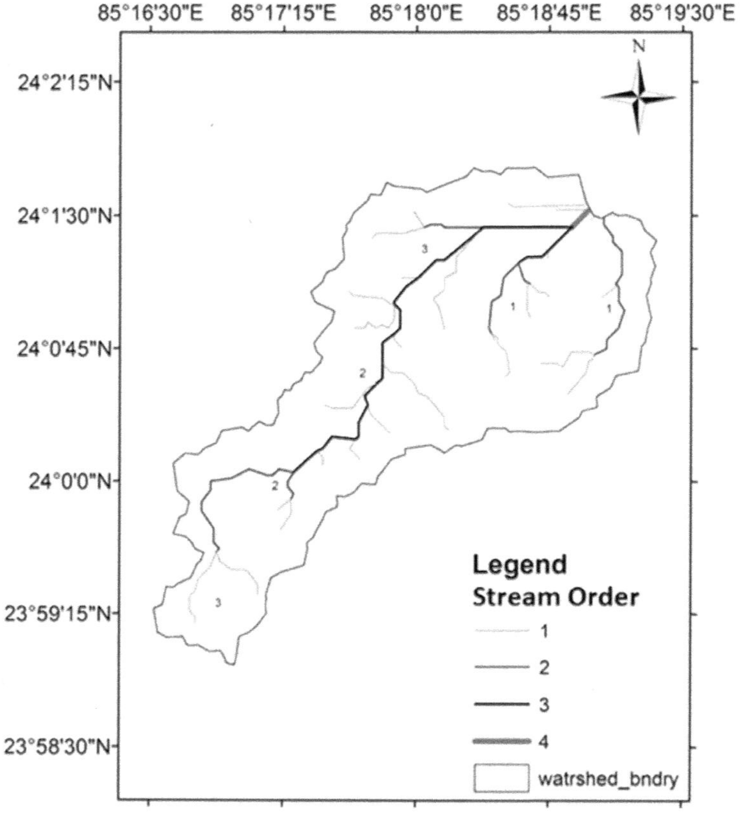

FIGURE 7.2 Stream network of study area.

7.5 CONCLUSION

In the present study, the quantitative morphometric analysis applied to the Charwa watershed in same river basin which consists of six mini watersheds using GIS and RS techniques which helps to understand the various functions of drainage, climate, physiography of the watershed, infiltration capacity, surface runoff, etc. In the present study, the watershed is classified as fourth-order basin which shows dendritic pattern, permeable soil, and coarse Dt. The domination of no. of streams is higher in lower order and total stream lengths. This parameter controls the erosion and run-off parameters.

The results of the Charwa watershed revealed that MSW-5 shows high prone to comparatively higher erosion and soil loss. The quantitate analysis of linear, aerial, and relief parameters found to be a great impact on watershed evaluation. The variation in bifurcation indicates inconsistency in geological nature. The Fs of all mini watersheds shows a positive interrelation with the Dd value. Re exhibits that some mini watersheds are classified under circular shape while the other in elongated shape. Circular basin is more efficient in run-off discharge compared with elongated patterns. After the verification of results, the study concludes that morphometric analysis applied to any region has the high potential to reveal the morphological and hydrological characteristics using RS and GIS which value for control of soil erosion, basin-level construction, flood-control planning, watershed management, and other hydrological morphological planning for any regions.

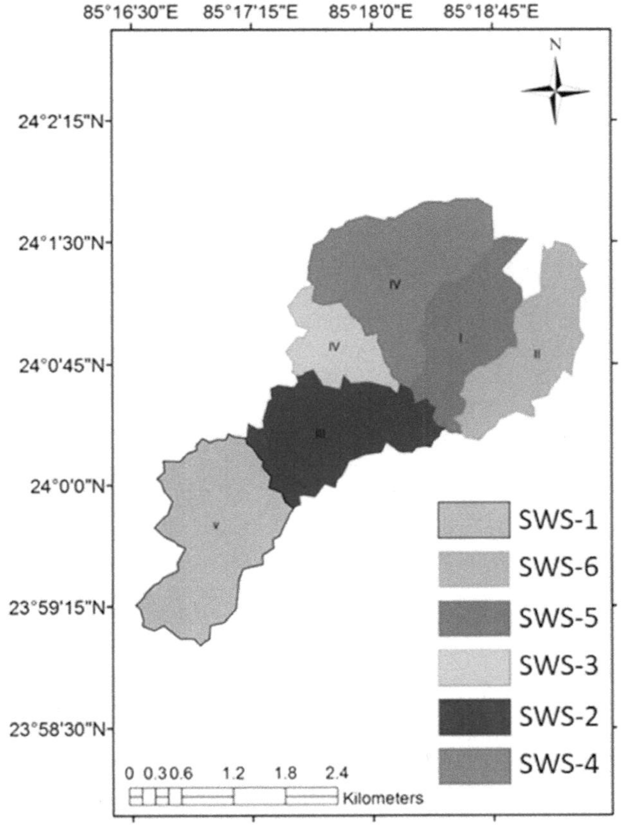

FIGURE 7.3 Prioritized subwatershed of study area.

KEYWORDS

- **geo-morphometric**
- **prioritization**
- **mini watershed**
- **geo-spatial-techniques**
- **GIS**

REFERENCES

Astras, T.; Soulankellis, N. Contribution of Digital Image Analysis Techniques on Landsat-5 TM Imageries for Drainage Delineation. A Case Study from the Olympus Mountain, West Macedonia, Greece. In *Proceedings of the 18th Annual Conference of Remote Sensing Society*; University of Dundee, Dundee, Scotland, 1992; pp 15–17.

Biswas, S.; Sudhakar, S.; Desai, V. Prioritisation of Subwatersheds Based on Morphometric Analysis of Drainage Basin: A Remote Sensing and GIS Approach. *J. Indian Soc. Remote Sens.* **1999,** *27*, 155–166.

Chopra, R.; Dhiman, R. D.; Sharma, P. Morphometric Analysis of Sub-Watersheds in Gurdaspur District, Punjab Using Remote Sensing and GIS Techniques. *J. Indian Soc. Remote Sens.* **2005,** *33*, 531–539.

Chow, V. T. Runoff. In *Handbook of Applied Hydrology*; McGraw-Hill: New York, 1964; pp 14-1–14-54.

Gajbhiye, S.; Mishra, S. K.; Pandey, A. Prioritizing Erosion-Prone Area through Morphometric Analysis: An RS and GIS Perspective. *Appl. Water Sci.* **2014,** *4* (1), 51–61.

Gregory, K. J.; Walling, D. E. *Drainage Basin form and Process: A Geomorphological Approach*; Edward Arnold: London, 1973.

Horton, R. E. Drainage-basin Characteristics. *Trans. Am. Geophys. Union* **1932,** *13*, 350–361.

Horton, R. E. Erosional Development of Streams and Their Drainage Basins; Hydrophysical Approach to Quantitative Morphology. *Geol. Soc. Am. Bull.* **1945,** *56*, 275–370.

Javed, A.; Khanday, M. Y.; Ahmed, R. Prioritization of Subwatersheds Based on Morphometric and Land Use Analysis Using Remote Sensing and GIS Techniques. *J. Indian Soc. Remote Sens.* **2009,** *37*, 261–274.

Langbein, W. B. *Topographic Characteristics of Drainage Basins*; US Government Printing Office: USA, 1947.

Leopold, L. B.; Miller, J. P. *Ephemeral Streams: Hydraulic Factors and Their Relation to the Drainage Net*; US Government Printing Office: USA, 1956.

Melton, M. A. An Analysis of the Relations among Elements of Climate, Surface Properties, and Geomorphology. DTIC Document, 1957.

Miller, V. C. A Quantitative Geomorphic Study of Drainage Basin Characteristics in the Clinch Mountain Area Virginia and Tennessee. DTIC Document, 1953.

Morisawa, M. Accuracy of Determination of Stream Lengths from Topographic Maps. *Trans. Am. Geophys. Union* **1957,** *38*, 86–88.

Nag, S. Morphometric Analysis Using Remote Sensing Techniques in the Chaka Sub-Basin, Purulia District, West Bengal. *J. Indian Soc. Remote Sens.* **1998,** *26*, 69–76.

Nag, S.; Chakraborty, S. Influence of Rock Types and Structures in the Development of Drainage Network in Hard Rock Area. *J. Indian Soc. Remote Sens.* **2003,** *31*, 25–35.

Patel, D. P.; Gajjar, C. A.; Srivastava, P. K. Prioritization of Malesari Mini Watersheds through Morphometric Analysis: A Remote Sensing and GIS Perspective. *Environ Earth Sci.* **2013,** *69*, 2643–2656.

Ratnam, K. N.; Srivastava, Y.; Rao, V. V.; Amminedu, E.; Murthy, K. Check Dam Positioning by Prioritization of Micro- Watersheds Using SYI Model and Morphometric Analysis—remote Sensing and GIS Perspective. *J. Indian Soc. Remote Sens.* **2005,** *33*, 25–38.

Registrar General, India. Census of India 2011: Provisional Population Totals-India Data Sheet. Office of the Registrar General Census Commissioner, India, Indian Census Bureau, 2011.

Rekha, B. V.; George, A. V.; Rita, M. Morphometric Analysis and Micro-Watershed Prioritization of Peruvanthanam Sub-Watershed, the Manimala River Basin, Kerala, South India. *Environ. Res. Eng. Manage.* **2011,** *57* (3), 6–14.

Scheidegger, A. E. The Algebra of Stream-Order Numbers. United States Geological Survey Professional Paper, 1965; pp 187–189.

Schumm, S. A. Evolution of Drainage Systems and Slopes in Badlands at Perth Amboy, New Jersey. *Geol. Soc. Am. Bull.* **1956,** *67*, 597–646.

Shreve, R. L. Infinite Topologically Random Channel Networks. *J. Geol.* **1967,** *75* (2), 178–186.

Singh, S.; Singh, M. Morphometric Analysis of Kanhar River Basin. *Natl. Geograph. J. India* **1997,** *43*, 31–43.

Strahler, A. N. Hypsometric (area-altitude) Analysis of Erosional Topography. *Geol. Soc. Am. Bull.* **1952,** *63*, 1117–1142.

Strahler, A. N. Quantitative Analysis of Watershed Geomorphology. *Trans. Am. Geophys. Union* **1957,** *38*, 913–920.

Strahler, A. N. Dimensional Analysis Applied to fluvially Eroded Landforms. *Geol. Soc. Am. Bull.* **1958,** *69*, 279–300.

Strahler, A. N. Quantitative Geomorphology of Drainage Basins and Channel Networks. In *Handbook of Applied Hydrology*; Chow, V. T., Ed.; McGraw-Hill: New York, 1964; pp 4-39–4-76.

Surendra Kumar, C.; Kansal, M. L. Springerlink.com. Prioritization of Sub-Watershed of Piperiya Watershed, India, 2014.

Survey of India, Ranchi (Toposheet).

Tarboton, D. G.; Bras, R. L.; Rodriguez-Iturbe, I. A Physical Basis for Drainage Density. *Geomorphology* **1992,** *5*, 59–76.

Verstappen, H. T. *Applied Geomorphology: Geomorphological Surveys for Environmental Development*; Elsevier: Amsterdam, 1983.

Vijith, H.; Satheesh, R. GIS Based Morphometric Analysis of Two Major Upland Sub-Watersheds of Meenachil River in Kerala. *J. Indian Soc. Remote Sens.* **2006,** *34*, 181–185.

Vittala, S. S.; Govindaiah, S.; Gowda, H. H. Morphometric Analysis of Sub-Watersheds in the Pavagada Area of Tumkur District, South India Using Remote Sensing and GIS Techniques. *J. Indian Soc. Remote Sens.* **2004,** *32*, 351–362.

Yadav, S. K.; Singh, S. K.; Gupta, M.; Srivastava, P. K. Morphometric Analysis of Upper Tons Basin from Northern, 2014a.

Yadav, S. K.; Singh, S. K.; Gupta, M.; Srivastava, P. K. Foreland of Peninsular India Using CARTOSAT satellite and GIS. *Geocarto Int.* **2014b,** *29* (8), 895–914.

CHAPTER 8

A Framework for Integrated Wetland Sustainability Index

ALKA YADAV and M. L. KANSAL*

Department of Water Resources Development & Management, Indian Institute of Technology, Roorkee 247667, Uttarakhand, India

**Corresponding author. E-mail: mlk@wr.iitr.ac.in*

ABSTRACT

Wetland ecosystems furnish human societies with various regulatory, provisioning, and cultural services. However, the overexploitation and environmental pressures created by anthropogenic activities have caused the deterioration of their health and functions. A sustainable wetland ensures the availability of its functions for both present and future generations. Therefore, there is a need for the assessment of wetland sustainability incorporating socioeconomic conditions and ecological baseline of the area. This chapter reviews the various sustainability indices suggested by the researchers and recommends an Integrated Wetland Sustainability Index (IWSI). The IWSI applies state-pressure–response function based on the Wetland Hydrology, Environment, Life, and Policy (HELP). A HELP Framework is a beneficial tool that obtains data on the current environment of wetland and integrates the four realms in one aggregated, dynamic, and quantitative indicator. The proposed indicator will help in understanding the wetland hydrology, environment (ecological health), life (socioeconomic) issues, and policies. It can be employed effectively as an analytical tool for policymaking in the context of sustainable management of wetland ecosystems.

Advances in Hydrology and Climate Change: Historical Trends and New Approaches in Water Resources Management. Surendra Kumar Chandniha, Anil Kumar Lohani, Gopal Krishan, & Ajay Krishna Prabhakar (Eds.)

8.1 INTRODUCTION

Wetlands are highly fragile, and dynamic ecosystems comprise of marshes, lakes, mangroves, floodplains, and peatlands, which are continually or periodically inundated with flowing or still fresh, brackish, or salty water (Syphard and Garcia, 2001). Wetland is a transitional zone between terrestrial and aquatic with rich biodiversity, but sensitive to the ecological atmosphere (Mitch and Gosselink, 2000). Historically, wetlands have played the most crucial role in the preliminary settlement and subsequent development of human civilization. Wetlands are usually present along the shorelines of lakes, rivers, oceans, and in local depressions. Natural wetlands derive their water from lakes, flooding rivers, tidal flows, or connections to groundwater. A typical wetland is depicted in Figure 8.1.

FIGURE 8.1 A typical mangrove wetland.

According to Ramsar Convention 1971, "Wetlands are the area of marsh, fen, peatland or water whether natural or artificial, permanent or temporary with water that is static or flowing, fresh, brackish, salt including the area of marine water, the depth of which at low tides does not exceed six meters" (Ramsar Convention Secretariat, 2006). Further, article 2.1 of the convention stated that wetlands "may incorporate riparian and coastal zones adjacent to the wetlands, and islands or bodies of marine water deeper than six meters at low tide lying within the wetlands."

The definition given by U.S. National Wetlands Inventory states wetlands as "land transitional between terrestrial and aquatic systems where the water table is usually at or near the surface or land covered by shallow water." Wetlands must have at least one of the three characteristics mentioned:

- The land lends predominant support to biota particularly hydrophytes, at least periodically;
- Soils formed mainly due to anaerobic processes, that is, hydric soil is the dominant substrate
- Presence of a nonsoil substrate that remains wet or flooded with water for some period in a year mostly the growing season.

Wetlands offer various services, for example, regulating, provisioning, supporting, and cultural services (Chopra et al., 2005). For example, wetlands act as a natural solution for mitigating climate change, which has become global stress (Swain, 2017; Swain et al., 2018; Dayal et al., 2019,). They also have a role in controlling major issues of urban flooding as constructed wetlands (Kansal et al., 2019). A list of services under different categories are shown in Table 8.1.

8.1.1 WETLAND CLASSIFICATION SYSTEM

The classification system introduced by Ramsar Convention delivered a comprehensive framework for delineating wetlands and is also acquiescent to remote sensing based data. Wetland refers to all aspects of the water mass found in wetlands, including the ecotones. The wetland classification system incorporates deepwater habitats and impoundments in addition to all wetlands (Table 8.2).

The main conditions that are taken in this system are given below:

- Wetland hydrology, that is, display of presence of water through the satellite images.
- Wetland vegetation—primarily aquatic plants and other hydrophytes in a portion or whole of the wetland as perceived on satellite.

TABLE 8.1 Wetlands Provide or Benefit from Ecosystem Services.

Ecosystem services	Examples
Provisioning	
Biochemical products	Pharmaceutical and other products from biota
Genetic materials	Plant pathogens resistant genes, ornamental species, and so on
Food, fiber, and fuel	Provide fish, fruits, grains, wild game, logs, peat, fuel wood, and fodder
Freshwater	Water storage and supply for a range of applications, including residential, agricultural, and irrigational
Regulating	
Erosion regulation	Soils and sediments are kept in place by wetlands
Prevention of natural hazard	Storm protection, flood control
Regulation of climate change	CO_2, CH_4 (greenhouse gases) source, and sink affect climate processes at a local and regional scales
Water purification and Waste treatment	Improvement of water quality by removal of excess nutrients and other toxic pollutants
Water regulation	Regulation by discharge and recharge of groundwater reservoirs
Cultural	
Aesthetic value	Aesthetic appeal in wetland ecosystems
Educational value	Training, education opportunities
Recreational	Opportunities for various recreational events
Spiritual and Inspirational	Inspirational source; attachment of spiritual and religious sentiments with various aspects of wetland ecosystems
Supporting	
Nutrient cycling	Serving as chemical sources, sinks, storage, recycling, and acquisition of nutrients.
Soil formation	Accumulation of rich organic matter, Hydric soil development, Sediment retention.

The wetlands classification system of the International Union for Conservation of Nature (IUCN) (Dugan, 1990), presented in Table 8.3, is broader than the Ramsar system. It can be summarized as a four-level hierarchical structure. Level 1 consists of two types of wetlands: inland and coastal. Level 2 comprises the information on geology (estuary, marine). Level 3 includes the information on hydrology (inundation and duration of inundation). Although, artificial wetlands are assembled in a single category.

TABLE 8.2 Wetland Classification System (SAC, 2017).

Level I	Level II	Level III
Inland wetlands	Natural	• High altitude wetland
		• Lake/pond
		• Ox-bow lake/cut-off meander
		• River/stream
		• Riverine wetlands
		• Waterlogged
	Man-made	• Reservoir/barrage
		• Salt pan
		• Tank/pond
		• Waterlogged
Coastal wetlands	Natural	• Coral reef
		• Creek
		• Intertidal mud flat
		• Lagoon
		• Mangrove
		• Salt marsh
		• Sand/beach
	Man-made	• Aquaculture Pond
		• Salt pan

8.1.2 STATUS OF WETLANDS IN WORLD

Wetlands are a prominent source of numerous essential ecosystem services (Chopra et al., 2005; Junk et al., 2013), having socioeconomic and ecological values besides acting as a lifeline for terrestrial ecosystems (Gardner and Finlayson, 2018). Of late, there has been substantial degradation of natural wetlands (coastal, marine, and inland) all over the world. According to a study, there has been an annual average decline of −0.78% in the global natural wetlands; The rate is alarming given that it is more than thrice the average annual rate of loss of natural forest cover during 1990 and 2015 (−0.24%), as estimated by the WET Index (Gardner and Finlayson, 2018).

Wetland cover around the globe spans over 12.1 million km^2 out of which 54% is permanently submerged and 46% remains inundated seasonally. Around 35% of wetlands vanished between 1970 and 2015. Since 1970, the world has witnessed a serious declination in the population of wetland-dependent species. Recent studies suggest that the figure is as high as 81% for inland wetland species and 36% for coastal and marine species. Consequently, the valuable ecosystem services delivered by wetlands have been badly hit across the globe (Dodds et al., 2013; Gardner and Finlayson, 2018). The major factors that have led to this continual degradation of wetlands over the years include usurping the wetlands to drive urbanization, development of catchment areas to promote agriculture activities. Besides other factors like pollution, disturbing climatic patterns also poses a severe threat to the wetlands (Finlayson et al., 2012). The regional distribution of wetland areas at a global level is given in Figure 8.2.

TABLE 8.3 IUCN Wetlands Classification System (Dugan, 1990).

Level I	Level II	Level III	Level IV
Saltwater	Marine	Tidal	Geomorphology, soil composition, land use, vegetation, and other variables are used to identify 39 distinct types of wetlands.
		Intertidal	
	Estuary	Tidal	
		Intertidal	
	Lagoon		
	Salt pans		
Freshwater	Rivers	Perennial	
		Temporary	
	Lacustrine	Permanent	
		Seasonal	
	Palustrine	Forest	
		Emergence	
Artificial	Agriculture		
	Aquaculture		
	Salt pans		
	Salt pans		
	Urban/industry		

8.1.3 STATUS OF WETLANDS IN INDIA

India boats of both inland and coastal wetlands, including natural and man-made. According to Wetlands International South Asia, the number of wetlands in India is estimated to be around 7.7 lakhs which accounts for 4.6%

of the country's geographical area, establishing India's credentials as a major contributor to the biodiversity in South Asia. As per the latest data available, 42 sites have been recognized for their global importance and designated as Ramsar sites. These wetlands cover an area of 10,814.38 km^2 and traverse 17 different Indian States and two Union Territories (UTs). They are home to at least 13% amphibians, 23% fish, 23% reptiles, 65% birds, and 26% of mammalian species, according to the Ministry for Environment. With 42 Ramsar wetland sites, India will have the maximum number of Ramsar sites in South Asia and next only to Japan and China in Asia.

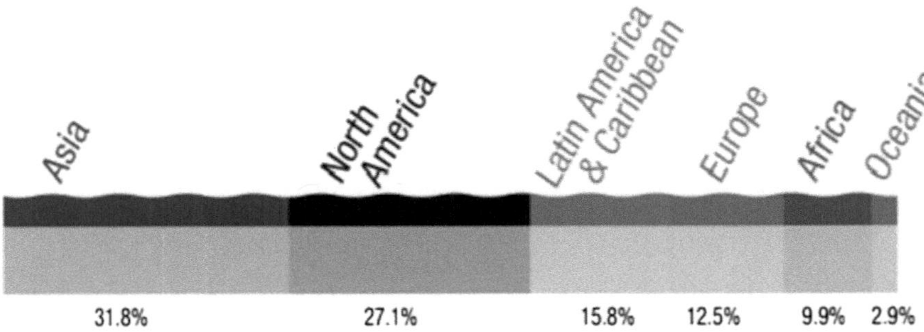

FIGURE 8.2 Wetland area regional distribution (%) (Davidson et al., 2018).

The Ramsar Convention, signed in Ramsar, Iran in 1971, is a global treaty aimed at managing and protecting valuable wetlands. India is among 170 signatory countries that agreed to pay attention and initiate needful actions toward the conservation of wetlands. Chilika Lake (Odisha) and Keoladeo National Park (Rajasthan) were the first wetlands to be identified as Ramsar sites of international significance a decade after signing the Ramsar convention in 1971. One of the main goals of the Ramsar Convention is to achieve and preserve the ecological character of wetlands, as well as to promote their sustainable and wise use for the benefit of people and the environment. It also mandates them to develop National Wetland Policies, establish wetland inventories, undertake and create monitoring, research and awareness, and prepare integrated management plans.

Out of 42 Ramsar sites in India now, just two states Uttar Pradesh (8) and Punjab (6) have 14 sites, while the large state of Andhra Pradesh, Gujarat, Tamil Nadu, and Madhya Pradesh each have one Ramsar site. The largest Ramsar sites are Sundarbans (4230 km^2) followed by Chilika (1165 km^2). According to the State of Forest Report of 2019, a number of wetlands located within forests is 62,466 that covers 3.8% of the recorded forest area. Many wetland conservation efforts have been placed on those that are present within forests or protected area networks. The list of 42 designated Ramsar sites is given in Table 8.4.

TABLE 8.4 Wetlands of International Importance Situated in India.

State/UT	Name of wetlands	Entitled on	Area (km^2)
Andhra Pradesh	Kolleru Lake	19.08.2002	901
Assam	Deepor Beel	19.08.2002	40
Bihar	Kabartal	21.07.2020	26.20
Gujarat	Nalsarovar Bird Sanctuary	24.09.2012	123
Himachal Pradesh	Chandertal	08.11.2005	0.49
	Pong Dam Lake	19.08.2002	156.62
	Renuka	08.11.2005	0.2
Jammu & Kashmir	Hokera	08.11.2005	13.75
	Surinsar-Mansar Lakes	08.11.2005	3.5
	Wular Lake	23.03.1990	189
Kerala	Sasthamkotta Lake	19.08.2002	3.73
	Vembanad-Kol	19.08.2002	1512.5
	Ashtamudi	19.08.2002	614
Ladakh	Tsomoriri	19.08.2002	120
	Tso Kar	17.11.2020	95.77
Madhya Pradesh	Bhoj	19.08.2002	32
Maharashtra	Nandur Madhameshwar	21.06.2019	14
	Lonar Lake	22.07.2020	4.27
Manipur	Loktak Lake	23.03.1990	266
Odisha	Chilika Lake	01.10.1981	1165
	Bhitarkanika Mangroves	19.08.2002	650
Punjab	Ropar	22.01.2002	13.65
	Harike Lake	23.03.1990	41
	Nangal Wildlife Sanctuary	26.09.2019	1
	Keshopur-Miani Community Reserve	26.09.2019	34
	Kanjli	22.01.2002	1.83
	Beas Conservation Reserve	26.09. 2019	64
Rajasthan	Keoladeo National Park	01.10.1981	28.73
	Sambhar Lake	23.03.1990	240
Tamil Nadu	Point Calimere Wildlife and Bird Sanctuary	19.08.2002	385

TABLE 8.4 *(Continued)*

State/UT	Name of wetlands	Entitled on	Area (km^2)
Tripura	Rudrasagar Lake	08.11.2005	2.4
Uttar Pradesh	Upper Ganga River	08.11.2005	265.9
	Saman Bird Sanctuary	02.12.2019	5
	Samaspur Bird Sanctuary	03.10.2019	8
	Sandi Bird Sanctuary	26.09.2019	3
	Sarsai Nawar Jheel	19.09.2019	2
	Nawabganj Bird Sanctuary	19.09.2019	2
	Parvati Aranga Bird Sanctuary	2.09.2019	7
	Sur Sarovar	21.08.2020	4.31
Uttarakhand	Asan Conservation Reserve	21.07.2020	4.44
West Bengal	East Calcutta	19.08.2002	125
	Sundarban	30.01.2019	4230

8.2 WETLANDS SUSTAINABILITY

Sustainable development has been defined in many ways, the commonly accepted being the adoption of a model of development which seeks to meet the needs of the current generation without stripping the resources of future generations (Chandniha et al., 2014; Chandniha and Kansal, 2018). The Brundtland report, published in 1987 by the World Commission on Environment and Development (WECD), popularized the concept of sustainable development. The report defines sustainable development as "Development that meets the needs of the present, without compromising the ability of future generations to meet their needs." This has necessitated the debate around the globe on how to strike a balance between economy, environment, and society.

There are substantial examples pointing to the environmental degradation of wetlands over the years mostly on account of drainage or erosion resulting in deranged water tables, the root cause being overexploitation of wetland ecosystem services to meet the ever-increasing demand of society. This undue stress has more or less rendered the wetlands as wastelands, ill effects of which are pollution, eutrophication, overexploitation of resources, and intrusion of alien species into wetland landscapes.

The scientific community all over the world has emphasized upon the need to achieve sustainable use of wetlands considering their importance for water supply amid increasing calls to put them to agricultural uses.

McFarlane and Whitlow (1990) have studied the loss of wetlands through inappropriate management, highlighting the effect of community activities on their wetlands. The focus is to be kept on their judicious utilization without affecting the ongoing environmental activities. This can be ensured by following and implementing effective management policies and plans at various levels of governance viz. national, regional, and local that seek to preserve the natural characteristics of wetlands while simultaneously allowing the activities catering to the socioeconomic needs of communities.

The environmental functioning of wetlands and exploitation for purposes has to go hand in hand. For a better understanding of wetland functionality and the establishment and achievement of sustainable development goals, dynamic responses in accordance with current market/economic conditions are needed.

In this regard, the need of the hour is to develop a sound wetland utilization strategy encompassing the four basic pillars of a holistic sustainable development model (Fig. 8.3), that is

- **Hydrological sustainability,** which focuses on hydrological aspects. Hydrological sustainability. It helps in balancing the water requirement of the ecosystem along with maintaining both water quantity and quality. It is crucial to maintain sustainable demand, supply and use of hydrological resources for the needs of upcoming generations.
- **Environment sustainability,** which emphasizes on ecological factors. It encourages the development of a sound wetland utilization plan whose implementation does not negatively impact or alter the wetland and also ensures the maintenance of ecosystem and life support systems from an intergenerational equity perspective. The ultimate aim is to protect the environment from being degraded or polluted by securing sustainable development goals.
- **Socioeconomic (Life) sustainability,** which focuses on utilizing wetland in such a way as provides both social and economic benefits to the users in the long run. It ensures the active participation of the local community and their collaborative efforts toward the development of a sustainable wetland model. It also contributes to improvement in social health and mitigation of conflict among stakeholders on resource utilization.
- **Policy-wise sustainability,** which facilitates the holistic management of wetlands that are constantly under stress for exploitation for various other needs. Wetlands are the source of numerous valuable ecosystem goods and services ranging from water storage and purification to

providing biodiversity hotspots providing nourishment to flora and fauna to protecting us from disasters. It is imperative that wetlands become a part and parcel of discussions and debates on the conservation and development of natural resources. Many international policies and environmental regulations have recognized the potential of wetlands for conservation hotspots and agriculture development.

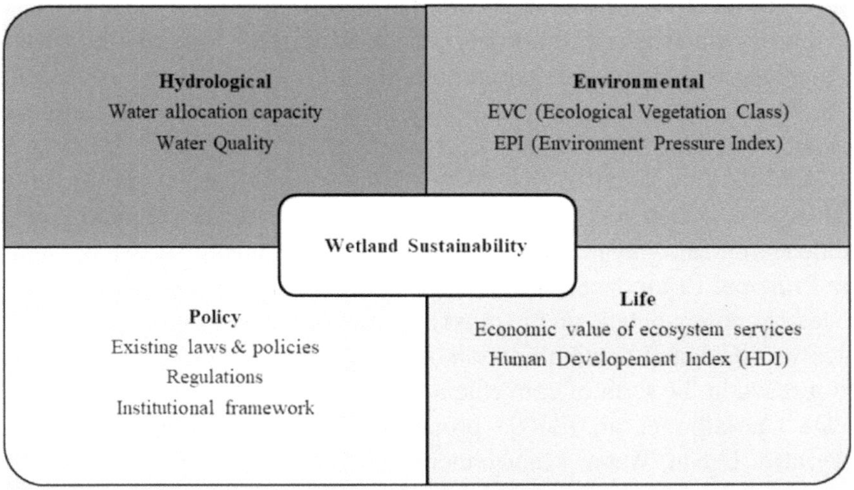

FIGURE 8.3 Components of wetland sustainability.

8.2.1 WATERSHED SUSTAINABILITY STUDIES

Various studies have been undertaken by different authors around the globe to develop models for determining water sustainability. The Delft Hydraulics Laboratory, Netherlands conducted a study on a water resources project and developed a procedure to establish the contribution to sustainable development (Baan, 1994). The basis of measurement was a combination of various indices derived from five criteria that are based on socioeconomic factors; utilization of environmental resources and waste discharge within the natural system carrying capacity; impact on resilience, growth, and stability; natural resources improvement and conservation and upgrading of their carrying capacity; flexibility and sustainability of infrastructure work safety, and well-being of public health; organizational opportunities for different functional use, and opportunistic adaptation to the changing situations. Each factor was further divided into subcriteria and each subcriterion was allocated equal emphasis. The sustainability index was derived from the addition of the numerical values allocated to each sub-criterion.

Another approach was devised by Loucks (1997) to identify sustainability indices in terms of reliability, resilience, and vulnerability with regards to various ecological, environmental, economic, and social conditions. To begin with suitable factors viz., ecological, environmental, economic, and social conditions were found out which could be represented quantitatively and determined from time-series variables of the water resource system.

Chaves and Alipaz (2007) proposed a Watershed Sustainability Index (WSI) that combined issues based on hydrology, climate, life, and politics, as well as current pressures and policy responses, into a single composite indicator using the pressure-state-response (PSR) framework.

Ali (2009) proposed an Arab Water Sustainability Index (AWSI) for the Arab region, which was a theoretical framework that incorporated various socioeconomical, physical, and ecological components of water status. The Principal Component Analysis has been used to design a multivariate model, to convey details on the most important factors. Configuration of the index was designed by combination of 8 indicators derived from 22 factors compressed in the form of convenient data sets.

De Carvalho et al. (2009) proposed "The Sustainability Index for Integrated Urban Water Management" (SIUWM) which comprises five components resulting from 64 variables and 20 indicators. It was deduced from the concept of the "triple bottom line," which shows environmental, social-economic sides.

A water sustainability index for West Java, Indonesia was proposed by Juwana et al. (2010a, b) in a two-part series. The first part was linked with the theoretical framework and consists of a set of 4 factors and 11 indicators. The application of the DELPHI technique was introduced in the second part to finalize the West Java Water Sustainability Index (WJWSI) framework on the basis of feedback coming from stakeholders.

In order to assess water resource sustainability, Gonzales and Ajami (2015) proposed a regional integration framework that computes the sustainability of the system. The framework was a mathematical index that comprises demand, supply, and adaptive capacity-based indicators. It takes in an advanced way to deal with the significance of having different sources of supply.

Kim et al. (2016) developed a Sustainable Water Index (SWI) from Organization for Economic Cooperation and Development (OECD) to determine the extent of sustainability of water use by the application of the PSR framework at the administrative level in Korea.

8.2.2 WETLAND SUSTAINABILITY STUDIES

Moctezuma-Malagón et al. (2008) proposed a sustainability assessment framework for inland wetlands that are based on agroecosystem. This methodology not only considers economic factors but also the environment and social factors. Because of its interdisciplinary and holistic approach, the methodology for the evaluation of management systems using indicators (MESMIS) was chosen because it provided for a comparison of conventional and alternative resource management systems. The construction involved, the consideration of technological, management, socioeconomic characteristics, and biophysical components.

A methodology was proposed by Wang et al. (2011) in which wetland ecosystems were studied in terms of their socioeconomic and environmental benefits, and an integrated assessment system of the wetland resources sustainability has been constructed by employing the PSR model, DELPHI method, and Analytic Hierarchy Process (AHP). Initially, an integrated assessment of wetland assets index system was built, and it includes the given five factors- reasonable utilization of wetlands, ecological environment, social culture, economic system, and safeguard system. With the corporations of 5 subsystems and 34 indicators, an index system of integrated assessment of wetland assets was developed.

The trans-disciplinary framework developed by Van Dam et al. (2011) explored the relationship between ecosystem functioning and livelihood outcomes in wetlands. A Bayesian Network model with 34 nodes was used to integrate the relationship between ecosystem structure, hydrology, and livelihood outcomes in wetlands. The findings of the socioeconomic analysis, eco-hydrological studies, and stakeholder consultations were used to make the decision.

Marambanyika et al. (2016) proposed the WET-Health framework to determine environmental sustainability. It was applied to various types of cultivation in wetlands. The method is focused on indicator and impact scores of the intensity, scope, and significance of various forms of land-use impacts on wetlands. The indicators used in this approach are used to assess the degree of anthropogenic impacts on three key correlated factors that characterize the ecological status of wetland, in particular hydrology, geomorphology, and flora composition.

Chen (2017) proposed the Sustainable Ecological Evaluation Indicators (SEEI) employed the various computable indicators: Ecological Deficit (ED), Environmental Sustainability Index (ESI), Ecological Footprint Index (EFI), Ecological Remainder (ER). These indicators were used to build an indicator system to assess the ecological sustainability of wetland and to develop standards in order to accurately assess ecological protection. The sustainability condition was measured using the two models of ecosystem services and ecological footprint. The value of ecological services was assessed after a sustainability assessment based on each of the ecological services available for the different wetlands land uses and land covers.

8.3 PROPOSED FRAMEWORK FOR INTEGRATED ASSESSMENT OF WETLAND SUSTAINABILITY

To determine the sustainability of wetland, an integrated framework is required. The framework can be designed by given steps: (1) Determining the boundary and analyze the relevant parameters; (2) quantifying, weighing of index, development of a model for integrated evaluation of sustainability; (3) computation of the degree of sustainability of the given study area. Figure 8.4 illustrates the flow diagram of the study.

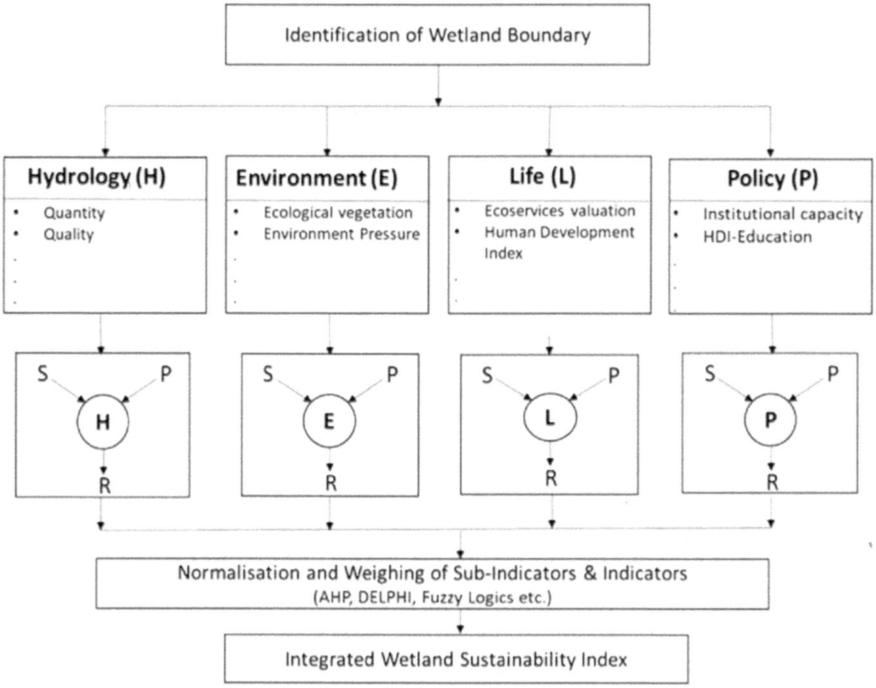

FIGURE 8.4 Evaluation process of wetland sustainability.

The task to build the assessment indexes is quite challenging. The indexes that have an impact on sustainability consists of not only ecological and environmental factors but also socioeconomic factors; also, the supporting condition needs to consider while applying (Cui and Yang, 2002). The State-Pressure-Response model is thus developed for the evaluation of wetlands, as shown in Figure 8.5.

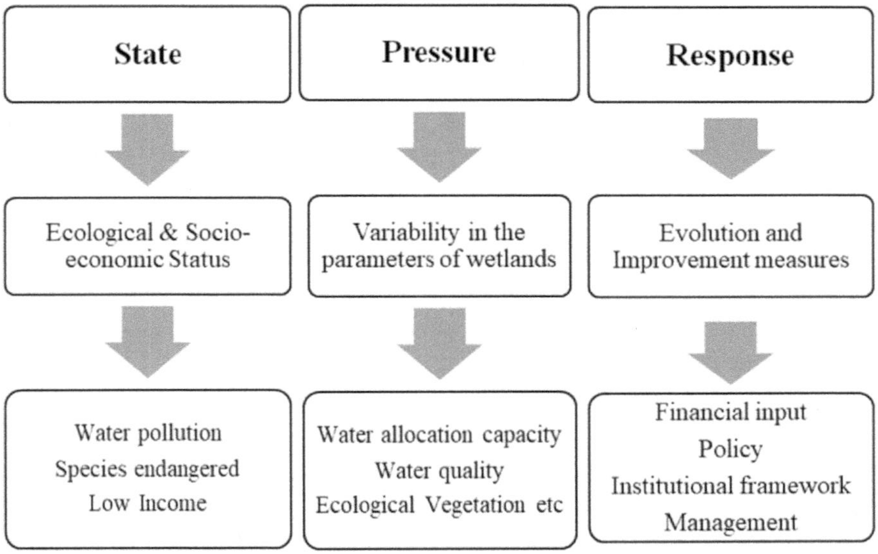

FIGURE 8.5 State-pressure-response model.
Source: Modified from Cheng (2007).

The pressures that affect the wetland status are mainly divided into four aspects: hydrological state, environment state, policy, and the socioeconomic state of the local society. Lastly, both pressures and state are considered, certain actions should be taken by the authority for the recovery & restoration of a healthy ecosystem, for example, policy-making, financial support, etc. (Yang et al., 2006).

8.3.1 INTEGRATED WETLAND SUSTAINABILITY INDEX BASED ON HYDROLOGY, ENVIRONMENT, LIFE, AND POLICY ISSUES

It is recognized that wetland sustainability is directly a function of its hydrologic factors, environmental factors, socioeconomic, and policy factors, but a few efforts have been taken to combine these factors in a single comprehensive and comparable number. Considering that management of wetland is a holistic and dynamic process, and suppose the wetland sustainability is a function of wetland hydrology (*H*), wetland environment (*E*), its socio-economic state (*L*), and water-based policy (*P*), a dynamic, State-Pressure-Response model based on four parameters (*H, E, L, P*) was applied. Hence, a Wetland Sustainability Index – IWSI has resulted. Mathematically, the IWSI is given by:

$$IWSI(0-1) = \frac{(H + E + L + P)}{4} \tag{8.1}$$

where, *IWSI* =Integrated Wetland Sustainability Index, *H* = Hydrology indicator, *E* = Environmental indicator, *L* = Life indicator and *P* = Policy indicator.

Table 8.5 illustrates the IWSI parameters with reference to the given four indicators (*H, E, L, P*). A score between 0 and 1 is given for each combination of parameters and indicators. A zero value is given to the poorest conditions, and one is given for the optimum conditions.

Chaves and Alipaz (2007) divided parameters into five scale scores in both the qualitative and quantitative parameters (0, 0.25, 0.50, 0.75, and 1.0) in order to simplify the determination of the parameter by the consumers. As mentioned in eq 8.1, all indicators have equal weightage as there is no evidence that it may be otherwise (Harr, 1987). By the use of same weight, skewing of outcomes can be avoided (Heathcote, 1998) and the input allowed from the diverse sectors and stakeholders (environmentalists, hydrologists, water users, and policymakers) as shown in eq 8.1. The scale score may vary, can be of 5 scale, 7 scale, or 10 scale. The Equation obtained can be a linear or nonlinear function. In addition, Wetland to wetland, indicator weights can vary, and they should be decided by stakeholders in agreement as given in eq 8.2.

$$IWSI = \frac{W_1 H + W_2 E + W_3 L + W_4 P}{W_1 + W_2 + W_3 + W_4} \tag{8.2}$$

Once the parameters of all four indicators are obtained, there is a selection of a definite epoch for the study (for example, a 6-year epoch, corresponding to the accessible data e.g. HDI), the IWSI is calculated, with the help of various methods, for example Fuzzy logics, analytic hierarchy process (AHP), DELPHI method, PSR model, etc. according to eqs 8.1 and 8.2.

TABLE 8.5 Parameters and Indicators of the Integrated Wetland Sustainability Index.

Indicators	State parameters	Pressure parameters	Response parameters
Hydrology	Water allocation capacity of basin (long-term average)	Variability in water allocation capacity of the basin in the period analyzed (Mm^3/year/km^2)	Enhancement of water allocation capacity in the period analyzed
	Basin Carlson's Trophic State Index (long-term average)	Variability in Trophic state of the basin in the period analyzed	Reduction of agriculture run off or improvement in treatment in the period analyzed
Environment	Basin Biota sub-index score for each EVC	Variation in EVC (Ecological Vegetation Class) in the period analyzed	Progress in conservation of biodiversity analyzed in the given time
	% of a wetland area with natural flora	Variability in EPI (Environment Pressure Index) in the period examined	Progress in wetland conservation analyzed (% of protected areas, BMPs) in the given time
Life	Basin monetary valuation	Variability in value of services of wetland in the given analyzed	Advancement in the awareness of the importance of basin services in the period analyzed
	Basin Human Development Index (weighed by population of country)	Variability in the wetland per capita income in the given time analyzed	Improvement in the wetland Human Development Index in specific time studied
Policy	Wetland institutional capacity in Conservation and Restoration	Variability in the HDI-Education of the wetland reviewed in the given time	Improvement in the wetland conservation and restoration through expenditure in the period analyzed

Source: Modified from Chaves and Alipaz (2007).

8.4 CONCLUSION

Several issues affect the status of wetland and its sustainable development, for example, environmental, economic issues. Normally they are considered individually, not as a composite, dynamic process. Evaluation of the overall condition of wetlands and their sustainable development should be done by using a holistic methodology that incorporates ecological and socioeconomic components. For the purpose of integrating the HELP components with the present pressures and policy aspects in a single measurable and composite indicator, Integrated Wetland Sustainability Index (IWSI) has been recommended for wetlands. This index is straightforward and utilizes easily procurable data. The results can be listed as follows: highly sustainable, fairly sustainable, moderate sustainable, poorly sustainable, and not sustainable. It is important to build an integrated understanding of the impact of natural and anthropogenic events on wetland functions and the impact of wetlands on livelihoods, for the formulation of effective sustainable wetland policies and management strategies. It is also crucial to understand how decision makers cause exploitation of wetland resource and how sociocultural and policy issues influence those decisions. There is an advantage of the use of a State-Pressure-Response way as it considers cause–effect relationships, allowing different managers, decision-makers, NGOs, stakeholders, etc. to recognize and comprehend the relation among the factors.

KEYWORDS

- **wetlands**
- **integrated wetland sustainability index**
- **hydrology**
- **environment**
- **life**
- **policy**

REFERENCES

Ali, H. M. M. Development of Arab Water Sustainability Index Using Principal Component Analysis. In *Proceedings of the Thirteenth International Water Technology Conference IWTC13*, Hurghada, Egypt, 2009.

Baan, J. A. Evaluation of Water Resources Projects on Sustainable Development. In *Proceedings of the International Symposium on Water Resources Planning in a Changing World*; Karlsruhe, Germany, 28–30, 1994; p 19.

Chandniha, S. K.; Kansal, M. L. Water Sustainability Assessment under Climatic Uncertainty—A Case Study of Chhattisgarh (India). In *Climate Change Impacts*; Springer: Singapore, 2018; pp 231–261.

Chandniha, S. K.; Kansal, M. L.; Anvesh, G. Watershed Sustainability Index Assessment of a Watershed in Chhattisgarh, India. *Curr. World Environ.* **2014,** *9* (2), 403.

Chaves, H. M. L.; Alipaz, S. An Integrated Indicator Based on Basin Hydrology, Environment, Life, and Policy: The Watershed Sustainability Index. *Water Resour. Manage.* **2007,** *21*, 883–895. doi:10.1007/s11269-006-9107-2

Chen, H. S. Establishment and Application of Wetlands Ecosystem Services and Sustainable Ecological Evaluation Indicators. *Water* **2017,** *9* (3), 197.

Cheng, S. *Study on Integrative Assessment for Conservation and Sustainable Utilization of Wetlands Resources*; Chinese Academy of Forestry, 2007.

Chopra, K.; Leemans, R.; Kumar, P.; Simons, H. *Ecosystems and Human Well-Being: Policy Responses*; Island Press, 2005.

Cui, B. S.; Yang Z. F. Establishing an Indicator System for Ecosystem Health Evaluation on Wetlands I. A Theoretical Framework. *Acta Ecologica Sinica* **2002,** *22* (7), 1005–1011.

Davidson, N. C.; Fluet-Chouinard, E.; Finlayson, C. M. Global Extent and Distribution of Wetlands: Trends and Issues. *Mar. Freshwater Res.* **2018,** *69* (4), 620–627.

Dayal, D.; Swain, S.; Gautam, A. K.; Palmate, S. S.; Pandey, A.; Mishra, S. K. Development of ARIMA Model for Monthly Rainfall Forecasting over an Indian River Basin. In *World Environmental and Water Resources Congress 2019: Watershed Management, Irrigation and Drainage, and Water Resources Planning and Management*; American Society of Civil Engineers: Reston, VA, 2019; pp 264–271.

De Carvalho, S. C. P.; Carden, K. J.; Armitage, N. P. Application of a Sustainability Index for Integrated Urban Water Management in Southern African Cities: Case Study Comparison–Maputo and Hermanus. *Water SA* **2009,** *35* (2).

Dixon, A. B. *Indigenous Management of Wetlands: Experiences in Ethiopia*; Ashgate Publishing, 2003.

Dodds, W. K.; Perkin, J. S.; Gerken, J. E. Human Impact on Freshwater Ecosystem Services: A Global Perspective. *Environ. Sci. Technol.* **2013,** *47* (16), 9061–9068.

Dugan, P.; Dugan, P. J. (Eds.). Wetland Conservation: A Review of Current Issues and Required Action. IUCN-The World Conservation Union, Gland, 1990.

Finlayson, C. M. Forty Years of Wetland Conservation and Wise Use. *Aquat. Conserv. Mar. Freshw. Ecosyst.* **2012,** *22* (2), 139–143.

Gardner, R. C.; Finlayson, C. Global Wetland Outlook: State of the World's Wetlands and Their Services to People. In *Ramsar Convention Secretariat*, 2018.

Gonzales, P.; Ajami, N. K. Urban Water Sustainability: An Integrative Framework for Regional Water Management. *Hydrol. Earth Syst. Sci. Discus.* **2015,** *12* (11).

Harr, M. E. *Reliability-based Design in Civil Engineering* (No. BOOK); McGraw-Hill, 1987.

Heathcote, I. W. *Integrated Watershed Management: Principles and Practice*; John Wiley & Sons, 2009.

Junk, W. J.; An, S.; Finlayson, C. M.; Gopal, B.; Kvˇet, J.; Mitchell, S. A.; Mitsch, W. J.; Robarts, R. D. Current State of Knowledge Regarding the World's Wetlands and Their Future under Global Climate Change: A Synthesis. *Aquat. Sci.* **2013,** *75* (1), 151–167.

Juwana, I.; Perera, B.; Muttil, N. A Water Sustainability Index for West Java-Part 1: Developing the Conceptual Framework. *Water Sci. Technol.* **2010a**, *62*, 1629–1640.

Juwana, I.; Perera, B.; Muttil, N. A Water Sustainability Index for West Java-part 2: Refining the Conceptual Framework Using Delphi Technique. *Water Sci. Technol.* **2010b**, *62*, 1641–1652.

Kansal, M. L.; Osheen; Tyagi, A. Hotspot Identification for Urban Flooding in a Satellite Town of National Capital Region of India. In *World Environmental and Water Resources Congress 2019: Emerging and Innovative Technologies and International Perspectives*. American Society of Civil Engineers, Reston, VA, 2019; pp 12–24.

Kim, I.; Kim, Y.; Kong, I. Assessing Sustainable Water Use and Local Economy in Korea. In *Proceedings of the 36 Annual Conference of the International Association for Impact Assessment*; Nagoya, Japan, May 11–14, 2016.

Linster, M. OECD Environmental Indicators: Development, Measurement and Use, 2003.

Loucks, D. P. Quantifying Trends in System Sustainability. *Hydrol. Sci. J.* **1997,** *42*, 513–530.

Marambanyika, T.; Beckedahl, H.; Ngetar, N. S.; Dube, T. Assessing the Environmental Sustainability of Cultivation Systems in Wetlands Using the WET-health Framework in Zimbabwe. *Phys. Geogr.* **2017,** *38* (1), 62–82.

McFarlane, M. J.; Whitlow, R. Key Factors Affecting the Initiation and Progress of Gullying in Dambos in Parts of Zimbabwe and Malawi. *Land Degrad. Develop.* **1990,** *2* (3), 215–235.

Mitsch, W. J.; Gosselink, J. G.; Zhang, L.; Anderson, C. J. *Wetland Ecosystems*; John Wiley & Sons, 2009.

Moctezuma-Malagón, A.; González-Esquivel, C. E.; De la Lanza-Espino, G.; Islas, C. G. R. A Methodology for Evaluating the Sustainability of Inland Wetland Systems. *Aquacult. Int.* **2008,** *16* (6), 525–537.

Ramsar Convention Bureau. The Ramsar Convention Manual: A Guide to the Convention on Wetlands, Ramsar, Iran, 1971. Ramsar Convention Bureau, 2006.

Space Applications Centre (SAC). National Wetland Inventory and Assessment (NWIA). SAC, Indian Space Research Organisation, Ahmedabad, 2017.

Swain, S. Hydrological Modeling through Soil and Water Assessment Tool in a Climate Change Perspective a Brief Review. In *2017 2nd International Conference for Convergence in Technology (I2CT)*; IEEE, 2017; pp 358–361.

Swain, S.; Verma, M. K.; Verma, M. K. Streamflow Estimation Using SWAT Model over Seonath River Basin, Chhattisgarh, India. In *Hydrologic Modeling*; Springer: Singapore, 2018; pp 659–665.

Syphard, A. D.; Garcia, M. W. Human-and Beaver-Induced Wetland Changes in the Chickahominy River Watershed from 1953 to 1994. *Wetlands* **2001,** *21* (3), 342–353.

van Dam, A. A.; Kipkemboi, J.; Rahman, M. M.; Gettel, G. M. Linking Hydrology, Ecosystem Function, and Livelihood Outcomes in African Papyrus Wetlands Using a Bayesian Network Model. *Wetlands* **2013,** *33* (3), 381–397.

Wang, Y.; Cheng, S.; Ju, H.; Zhang, H.; Jiang, D.; Zhuang, D. Evaluating Sustainability of Wetland System: A Case Study in East Dongting Lake Area, China. *Afr. J. Agric. Res.* **2011,** *6* (28), 6167–6176.

Yang, S. M.; Li, Y. W.; Lv, Y. Z. The Assessment Index System of Zhalong Wetland Eco-security. *Scientia Silvae Sinicae* **2006,** *42*, 127–132.

CHAPTER 9

Long-Term Trend Analysis of Rainfall Series (1901–2010) over Southern Bihar, India

SAURABH SUMAN, PRATIBHA WARWADE*, and PRITI KUMAR

Department of Water Engineering and Management Central University of Jharkhand, India

**Corresponding author. E-mail: pratibhawarwade@gmail.com*

ABSTRACT

In the present study, Mann–Kendall (MK)/Modified Mann–Kendall (MMK) test, Innovative Trend Method (ITM), and Sen's slope test were used to determine the trend over the Southern part of Bihar at monthly, seasonal, and annual time scales for 11 stations using 110 years data (1901–2010). In the MMK test, all stations in the annual and monsoon seasons showed a significant decrease in rainfall trend at the 10% significance level which was also confirmed by the ITM test. The maximum decrease in the magnitude of rainfall observed for the annual season was 2.861 mm/year, for monsoon rainfall was 2.792 mm/year, and for winter rainfall it was 0.085 mm/year. On the other hand, the maximum increase in rainfall observed for Post monsoon season was 0.225 mm/year and for the summer season it was 0.138 mm/year during the study period. All stations showed a significant decreasing trend in the month of February, June, July, and August while the significant increasing trend in the month of January and May although October, November, and December also showed a significant increasing trend for 3, 9, and 5 stations, respectively. The spatial trend depicted that the Nalanda

Advances in Hydrology and Climate Change: Historical Trends and New Approaches in Water Resources Management. Surendra Kumar Chandniha, Anil Kumar Lohani, Gopal Krishan, & Ajay Krishna Prabhakar (Eds.)

and Jamui have the greatest reduction of rainfall from the last 110 years. This study provided very interesting and informative results to take immediate initiatives on water conservation measures and irrigation practices to mitigate the rainfall variability on agricultural yield. This study will be very helpful for policymakers, agriculturists, and framers to take a decision on water management, crop scheduling, and patterns over this region.

9.1 INTRODUCTION

Climate change is a regular process that results a new weather pattern in which rainfall is a very important factor that determines the availability of water on the surface of the earth. During the last few decades, there is a large variation in rainfall patterns leading to the formation of floods and drought due to which there is a scarcity of availability of food and fresh water. Changes in frequency and quantity of rainfall may vary the stream flows pattern and especially agricultural demand, the temporal and spatial distribution of runoff, groundwater, and soil moisture.

Rainfall, being the prime source of water on the surface of the Earth, studies of rainfall is very essential for the deeper understanding of the hydrological cycle and its importance for the existence of all lives on the earth. Global warming directly affects the rainfall quantity, intensity, and frequency; hence, trend analysis plays crucial role in identifying long-term changes in rainfall at various temporal and spatial scales. IPCC (2007) reported global surface temperature variation was 0.74 ± 0.18°C from 1906 to 2005, glaciers, freshwater availability, and living being may get affected by variation of future climate change. The most susceptible (climate change) area is agriculture as it is highly dependent on rainfall and temperature variation (Philip et al., 2014). In developing countries like India, agriculture is the main source of income and any noticeable changes in the monsoon rainfall (pattern, intensity, and frequency) directly affect the livelihood of the major population (Kumar et al., 2010).

For assessing the spatial and temporal variation of annual and seasonal rainfall trend over India, several studies were conducted (Basistha et al., 2009; Kumar et al., 2010; Joshi and Pandey, 2011; Duhan and Pandey, 2013; Mondal et al., 2014; Pingale et al., 2014; Kundu et al., 2015; Gajbhiye et al., 2016; Sanikhani et al., 2018; Namitha et al., 2019). There have been attempts also made to study the rainfall trend of Bihar, Warwade et al. (2018) analyzed spatiotemporal variation of annual and seasonal rainfall over Bihar for the period between 1901 and 2002. They found that annual and monsoon rainfall trends decreased significantly over the area and no significant trends in pre-monsoon, post-monsoon, and winter rainfall. The average decline in rainfall rate was −2.17 mm/year and −2.13 mm/year for annual and monsoon periods.

Trend analysis of a time series is done to check its statistical significance and get the magnitude of the trend. Many parametric and non-parametric methods can be applied for long-term rainfall trend analysis. In the case of a parametric test, the data should be followed to a particular distribution. While in a non-parametric trend analysis, the time series data used should be independent and allow the outliers present in the data (Lanzante, 1996). Therefore, in the present study various non-parametric statistical techniques namely the Mann Kendall (MK)/Modified Mann–Kendall (MMK) test (Hamed and Rao, 1998) was used to analyze the statistical significance of the trend and Sen's slope estimator (Sen, 1968), was used to determine the magnitude of trend in the rainfall series. However, to support the result of MK/MMK test, the Innovative Trend Method (ITM) was also applied ITM. The newly developed nonparametric ITM method (Sen, 2011) is very popular in recent years for the identification of trend in time series. ITM compares the data points in a Cartesian coordinate system with help of 45° line as a benchmark (1:1 line). It is an effortless method free of preliminary tests like autocorrelation, non-normality, and large data requirement, and it is flexible as per data length. Due to its simplicity and flexibility, attracted many researcher's attention to analyze the time series (Tabari and Willems, 2015; Demir and Kisi, 2016; Dabanl et al., 2016; Tosunoglu and Kisi, 2017; Wu and Qian, 2017; Öztopal and Sen, 2017; Sanikahani et al., 2018).

All the studies, as mentioned above, dealt with the trend analysis of rainfall in different parts of India as well as the globe. However, information and analysis of the trend of rainfall both in temporal and spatial scales on a seasonal and annual basis and by ITM for southern Bihar is not available in the literature. Bihar has been divided into two unequal parts by the Ganga River; that is, the Northern part and the Southern part. It is generally seen that the Northern part is affected by flood whereas the Southern part is affected by drought. Many of studies have been done in the Northern part due to the flood of the Kosi River, but there is a lack of study in the Southern part of Bihar. Therefore, the present study has been carried with the main objectives as: (1) to detect the magnitude and direction of rainfall trend of Southern Bihar over a period of 110 years (1901–2010). (2) to identify the spatial distribution of rainfall trend over the study area and (3) to identify the possible trends of rainfall time series of Southern Bihar using recently developed ITM.

9.2 MATERIALS AND METHODS

9.2.1 *STUDY AREA*

The study area is the southern part of Bihar state in India with an area of 41,255 km^2 and lies between 83°19′50″ E to 88°17′40″ E and 24°20′10″N to 26°31′15″E. The location of the study area and considered stations were shown in Figure 9.1. The average rainfall in the area is 990–1300 mm. The average minimum temperature is between 7.56°C and 10.56°C and the maximum temperature of about 35°C to 42°C.

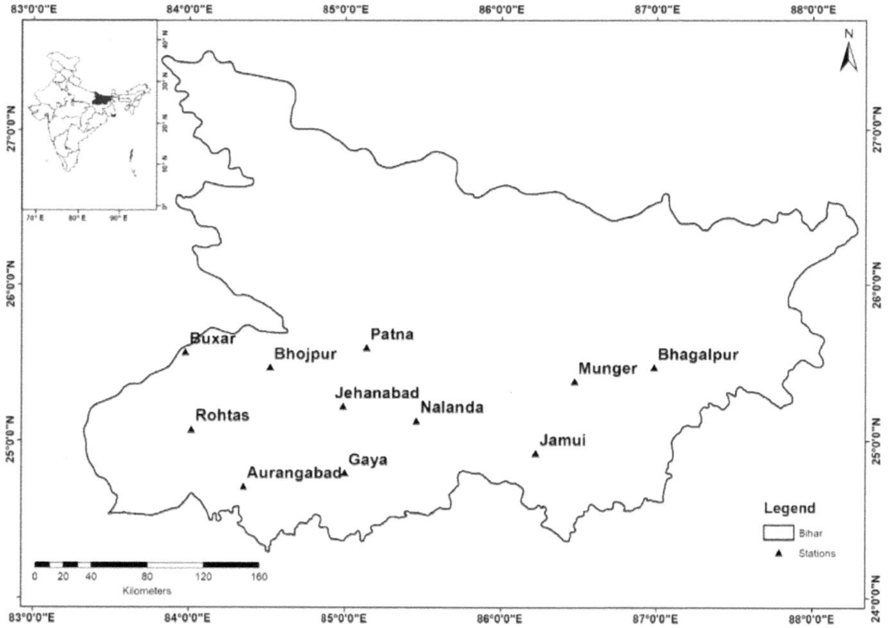

FIGURE 9.1 Location map of the study area and location of different stations.

9.2.2 *DATA USED*

Monthly rainfall data from 1901 to 2010 (110 years) for 11 stations of southern Bihar were used for the present study obtained from India Meteorological Department (IMD). The whole rainfall time series classified into four seasons as summer (March–May), monsoon (June–September), post monsoon (October and November), and winter (December–February).

9.2.3 HOMOGENEITY TEST

To assess the homogeneity of the precipitation series, the Standard Normal Homogeneity Test (SNHT) (Alexandersson, 1986; Alexandersson and Moberg, 1997) was applied at 5% significant level. The series is homogenous if the critical values of the SNHT statistic T_0 for 102 sample sizes at 95% critical levels is less than 9.17 (Khaliq and Ouarda, 2007).

9.2.4 MAGNITUDE OF THE TREND

For determining the magnitude of trend in a time series, Sen's estimator was widely used by the various researchers in a hydro-meteorological time series. In this method, the slopes (T_i) of all data pairs are first calculated by the given eq 9.1

$$T_i = \frac{x_j - x_i}{j - i} \tag{9.1}$$

where x_j and x_i are data values at the time j and i ($j > i$) respectively. The median of these values of T_i is Sen's estimator of slopes which is calculated as eq 9.2

$$\beta = \begin{cases} T_{\frac{N+1}{2}} & N \text{ is odd} \\ \frac{1}{2}\left(T_{\frac{N}{2}} + T_{\frac{N+2}{2}}\right) & N \text{ is even} \end{cases} \tag{9.2}$$

If N is Odd, the Sen's estimator is computed as $\beta = T_{(N+1)/2}$ but in case of N is even, $\beta = [T_{N/2} + T_{(N+2)/2}]/2$. Finally, β is computed by a two-sided test at 100 (1–α) % confidence interval, and then the true slope can be computed by using the non-parametric test. A positive value of β indicates an upward (increasing) trend and a negative value indicates a downward (decreasing) trend in the time series.

9.2.5 SIGNIFICANCE OF THE TREND

Nonparametric MK/MMK tests were used for trend detection in annual and seasonal rainfall series (Mann, 1945; Kendall, 1975). Various researchers applied the MK test to examine the trend in hydrometeorological series (e.g., Yu et al., 1993; Douglas et al., 2000; Yue et al., 2003; Burn et al.,

2004; Singh et al., 2008). It is essential to check the significance of autocorrelation in time series as a preliminary examination in the MK test. For auto-correlated series, Hamed and Rao (1998) provided MMK to remove the effect of autocorrelation.

9.2.6 INNOVATIVE TREND METHOD

Sen (2011) introduced ITM for observation of the trend in various hydrological and meteorological parameters (Sen, 2013; Kisi, 2015). In this method, data divided into two sub-series (first half and second half), and both the series are arranged in ascending order. 1:1 line (no trend line) was drawn by putting the first half on *X*-axis and the second half on *Y*-axis. Data accumulated on the upper region show an increasing trend, and data on the lower region as decreasing trend, while data accumulated on the 1:1 line show no trend.

9.3 RESULT AND DISCUSSIONS

9.3.1 GENERAL STATISTICS

Geographical location, mean, minimum (min), maximum (max), standard deviation (SD), coefficient of variation (CV), and Skewness Coefficient (SC) of all the stations are presented in Table 9.1. Basic statistics were analyzed for annual rainfall from 1901 to 2010. Mean min, max and SD varied from 1007.53 to 1203.68 mm, 366.30 to 765.20 mm, 1428.64 to 1961.87, and 194.06 to 240.74 mm, respectively. Coefficient of variation varied from 0.17 to 0.22 and the skewness coefficient from −0.40 to 0.40, respectively.

9.3.2 HOMOGENEITY OF ANNUAL RAINFALL SERIES

Homogeneity test results are shown in Table 9.2.

9.3.3 MONTHLY, ANNUAL AND SEASONAL TREND OF RAINFALL SERIES USING MANN–KENDALL TEST

The MMK test, Sen's slope estimator test, and ITM were used for identifying possible trends in monthly, seasonal, and annual rainfall data obtained from

11 stations in the Southern part of Bihar, India for the period 1901–2010. The Z statistics of the MMK test for the monthly data of each station at a 10% significance level are reported in Table 9.3. A significant increasing trend in the month of January and a significant decreasing trend in the month of February were observed for all stations. In March, six stations (Aurangabad, Bhagalpur, Munger, Nalanda, Patna, and Rohtas) have experienced a significant decreasing trend. In April, only 2 stations out of 11 showed a significant increasing trend. In the month of May, all stations showed a significant increasing trend except Rohtas. In the month of June, seven stations showed a significant decreasing trend. In July, nine stations showed a significant decreasing trend while two stations showed a significant increasing trend. Ten stations out of 11 showed a significant decreasing trend in the month of August. In the month of September Buxar, Nalanda and Patna showed a significant decreasing trend and other stations did not show any significant trend. In October, Bhagalpur, Jamui, and Munger showed a significant increasing trend whereas Bhojpur showed a significant decreasing trend. In month of November, nine stations showed a significant increasing trend except Gaya and Rohtas that did not show any significant trend. Five stations showed a significant increasing trend in the month of December, and rest stations did not show any significant trend.

TABLE 9.1 Rainfall (1901–2010) Statistics of Various Stations Considered for Present Study.

Stations	Latitude °N	Longitude °E	Mean (mm)	Minimum (mm)	Maximum (mm)	Standard deviation	CV	SC
Aurangabad	24.7	84.3	1083.29	492.62	1691.99	240.74	0.22	−0.13
Bhagalpur	25.5	86.9	1200.64	765.20	1664.64	200.04	0.17	0.09
Bhojpur	25.5	84.5	1022.40	439.89	1464.84	211.54	0.21	−0.6
Buxar	25.6	83.9	1007.53	460.50	1428.64	194.06	0.19	−0.40
Gaya	24.8	85.0	1125.79	551.42	1771.52	238.16	0.21	0.03
Jamui	24.9	86.2	1203.68	749.92	1892.44	216.55	0.18	0.38
Jehanabad	25.2	84.9	1065.751	537.70	1548.48	207.99	0.20	−0.22
Munger	25.3	86.5	1166.36	695.00	1961.87	229.09	0.20	0.40
Nalanda	25.1	85.4	1083.85	366.30	1655.38	223.45	0.21	−0.29
Patna	25.6	85.1	1092.04	589.92	1761.20	213.02	0.20	0.30
Rohtas	25.1	84.0	1053.61	500.95	1599.00	207.86	0.20	−0.15

TABLE 9.2 Results of Standard Normal Homogeneity Test (SNHT).

S. no.	Stations name	SNHT statistics (T_o)	Trend
1	Aurangabad	7.14	H_o
2	Bhagalpur	20.72	H_a
3	Bhojpur	7.68	H_o
4	Buxar	6.08	H_o
5	Gaya	5.78	H_o
6	Jamui	8.48	H_o
7	Jehanabad	8.27	H_o
8	Munger	5.80	H_o
9	Nalanda	9.09	H_o
10	Patna	19.63	H_o
11	Rohtas	7.06	H_o

Series is homogeneous at 5% significant level, if T_o is less than 9.214, H_0 Homogenous and H_a Heterogeneous

A summary of the MMK test results is provided in Table 9.4 for the annual and seasonal rainfall data of 11 stations. At the annual and monsoon season time scale, all stations showed a significant decreasing trend at a significance level of 10%. In the summer season, six stations (Bhagalpur, Bhojpur, Buxar, Gaya, Patna, and Rohtas) showed a significant increasing trend. In the winter season, four stations showed a significant decreasing trend and one station (Jamui) showed a significant increasing trend. In post-monsoon season, a significant increasing trend was observed in three stations and a significant decreasing trend in one station.

9.3.4 MAGNITUDE OF THE MONTHLY, ANNUAL, AND SEASONAL TREND SLOPE

The magnitude of the annual and seasonal rainfall trends for the study period was calculated from Sen's slope estimator and presented in Table 9.5. Annual and monsoon rainfall clearly showed a decreasing rate of rainfall in all stations, while in summer all stations showed a rising rate of rainfall except Aurangabad. The maximum decrease in the magnitude of rainfall was observed for annual rainfall (−2.861 mm/year) and monsoon rainfall showed (−2.792 mm/year) decrease in rainfall, while the maximum decrease of winter rainfall was (−0.085 mm/year). The maximum increase in post

TABLE 9.3 The *Z* Statistic Values of Monthly Rainfall Using MK/MMK for Study Region (1901–2010).

Stations	Jan	Feb	Mar	Apr	May	Jun	Jul	Aug	Sep	Oct	Nov	Dec
Aurangabad	**7.873**	**−2.137**	**−2.676**	0.071	**3.832**	−2.122	**−2.820**	**−4.915**	−1.303	−0.920	**3.444**	**1.758**
Bhagalpur	**2.353**	**−4.441**	**−2.729**	−0.103	**3.377**	**−11.67**	**−3.515**	**−5.782**	−1.142	**5.083**	**3.633**	0.984
Bhojpur	**7.801**	**−1.854**	−0.264	**1.763**	**4.139**	**−3.331**	**−3.554**	**−5.785**	−1.378	**−2.583**	**4.416**	**2.821**
Buxar	**9.505**	**−2.161**	−0.162	**3.047**	**3.591**	**−3.616**	**−3.520**	**−5.194**	**−2.311**	−1.558	**4.470**	**2.230**
Gaya	**6.548**	**−3.424**	**−3.252**	1.157	**4.160**	−1.619	**2.860**	**−5.555**	−1.244	−0.368	1.026	−0.290
Jamui	**3.732**	**−1.985**	−0.538	1.915	**3.730**	**−6.376**	**−4.118**	**−5.068**	1.004	**3.416**	**7.621**	**1.773**
Jehanabad	**7.800**	**−2.041**	−1.145	1.638	**4.330**	**−3.049**	**3.034**	**5.328**	−1.081	1.374	**3.263**	**1.696**
Munger	**3.559**	**−`1.733**	**−2.179**	1.639	**3.124**	**−9.380**	**−2.611**	**−4.453**	−1.097	**3.884**	**4.150**	1.257
Nalanda	**5.548**	**−2.673**	**−2.010**	−0.376	**3.563**	0.541	**−3.290**	**−6.085**	**−2.231**	1.552	**4.041**	1.449
Patna	**5.659**	**−3.159**	**−5.120**	1.116	**3.259**	**−5.124**	**−2.729**	**−5.492**	**−1.791**	−0.810	**1.768**	0.706
Rohtas	**4.812**	**−4.372**	**−2.443**	1.339	1.540	−0.852	**−3.136**	**−4.712**	−1.363	−0.131	−0.513	−0.318

Numbers in bold indicate significant values at the 10% level ($Z_{10\%} = \pm 1.645$).

monsoon rainfall was observed as (0.225 mm/year), and summer rainfall was increased by (0.138 mm/year) during the study period. The central part of the study area has experienced a greater decreasing rate in annual rainfall than the eastern and western parts (Fig. 9.2). Figure 9.2 illustrates significant and non-significant increasing and decreasing trend with different symbols. An innovative trend plot of rainfall data has been illustrated in Figure 9.3. The monsoon season also almost showed the same decreasing pattern over the area as annual rainfall, and the condition is not good as it contributes to maximum rainfall. For the summer season, the south-western part of the study area showed a minimum decreasing rate and the maximum increase in the magnitude of rainfall was observed in the eastern part; however, the eastern part experienced maximum decrease in winter rainfall. The post-monsoon seasons showed a maximum decreasing slope in the western part and an increasing rate in the eastern part.

TABLE 9.4 Values of *Z* Statistics for Seasonal and Annual Rainfall Using MK/MMK for Study Region.

Stations	Annual	Summer	Winter	Monsoon	Post-monsoon
Aurangabad	**−3.693**	−0.983	−0.421	**−3.494**	−1.084
Bhagalpur	**−4.480**	**2.472**	**−3.372**	**−4.903**	**5.022**
Bhojpur	**−3.694**	**1.873**	−0.225	**−3.511**	**−2.193**
Buxar	**−3.740**	**1.786**	0.793	**−3.521**	−0.692
Gaya	**−4.084**	**1.680**	**−1.754**	**−3.765**	−0.536
Jamui	**−4.805**	1.621	**2.861**	**−4.921**	**3.378**
Jehanabad	**−4.058**	1.103	−0.025	**−3.866**	−0.698
Munger	**−4.085**	1.118	**−3.167**	**−4.691**	**3.627**
Nalanda	**−4.019**	0.094	−1.163	**−4.178**	1.353
Patna	**−4.369**	**2.703**	1.254	**−4.696**	−1.183
Rohtas	**−3.369**	**1.884**	**−2.395**	**−3.281**	−0.861

Numbers in bold indicate significant values at the 10% level ($Z_{10\%} = \pm 1.645$).

Sen's slope value of monthly rainfall of all the stations is shown in Table 9.6. In January maximum increase rate was found to be (0.118 mm/year) while in February and March maximum decrease rate was found to be (−0.109 mm/year). Again, the increasing magnitude of trends was observed for months of April and May with a maximum increasing rate as (0.168 mm/year). The maximum decreasing trend slope for the month of June and July was found to be (−0.962 mm/year) and (−0.957 mm/year) respectively. The maximum decrease in magnitude for September rainfall was observed as (−0.339 mm/year). In the month of October, maximum increasing rate was (0.223 mm/year). Most of the stations have no trend in rainfall for November and December months.

TABLE 9.5 Values of Sen's Slope for Seasonal and Annual Rainfall (1901–2010).

Stations	Annual	Summer	Winter	Monsoon	Post-monsoon
Aurangabad	−2.135	−0.027	−0.007	−2.003	−0.048
Bhagalpur	−1.779	0.138	−0.085	−1.936	0.225
Bhojpur	−2.389	0.050	−0.006	−2.269	−0.111
Buxar	−2.209	0.041	0.015	−2.132	−0.026
Gaya	−2.250	0.051	−0.040	−2.180	−0.028
Jamui	−2.516	0.048	0.074	−2.470	0.220
Jehanabad	−2.137	0.029	0	−2.092	−0.029
Munger	−2.134	0.037	−0.071	−2.288	0.150
Nalanda	−2.861	0.004	−0.036	−2.792	0.056
Patna	−2.149	0.112	0.033	−1.952	−0.064
Rohtas	−1.769	0.054	−0.078	−1.71	−0.026

Sharma and Singh (2017) found decreasing rainfall trends in winter, monsoon, and annual rainfall series during the time period of 102 years over a neighboring state of Bihar, that is; Jharkhand while an increasing trend in pre-monsoon and post-monsoon rainfall. Warwade et al. (2018) analyzed spatiotemporal variation of annual and seasonal rainfall over Bihar for the period between 1901 and 2002. They also found that annual and monsoon rainfall trends decreased significantly over the state; however, no significant trends were observed in pre-monsoon, post-monsoon, and winter rainfall.

The result of ITM test at the 10% level of significance is presented in Table 9.7, for monthly, seasonal and annual rainfall of Bhagalpur station (as an example). If the slope values are lies between upper and lower confidence limits (CL), indicates an insignificant trend while slope values out of range of both the limits indicate a significant trend.

9.3.5 MONTHLY, SEASONAL, AND ANNUAL TREND OF RAINFALL FOR 11 STATIONS FROM 1901–2010 USING INNOVATIVE TREND METHOD

The rainfall data generally showed a significantly decreasing trend (Jan, Feb, June, July, Aug, and Sept) and a significantly increasing trend (April,

TABLE 9.6 Change in Rate of Magnitude of Monthly Rainfall (1901–2010).

Stations	Jan	Feb	Mar	Apr	May	Jun	Jul	Aug	Sep	Oct	Nov	Dec
Aurangabad	0.118	−0.068	−0.023	0	0.037	−0.147	−0.726	−1.021	−0.170	−0.026	0	0
Bhagalpur	0.031	−0.078	−0.027	0	0.168	−0.937	−0.595	−0.551	−0.166	0.223	0.006	0
Bhojpur	0.105	−0.055	−0.002	0.010	0.050	−0.351	−0.781	−1.263	−0.172	−0.099	0	0.005
Buxar	0.119	−0.055	−0.001	0.014	0.041	−0.301	−0.774	−0.972	−0.261	−0.037	0	0.003
Gaya	0.081	−0.106	−0.032	0.008	0.072	−0.126	−0.680	−1.257	−0.175	0.019	0	0
Jamui	0.041	−0.074	−0.008	0.025	0.092	−0.595	−0.811	−0.652	−0.111	0.21	0.012	0
Jehanabad	0.106	−0.065	−0.010	0.010	0.050	−0.282	−0.643	−0.891	−0.154	0.061	0	0
Munger	0.053	−0.045	−0.024	0.020	0.096	−0.962	−0.455	−0.440	−0.141	0.143	0.008	0
Nalanda	0.085	−0.080	−0.016	0.004	0.051	−0.440	−0.957	−0.979	−0.339	0.079	0	0
Patna	0.078	−0.088	−0.023	0.007	0.092	−0.329	−0.577	−0.726	−0.214	−0.042	0	0
Rohtas	0.088	−0.109	−0.021	0.005	−0.035	−0.053	−0.692	−0.979	−0.155	−0.004	0	0

TABLE 9.7 Detailed Trend Parameters of ITM Test for the Monthly Rainfall of Bhagalpur Station (Significant at the 10% Level).

Time	Slope, s	Intercept, a	Std. dev.	Correlation	Slope std. dev.	Lower CL	Upper CL	Hypothesis	Decision	Types of trend
Jan	−0.025	16.025	14.654	0.951	0.0080	−0.0131	0.0131	Ha	Yes	Decreasing
Feb	−0.123	20.923	13.675	0.945	0.0079	−0.0130	0.0130	Ha	Yes	Decreasing
Mar	−0.053	16.483	12.489	0.989	0.0032	−0.0053	0.0053	Ho	No	No trend
Apr	0.024	15.738	14.893	0.955	0.0077	−0.0127	0.0127	Ha	Yes	Increasing
May	0.177	42.650	33.424	0.985	0.0102	−0.0167	0.0167	Ha	Yes	Increasing
Jun	−0.966	239.497	92.282	0.983	0.0298	−0.0490	0.0490	Ha	Yes	Decreasing
Jul	−0.899	351.324	93.302	0.977	0.0348	−0.0572	0.0572	Ha	Yes	Decreasing
Aug	−0.554	309.407	85.114	0.989	0.0223	−0.0368	0.0368	Ha	Yes	Decreasing
Sep	−0.145	238.109	92.758	0.981	0.0318	−0.0522	0.0522	Ha	Yes	Decreasing
Oct	0.228	70.874	60.64	0.986	0.0196	−0.0290	0.0290	Ha	Yes	Increasing
Nov	0.024	4.653	7.533	0.983	0.0024	−0.0040	0.0040	Ha	Yes	Increasing
Dec	0.025	0.912	4.201	0.971	0.0018	−0.0029	0.0029	Ha	Yes	Increasing
Annual	−2.290	1326.593	200.041	0.973	0.0811	−0.1334	0.1334	Ha	Yes	Decreasing
Summer	0.148	74.871	39.288	0.966	0.0179	−0.0294	0.0294	Ha	Yes	Increasing
Winter	−0.126	37.860	19.836	0.989	0.0052	−0.0085	0.0085	Ha	Yes	Decreasing
Monsoon	−2.263	1138.282	178.819	0.981	0.0605	−0.0995	0.0995	Ha	Yes	Decreasing
Post Monsoon	0.252	75.527	62.215	0.989	0.0157	−0.0258	0.0258	Ha	Yes	Increasing

May, Oct, Nov, and Dec), except May (no trend), at the level of 10%. Summary of the ITM results for the monthly rainfall of all stations at a different significant levels (10, 5, and 1%) is provided in Table 9.8. Unlike MK/MMK, no significant trend was observed in any of the months in all the stations. There is a slight difference between results obtained from both the methods for monthly rainfall (MMK and ITM), which may be due to the fact that ITM is based on the division of series into halves (first half and second half) and MK/MMK based on the whole series. However, both significant and insignificant trends were observed for annual and seasonal rainfall presented in Table 9.9. A significant negative trend was prominently found in all the stations in both annual and monsoon rainfall whereas an insignificant negative trend was observed in winter rainfall. The summer season showed an insignificant increasing trend in all the stations, while the mixed trend recorded in post-monsoon rainfall. ITM provides a clear picture of the trend by a graphical presentation. From the graph also it can be seen that rainfall data of most of the stations are accumulated in the lower triangular region thereby clearly indicating decreasing trend over the time period.

Agricultural production, water resource availability, urban water supply, water uses in industrial, residential, and agricultural purposes are directly affected by rainfall. From the present study, it was observed that most of the considered stations have a significant decreasing trend in annual, monsoon, and winter season (few stations) and an increasing trend in summer (all) and post-monsoon season (very few). About 80% of the rainfall is contributed by the monsoon season and this reduction of rainfall in such a manner may imbalance the water availability on the earth. This imbalance of water causes greater extraction of groundwater for irrigating crops and decreasing of groundwater levels, facing drought problems, and reducing soil moisture (Mirabbasi and Eslamian, 2010).

Availability of water at the right time, right place, and right amount are very essential for crop growth to maximize the yield. Summer season indicates an increase in rainfall from the last 110 years, and this increase in rainfall can be utilized through the suitable design of conservation measures and may be one of the supplements of monsoon rainfall reduction. It can also help in fulfilling the moisture requirement for land preparation and tillage practices for the sowing of crops. Some of the areas showed an increase in post-monsoon rainfall, which may lead to the destruction of major Kharif crops, as most of the crops are at the harvesting stage during that time. This may affect the agricultural contribution to the Indian economy. Therefore,

TABLE 9.8 Slope, *S*, Values of Monthly Rainfall Using ITM Test for Study Region (1901–2010).

Stations	Jan	Feb	Mar	Apr	May	Jun	Jul	Aug	Sep	Oct	Nov	Dec
Aurangabad	0.121	−0.202	−0.094	0	0.058	−0.138	−0.459	−1.268	−0.081	−0.053	−0.011	0.036
Bhagalpur	−0.025	−0.126	−0.053	0.024	0.177	−0.966	−0.899	−0.554	−0.145	0.228	0.024	0.025
Bhojpur	0.108	−0.181	−0.050	0.002	0.062	−0.360	−0.733	−1.390	−0.100	−0.132	0.021	0.026
Buxar	0.117	−0.160	−0.048	0.006	0.056	−0.232	−0.797	−1.212	−0.167	−0.056	0.027	0.024
Gaya	0.120	−0.220	−0.101	0.014	0.108	−0.053	−0.524	−1.376	−0.143	−0.015	−0.018	−0.027
Jamui	0.030	−0.180	−0.067	0.022	0.086	−0.719	0.944	0.902	0.129	0.161	0.016	0.039
Jehanabad	0.111	−0.193	0.062	0.005	0.058	−0.354	−0.647	−1.139	−0.079	−0.016	0.005	0.020
Munger	0.004	−0.133	−0.060	0.017	0.083	−0.924	−0.733	−0.573	−0.095	0.189	0.028	0.020
Nalanda	0.092	−0.189	−0.063	0.004	0.057	−0.521	−0.978	−1.161	0.288	0.029	0.014	0.018
Patna	0.097	−0.194	−0.052	0.018	0.131	−0.337	−0.554	−0.942	−0.131	−0.063	0.007	0.017
Rohtas	0.104	−0.021	−0.077	0.016	0.131	−0.085	−0.545	−1.247	−0.041	−0.002	−0.017	0.021

changes in rainfall should be carefully taken into consideration for the water management in the study area as the major population of Bihar is dependent on agriculture for their livelihood.

TABLE 9.9 Slope, *S*, Values of Seasonal, and Annual Rainfall Using ITM Test for Study Region.

Stations	Annual	Summer	Winter	Monsoon	Post monsoon
Aurangabad	**−2.092**	-0.038	−0.045	**−1.945**	−0.064
Bhagalpur	**−2.290**	0.148	−0.126	**−2.563**	0.252
Bhojpur	**−2.727**	0.015	−0.046	**−2.584**	−0.112
Buxar	**−2.441**	0.0142	−0.019	**−2.41**	−0.028
Gaya	**−2.183**	0.018	−0.073	**−2.095**	0.033
Jamui	**−2.586**	0.041	−0.111	**−2.693**	0.177
Jehanabad	**−2.291**	0	−0.062	**−2.219**	−0.011
Munger	**−2.177**	0.040	−0.109	**−2.325**	0.217
Nalanda	**−2.986**	−0.003	−0.079	**−2.947**	0.043
Patna	**−2.003**	0.097	0.080	**−1.964**	−0.057
Rohtas	**−1.951**	0.070	−0.084	**−1.917**	−0.019

Number in bold indicates a significant value at the 10% significant level.

Monthly trend analysis is very helpful to manage the water requirement for different crops during different-different seasons. December (45% stations) and January indicate the increasing trend of rainfall during this study period, which may encourage the farmer to take the Rabi crops like gram, mustard, pea, etc. (having less water requirement). During July and August, the strongly negative trend of rainfall was recorded over the study area, which is the major concern for water management; these months are highly important for major production of agricultural crops. The author suggested seasonal irrigation system must be developed to utilize excess water and to mitigate drought impacts on agriculture production. Annual rainfall was decreasing from the last 110 years, monsoon rainfall has a major contribution to annual rainfall was also decreasing in the same manner as annual rainfall. This is an alarming situation for the agricultural sector, most of the agricultural practices are dependent on monsoon rainfall. This shows the adaptation of an irrigation system as a supportive agent to maintain and enhance major agricultural production.

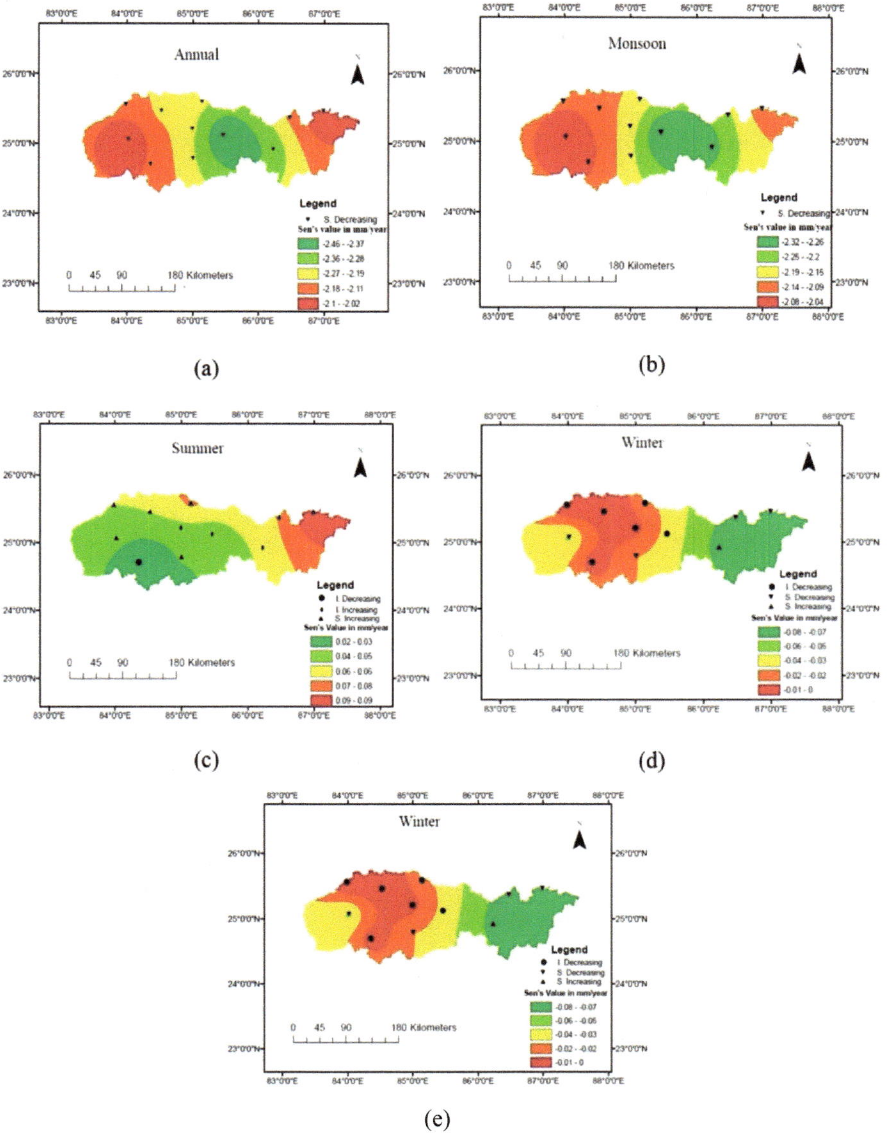

FIGURE 9.2 (a–e) Spatial distribution of the magnitude of annual and seasonal rainfall trend determined by Sen's slope test.

9.4 CONCLUSIONS

In this study, the trends of rainfall over Southern Bihar using 110-year data were evaluated by MK/MMK, Sen's slope estimator, and ITM. Similar trends

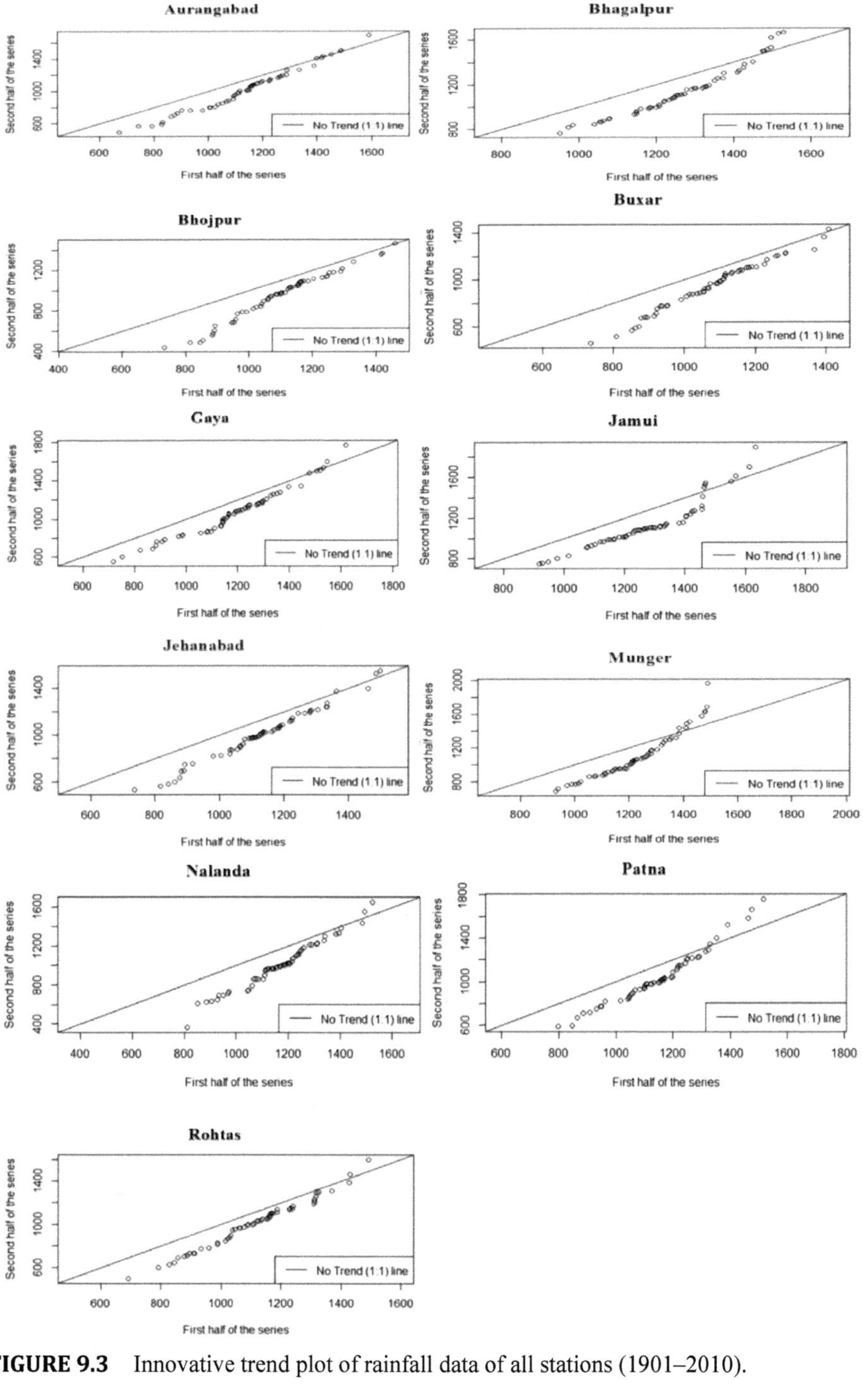

FIGURE 9.3 Innovative trend plot of rainfall data of all stations (1901–2010).

were obtained for monthly, annual, and seasonal rainfalls from both MK/MMK. All stations in the annual and monsoon season showed a significant decrease in rainfall. The maximum decrease in the magnitude of rainfall was observed for annual rainfall (−2.861 mm/year), and monsoon rainfall showed (−2.792 mm/year) decrease in rainfall. In the summer season, both significant and nonsignificant increasing trends were observed. Post-monsoon season also showed a significant increasing in some of the stations while some stations showed an insignificant decreasing trend. The winter season also showed a clear decreasing trend. Unlike ITM which showed an insignificant trend in all months, the MMK test showed a significant increasing trend in the month of January and a significant decreasing trend in month of February. Therefore, the present study of trend analysis provided researchers with very interesting and informative findings that will be very useful for irrigation engineer, climate change experts, and policymakers to take a decision for adaptation of water management practice especially agricultural purposes.

ACKNOWLEDGMENTS

The authors thank India Meteorological Department for providing climatological data.

KEYWORDS

- **trend analysis**
- **modified Mann-Kendall**
- **Sens's slope**
- **innovative trend method**

REFERENCES

Alexandersson, H. A Homogeneity Test Applied to Precipitation Data. *Int. J. Climatol.* **1986,** *6*, 661–675.

Alexandersson, H.; Moberg, A. Homogenization of Swedish Temperature Data. Part I: A Homogeneity Test for Linear Trends. *Int. J. Climatol.* **1997,** *17*, 25–34.

Basistha, A.; Arya, D. S.; Goel, N. K. Analysis of Historical Changes in Rainfall in the Indian Himalayas. *Int. J. Climatol.: J. Royal Meteorol. Soc.* **2009,** *29* (4), 555–572.

Burn, D. H.; Cunderlik, J. M.; Pietroniro, A. Hydrological Trends and Variability in the Liard River basin/Tendances hydrologique set variabilitédans le basin de la rivière Liard. *Hydrol. Sci. J.* **2004,** *49* (1), 53–67.

Dabanlı, I.; Sen, Z.; Yeleğen, M. O.; Şişman, E.; Selek, B.; Güçlü, Y. S. Trend Assessment by the Innovative-Şen Method. *Water Resour. Manage.* **2016,** *30* (14), 5193–5203.

Demir, V.; Kisi, O. Comparison of Mann-Kendall and Innovative Trend Method (Şen trend) for Monthly Total Precipitation (Middle Black Sea Region, Turkey). In *3rd International Balkans Conference on Challenges of Civil Engineering, 3-BCCCE*; Epoka University, Tirana, Albania, May 19–21, 2016; pp 344–351.

Douglas, E. M.; Niyogi, D.; Frolking, S.; Yeluripati, J. B.; PielkeSr, R. A.; Niyogi, N.; Mohanty, U. C. Changes in Moisture and Energy Fluxes due to Agricultural Land Use and Irrigation in the Indian Monsoon Belt. *Geophys. Res. Lett.* **2006,** *33* (14), 1–5.

Gajbhiye, S.; Meshram, C.; Mirabbasi, R.; Sharma S. K. Trend Analysis of Rainfall Time Series for Sindh River Basin in India. *Theor. Appl. Climatol.* **2016,** *125* (3–4), 593–608.

Hamed, K. H.; Rao, A. R. A Modified Mann-Kendall Trend Test for Autocorrelated Data. *J. Hydrol.* **1998,** *204* (1), 182–196.

Joshi, M. K.; Pandey, A. C. Trend and Spectral Analysis of Rainfall over India during 1901–2000. *J. Geophys. Res.: Atmospheres* **2011,** *116* (D6), 1–13.

Khaliq, M. N.; Ouarda, T. B. M. J. Short Communication on the Critical Values of the Standard Normal Homogeneity Test (SNHT). *Int. J. Climatol.* **2007,** *27*, 681–687.

Kumar, V.; Jain, S. K.; Singh, Y. Analysis of Long-Term Rainfall Trends in India. *Hydrol. Sci. J.* **2010,** *55* (4), 484–496.

Kundu, A.; Chatterjee, S.; Dutta, D.; Siddiqui, A. R. Meteorological Trend Analysis in Western Rajasthan (India) Using Geographical Information System and Statistical Techniques. *J. Environ. Earth Sci.* **2015,** *5* (5), 90–99.

Lanzante, J. R. Resistant, robust and non-parametric techniques for the analysis of climate data: Theory and examples, including applications to historical radiosonde station data. *Int. J. Climatol.: J. Royal Meteorol. Soc.* **1996,** *16* (11), 1197–1226.

Mondal, A.; Khare, D.; Kundu, S. Spatial and Temporal Analysis of Rainfall and Temperature Trend of India. *Theor. Appl. Climatol.* **2015,** *122* (1–2), 143–158.

Namitha, M. R.; Vinothkumar, K. Derivation of the Intensity –Duration –Frequency Curve for Annual Maxima Rainfall Using Generalized Extreme Value Distribution. *Int. J. Curr. Microbiol. Appl. Sci.* **2019,** *8* (1), 2626–2632.

Öztopal, A.; Şen, Z. Innovative Trend Methodology Applications to Precipitation Records in Turkey. *Water Resour. Manage.* **2017,** *31* (3), 727–737.

Philip, A.; Augistine, Y.; Abindaw, B. Impact of Climate Variability on Smallholder Households and Indigenous Coping Strategies in Bonga District. *Int. J. Develop. Res.* **2014,** *4* (3), 693–699.

Pingale, S. M.; Khare, D.; Jat, M. K.; Adamowski, J. Spatial and Temporal Trends of Mean and Extreme Rainfall and Temperature for the 33 Urban Centers of the Arid and Semi-Arid State of Rajasthan, India. *Atmos. Res.* **2014,** *138*, 73–90.

Sanikhani, H.; Kisi, O.; Mirabbasi, R.; Meshram, S. G. Trend Analysis of Rainfall Pattern over the Central India during 1901–2010. *Arab. J. Geosci.* **2018,** 437, 1–14.

Sen, P. K. Estimates of the Regression Coefficient Based on Kendall's Tau. *J. Am. Stat. Assoc.* **1968,** 324, 1379–1389.

Şen, Z. Innovative Trend Analysis Methodology. *J. Hydrol. Eng.* **2011,** *17* (9), 1042–1046.

Singh, P.; Kumar, V.; Thomas, T.; Arora, M. Basin-wide Assessment of Temperature Trends in Northwest and Central India/Estimation par bassin versant de tendances de température au nord-ouestet au centre de l'Inde. *Hydrol. Sci. J.* **2008,** *53* (2), 421–433.

Srivastava, P. K.; Han, D.; Rico Ramirez, M. A.; Islam T. Sensitivity and Uncertainty Analysis of Mesoscale Model Downscaled Hydro-Meteorological Variables for Discharge Prediction. *Hydrol. Proces.* **2014,** *28* (15), 4419–4432.

Tabari, H.; Willems, P. Investigation of Streamflow Variation Using an Innovative Trend Analysis Approach in Northwest Iran. In *The 36th IAHR World Congress*; The Hague, the Netherlands, June 28–July 3, 2015.

Tosunoglu, F.; Kisi, O. Trend Analysis of Maximum Hydrologic Drought Variables Using Mann–Kendall and Şen's Innovative Trend Method. *River Res. Appl.* **2017,** *33* (4), 597–610.

Warwade, P.; Tiwari, S.; Ranjan, S.; Chandniha, S. K.; Adamowski, J. Spatio-temporal Variation of Rainfall over Bihar State, India. *J. Water Land Develop.* **2018,** *36* (1), 183–197.

Wu, H.; Qian, H. Innovative Trend Analysis of Annual and Seasonal Rainfall and Extreme Values in Shaanxi, China, since the 1950s. *Int. J. Climatol.* **2017,** *37* (5), 2582–2592.

Yu, Y. S.; Zou, S.; Whittemore, D. Non-parametric Trend Analysis of Water Quality Data of Rivers in Kansas. *J. Hydrol.* **1993,** *150* (1), 61–80.

Yue, S.; Wang, C. The Mann-Kendall Test Modified by Effective Sample Size to Detect Trend in Serially Correlated Hydrological Series. *Water Resour. Manage.* **2004,** *18* (3), 201–218.

CHAPTER 10

Groundwater Flow Modeling of a Hypothetical Groundwater System

SUMANT KUMAR*, RAVI K. SAINI, and VINOD KUMAR

Groundwater Hydrology Division, National Institute of Hydrology, Roorkee, India

Corresponding author. E-mail: sumantks@gmail.com

ABSTRACT

Groundwater modeling has proved to be a powerful tool for understanding the behavior of the groundwater system. Generally, a groundwater model is a conceptual description or mathematical approximations that describe the physical system. Groundwater models are applied in groundwater resource management for sustainable development. It is important to understand general aspects of the groundwater flow models so that the application or the evaluation of these models are performed correctly. In this case, the governing equation, initial and boundary conditions, solving method, and modeling procedures are elaborated. The modeling procedures include constructing the conceptual model, designing the numerical model, providing inputs to the model, calibration and validation of the model. For illustrating the concept of this groundwater model, a simple hypothetical case study is presented. The hypothetical case is formulated by assuming a confined aquifer that is bound by two parallel rivers in the east and west. The north and south sides of the model domain are considered as an impermeable boundary. The first case was when there was no pumping and hence the model was run to simulate natural groundwater flow. It was observed that the groundwater level decreases approximately 0.5 m/200 m from the

Advances in Hydrology and Climate Change: Historical Trends and New Approaches in Water Resources Management. Surendra Kumar Chandniha, Anil Kumar Lohani, Gopal Krishan, & Ajay Krishna Prabhakar (Eds.)

west river and the inflow from the river was 1728.46 m^3/day. In the second case, where pumping was considered, the maximum drop of groundwater level was 2.7 m in the center of the cone of depression and inflow from the left river increased to 2241 m^3/day. The output of the model demonstrated groundwater head distribution and the impact of groundwater pumping on the aquifer response. The software used to demonstrate the hypothetical case was Visual MODFLOW flex. The case studies helped in understanding the concept of the groundwater model which is essentially required for sustainable groundwater resources management.

10.1 INTRODUCTION

Groundwater is one of the major sources for domestic, irrigation, and industry use; however, it is not uniformly distributed in space and time. In India, about 85% of rural drinking water supply is met from groundwater sources. Groundwater requirement is increasing rapidly and approximately 60% of total irrigated area is fed from groundwater (Shah, 2007). With growing population and developmental activities, problems such as groundwater head depletion and groundwater quality deterioration have been observed in many places in the last decades (Singh and Singh, 2002). Therefore, integrated groundwater resources management must be adopted for sustainable development. Groundwater Modeling helps in formulating technically sound groundwater resources management policies by the decision makers (Jakeman et al., 2016). The groundwater model is the best tool for groundwater hydrologists to answer few questions such as: (1) How long aquifer can sustain with the present rate of withdrawal? (2) What will be the effect on groundwater system if abstraction rate/recharge changes from the present situation? (3) What is the capture zone of a water supply well? (4) What are the most likely pathways for contaminants if it leaches from solid waste disposal site or domestic waste, etc.? The groundwater model is used for understanding the behaviour of the groundwater flow system and prediction of the system's response due to forcing function (Elango, 2005). It also acts as an assessment tools for evaluating recharge and discharge, and for quantifying sustainable yield of aquifer. Typically, a groundwater flow model comprises an equation (algebraic or differential) governing the flow and numerical solution to compute the head distribution in space and time (Dou et al., 1995). Computer-based numerical groundwater models are being used to simulate groundwater flow system for various multiple scenarios of groundwater development and land uses, etc. The most popular numerical model is USGS MODFLOW (McDonald and

Harbaugh, 1988) which is being used by several researchers (Majumdar et al., 2009; Szczepiński, 2019; Sundararajan and Sankaran, 2020) for solving real-life problems related to groundwater quantity and quality. This chapter presents an overview of groundwater flow modeling concept and procedure. A hypothetical case study has been presented to display how to construct numerical groundwater flow model in MODFLOW. The methodologies and results of the hypothetical case study are discussed and summarized in the subsequent sections.

10.2 MATHEMATICAL MODEL

A mathematical model of the groundwater system is defined by a governing equation (partial differential equation), initial and boundary conditions. The governing equation is essentially an expression of continuity equation, that is, the difference between inflow and outflow rates equals the rate of change of storage in any selected domain of saturated flow (Fitts, 2002). The governing equation is derived in one dimension and then extended to three dimensions. In a typical mass balance analysis, the net flux of mass through the boundary of an element is equated to the rate of change of mass within the element. Consider the mass balance for a small rectangular element within the saturated zone as shown in Figure 10.1.

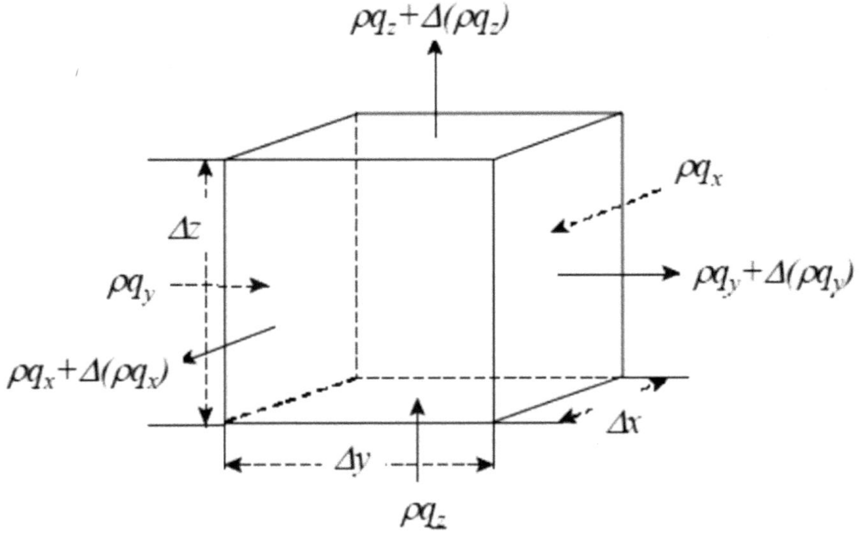

FIGURE 10.1 Elementary volume in saturated flow with dimension Δx, Δy, and Δy.

The mass flux (mass/time) in through back face of *X*-axis is given by

$$\rho_w\ (x)\ q_x(x)\Delta_y\Delta_z \tag{10.1}$$

where $\rho_w(x)$ is the water density at coordinate x and $q_x(x)$ is the specific discharge at coordinate x.

The corresponding flux out through the front face of the element is

$$\rho_w\ (x+\Delta x)\ q_x(x+\Delta x)\Delta y\Delta z \tag{10.2}$$

When these two fluxes are equal, then the flow is in steady state. When they differ, it means that the flow is in transient condition and there must be a change in mass of water stored in the element. The change in the volume of water stored in the element volume V_t may be defined by eq 10.3 as per definition of specific storage.

$$dV_w = S_s dhV_t \tag{10.3}$$

For the element illustrated in Figure 10.1 and in time interval *dt*,

$$\frac{\partial V_w}{\partial t} = S_s \frac{\partial h}{\partial t}\Delta_x\Delta_y\Delta_z \tag{10.4}$$

The rate of change of mass is given by

$$\frac{\partial m}{\partial t} = \rho_w \frac{\partial V_w}{\partial t} = \rho_w S_s \frac{\partial h}{\partial t}\Delta_x\Delta_y\Delta_z \tag{10.5}$$

The rate $\frac{\partial h}{\partial t}$ is expressed as partial derivative as h is a function of two variables x and t.

As per mass balance, the mass flux in by the element minus mass flux out is the rate of change of mass stored in the element.

$$\rho_w\ (x)\ q_x(x)\Delta_y\Delta_z - \rho_w\ (x+\Delta x)\ q_x(x+\Delta x)$$
$$\Delta_y\Delta_z = \rho_w S_s \frac{\partial h}{\partial t}\Delta_x\Delta_y\Delta_z \tag{10.6}$$

Dividing eq 10.6 by Δx, Δx, and Δz, gives

$$-\left[\frac{\rho_w\ (x+\Delta x)q_x(x+\Delta x) - \rho_w\ (x)\ q_x(x)}{\Delta x}\right] = \rho_w S_s \frac{\partial h}{\partial t}\Delta_x\Delta_y\Delta_z \tag{10.7}$$

L.H.S of eq 10.7 is derivative and may be rewritten as eq 10.8

$$-\frac{\partial(\rho_w q_x)}{\partial x} = \rho_w S_s \frac{\partial h}{\partial t} \tag{10.8}$$

Expanding the derivatives of L.H.S., it becomes

$$-\rho_w \frac{\partial q_x}{\partial x} - q_x \frac{\partial \rho_w}{\partial x} = \rho_w S_s \frac{\partial h}{\partial t} \tag{10.9}$$

The second term of LHS of eq 10.9 is generally smaller than first term

$$\rho_w \frac{\partial q_x}{\partial x} \gg q_x \frac{\partial \rho_w}{\partial x} \tag{10.10}$$

Finally, the continuity equation may be written as eq 10.11

$$-\frac{\partial q_x}{\partial x} = S_s \frac{\partial h}{\partial t} \tag{10.11}$$

Substituting the q_x of the Darcy equation in eq 10.12, it gives one-dimensional groundwater saturated flow equation

$$\frac{\partial}{\partial x}\left(K_x \frac{\partial h}{\partial x}\right) = S_s \frac{\partial h}{\partial t} \tag{10.12}$$

For three-dimensional flow, the general equation may be written as eq 10.13

$$\frac{\partial}{\partial x}\left(K_x \frac{\partial h}{\partial x}\right) + \frac{\partial}{\partial y}\left(K_y \frac{\partial h}{\partial y}\right) + \frac{\partial}{\partial z}\left(K_z \frac{\partial h}{\partial z}\right) = S_s \frac{\partial h}{\partial t} \tag{10.13}$$

If the hydraulic conductivity is assumed to be homogeneous (independent of x, y and z), the general equation can be written as

$$K_x \frac{\partial^2 h}{\partial x^2} + K_y \frac{\partial^2 h}{\partial y^2} + K_z \frac{\partial^2 h}{\partial z^2} = S_s \frac{\partial h}{\partial t} \tag{10.14}$$

If the hydraulic conductivity are assumed to be homogeneous (independent of x, y, and z) and isotropic ($K_x = K_y = K_z$), the general equation can be written as

$$\frac{\partial^2 h}{\partial x^2} + \frac{\partial^2 h}{\partial y^2} + \frac{\partial^2 h}{\partial z^2} = \nabla^2 h = \frac{S_s}{K} \frac{\partial h}{\partial t} \tag{10.15}$$

If the flow is steady state ($\frac{\partial h}{\partial t} = 0$), the general equation for steady-state flow with homogeneous and isotropic K is written as

$$\nabla^2 h = 0 \tag{10.16}$$

This is a common partial differential equation known as the Laplace equation. The symbol ∇^2 is the Laplacian operator.

10.2.1 INITIAL AND BOUNDARY CONDITIONS

Initial conditions, as the name implies, refer to the spatial distribution of heads everywhere in the model domain at the beginning of a reference time. The initial condition is necessary to arrive at a unique solution of the governing differential equation. Conceptually, they can be visualized as the initial condition before the application of external forces such as pumping recharge, etc. (Franke et al., 1987).

Boundary conditions specify the head or flux of the groundwater flow at the boundary of model domain. Physical boundaries of groundwater flow systems are formed by the physical presence of an impermeable body of rock or a large body of surface water. Other boundaries form as a result of the hydrologic conditions. These invisible boundaries are called hydraulic boundaries which include groundwater divides and streamlines (Ge and Garven, 1992).

Hydrogeological boundaries are represented by three types of mathematical conditions. The first type is called specified head boundaries (Dirichlet conditions). Along these boundaries, groundwater heads are known and are specified as:

$$h|_{\Gamma_1} = h_1(x,y,z) \quad (x,y,z) \in \Gamma s_1$$

Γ_1 represents location of the first type of boundary. Rivers may be treated as specified head boundary.

Second type is called specified flow boundaries (Neumann conditions). Across these boundaries', groundwater flow is known and specified as:

$$K\frac{\partial h}{\partial n}|_{\Gamma_2} = q(x,y,z) \quad (x,y,z) \in \Gamma_2$$

Γ_2 represents the location of the second type boundary. A special case of the specified flow boundaries is no-flow boundary where flux is specified to be zero.

The third type is called head-dependent flow boundaries (Cauchy or mixed boundary conditions). Groundwater flow across these boundaries depends on the head at boundaries. These boundaries are specified as:

$$q(x,y,z)|_{\Gamma_3} = K'\frac{h-h_0}{B'} \quad (x,y,z) \in \Gamma_3$$

Γ_3 represents location of the third type of boundary. Rivers may also be treated as head dependant flow boundaries if the river bed deposits are different from aquifer medium.

10.3 FINITE DIFFERENCE METHOD

Finite difference method transforms governing differential equation into a determinate system of linear equations (Igboekwe and Achi, 2011). The first step of the modeling process is to discretize the model domain into finite number of points that are usually denoted by nodes (typically the nodes may lie at the intersections of rows and columns). Similarly, the time domain, that is, the period over which the system response is to be simulated, is discretized by a finite number of discrete times. Thus, a spatial distribution of any variable (say hydraulic conductivity) implies data comprising the values of the variable at each node. Similarly, a spatial and temporal distribution of any variable (say, piezometric head) implies data comprising nodal values of the variable at the selected discrete times

10.3.1 SPATIAL DISCRETIZATION

Grids: rectangular network consisting of gridlines in x and y directions as shown in Figure 10.2;

- Cells: rectangular blocks divided by gridlines;
- Nodes: centre points of cells;
- Δ_x: column width; Δ_y: row width;
- Δ_z: layer thickness;
- First type boundary: nodes;
- Second type boundary: grid lines.

10.3.2 FINITE DIFFERENCE APPROXIMATION

Derivative of a function is represented by the change of variables with finite interval.

10.3.2.1 TAYLOR SERIES

Expansion of $h\,(x)$ in positive x direction:

$$h(x+\Delta_x) = h(x) + \Delta x \frac{\partial h}{\partial x} + \frac{(\Delta x)^2}{2!}\frac{\partial^2 h}{\partial x^2} + \frac{(\Delta x)^3}{3!}\frac{\partial^3 h}{\partial x^3} + \ldots \tag{10.17}$$

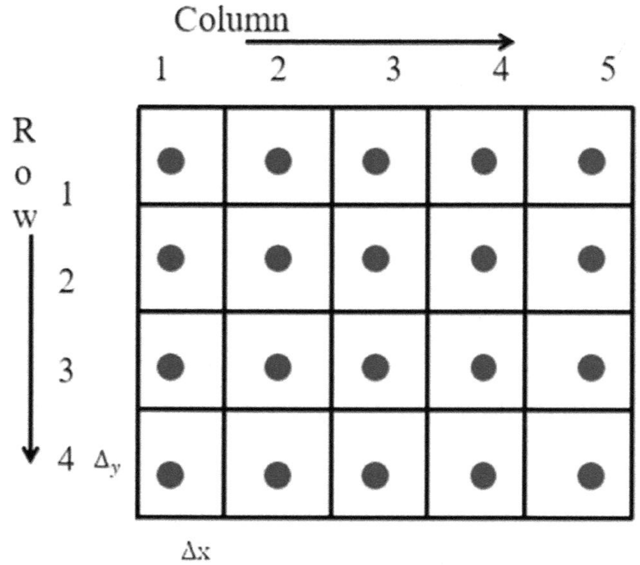

FIGURE 10.2 Finite difference discretization.

$$h(x+\Delta_x) = h(x) + \Delta x \frac{\partial h}{\partial x} + 0(\Delta x) \tag{10.18}$$

$$\frac{\partial h}{\partial x} = \frac{h(x+x) - h(x)}{\Delta x} + 0(\Delta x) \tag{10.19}$$

0 (Δx) is the first-order truncation error

Forward-difference approximation

$$\frac{\partial h}{\partial x} \approx \frac{h(x+\Delta x) - h(x)}{\Delta x} \tag{10.20}$$

Expansion of $h(x)$ in negative x direction:

$$h(x-\Delta_x) = h(x) - \Delta x \frac{\partial h}{\partial x} + \frac{(\Delta x)^2}{2!} \frac{\partial h^2}{\partial x^2} - \frac{(\Delta x)^3}{3!} \frac{\partial^3 h}{\partial x^3} + \ldots \tag{10.21}$$

Backward-difference approximation

$$\frac{\partial h}{\partial x} \approx \frac{h(x) - h(x-\Delta x)}{\Delta x} \tag{10.22}$$

Central-difference approximation

Equations 10.17–10.21 yield

$$h(x+\Delta_x)-h(x-\Delta_x)=2\Delta x\frac{\partial h}{\partial x}+2\frac{(\Delta x)^3}{3!}\frac{\partial^3 h}{\partial x^3}+\ldots \quad (10.23)$$

$$\frac{\partial h}{\partial x}=\frac{h(x+\Delta_x)-h(x-\Delta_x)}{2\Delta x}+0\left((\Delta x)^2\right) \quad (10.24)$$

$$\frac{\partial h}{\partial x}\cong\frac{h(x+\Delta x)-h(x-\Delta x)}{2\Delta x} \quad (10.25)$$

$o[(\Delta x)^2]$ is the second-order truncation error.

Finite-difference approximation of second-order derivative

Equations 10.17 and 10.21 yield

$$h(x+\Delta_x)+h(x-\Delta_x)=2h(x)+2\frac{(\Delta x)^2}{2!}\frac{\partial h^2}{\partial x^2}+o[(\Delta x)^4]\ldots \quad (10.26)$$

$$\frac{\partial^2 h}{\partial \mathrm{x}^2}\approx\frac{h(x+\Delta x)-2h(x)+h(x-\Delta x)}{(\Delta x)^2} \quad (10.27)$$

$o[(\Delta x)^4]$ is the fourth-order truncation error.

10.4 MODELING PROCESS

The modeling process involves several steps which are shown in Figure 10.3. The first step in the modeling process is to define purposes. The model needs to answer the questions: for example, is the model used for prediction, system understanding, or generic exercises, etc. before starting the modeling.

10.4.1 CONSTRUCTING CONCEPTUAL MODEL

A conceptual model is the base for model analysis. It is a representation of groundwater systems in terms of hydrogeological units, boundary conditions, hydrogeological parameters, hydrological stresses, flow patterns, etc. Field visits to the modeling area are necessary to visualize the problems and for getting impression about the area to be modeled.

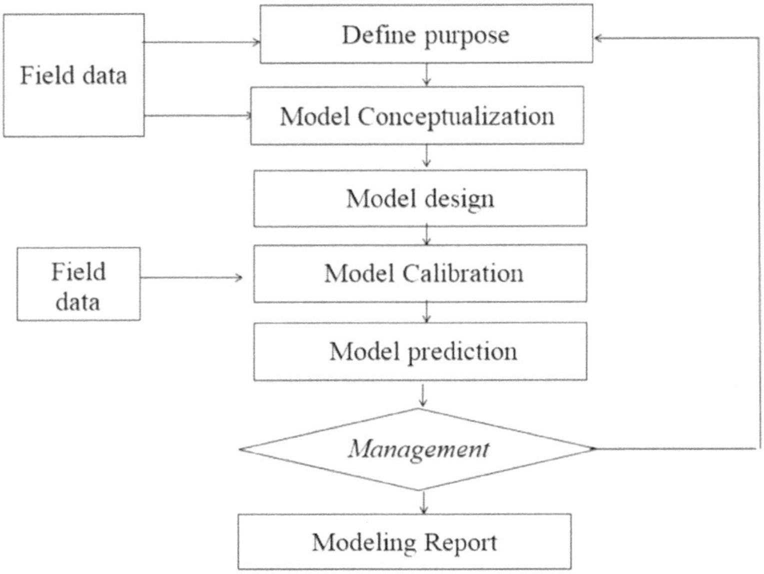

FIGURE 10.3 Flow chart of processes involved in groundwater modeling.

10.4.2 DESIGNING NUMERICAL MODEL

A numerical model is designed based on the developed conceptual model. The design of numerical model comprises the size of model grids, defining stress periods and time steps, and setting model initial and boundaries conditions. The grid size and time steps depend on the purpose of the modeling as the memory and computing time of computers/computer code may have limitations.

10.4.3 MODEL INPUTS

The inputs to the model include initial (initial heads) and boundary conditions, hydrogeological parameters such as transmissivity/conductivity and storativity/ specific yield, and hydrologic stresses. The input data have to be entered into all grid points for all stress periods.

10.4.4 CALIBRATION OF THE MODEL

The model is calibrated by tweaking the input parameters in such a way that it can reproduce the field-measured groundwater heads or concentrations.

The alteration of aquifer parameters or stresses is done by a trial-and-error method or automated parameter estimation method. Sensitivity analysis may also be carried out to evaluate the influences of uncertainties of model parameters and boundary conditions on the model results.

10.4.5 VERIFICATION OF THE MODEL

The model is verified with another set of field data (with another period of time) to check whether the calibrated model has the predictive power. The model should also be able to reproduce the field-measured values of groundwater heads or concentrations with hydrologic stresses in this process.

10.4.6 APPLICATION OF THE MODEL

The calibrated and validated model is used to predict the aquifer system response to a future scenario. In the prediction, the model is run with the calibrated aquifer parameters and future hydrologic stresses. Future hydrologic stresses may be the proposed actions such as abstraction or natural uncontrolled stresses such as recharge from precipitation.

Groundwater modeling comprises steps that are outlined above and these have to be repeated as groundwater modeling is an iterative process. Assumptions and even simplifications are required in the modeling process because of the (1) complexity involved in the hydrogeological features and (2) lack of sufficient field data. It should be kept in mind that models are only approximations of reality, but not reality itself. Therefore, groundwater modeling is not only a science but also an art. The science behind modeling can relatively be learned easily from many standard text books; however, the art of modeling can only be learned from practicing how to apply models. A successful modeler will know the science of the modeling and practice the art of modeling.

10.5 A HYPOTHETICAL GROUNDWATER MODEL

A hypothetical model is illustrated below. A confined aquifer is bound by two parallel rivers (as constant head boundaries) in east and west sides. The north and south sides of the model domain is considered as an impermeable boundary (Fig. 10.4). The water level is 15 m in the west river and 10 m in the east river. The aquifer is homogeneous with thickness of 10 m and having hydraulic conductivity of 30 m/day. These examples demonstrate (1) how groundwater head distributes in the aquifer and how much groundwater flows through the aquifer; (2) what is the impact of groundwater pumping on the response of aquifer if a pumping well is installed in the centre with capacity of 1000 m^3/day.

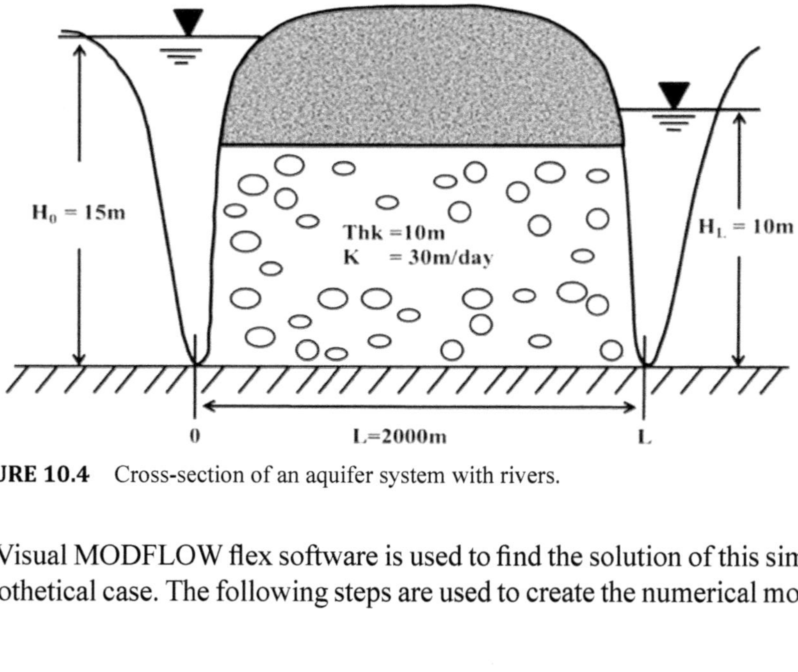

FIGURE 10.4 Cross-section of an aquifer system with rivers.

Visual MODFLOW flex software is used to find the solution of this simple hypothetical case. The following steps are used to create the numerical model.

10.5.1 MODEL GRID

In the first step, the area is divided into cells with grid lines in the *X* and *Y* directions. A uniform grid size of 50 m is taken and hence 40 rows and 40 columns are created (Fig. 10.5a). The first column and last column are represented by constant head boundaries where water levels are known.

10.5.2 AQUIFER GEOMETRY

The following properties of aquifer are defined:

Aquifer type: Confined
Top elevation of the aquifer: 10 m
Bottom elevation of the aquifer: 0 m

10.5.3 MODEL PARAMETERS

The following parameters are specified:

Time unit: day
Simulation time: 1 day
Initial head: 15 m in the first column, 10 m in the last column and 12 m in other cells
Hydraulic conductivity: 30 m/day
Evapotranspiration rate: Maxm. ET rate 0.0006 m/day, distinction depth 3 m.
Recharge rate:0.5 mm/day, ponding depth 0.1 m
Pumping rate: 1000 m^3/day

10.5.4 SIMULATION OF GROUNDWATER FLOW

The model is run to simulate natural groundwater flow. The contour map of groundwater level is shown in Figure 10.5b. It may be observed that level decreases approx. 0.5 m/200 m from the left river. The water balance shows that inflow to the system is 3628.46 m^3/day which comprises of 1728.46 m^3/day from river and recharge is 1900 m^3/day. The total outflow is also 3628.46 m^3/day (river flow 1348.45 m^3/day and ET is 2280 m^3/day) which holds true for the steady state conditions (Fig. 10.6).

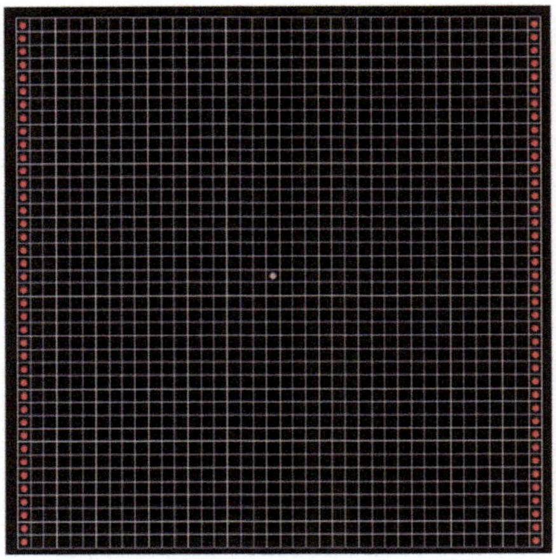

FIGURE 10.5(a) Model grid with boundary conditions.

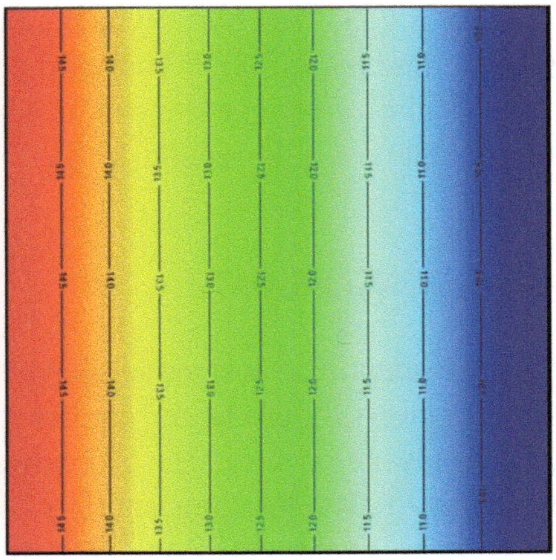

FIGURE 10.5(b) Contour map of groundwater level without pumping.

```
VOLUMETRIC BUDGET FOR ENTIRE MODEL AT END OF TIME STEP     1, STRESS PERIOD   1
----------------------------------------------------------------------------------

   CUMULATIVE VOLUMES          L**3        RATES FOR THIS TIME STEP        L**3/T
   ------------------                      ------------------------

           IN:                                      IN:
           ---                                      ---
             STORAGE =          0.0000                STORAGE =          0.0000
       CONSTANT HEAD =       1728.4630          CONSTANT HEAD =       1728.4630
                  ET =          0.0000                     ET =          0.0000
            RECHARGE =       1900.0000               RECHARGE =       1900.0000

            TOTAL IN =       3628.4629               TOTAL IN =       3628.4629

          OUT:                                     OUT:
          ----                                     ----
             STORAGE =          0.0000                STORAGE =          0.0000
       CONSTANT HEAD =       1348.4591          CONSTANT HEAD =       1348.4591
                  ET =       2280.0000                     ET =       2280.0000
            RECHARGE =          0.0000               RECHARGE =          0.0000

           TOTAL OUT =       3628.4590              TOTAL OUT =       3628.4590

            IN - OUT =      3.9062E-03               IN - OUT =      3.9062E-03

 PERCENT DISCREPANCY =            0.00    PERCENT DISCREPANCY =            0.00
```

FIGURE 10.6 Volumetric water budgeting of the model.

10.5.5 SIMULATION OF GROUNDWATER FLOW WITH IMPACT OF PUMPING

The pumping rate of 1000 m^3/day is distributed in the center of model domain area. The model again run to calculate groundwater level and budget under the pumping conditions. Figure 10.7a and b show the contour map of groundwater level and drawdowns caused by pumping and the maximum drop down of groundwater level is around 2.7 m in the center of cone of depression. The water balance shows that inflow from the left boundary is 2241.19 m^3/day and recharge is 1900 m^3/day and therefore total inflow is 4141.19 m^3/day. The outflow to the right river is 861.31 m^3/day, Evapotranspiration loss is 2279.87 m^3/day and pumping withdrawal is 1000 m^3/day and sum of all outflow components is 4141.19 m^3/day which holds true for the steady state conditions (Fig. 10.8).

10.6 CONCLUSION

Groundwater numerical models have become a commonly used tool by scientists/engineers for predicting aquifer response and estimating water budgeting of groundwater system. The groundwater models basically calculate rate and direction of flow through aquifers. A groundwater flow model was developed using the visual Modflow software under steady-state conditions by constructing a hypothetical case. The hydrogeological system consists of homogeneous aquifer with thickness of 10 m (hydraulic conductivity 30 m/

day) and bound by two rivers which represents constant head boundary. The model indicates the average inflow to the system (flow direction is from west to east) is 3628.46 m^3/day. The results also suggest that drawdown of 2.7 m occurs when pumping (rate 1000 m^3/day) is allowed in the center of model domain area and the inflow increases to 4141.19 m^3/day. The present exercise may help readers in understanding the concept of groundwater flow modeling.

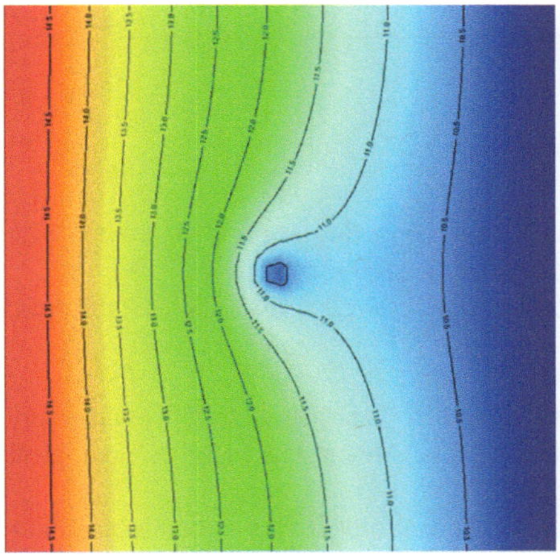

FIGURE 10.7(a) Contour map of groundwater level without pumping.

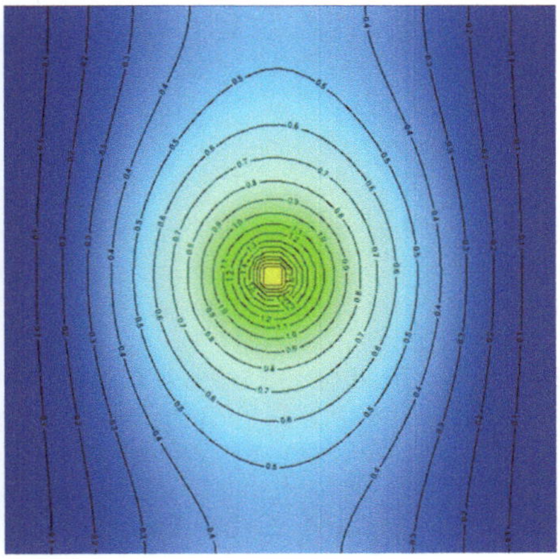

FIGURE 10.7(b) Contour map of drawdown caused by pumping.

```
VOLUMETRIC BUDGET FOR ENTIRE MODEL AT END OF TIME STEP    1, STRESS PERIOD   1
------------------------------------------------------------------------------

  CUMULATIVE VOLUMES          L**3       RATES FOR THIS TIME STEP        L**3/T
  ------------------                     ------------------------

        IN:                                      IN:
        ---                                      ---
          STORAGE =           0.0000               STORAGE =           0.0000
    CONSTANT HEAD =        2241.1909         CONSTANT HEAD =        2241.1909
            WELLS =           0.0000                 WELLS =           0.0000
               ET =           0.0000                    ET =           0.0000
         RECHARGE =        1900.0000              RECHARGE =        1900.0000

         TOTAL IN =        4141.1909              TOTAL IN =        4141.1909

       OUT:                                     OUT:
       ----                                     ----
          STORAGE =           0.0000               STORAGE =           0.0000
    CONSTANT HEAD =         861.3183         CONSTANT HEAD =         861.3183
            WELLS =        1000.0000                 WELLS =        1000.0000
               ET =        2279.8728                    ET =        2279.8728
         RECHARGE =           0.0000              RECHARGE =           0.0000

        TOTAL OUT =        4141.1914             TOTAL OUT =        4141.1914

         IN - OUT =      -4.8828E-04              IN - OUT =      -4.8828E-04

PERCENT DISCREPANCY =          -0.00     PERCENT DISCREPANCY =          -0.00
```

FIGURE 10.8 Volumetric water budget of the model with pumping.

KEYWORDS

- **groundwater**
- **modeling**
- **visual MODFLOW flex**
- **aquifer**

REFERENCES

Dou, C.; Woldt, W.; Bogardi, I.; Dahab, M. Steady State Groundwater Flow Simulation with Imprecise Parameters. *Water Resour. Res.* **1995**, *31* (11), 2709–2719.

Elango, L. *Numerical Simulation of Groundwater Flow and Solute Transport*; Allied Publishers: Chennai, 2005.

Fitts, C. R. *Groundwater Science*; Elsevier, 2002.

Franke, O. L.; Reilly, T. E.; Bennett, G. D. Definition of Boundary and Initial Conditions in the Analysis of Saturated Ground-Water Flow Systems: An Introduction. US Geological Survey, US Govt. printing office, Washington, DC, 1987.

Ge, S.; Garven, G. Hydro Mechanical Modelling of Tectonically Driven Groundwater Flow with Application to the Arkoma Foreland Basin. *J. Geophys. Res.: Solid Earth* **1992**, *97* (B6), 9119–9144.

Igboekwe, M. U.; Achi, N. J. Finite Difference Method of Modelling Groundwater Flow. *J. Water Resour. Protect.* **2011**, *3* (3), 192–198.

Jakeman, A. J.; Barreteau, O.; Hunt, R. J.; Rinaudo, J. D.; Ross, A.; Arshad, M.; Hamilton, S. Integrated Groundwater Management: An Overview of Concepts and Challenges. In *Integrated Groundwater Management*; Springer: Cham, 2016; pp 3–20.

Majumdar, P.; Ram, S.; Rao, P. Artificial Recharge in Multi Aquifers of a Mountainous Watershed. *J. Hydrol. Eng.* **2009**, *14* (3), 215–222.

McDonald, M. G.; Harbaugh, A. W. A *Modular Three-Dimensional Finite-Difference Ground-Water Flow Model*; US Geological Survey, 1988.

Shah, T. The Groundwater Economy of South Asia: An Assessment of Size, Significance and Socio-Ecological Impacts. In *The Agricultural Groundwater Revolution: Opportunities and Threats to Development*, 2007; pp 7–36.

Singh, D. K.; Singh, A. K. Groundwater Situation in India: Problems and Perspective. *Int. J. Water Resour. Develop.* **2002**, *18*, 563–580. 10.1080/0790062022000017400

Sundararajan, N.; Sankaran, S. Groundwater Modeling of Musi Basin Hyderabad, India: A Case Study. *Appl. Water Sci.* **2020**, *10* (1), 14.

Szczepiński, J. The Significance of Groundwater Flow Modeling Study for Simulation of Opencast Mine Dewatering, Flooding, and the Environmental Impact. *Water* **2019**, *11* (4), 848.

CHAPTER 11

Classification and Change Detection of an Urban Built-Up Area: A Case Study of Gurugram

OSHEEN* and M. L. KANSAL

Department of Water Resources Development & Management, Indian Institute of Technology, Roorkee 247667, Uttarakhand, India

**Corresponding author. E-mail: osheen@wr.iitr.ac.in*

ABSTRACT

Numerous built-up indices have gained wide attention for classifying urban areas using satellite imageries since the mid-1990s. In this study, three spectral indices performance to classify images and detect changes in the built-up area have been examined using Level 1C multispectral instrument Sentinel-2A imageries and compared. The following three mid-infrared-based indices are used: normalized difference built-up index (NDBI), index-based built-up index (IBI), and urban index (UI). These indices are used to classify and detect the temporal changes in urban areas in Gurugram using the image thresholding technique. For each image, 100 reference points from Google Earth were used to assess the classification. The built-up area's evaluation revealed that most of the urbanization is concentrated in the eastern part and is expanding in the central part of the Gurugram city. The total geographical area of the city is 473 km^2, out of which the built-up area for the year 2016 is estimated as 357, 350, and 349 km^2 using IBI, UI, and NDBI, which increased to 373, 365, and 369 km^2 in 2019. The increase in the built-up area from 2016 to 2019 is estimated as 4.48, 4.28, and 5.73% using IBI, UI, and

Advances in Hydrology and Climate Change: Historical Trends and New Approaches in Water Resources Management. Surendra Kumar Chandniha, Anil Kumar Lohani, Gopal Krishan, & Ajay Krishna Prabhakar (Eds.)

NDBI, respectively. The findings show the significance of built-up indices in the rapid, spatial and temporal assessment of larger city areas. Such studies would enable the urban planners and policymakers in urban cool/heat island modeling, rainfall-runoff modeling, urban sprawl model parameterization, and climate-change mitigation.

11.1 INTRODUCTION

The world is going through the hardship of urbanization. Urbanization leads to switching of natural landscapes by built-up areas (Xu, 2010). The temporal and spatial distribution of built-up area plays an essential role in the global environmental change and ecosystem services (Xi et al., 2019). The three indicators that specify urban ecological environment are density of vegetation, biodiversity, and imperviousness (Behling et al., 2015; Arnold and Gibbons, 1996; Jaeger and Schwick, 2014; Savard et al., 2000). Impervious surface is defined as a surface that does not allow water to penetrate into the soil (Weng, 2012; Arnold and Gibbons, 1996). An increased imperviousness leads to many problems such as urban flooding and; urban heat island effect (Bisht et al., 2016; Chandniha et al., 2014; Osheen and Singh, 2019). Consequently, it becomes essential to monitor the pattern of urban sprawl precisely (Masek et al., 2000; Grimm et al., 2008; Chandniha et al., 2017; Xi et al., 2019; Buo et al., 2020).

Urban remote sensing has gain prodigious attention in the recent years (Estoque and Murayama, 2013). Knowledge of the extent of urban sprawl area is necessary for urban planning, environmental management, and disaster risk assessment (Buo et al., 2020). The satellite imageries are cost-effective and time-saving tools to map the built-up areas (Xu, 2008). Many Earth observational satellites having a medium resolution of 10–30 m have become operational in the recent years. The most commonly used satellite imagery products are from Landsat satellites having a spatial resolution of 15–60 m. Therefore, presently, the increasing trend is toward the use of Sentinel imagery having a high resolution of 10–60 m (Adamu et al., 2021; Zhang et al., 2021). The Sentinel-2A MSI—Multi-Spectral Instrument has a total of 13 bands in the visible, near-infrared (NIR), and short-wave infrared (SWIR) part of the spectrum (Drusch et al., 2012). The various bands have a spatial resolution of 10 m, 20 m, and 60 m (see Table 11.1 for details).

To rapidly detect the built-up area, instead of supervised classification of satellite imagery, various researchers have introduced built-up indices. They are based on the difference in reflectance of the built-up area in various channels of the electromagnetic spectrum. The built-up index rapidly and accurately detects the built-up area in the satellite imagery. The various built-up indices determined are Urban Index (UI) in the mid-1990s by Kawamura (1996); Normalized Difference Built-up Index (NDBI) in the early 2000s by Zha et al. (2003), Index-based built-up index (IBI) in the late 2000s by Xu (2008); the normalized difference impervious surface index (NDISI) by Xu (2010) and VgNIR-BI and VrNIR-BI by Estoque and Murayama (2015). The IBI, NDBI and UI, VgNIR-BI, and VrNIR-BI use the visible NIR, and MIR VrNIR-BI and VgNIR-BI NDISI use the thermal infrared (TIR), MIR, and NIR portion of the electromagnetic spectrum. The objective of this study was to perform the comparative assessment of three built-up indices, UI, NDBI, and IBI, using the satellite imagery of Sentinel-2A to detect the change in the built-area of Gurugram for four consecutive years after 2016.

TABLE 11.1 Features of a Sentinel-2A Imagery.

Electromagnetic region	Band	Spatial resolution (m)	Central wavelength (nm)
Coastal aerosol	1	60	442.7
Blue	2	10	492.4
Green	3	10	559.8
Red	4	10	664.6
Vegetation Red Edge	5	20	704.1
Vegetation Red Edge	6	20	740.5
Vegetation Red Edge	7	20	782.8
Near-Infrared (NIR)	8	10	832.8
Narrow NIR	8a	20	864.7
Water Vapor	9	60	945.1
Short-Wave Infrared-Cirrus	10	60	1373.5
Short-Wave Infrared (SWIR)	11	20	1613.7
Short-Wave Infrared (SWIR)	12	20	2202.4

11.2 STUDY AREA

Gurugram, a city situated in the National Capital Region of India, is opted as a study area in this study. The city is located in the Gurugram district of Haryana under 28.4595° N and 77.0266° E, as shown in Figure 11.1, having an area of 473 km^2 and a population of 1.4 million (Kansal et al., 2019). The city has a subtropical continental monsoonal climate having unreliable rainfall and significant temperature variation. The average annual rainfall of Gurugram is 762 mm, out of which 77% of rainfall is from the southwest monsoon.

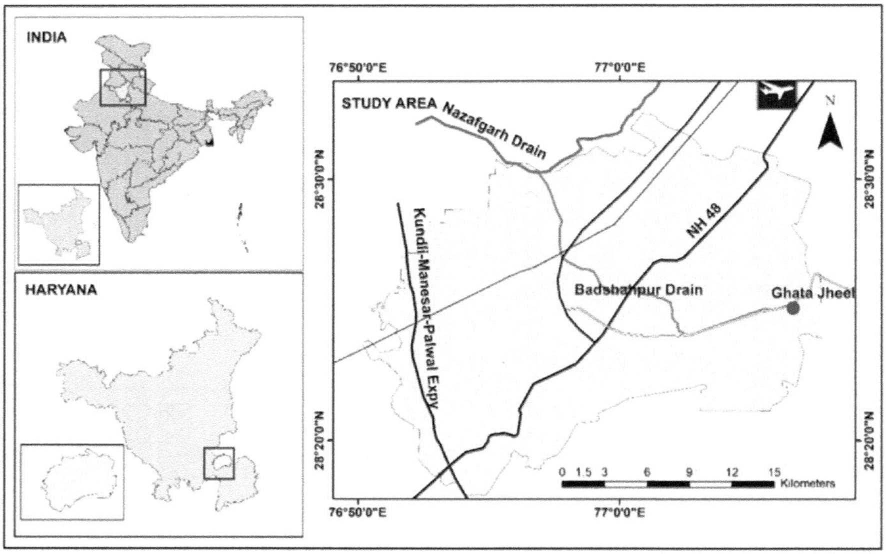

FIGURE 11.1 Location map of the study area.

11.3 DATA USED

In this study, Level 1C multispectral instrument (MSI) Sentinel-2A data has been used obtained from the cloud-based Google Earth engine platform. The datasets are filtered for the cloud pixel percentage of 10% and the date range for February for the years 2016–2019. The standard false-color composite of the study area and its surrounding is as shown in Figure 11.2. The image contains 13 bands and has been resampled to 10 m resolution in the GIS environment. Built-up areas appear in cyan to turquoise color and vegetation in red. The shrubland or wasteland appears in moss color, while the water body appears in dark blue. The black line marks the metropolitan boundary of the city.

11.4 METHODOLOGY

This study aims to determine the change in the built-up area each year from 2016 to 2019 using built-up indices. A comparative assessment of three built-up indices, that is, IBI, NDBI, and UI, is also done to identify the best suitable built-up index for the urban area. The validation of indices is done by using ground-truthing with 100 random sample points on Google Earth Pro. The step-by-step procedure followed for the work is as shown in Figure 11.3. The present work is carried out in Arc-GIS 10.4

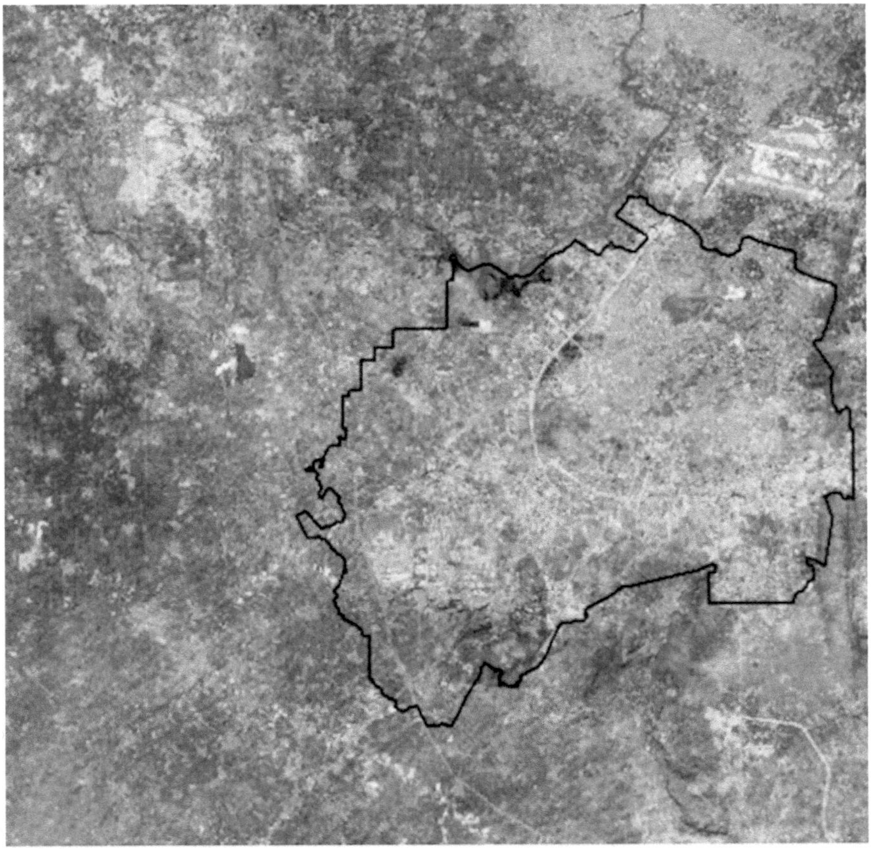

FIGURE 11.2 Sentinel-2A image of the study area: Gurugram and its neighboring areas (February 24, 2019).

11.4.1 URBAN INDEX

Kawamura (1996) in the mid-1990s proposed the UI to overcome the costly and time-consuming process of data collection to access updated information on built-up areas using satellite imageries. This index is based on the inverse relation of the brightness of urban areas in the MIR and NIR region. The UI utilizes the Near-Infrared channel (Band 4) and Mid-Infrared channel (Band 7) of Landsat TM (eq. 11.1). Kawamura (1996) validated the UI by developing the relation with NDVI and land cover data from the case area Colombo, Sri Lanka. This study uses Sentinel-2 data for calculating the UI of case area, so using the NIR and MIR bands of Sentinel-2, the formula for UI is:

$$UI = \frac{R_{SWIR2} - R_{NIR}}{R_{SWIR2} + R_{NIR}} \tag{11.1}$$

where R_{NIR} and R_{SWIR2} represent the surface reflectance values of the atmospherically corrected band 8 and band 12 of Sentinel-2A, respectively.

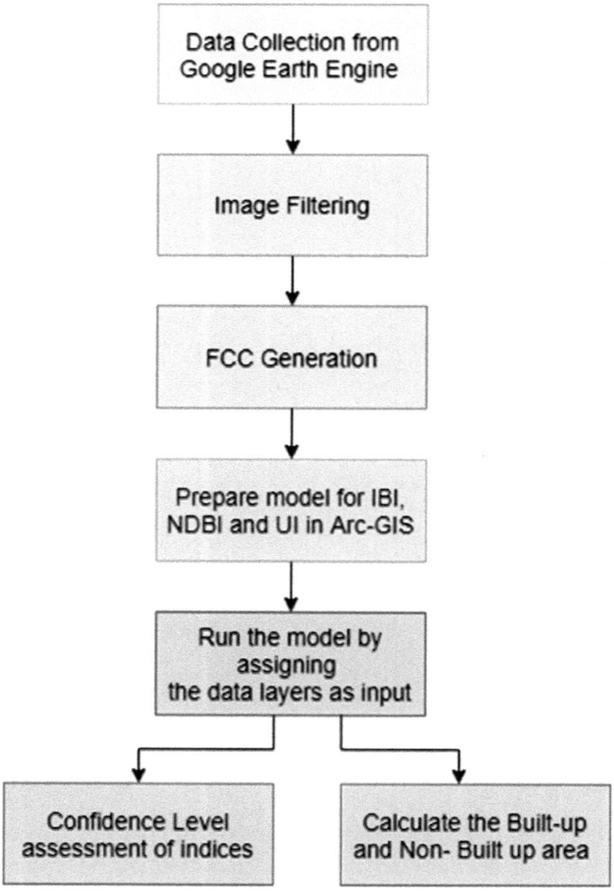

FIGURE 11.3 Methodology flowchart for the study.

11.4.2 NORMALIZED DIFFERENCE BUILT-UP INDEX

Zha et al. (2003) in the early 2000s used NDBI over Nanjing, China, to automate the method of mapping built-up areas. NDBI uses the NIR channel (Band 4) and Mid- Infrared channel (Band 5) of Landsat TM. The index is based on different spectral responses of the built-up areas in the MIR and NIR regions of the spectrum. Zha et al. (2003) assessed the accuracy of NDBI spatially, taking 68 random points from NDBI classified and ground-truthing of them with a global positioning system (GPS). Spatial validation is done by comparing the overall built-up area derived from the NDBI classified map and manual measurement of the area from the topographic map. This study uses Sentinel-2 data for calculating the UI of case area, so using the NIR and MIR bands of Sentinel-2, the formula for NDBI is:

$$NDBI = \frac{R_{SWIR} - R_{NIR}}{R_{SWIR} + R_{NIR}} \tag{11.2}$$

where R_{NIR} and R_{SWIR} represent the surface reflectance values of the atmospherically corrected band 8 and band 11 of Sentinel-2A eq 11.2. The only difference between UI and NDBI is that the NDBI utilizes the SWIR 1 band (i.e., first band) of the MIR spectrum channel, and UI uses the SWIR 2 (i.e., second band) of the MIR spectrum channel of Landsat-TM. The limitation of NDBI is its inability to differentiate between built-up areas and bare lands.

11.4.3 INDEX-BASED BUILT-UP INDEX

Xu (2008) in the late 2000s proposed the IBI to expedite the process to detect built-up areas using satellite imageries. The IBI is derived using three thematic indices, that is, the soil adjusted vegetation index (SAVI), NDBI, and modified normalized difference water index (MNDWI). The validation of IBI has been done on Fuzhou City in China using the Landsat ETM+ imagery.

$$\text{IBI} = \frac{\dfrac{2R_{\text{SWIR}}}{R_{\text{SWIR}} + R_{\text{NIR}}} - \left[\dfrac{R_{\text{NIR}}}{R_{\text{NIR}} + R_{\text{Red}}} + \dfrac{R_{\text{Green}}}{R_{\text{Green}} + R_{\text{SWIR}}}\right]}{\dfrac{2R_{\text{SWIR}}}{R_{\text{SWIR}} + R_{\text{NIR}}} + \left[\dfrac{R_{\text{NIR}}}{R_{\text{NIR}} + R_{\text{Red}}} + \dfrac{\text{R}_{\text{Green}}}{R_{\text{Green}} + R_{\text{SWIR}}}\right]} \tag{11.3}$$

where R_{Green}, R_{Red}, R_{NIR}, and R_{SWIR} are referred as the surface reflectance values of atmospherically corrected band 3, band 4, band 8, and band 11 of Sentinel-2A eq 11.3. These wavelength bands selected for the index were based on the absorption and reflection range of nonbuilt-up and built-up areas. IBI is derived to overcome the limitation of NDBI of the background noise of plants as it is observed that when the water content in vegetation decreases, its reflectance increases in the MIR channel. Built-up areas have higher reflectance values as their sensor wavelengths are longer (Herold et al., 2003). For the Sentinel 2A, the reflectance of built-up land is low, and vegetation is high in Band 8. On the contrary, the reflectance of built-up land is high, and vegetation is low in Band 12 (Herold et al., 2004; As-syakur et al., 2012).

11.5 RESULTS

The built-up indices are used to compute the built-up area in the city and detect the change in the built-up area for the four consecutive years from 2016 to 2019. Land cover maps of Gurugram classified from Sentinel 2A imagery using the three built-up indices, namely IBI, UI, and NDBI, are shown in Figure 11.4. Most of the urbanization is concentrated in the eastern part of the city, and the region shows an increasing pace of urbanization in the central part of the city. In the last 10 years, most of the agricultural land in the central part of the city is converted into built-up land. In this study, we classified the land cover into built-up and nonbuilt-up areas. The built-up class includes houses, buildings, impervious surfaces such as parking lots, roads, etc., and the non-built-up group includes grasslands, forests, croplands (including dry croplands), that is, all vegetated areas and non-impervious areas. Table 11.2 summarizes the extent of built-up and nonbuilt-up areas in the city for the years 2016–2019 and yearly percentage change in the built-up area.

The result shows that the proportion of urban built-up area for case area in 2016 varied from 73.7 to 75.4% and from 77.2 to 78.8% in 2019. Accordingly, the increase detected of the built-up area between 2016 and 2019 is observed as 4.48, 4.29, and 5.73% using IBI, UI, and NDBI, respectively.

11.5.1 CONFIDENCE LEVEL OF ASSESSMENT

The validation of the three built-up indices is done by comparing the derived maps and Google Earth imageries. For this, 100 random points sampling over the classified image of all three indices were used for the validation. The Google Earth images of February 2016–2019 were used for ground-truthing. The confidence level achieved is 84%, 76%, and 80%, using UI, NDBI, and IBI, respectively.

TABLE 11.2 The Detected Extent of Built-Up Area on the Basis of Three Spectral Indices.

	Non-Built-up area (in km^2)	Built-up area (in km^2)	% change (year-wise)
IBI			
2016	116	357	
2017	110	363	1.68
2018	104	367	1.10
2019	100	373	1.63
UI			
2016	123	350	
2017	117	356	1.71
2018	111	363	1.96
2019	108	365	0.55
NDBI			
2016	124	349	
2017	118	355	1.72
2018	112	361	1.69
2019	104	369	2.22

Further, it was observed that at some points, the bare land was misclassified into built-up lands. The incidences of nonbuilt-up areas classified as built-up were more than incidences of the built-up area classified into the nonbuilt-up. Grasslands, wastelands, and bare lands have relatively higher reflectance in the SWIR channel of the spectrum than NIR, which leads to spectral confusion in indices (Cibula et al., 1992; Gao, 1996). The results obtained from UI are comparatively better as UI uses the SWIR-2 band of the electromagnetic spectrum (Estoque and Murayama, 2015).

11.6 CONCLUSION

The continuous developments in the urban remote sensing promote the use of built-up indices for monitoring the pattern of urban sprawl in a simplified way. Detecting the built-up area using the built-up index is a cost-effective and time-saving tool in the present scenario. This study compares the performance of three built-up indices, UI, NDBI, and IBI, in classifying the built-up and nonbuilt-up area of the city using the imagery of Sentinel-2A. All the three indices use NIR, and SWIR channels of the electromagnetic spectrum as built-up areas have higher reflectance in the SWIR channel. The limitation of these indices is that they cannot differentiate the bare lands from built-up areas, which is tried to overcome by using the SWIR-2 band in the UI index. Based on the study area analysis, an average increase in the built-up area by 5.67 km^2/year is observed. Consequently, it may lead to the increased land surface temperature, environmental pollution, groundwater stress, and water scarcity condition in the city. Therefore, urban monitoring and planning are vital for the sustainable development of the city.

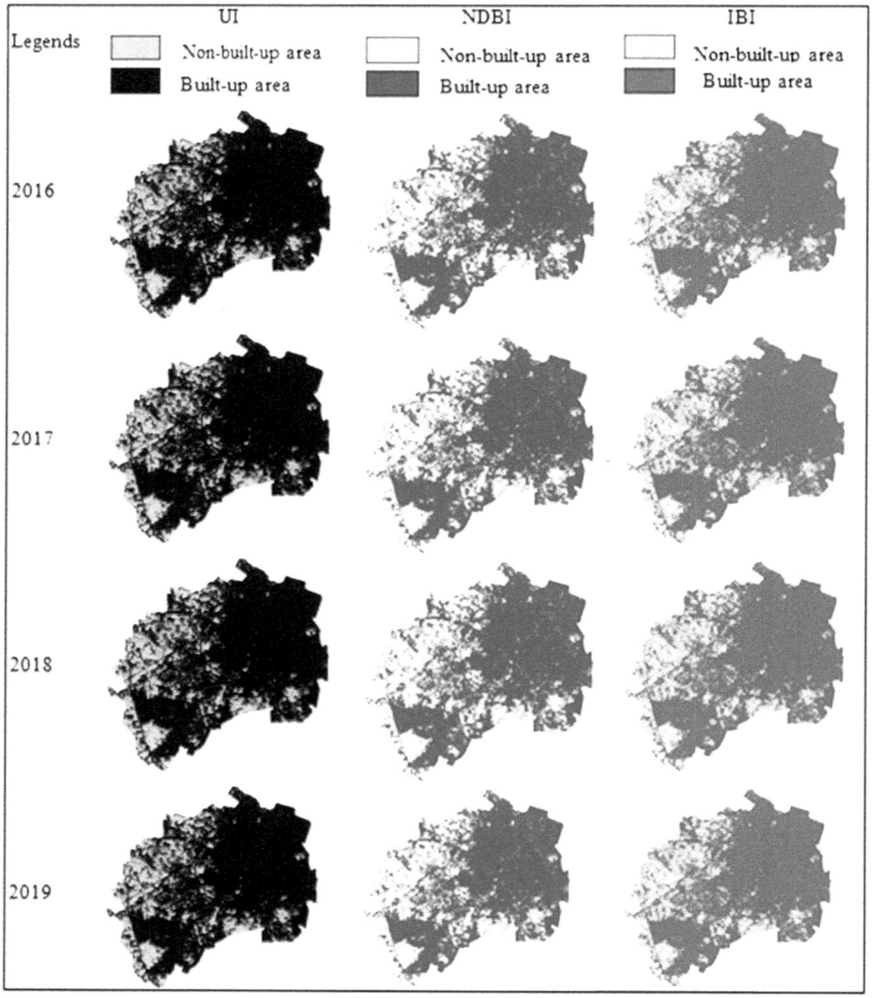

FIGURE 11.4 Land use cover maps of Gurugram using UI, NDBI, and IBI for the year 2016–2019.

KEYWORDS

- **Sentinel 2 imagery**
- **built-up index**
- **index-based built-up index**
- **normalized difference built-up index**
- **urban index**

REFERENCES

Adamu, B.; Rasul, A.; Whanda, S. J.; Headboy, P.; Muhammed, I.; Maiha, I. A. Evaluating the Accuracy of Spectral Indices from Sentinel-2 Data for Estimating Forest Biomass in Urban Areas of the Tropical Savanna. *Remote Sens. Appl.: Soc. Environ.* **2021,** *22*, 100484.

Arnold Jr, C. L.; Gibbons, C. J. Impervious Surface Coverage: The Emergence of a Key Environmental Indicator. *J. Am. Plan. Assoc.* **1996,** *62* (2), 243–258.

As-syakur, A.; Adnyana, I.; Arthana, I. W.; Nuarsa, I. W. Enhanced Built-up and Bareness Index (EBBI) for Mapping Built-up and Bare Land in an Urban Area. *Remote Sens.* **2012,** *4* (10), 2957–2970.

Behling, R.; Bochow, M.; Foerster, S.; Roessner, S.; Kaufmann, H. Automated GIS-based Derivation of Urban Ecological Indicators Using Hyperspectral Remote Sensing and Height Information. *Ecol. Indicat.* **2015,** *48*, 218–234.

Bisht, D. S.; Chatterjee, C.; Kalakoti, S.; Upadhyay, P.; Sahoo, M.; Panda, A. Modeling Urban Floods and Drainage Using SWMM and MIKE URBAN: A Case Study. *Nat. Haz.* **2016,** *84* (2), 749–776.

Buo, I.; Sagris, V.; Burdun, I.; Uuemaa, E. Estimating the Expansion of Urban Areas and Urban Heat Islands (UHI) in Ghana: A Case Study. *Nat. Haz.* **2020,** *105,* 1299–1321.

Chandniha, S. K.; Kansal, M. L.; Anvesh, G. Watershed Sustainability Index Assessment of a Watershed in Chhattisgarh, India. *Curr. World Environ.* **2014,** *9* (2), 403.

Chandniha, S. K.; Meshram, S. G.; Adamowski, J. F.; Meshram, C. Trend Analysis of Precipitation in Jharkhand State, India. *Theor. Appl. Climatol.* **2017,** *130* (1–2), 261–274.

Cibula, W. G.; Zetka, E. F.; Rickman, D. L. Response of Thematic Mapper Bands to Plant Water Stress. *Int. J. Remote Sens.* **1992,** *13* (10), 1869–1880.

Drusch, M.; Del Bello, U.; Carlier, S.; Colin, O.; Fernandez, V.; Gascon, F.; Hoersch, B.; Isola, C.; Laberinti, P.; Martimort, P.; Meygret, A. Sentinel-2: ESA's Optical High-Resolution Mission for GMES Operational Services. *Remote Sens. Environ.* **2012,** *120*, 25–36.

Estoque, R. C.; Murayama, Y. Landscape Pattern and Ecosystem Service Value Changes: Implications for Environmental Sustainability Planning for the Rapidly Urbanizing Summer Capital of the Philippines. *Landsc. Urban Plann.* **2013,** *116*, 60–72.

Estoque, R. C.; Murayama, Y. Classification and Change Detection of Built-up Lands from Landsat-7 ETM+ and Landsat-8 OLI/TIRS Imageries: A Comparative Assessment of Various Spectral Indices. *Ecol. Indicat.* **2015,** *56*, 205–217.

Gao, B. C. NDWI—A Normalized Difference Water Index for Remote Sensing of Vegetation Liquid Water from Space. *Remote Sens. Environ.* **1996,** *58* (3), 257–266.

Grimm, N. B.; Faeth, S. H.; Golubiewski, N. E.; Redman, C. L.; Wu, J.; Bai, X.; Briggs, J. M. Global Change and the Ecology of Cities. *Science* **2008,** *319* (5864), 756–760.

Herold, M.; Gardner, M. E.; Roberts, D. A. Spectral Resolution Requirements for Mapping Urban Areas. *IEEE Trans. Geosci. Remote Sens.* **2003,** *41* (9), 1907–1919.

Herold, M.; Roberts, D. A.; Gardner, M. E.; Dennison, P. E. Spectrometry for Urban Area Remote Sensing—Development and Analysis of a Spectral Library from 350 to 2400 nm. *Remote Sens. Environ.* **2004,** *91* (3–4), 304–319.

Jaeger, J. A.; Schwick, C. Improving the Measurement of Urban Sprawl: Weighted Urban Proliferation (WUP) and Its Application to Switzerland. *Ecol. Indicat.* **2014,** *38*, 294–308.

Kansal, M. L.; Osheen; Tyagi, A. Hotspot Identification for Urban Flooding in a Satellite Town of National Capital Region of India. In *World Environmental and Water Resources Congress 2019: Emerging and Innovative Technologies and International Perspectives*; American Society of Civil Engineers: Reston, VA, 2019; pp 12–24.

Kawamura, M. Relation between Social and Environmental Conditions in Colombo Sri Lanka and the Urban Index Estimated by Satellite Remote Sensing Data. In *Proceedings of the 51st Annual Conference of the Japan Society of Civil Engineers*, 1996; pp 190–191.

Masek, J. G.; Lindsay, F. E.; Goward, S. N. Dynamics of Urban Growth in the Washington DC Metropolitan Area, 1973–**1996,** from Landsat Observations. *Int. J. Remote Sens.* **2000,** *21* (18), 3473–3486.

Osheen; Singh, K. K. Rain Garden—A Solution to Urban Flooding: A Review. In *Sustainable Engineering. Lecture Notes in Civil Engineering*; Springer: Singapore, 2019; 30, pp 27–35.

Savard, J. P. L.; Clergeau, P.; Mennechez, G. Biodiversity Concepts and Urban Ecosystems. *Landsc. Urban Plann.* **2000,** *48* (3–4), 131–142.

Weng, Q. Remote Sensing of Impervious Surfaces in the Urban Areas: Requirements, Methods, and Trends. *Remote Sens. Environ.* **2012,** *117*, 34–49.

Xi, Y.; Thinh, N. X.; Li, C. Preliminary Comparative Assessment of Various Spectral Indices for Built-Up Land Derived from Landsat-8 OLI and Sentinel-2A MSI imageries. *Eur. J. Remote Sens.* **2019,** *52* (1), 240–252.

Xu, H. A New Index for Delineating Built-up Land Features in Satellite Imagery. *Int. J. Remote Sens.* **2008,** *29* (14), 4269–4276.

Xu, H. Analysis of Impervious Surface and Its Impact on Urban Heat Environment Using the Normalized Difference Impervious Surface Index (NDISI). *Photogram. Eng. Remote Sens.* **2010,** *76* (5), 557–565.

Zha, Y.; Gao, J.; Ni, S. Use of Normalized Difference Built-up Index in Automatically Mapping Urban Areas from TM Imagery. *Int. J. Remote Sens.* **2003,** *24* (3), 583–594.

Zhang, T.; Su, J.; Xu, Z.; Luo, Y.; Li, J. Sentinel-2 Satellite Imagery for Urban Land Cover Classification by Optimized Random Forest Classifier. *Appl. Sci.* **2021,** *11* (2), 543.

CHAPTER 12

Hydrological Water Balance Study in Champua Watershed of the Baitarani River Basin Using SWAT Model

AJAY KRISHNA PRABHAKAR[1*], K. K. SINGH[1], ANIL KUMAR LOHANI[2], and SURENDRA KUMAR CHANDNIHA[3]

[1]*Department of Civil Engineering, National Institute of Technology, Kurukshetra, India*

[2]*Surface Water Hydrology Division, National Institute of Hydrology Roorkee, Uttarakhand, India*

[3]*Department of Soil and Water Engineering, BRSM CAET & RS, Indira Gandhi Krishi Vishwavidyalaya Raipur, India*

Corresponding author. E-mail: ajayprabhakar2@gmail.com

ABSTRACT

The soil and water assessment tool (SWAT), a semidistributed model, has been employed at Champua watershed (Odisha) of Baitarani river basin to predict the streamflow and water balance. The sequential uncertainty fitting (SUFI-2) technique has been used to identify the most sensitive parameters that define well the hydrological process. The model performance has been evaluated by using P-factor (Pf), R-factor (Rf), the coefficient of determination (R^2), root mean square error (RSR), observation standard deviation, and Nash–Sutcliffe coefficient (NS). The SWAT model performs well for the monthly streamflow as Pf = 0.80, Rf = 0.80, R^2 = 0.84, NS = 0.84, and RSR = 0.40 for the calibration period 1991–2001, while Pf = 0.82, Rf = 0.78, R^2 = 0.91, NS = 0.91, and

Advances in Hydrology and Climate Change: Historical Trends and New Approaches in Water Resources Management. Surendra Kumar Chandniha, Anil Kumar Lohani, Gopal Krishan, & Ajay Krishna Prabhakar (Eds.)

RSR = −0.30, respectively for validation period 2002–2011. The water balance study is investigated to evaluate the water balance component of the Champua watershed for 21 years (1991–2011). From the water balance results, it has been revealed that the averages of evapotranspiration and surface runoff are about 57% and 11% of annual precipitation, surface runoff generation is about 90% of the monsoon precipitation, and annual precipitation in the Champua basin is about 1381 mm. The planning and management are possible with the SWAT model outputs for the current and future need of the people.

12.1 INTRODUCTION

The water utilization is increasing at a speedy pace and water availability is decreasing or remains constant with time. This causes massive gap between water demand of the growing population and water availability. The percentage decadal population growth rate of India for the period 2001–2011 was 17.64% (census 2011). The average annual potential flow in the rivers of India is 1869 km^3, out of which, only 1123 km^3 of water (surface water 690 km^3 and groundwater—433 km^3) is utilizable. The water demand was 634 km^3 in the year 2000, and it is likely to be 1093 km^3 by the year 2025 (India-WRIS website). The developing country like India whose 68% of the population is directly or indirectly dependent upon agricultural sector. The water required for the sustenance of population is high but the availability of water varies due to climate and erratic distribution of rainfall (O'Brien et al., 2004).

If the available water is appropriately managed, then it would increase agricultural productivity, however, it is a significant concern at local and regional scale for many countries (De Fraiture and Wichelns, 2010). The water availability from the Indra Sagar canal command area of Madhya Pradesh is one of the best example of the managed watershed which ensured irrigation throughout the year and witnessed increased the socioeconomic condition of people (Prabhakar and Tiwari, 2015). Distributed hydrological models are the efficient and powerful tool for assessing the impact of climate change and land use/land cover on water resources management (Pandey et al., 2017; Meng et al., 2017; Narsimlu et al., 2013). Among semidistributed hydrological models, SWAT model was initially developed for simulation and prediction of discharge from ungauged basins (Arnold et al., 1998). The SWAT model has been successfully applied at Nagwan watershed in Eastern India (Tripathi et al., 1999) to estimate the daily and monthly discharge. The SWAT model has been used to analyze the water balance of Ken basin (Central India) from 1985 to 2009 (Murty et al., 2014). To analyze the effect of watershed division on water balance components at Nagwan watershed (Tripathi et al., 2006), the SWAT model was applied. The results indicated that a lot of uncertainties in the SWAT model exist in the streamflow prediction at a watershed.

The spatial distribution of water (Rainfall) can be understood by the analysis of hydrological process through the water balance study. A significant amount of uncertainty exists while analyzing the hydrological process which are highly complex in nature, for example, water cycle cannot be understood through experiments. However, hydrological modeling can be a useful method to define the complicated hydrological process over any watershed (Praskievicz and Chang, 2009). Watershed models can investigate the surface and sub-surface hydrology, soil erosions, water use, conjunctive use management and water quality (chemical constituents) in the watersheds and also access the change in hydrologic processes due to land use, agricultural activities, and their consequences (Singh et al., 2005). The slope variation is also important factor that will affect runoff generation over the watershed (Xin et al., 2019) The climate, tropical condition, soil properties, rainfall distribution, and agrarian activities are different from the other parts of the world (Pandey et al., 2008). The runoff generation has been drastically reduced for Miyun Reservoir because of the climate change and human activities (Pang and Wang, 2016). The above studies suggest that the application of the SWAT model is quite useful to Indian basin. Hence, in the present study, SWAT model applicability has been explored for streamflow prediction and water balance study of Champua watershed of Baitarani river basin in Odisha along with planning and management of the resources to maintain healthy living environment condition at the watershed level.

12.2 STUDY AREA

The Champua watershed area is extended from 85°9′42.66″ to 85°44′10.42″ E longitude and 21°6′52.92″ to 22°11′51.65″ N latitude. The study area (1815 km^2) is mostly covered in Keonjhar district of Odisha. The annual average long-term rainfall of the catchment is about 1425 mm. Agriculture is the primary source of livelihood for the most tribal community living in the study area. The average mean monthly maximum temperature varies from 26°C in December to 34°C in May, and a mean monthly minimum temperature varies from 11°C in December–January to 27°C in May (Central Water Commission (CWC), 2009). The overall weather of the Odisha state is humid tropical, medium-to-high rainfall, short and mild winter. Figure 12.1 shows location and subwatersheds of Champua watershed in the state of Odisha.

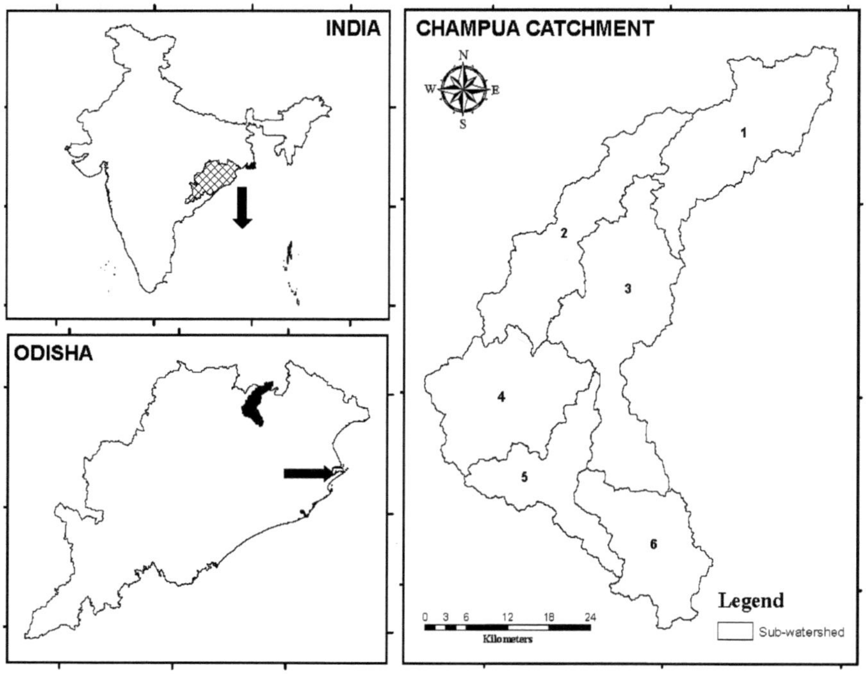

FIGURE 12.1 Location and sub-watersheds of Champua watershed.

12.3 MATERIALS AND METHODS

The performance of the SWAT model depends highly upon the precision of input data. The four types of data, for example, topographical, soil map, land use maps, and hydrometeorological data must comply in a specific format for SWAT model. The model runs on daily time steps. Therefore, all meteorological data (Shown in column 1 of Table 12.1) are arranged in a daily specific format at the basin scale. The three grids of size 0.25° × 0.25° of rainfall as provided by the India Meteorological Department (IMD), Pune lies in the study area, which have been used for simulation from 1991 to 2011. The data with their source (s) used in the study are arranged in Table 12.1.

Digital elevation model (DEM) plays an active role in deciding the streamflow network characteristics, that is, slope and slope length over the watershed area. The minimum and maximum elevation of the watershed are 340–1109 m. Satellite imagery is one of the important input which affects the water balance. Therefore, the highly rigorous unsupervised classification from the satellite imagery of the year 2011 has been used to classify the study area into number of important land use/land cover classes like agricultural land (22.54%), fallow land (18.66%), water (0.67%), shrub land (14.05%), settlement (1.45%), forest land (41.08%), wasteland (1.54%) by using the ERDAS 2014 image software. The model required the certain soil properties for the perfect water balance. Therefore, the study area soil series map was derived like Soil-48 (5.03%), Soil-46 (1.26%), Soil-52 (1.77%), Soil-51 (7.32%), Soil-53 (19.85%), Soil-43 (64.77%) and it was extracted from the digitized soil map of Odisha state in GIS Software and the layers wise the data have been input in the SWAT model according to the NBSS soil report. All the maps have been projected to common coordinates WGS-84 north system and projection system in polyconic with the spheroid of Everest-1956 and reference datum India (India, Nepal). The DEM, LULC, and soil map of the study area are shown in Figure 12.2.

TABLE 12.1 Data and Their Sources Used in This Study.

Data	Source(s)
DEM	Aster Digital Elevation Model (ASTER-DEM) (Resolution of 30 m) Source: http://gdem.ersdac.jspacesystems.or.jp/
Imageries	Landsat-5 TM and Landsat-7 ETM+ (Resolution of 30 m) Source: https://earthexplorer.usgs.gov/
Soil Series Map	National Bureau of Soil Survey and Land Use Planning (NBSS-LUP), Nagpur, India
Rainfall	Indian Meteorological Department (IMD), Pune (India) (Grid Resolution of $0.25° \times 0.25$)
Minimum/Maximum Temperature	Indian Meteorological Department (IMD), Pune (Grid Resolution of $1° \times 1$)
Relative Humidity Solar Radiation Wind Speed Dew Point	*NASA Prediction of Worldwide Energy Resource (POWER)* *Source:* http://power.larc.nasa.gov/cgi-bin/cgiwrap/solar/agro.cgi?email=agroclim@larc.nasa.gov
Discharge Data	India-Water Resources Information System (India-$WRIS_{webGIS}$) http://www.india-wris.nrsc.gov.in/wris.html

The whole study area has been delineated (using threshold values of 14,000 ha) into six numbers of the subwatersheds with GIS Software. All three maps of the study area have been overlaid with defining the multiple slopes (0–4; 4–8; 8–12 and 12-∞) and default threshold values are used in most of the studies to define the hydrological response units (HRUs) definition, for example, 20% of the land use, 10% of the soil class and 20% of the slopes (Winchell et al., 2007). All input data interface with the geographic

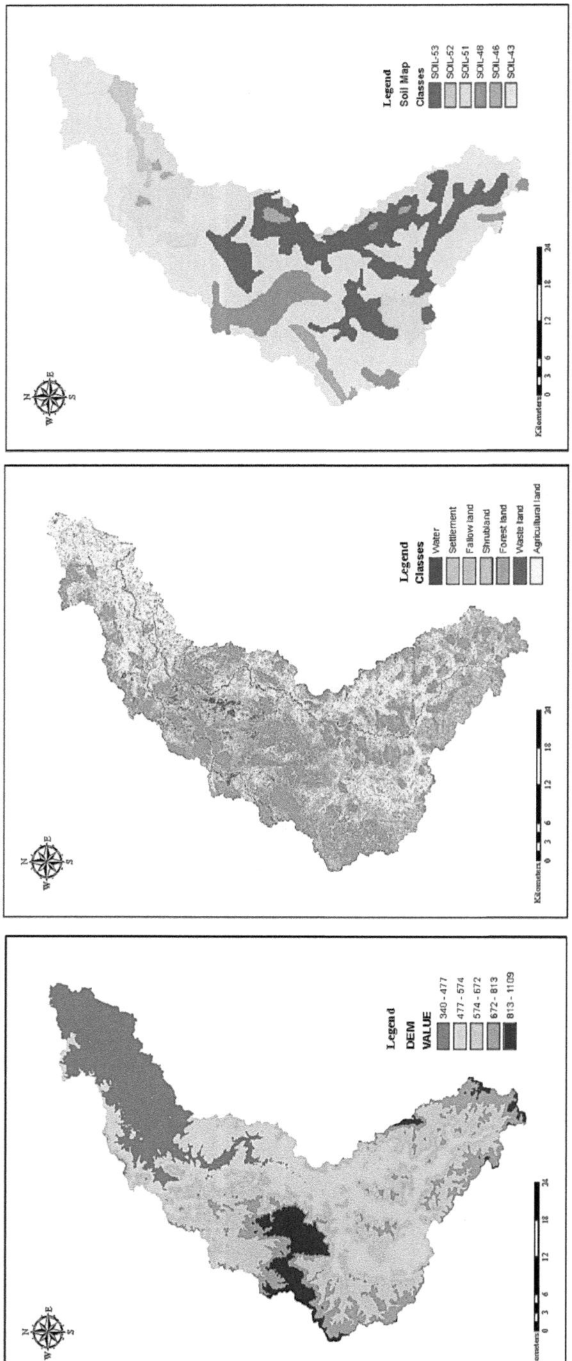

(a) Digital Elevation Model (DEM) (b) Land use Land Cover Map (LULC) (c) Soil Map

FIGURE 12.2 The DEM, LULC and soil map of study area.

information system and about 122 HRUs spread over the Champua watershed have been used for running the SWAT model.

All weather data, viz., rainfall, temperature, relative humidity, solar radiations, and wind are used as input for the weather generation. The SWAT model simulation period has been selected from 1991 to 2011 with 5 years of warm-up period (1986–1990), calibration period (1991–2001), sensitivity analysis, and validation period (2002–2011). The water balance relationship given in eq 12.1 creates the relationship among the various hydrological process of each HRUs of SWAT model (Arnold et al., 1998; Neitsch et al., 2011) which is given here:

$$SW_t = SW_0 + \sum_{i=1}^{t} \left(P_i - R_i - ET_i - S_i - GW_i \right) \tag{12.1}$$

where the SW_t and SW_0 denote the initial and final soil moisture contents (mm) at time (t) days. The amounts of daily overland surface runoff (mm) due to daily rainfall distribution (mm) are Q_{surf} and $R_{\text{day;}}$ The water seepage (mm) and return flow (mm) through the vadose zone of soil profile are W_{seep} and Q_{gw} and the evapotranspiration (mm) and potential evapotranspiration (mm) are represented as ET_a and PET.

The methodology used by the SWAT model to compute the runoff generation (Q) due to rainfall (R) uses the curve number (CN) method (USDA, 1972). The evapotranspiration (ET) and potential evapotranspiration (PET) are computed by the Penman–Monteith method and channel routing by the nonlinear Muskingum method under daily time steps input data.

The sequential uncertainty fitting (SUFI-2) algorithm is an efficient technique to identify the most sensitive parameters while equating the values of predicted flow output of the model and observed flow (Rostamian et al., 2008). All uncertainties of the model have measured data bracketed under significance of sensitivity of 95% prediction Uncertainty (95PPU). In an ideal condition, a p-factor (0–1) value should be 1 and an r-factor (0–1) value should be near to zero (Abbaspour, 2007). However, the higher p-factor value could be achieved only at higher r-factor values as both values are closely related to each other. The parameter uncertainties are under desired parameters range when the values of p-factor and r-factor have reached the acceptable range. In the present study, the calibration and validation are performed on monthly observed data of the Baitarani basin of Champua watershed. The model performance has been evaluated by using P-factor (Pf), R-factor (Rf), a coefficient of determination (R^2), Nash–Sutcliffe coefficient (NS), percent bias (Pbias), RSR and observation standard deviation (Moriasi et al., 2007).

12.4 RESULTS AND DISCUSSION

12.4.1 EVALUATION OF SOIL AND WATER ASSESSMENT TOOL MODEL FOR CHAMPUA WATERSHED (BAITARANI BASIN)

Sensitivity analysis was performed to examine the relative changes in model output concerning to the model input. The model input and the hydrological process over the watershed are dependent on the sensitive parameter. SUFI-2 algorithm was established to calculate the best-fitted parameter values for minimizing the gap between observed values and model output values. A large number of iterations have been performed to select most sensitive parameter with the help of sensitivity analysis method (one at a time Latin and hypercube) to define the natural hydrological process on the watershed. Ten hydrologic parameters were manually adjusted and their desired ranges and best-fitted values for the study area are given in Table 12.2.

TABLE 12.2 Sensitively Analysis of Parameters (Study Area).

Rank	Parameters	Descriptions	Parameter range	Parameter-calibrated values
1	CN2	Initial SCS runoff curve number for moisture condition II	35–98	58.46
2	CH_K2	Hydraulic conductivity in main channel	−0.01–150	23.44
3	SOL_K	Saturated hydraulic conductivity	0–1000	442.7
4	ALPHA_BF	Baseflow alpha factor	0–1	0.71
5	GW_DELAY	Ground water delay time	0–50	33.19
6	REVAPMN	Threshold depth of water in the shallow aquifer for revap to occur	0–500	221.14
7	ESCO	Soil evaporation compensation factor	0–1	1
8	CH_N2	Manning's n value for the main channel	0.01–0.0.5	0.12
9	EPCO	Plant uptake compensation factor	0–1	0.63
10	SOL_AWC	Available water capacity of the soil layer	0–1	0.13

 # R, relative value (existing value(1+R-value) and V, value replaces the existing value.*

Watershed model has been calibrated and validated using the monthly observed data to compute the sensitivity of the streamflow. In Table 12.2, the sensitivity of the parameters are arranged in descending order according to the higher ± values of t-stat from the SWAT-CUP. The top five parameters, that is, CN2, CH_K2, SOL_K, ALPHA_BF, and GW_DELAY are more sensitive and the remaining parameters such as REVAPMN, ESCO, CH_N2, EPCO, and SOL_AWC are less vulnerable for streamflow prediction. All these sensitive parameters alter the runoff, rate of evaporation, recharge rate, available soil moisture, etc. that would ultimately influence the whole water cycle. After sensitivity analysis, the monthly average observed and simulated/calculated discharge for the calibration period 1991–2001 are expressed with the help of Figure 12.3.

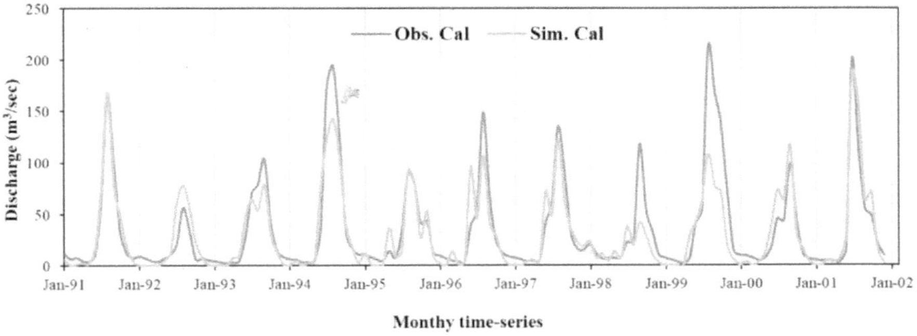

FIGURE 12.3 Average monthly observed and simulated discharge before sensitivity analysis.

The monthly average discharge simulated by the model for the year 1998 and 1999 are varied much compared with the observed average monthly discharge. On analysis of input rainfall, the vast annual rainfall deficits are found, that is, 12.88% for the year 1998 and 7.13% for the year 1999. The year 1998 has approximately half deficit than the previous year but the model has underpredicted discharge, this may be due to monthly rainfall distribution, the consecutive rainy months from January to October have less than 65 mm of rainfall except for the month of February which has only 6.3 mm rainfall (less CN2). The increased baseflow constant values (ALPHA_BF) would increase the rate of groundwater flow (Smedema and Rycroft, 1983). The best-fitted calibrated value of baseflow is 0.81. With the increase in ALPHA_BF, there is a decrease in discharge generally during the low flow. In the year 1999, more gap between observed and simulated monthly average discharge is may be due to less monthly average rainfall from May to October (varies from 214 to 353.2 mm) and less continuous rainy days. The other possibility could be the lesser density of meteorological stations in the study area. The monthly average discharge prediction shows good relationship with the observed data except for the extreme and low rain events (Gassman et al., 2007; Murty et al., 2014). Generally, the discharge increases when there is more soil moisture and less evaporation, more intensity of rainfall and rainier days. The agreement diagram of monthly average observed and simulated discharge for calibration period are shown in Figure 12.4.

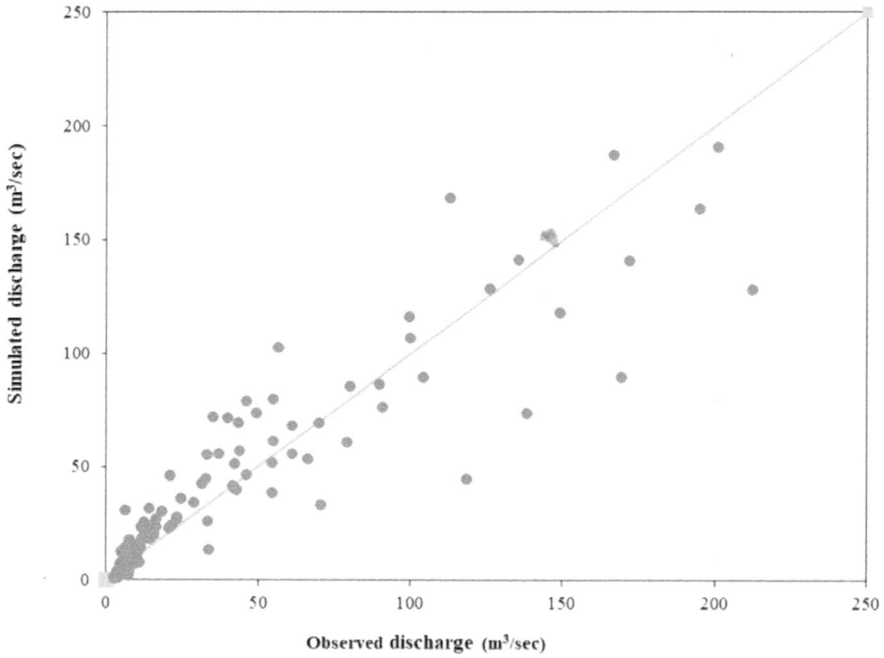

FIGURE 12.4 Agreement diagram of average monthly observed and simulated discharges.

The scattered plot between the monthly averages observed and simulated discharges around the perfect agreement line. This represents that most of the predicted values near to perfect agreement line. After sensitivity analysis, the monthly average observed and simulated/calculated discharge for the validation period 2002–2011 are expressed with the help of Figure 12.5. The distributed comparison plot of monthly average observed and simulated discharge for calibration period are shown in Figure 12.6.

After sensitivity analysis, the evaluation of SWAT model performance is found for the calibration, the values of R^2, NS, Pbias, and RSR are 0.84, 0.84, −4.4, and 0.4 and the rest of the values are given in Table 12.3 and for the validation periods values of R^2, NS, Pbias, and RSR are 0.91, 0.91, −0.3, and 0.3 and the rest of the values are given in Table 12.4, which shows a very good performance rating for the prediction of monthly average discharge in both calibration and validation periods. The comparison of mean monthly rainfall, standard deviation, maximum monthly discharge of observed and simulated for calibration and validation are also depicted in Tables 12.3 and 12.4.

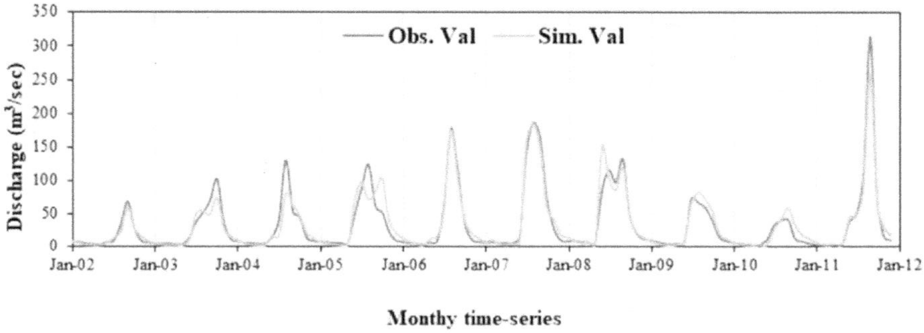

FIGURE 12.5 Average monthly observed and validated discharge during 2002–2011.

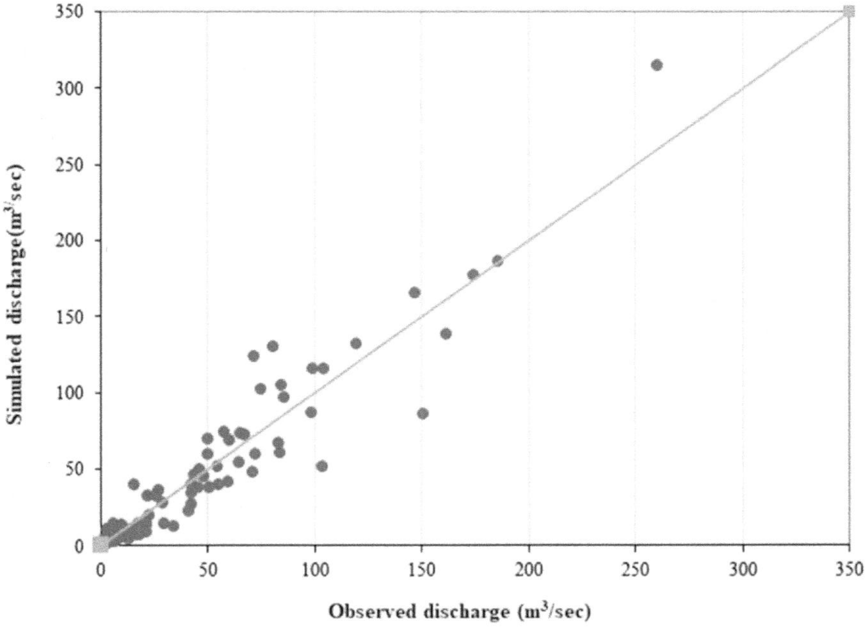

FIGURE 12.6 Agreement diagram of average monthly observed and simulated discharges.

The P-factor values in both calibration (0.80) and validation (0.82) are quite near to 1 and the R-factor values for the calibration and validation are

0.80 and 0.78, respectively. This shows that the graph between simulated and observed average monthly discharge is under bracket of 95% of significance prediction uncertainty (95PPU).

TABLE 12.3 The Statistical Parameters of the Study Area for the Calibration (1991–2001).

Statistical parameters (study area)	Avg. monthly discharge (m^3/s)		
	Observed		Simulated
Mean	34.6		36.1
Standard deviation, SD	46.7		42.0
Maximum	212.1		191.3
Count	132.0		132.0
Coefficient of determination, R^2		0.84	
Nash–Sutcliffe coefficient (NS)		0.84	
Percent bias (Pbias)		−4.4	
Root mean square error (RSE) observation standard deviation		0.4	
P-factor (Pf)		0.80	
R-factor(Rf)		0.80	

TABLE 12.4 The Statistical Parameter of the Study Area for the Validation (2002–2011).

Statistical parameter (study area)	Avg. monthly discharge (m^3/s)		
	Observed		Simulated
Mean	33.6		33.7
Standard deviation, SD	48.5		45.0
Maximum	314.9		260.0
Count	120.0		120.0
Coefficient of determination, R^2		0.91	
Nash–Sutcliffe coefficient (NS)		0.91	
Percent bias(Pbias)		−0.3	
Root mean square error (RSR) observation standard deviation		0.3	
P-factor (Pf)		0.82	
R-factor (Rf)		0.78	

The model results express that model was accurately validated for the Champua site of Baitarani river basin for predicting the average monthly streamflow. The SWAT model can be used for further study or the future variation of flow can be analyzed by using the future climate change scenarios.

12.4.2 WATER BALANCE STUDY FOR THE STUDY AREA

After the sensitivity analysis, the manual calibration was performed in SWAT model for six subwatershed. For each subwatersheds, the water balance components, viz., precipitation (PRECIP), evapotranspiration (ET), percolation (PERCO), surface runoff (SURQ), and lateral flow (LAT-Q) were evaluated by the model and are given in Table 12.5. The water balance (PRECIP- (ET+ PERCO+ SURQ+ LAT-Q*),* the average component values after calculation of all subwatersheds and Champua watershed (*WS*) are approximately equal to zero. The average evapotranspiration and average surface runoff generation are about 57% and 11% of the annual precipitation. The lesser runoff generation and more recharge of the groundwater may be due to the less urbanized area and uneven topographical area conditions which exist in the study area.

TABLE 12.5 Water Balance Components Study of Six Subwatershed (1991–2011).

S.no (basin)	Area (km^2)	PRECIP (mm)	ET (mm)	PERCO (mm)	SURQ (mm)	LAT-Q (mm)
1	332.7	1334.1	641.1	562.5	63.3	67.5
2	306.4	1334.1	876.1	247.1	197.4	13.7
3	426.3	1334.1	716.9	362.3	117.6	137.5
4	312.2	1482.2	833.9	323.3	172.8	152.6
5	204.6	1482.2	810.1	364.2	180.3	127.8
6	233.5	1373.3	900.1	245.2	218.0	10.2
Champua-WS	1815.8	1390.0	796.4	350.8	158.2	84.9

The monsoon precipitation contributes about 78% of the annual precipitation and the remaining 22% of the precipitation falls in the rest of the months. The surface runoff generation is about 90% of the monsoon precipitation. The minimum surface runoff generation is in the month of April and December. The water balance components (PRECIP- (ET+ PERCO+ SURQ+ LAT-Q) show maximum values in the month of August and also, most of the evapotranspiration occurred in the months of the monsoon season are presented in Table 12.6.

The graphical representation of various water balance components (PRECIP- (ET+ PERCO+ SURQ+ LAT-Q)) of Champua watershed on average monthly basis for the period of 21 years is shown in Figure 12.7.

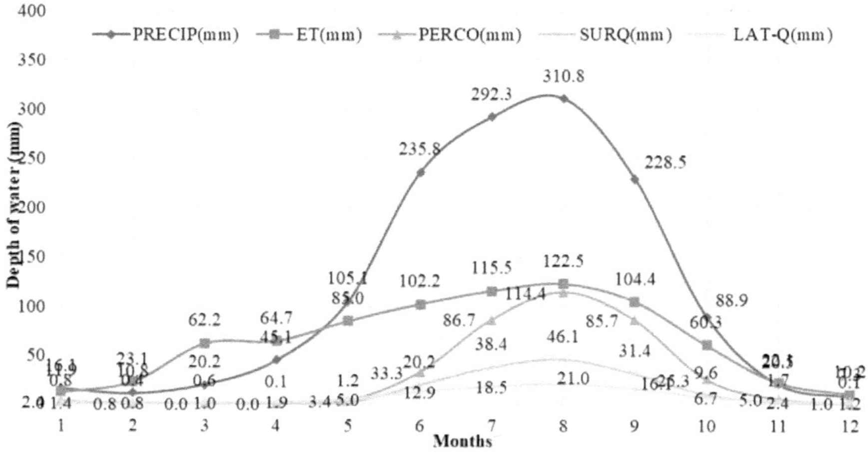

FIGURE 12.7 Graphical representation of average monthly (1991–2011) water balance components of the study area.

TABLE 12.6 Monthly Average Water Balance Study of Champua Watershed (1991–2011).

Months	Area (km²)	PRECIP (mm)	ET (mm)	PERCO (mm)	SURQ (mm)	LAT-Q (mm)
1	1815.8	16.1	11.9	2.4	0.8	1.4
2	1815.8	10.8	23.1	0.8	0.4	0.8
3	1815.8	20.2	62.2	0.0	0.6	1.0
4	1815.8	45.1	64.7	0.0	0.1	1.9
5	1815.8	105.1	85.0	3.4	1.2	5.0
6	1815.8	235.8	102.2	33.3	20.2	12.9
7	1815.8	292.3	115.5	86.7	38.4	18.5
8	1815.8	310.8	122.5	114.4	46.1	21.0
9	1815.8	228.5	104.4	85.7	31.4	16.1
10	1815.8	88.9	60.3	25.3	9.6	6.7
11	1815.8	20.5	22.1	5.0	1.7	2.4
12	1815.8	7.3	10.2	1.0	0.1	1.2
Champua-WS	1815.8	1381.3	784.1	358.0	150.6	88.9

The total 68 HRU's spread over the study area and out of this, 23 HRU's are spread over the forest area. Average CN is the HRU's spread over the forest area and is computed as 54 and the average surface runoff generation and evapotranspiration are about 8% and 56% of the annual average rainfall. The groundwater recharge is about 25% of annual rainfall. 13 HRU's are spread over the agricultural area, found average CN as 70 and average surface runoff generation and evapotranspiration are about 15% and 55% of the annual average rainfall. The groundwater recharge is about 29% of annual rainfall. Therefore, there is need to apply the best management practices (BMP) to reduce the surface runoff on the agricultural land. The strip cropping, grassed area, bunds and vegetative ground are some of the management practices which can diminish the runoff and protect the upper fertile soil.

12.5 CONCLUSIONS

In this study, SWAT model was applied to the champua watershed (Baitarani basin) to simulate the streamflow for the years from 1991 to 2011 after the rigorous calibration and validation analysis by using the SUFI-2 method. For calibration and validation periods, the values of R^2, NS, Pbias, and RSR were found to be 0.84, 0.84, −4.4 and 0.4, respectively and 0.91, 0.91, −0.3, and 0.3, respectively. These results show "very good" streamflow prediction in the study area. The water balance study indicates that the average evapotranspiration and surface runoff generation are 57% and 11% of the annual precipitation. The monsoon precipitation contributes about 78% of the annual precipitation and surface runoff generation has been observed as 90% of the monsoon precipitation. According to the HRUs, in the agricultural land, average surface runoff generation and evapotranspiration have been found as 15% and 55% of the annual average rainfall and in the forest area, the average surface runoff generation and evapotranspiration have been found as 8% and 56% of the average annual rainfall. BMP such as strip cropping, bunds, grassed land, and vegetative land, etc. can be used for reducing the surface runoff and soil erosion. It has been also noticed that sudden falling recession curve at the falling limp in model discharge hydrograph can be matched with the observed discharge hydrograph by using BMP method. Therefore, the development and need of people can be done sustainably with fewer effects on water availability, soil, land use and environment. The planning and management become possible with the SWAT model and could enrich the watershed for the current and future need of the people, such as food, clothing, and shelter as the hydrological modeling diminishes the uncertainty in decision-making.

ACKNOWLEDGMENT

The authors thank National Institute of Hydrology, Roorkee, and National Institute of Technology, Kurukshetra, for providing valuable support and their resource during the study.

KEYWORDS

- **SWAT**
- **hydrological modeling**
- **water balance**
- **SUFI-2 algorithm**

REFERENCES

Abbaspour, K. C. User Manual for SWAT-CUP, SWAT Calibration and Uncertainty Analysis Programs. Swiss Federal Institute of Aquatic Science and Technology, Eawag, Duebendorf, Switzerland, 2007.

Arnold, J. G.; Srinivasan, R.; Muttiah, R. S.; Williams, J. R. Large Area Hydrologic Modeling and Assessment Part I: Model Development. *JAWRA J. Am. Water Resourc. Assoc.* **1998,** *34* (1), 73–89.

De Fraiture, C.; Wichelns, D. Satisfying Future Water Demands for Agriculture. *Agric. Water Manage.* **2010,** *97* (4), 502–511.

Gassman, P. W.; Reyes, M. R.; Green, C. H.; Arnold, J. G. The Soil and Water Assessment Tool: Historical Development, Applications, and Future Research Directions. *Trans. ASABE* **2007,** *50* (4), 1211–1250.

Meng, X. Y.; Wang, H.; Cai, S. Y.; Zhang, X. S.; Leng, G. Y.; Lei, X. H.; Shang, Y. The China Meteorological Assimilation Driving Datasets for the SWAT Model (CMADS) Application in China: A Case Study in Heihe River Basin, 2017.

Moriasi, D. N.; Arnold, J. G.; Van Liew, M. W.; Bingner, R. L.; Harmel, R. D.; Veith, T. L. Model Evaluation Guidelines for Systematic Quantification of Accuracy in Watershed Simulations. *Trans. ASABE* **2007,** *50* (3), 885–900.

Murty, P. S.; Pandey, A.; Suryavanshi, S. Application of Semi-Distributed Hydrological Model for Basin Level Water Balance of the Ken Basin of Central India. *Hydrol. Process.* **2014,** *28* (13), 4119–4129.

Narsimlu, B.; Gosain, A. K.; Chahar, B. R. Assessment of Future Climate Change Impacts on Water Resources of Upper Sind River Basin, India Using SWAT Model. *Water Resour. Manage.* **2013,** *27* (10), 3647–3662.

Neitsch, S. L.; Arnold, J. G.; Kiniry, J. R.; Williams, J. R. *Soil and Water Assessment Tool Theoretical Documentation Version 2009*; Texas Water Resources Institute, 2011.

O'Brien, K.; Leichenko, R.; Kelkar, U.; Venema, H.; Aandahl, G.; Tompkins, H.; West, J. Mapping Vulnerability to Multiple Stressors: Climate Change and Globalization in India. *Global Environ. Change* **2004,** *14* (4), 303–313.

Pandey, A.; Chowdary, V. M.; Mal, B. C.; Billib, M. Runoff and Sediment Yield Modeling from a Small Agricultural Watershed in India Using the WEPP Model. *J. Hydrol.* **2008,** *348* (3), 305–319.

Pandey, B. K.; Gosain, A. K.; Paul, G.; Khare, D. Climate Change Impact Assessment on the Hydrology of a Small Watershed Using Semi-Distributed Model. *Appl. Water Sci.* **2017,** *7* (4), 2029–2041.

Pang, S.; Wang, X. The Characteristics and Attribution of Runoff Change in the Miyun Reservoir Watershed. *J. Arid Land Resourc. Environ.* **2016,** *30* (9), 144–148.

Prabhakar, A.; Tiwari, H. Land Use and Land Cover Effect on Groundwater Storage. *Model. Earth Syst. Environ.* **2015,** *1* (4), 45.

Praskievicz, S.; Chang, H. A Review of Hydrological Modelling of Basin-Scale Climate Change and Urban Development Impacts. *Prog. Phys. Geogr.* **2009,** *33* (5), 650–671.

Rostamian, R.; Jaleh, A.; Afyuni, M.; Mousavi, S. F.; Heidarpour, M.; Jalalian, A.; Abbaspour, K. C. Application of a SWAT Model for Estimating Runoff and Sediment in Two Mountainous Basins in Central Iran. *Hydrol. Sci. J.* **2008,** *53* (5), 977–988.

Singh, J.; Knapp, H. V.; Arnold, J. G.; Demissie, M. Hydrological Modeling of the Iroquois River Watershed Using HSPF and SWAT. *JAWRA J. Am. Water Resourc. Assoc.* **2005,** *41* (2), 343–360.

Smedema, L. K.; Rycroft, D. W. *Land Drainage: Planning and Design of Agricultural Systems*; Batsford Academic and Educational Ltd, 1983.

Tripathi, M. P.; Panda, R. K.; Raghuwanshi, N. S. Runoff Estimation from a Small Watershed Using SWAT Model. In *Proceedings of the International Conference on 'Water, Environment, Ecology, Socio-Economics and Health Engineering*; Seoul, Koreal, October 18–21, 1999; pp 43–152.

Tripathi, M. P.; Raghuwanshi, N. S.; Rao, G. P. Effect of Watershed Subdivision on Simulation of Water Balance Components. *Hydrol. Proces.* **2006,** *20* (5), 1137–1156.

Winchell, M.; Srinivasan, R.; Di Luzio, M.; Arnold, J. ArcSWAT Interface for SWAT 2005. User's Guide, 1-436, 2007.

Xin, Y.; Liu, G.; Xie, Y.; Gao, Y.; Liu, B.; Shen, B. Effects of Soil Conservation Practices on Soil Losses from Slope Farmland in Northeastern China Using Runoff Plot Data. *Catena* **2019,** *174*, 417–424.

CHAPTER 13

Long-Term Rainfall Trend Variability Analysis for Rampur Watershed: A Case Study of Mahanadi River Basin

PRIYANKA GUNJAN[1*], S. K. MISHRA[1], ANIL KUMAR LOHANI[2], SURENDRA KUMAR CHANDNIHA[3], and NITY TIRKEY[4]

[1]*Department of Water Resources Development & Management, Indian Institute of Technology, Roorkee, India*

[2]*Surface Water Hydrology Division, National Institute of Hydrology, Roorkee, India*

[3]*Department of Soil and Water Engineering, BRSM CAET & RS, Mungeli, IGKV Raipur, India*

[4]*Department of Water Engineering and Management, Central University of Jharkhand, India*

**Corresponding author. E-mail: pri.gunjan@gmail.com*

ABSTRACT

Precipitation is one of the pronounced indicators of climate change. The behavior of historical rainfall pattern becomes necessary to study the climate change, because of the complex phenomenon of the rainfall variability. The study of spatial and temporal variability of precipitation in the Rampur watershed of Mahanadi river basin has been carried out in this paper. Mann–Kendall (MK) and modified Mann–Kendall (MMK) tests have been used to determine the possible trends of daily rainfall data for the period 1983–2015.

Advances in Hydrology and Climate Change: Historical Trends and New Approaches in Water Resources Management. Surendra Kumar Chandniha, Anil Kumar Lohani, Gopal Krishan, & Ajay Krishna Prabhakar (Eds.)

Further, the Sen slope estimator (SSE) has been used to assess the magnitude of the whole time series. Arc-Map 10.1 software has been used to determine the spatial patterns of the trends over the whole watershed by applying the inverse distance weighting (IDW) interpolation technique. Further, the MK and MMK tests have been used to identify the possible trend analysis of the entire time series with 95% level of significance. The premonsoon season for both MK and MMK tests gave a positive significant trend for only SW 1. In monthly trends (using MK and MMK), 86% of the values has been indicated by a negative trend (either statistically significant or nonsignificant) and the month of September indicated a positive trend for all the subwatersheds (SWS). In MMK statistics, a positive significant trend has been observed in the month of August in only SW 1. The magnitude in the postmonsoon season gave all negative values for all the SWS. In monthly precipitation data, the months January, November, and December showed no trend values for all the 9 SWS. The months September and June showed all positive and negative values, respectively. The highest value of the coefficient of variation has been detected in postmonsoon season (146%) and the lowest has been detected in monsoon season (19.93%).

13.1 INTRODUCTION

Water apart from being the most essential natural resources is also a primary requirement for all living creatures. Any vital alteration in the water budget will make the hydrological process to suffer from a pivotal consequence which will later influence the welfare and economy of the country (Chattopadhyay and Hulme, 1997). The estimation has been done that by the year 2050, the availability of water and the annual mean runoff at higher latitudes and in few wet tropical areas will get increased by 10–40%, while 10–30% will get decreased in the dry tropics and at mid-latitudes over certain dry regions (IPCC, 2007). In India, agriculture is mostly dependent upon monsoon rainfall (June–October). Thus, the study of changes in the incidence and distribution of rainfall is essential for managing the water sustainably. Precipitation is a major component which controls the hydrology, vegetation, water quality, and agricultural production. The precipitation and temperature occurrence and variability determine the crop growth and its type to a large extent in distinct regions throughout the world. For sustainably optimizing the farm production, it is necessary to have a precise knowledge about the precipitation characteristics and variability of soil (Gajbhiye et al., 2015, 2016; Meshram et al., 2016). In India, the uneven water distribution is among the major problems that water resource management faces. The major constraints to crop growth in an area are uncertain and erratic distribution of precipitation and lack of water resource planning. The occurrence of these constraints are because of the natural precipitation pattern (spatial and temporal) which differs extensively in space and time (Bandyopadhyay and Perveen, 2006).

Trend analyses of precipitation are beneficial for the exploration of the climate change impact on water resources. Therefore, numerous studies on trend analysis and rainfall trend detection have been conducted globally (Bradley et al., 1987; Hulme et al., 1998; Hennessy et al., 1999; Fawcett, 2004; Smith, 2004; Guhathakurta and Rajeevan, 2006; Joshi and Rajeevan, 2006; Gallant et al., 2007; Fu et al., 2008; Murphy and Timbal, 2008; Taschetto and England, 2009; Chowdhury and Beecham, 2010; Kiem and Verdon-Kidd, 2010). Factors that have major impacts on climate change are increased deviation in the frequency and intensity of precipitation and higher temperatures (Dinpashoh et al., 2014). Praveen et al. (2020) stated that for achieving a high precision microlevel management plan, microlevel rainfall data like district wise data should be used. Further, rise in temperature increases the crop water requirement and reduces the precipitation and this results into shortage of crop water. Hence, trend analyses of precipitation is essential for facilitating the water use and planning more sustainably. For determining the supplemental requirements of water needed during the critical crop growth period, it is essential to have knowledge about the precipitation trend and its variability. High temperatures and uncommon space–time and frequency of rainfall distribution are the major climate change consequences taking place over decades in the country. The biggest sufferer are the agricultural sector who has to face the unusual occurrence of rainfall, that is, rainfall not occurring when expected and occurring when not expected. This unusual occurrence of rainfall along with high winds and hails causes huge loss to crops and to the farmers who are completely dependent on their crops. There is a satire that at the same time, a place may experience drought while the other place may experience flood. It is therefore important to determine whether the precipitation follows any trend in its variability. The studies carried out in India have not shown any evident trend of increasing or decreasing annual mean rainfall (Mooley and Parthasarathy, 1984; Lal, 2001; Chakraborty et al., 2013; Swain et al., 2015). However, in India, there has been no long-term trend for monsoon rainfall on a national scale. Although identification of significant long-term rainfall changes have been carried out by some researcher (Koteswaram and Alvi,

1969; Jagannathan and Parthasarathy, 1973; Raghavendra, 1974; Chaudhary and Abhyankar, 1979; Kumar et al,. 2005; Dash et al., 2007; Kumar and Jain, 2011; Gajbhiye et al., 2015).

The most commonly used tool for assessing the changes in the behavior of hydrologic and climatic time series is trend analysis. Commonly used tests are both parametric and nonparametric. Among both the tests, parametric trend test is more robust, but it requires independent and normally distributed data while the nonparametric trend test requires independent data. The nonparametric test can even tolerate the outliers in the data. The Mann–Kendall (MK) test is the most common and a well-known nonparametric test used for analyzing trend in the time series (Mann, 1945; Kendall, 1955). The MK trend test was proposed by Kendall (1955) and its derivation was done from a rank correlation test for two groups of observations. The MK trend test considers the correlation between the rank order of the observed value and their order in time. In the MK test, the null hypothesis is that the data are independent and ordered randomly, that is, the data do not follow any trend or any structural serial structure among the observations. Even in the nonexistence of a trend, the chances of significant answer is increased in the case of positive serial correlation among the observations. The elimination of the effects of seasonality can be performed by distinguishing the observations into different classes as per the seasons and then the MK trend test can be performed on the sum of the statistics from each season. This alteration is known as the seasonal Kendall test (Hirsch et al., 1982; Hirsch and Slack, 1984; Zetterqvist, 1991). However, the effect of dependence between seasons is eliminated by the seasonal test, but the series correlation within seasons is not accounted by it (Hirsch and Slack, 1984). The similar problem can be seen during the analysis of the yearly data, since they are often significantly auto correlated. Therefore, understanding the precipitation trend and variability is essential for determining the supplemental water required by the crops during their critical growth phases because the future of rainfed agriculture is becoming unpredictable due to climate change. From the above-mentioned issues, the main objective of the study is to investigate precipitation variability in Rampur watershed of Mahanadi river basin, using 33 years (1983–2015) of precipitation data.

13.2 STUDY AREA AND DATA USED

The origination of Rampur watershed takes place from Jonk river catchment which is a tributary of Mahanadi basin. Raipur and Mahasamund districts of

Chhattisgarh state occupies major part of the watershed and the minor part is occupied by Nuapada and Bargarh districts of Odisha state (Fig. 13.1). The watershed area is about 3424 km^2 and is situated between 81° 28′16″ to 83° 18′42″ eastern longitudes and 20° 27′53 to 21° 47′49″ northern latitudes. The region mostly consists of tropical wet and dry type of climate and the average temperature ranges from 15°C to 35°C. However, temperature remains moderate for the whole year, but April and May months can be extremely hot as temperature sometimes rises above 48C. Mean annual rainfall in the region ranges from 800 to 1200 mm. Forest and agricultural land occupy more than 50% of the watershed. The main culture in the region is paddy. Geologically, the district includes the Archean Rock Formation, the Chhattisgarh Supergroup, the Gondwana Supergroup, and the Deccan Basaltic Trap Formations. The height of the watershed varies between 205 and 875 m. The watershed comprises three classes of soil: clay loam, sandy loam, and sandy clay loam. The soil clay loam covers the maximum area and followed by sandy loam.

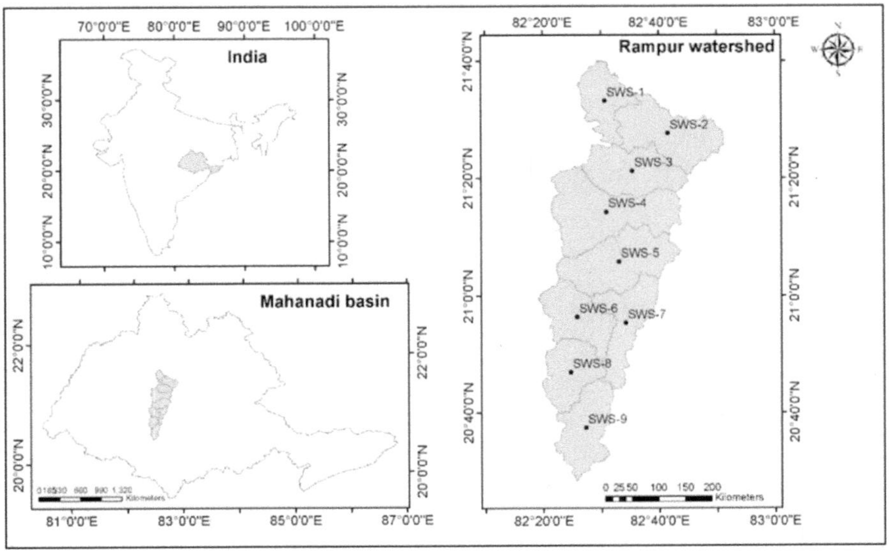

FIGURE 13.1 Location map of the watershed and all the nine subwatersheds.

Daily rainfall data of five grid points with resolution of 0.25° × 0.25° that falls within the catchment has been used in the study for the period 1983–2015. The data have been collected from the India Meteorological Department (IMD), Pune, acquired from National Institute of Hydrology (NIH), Roorkee.

13.3 METHODOLOGY

The watershed has been divided into nine subwatersheds (SWS) and the thiessen polygon method has been used for computing the rainfall in each SWS for the period 1983–2015. The computed daily rainfall for each watershed is then converted into monthly, seasonal, and annual rainfall series for performing trend analysis. The seasonal rainfall has been classified into four seasons: premonsoon (March–May), monsoon (June–September), postmonsoon (October–November), and winter (December–February). The trend detection of the precipitation has been performed by the following tests.

13.3.1 STATISTICAL TESTS FOR TREND ANALYSIS

Several methods are being emerged for analyzing the rainfall trend. The trend analysis is a well-known method which is performed through the tests like parametric and nonparametric. However, the nonparametric test holds a greater advantage than the parametric test as the former is more appropriate approach for non-normally distributed, missing, outlier and censored data, which are often experienced in hydrological time series. Consequently, the tests that are used widely for trend detection of the meteorological data are the MK and the modified Mann-Kendall (MK) tests.

13.3.2 MANN-KENDALL TEST

The MK test being a nonparametric test, is generally used for determining the changes in climatic and hydrological time series (Kendall, 1975; Hamed and Rao, 1998). The data which do not pursue a normal distribution (Tabari et al., 2011e) and permit multiple perceptions (Kampata et al., 2008) are suitable for this test. It identifies monotonic trends (either increasing or decreasing) and mostly determines trends as part of the time series. It is robust, simple, easy, and applicable where there are some missing values, and the data are not suitable with any statistical distribution (Libiseller and Grimval, 2002; Gilbert, 1987). The dependency of the statistics on plus or minus signs makes the test superior than the other nonparametric tests (Helsel and Hirsch, 1992; Birsan et al., 2005).

The description of the MK statistics S is given as:

$$S = \sum_{i=1}^{n-1} \sum_{j=i+1}^{n} \text{sign}\,(x_j - x_i) \tag{13.1}$$

where x_i and x_j are sequential data for *i*th and *j*th terms, *n* is the sample size, and

$$\text{sign}\,(x_j - x_i) = \begin{cases} +1, \text{ if } x_j - x_i > 1 \\ 0, \text{ if } x_j - x_i = 1 \\ -1, \text{ if } x_j - x_i < 1 \end{cases} \tag{13.2}$$

The variance, *Var*(*s*) of the *S* statistic is expressed as:

$$E(S) = 0,\ \text{Var}(S) = \frac{n(n-1)(2n+5)}{18} \tag{13.3}$$

Although, if there is an existence of tie in the data set, then the *Var*(*S*) expression is adjusted and it becomes:

$$\text{Var}(S) = \frac{\left\{ n(n-1)(2n+5) - \sum_{p=1}^{q} t_p(t_p-1)(2t_p+5) \right\}}{18} \tag{13.4}$$

where the variable q is the number of tied groups and the variable t_p is the number of data values in the *p*th group. The *Z* standardized statistic is expressed as:

$$Z_{\text{mk}} = \begin{cases} \dfrac{S-1}{\sqrt{\text{Var}(S)}} \text{ if } S > 0 \\ 0 \text{ if } S = 0 \\ \dfrac{S+1}{\sqrt{\text{Var}(S)}} \text{ if } S < 0 \end{cases} \tag{13.5}$$

The positive Z_{mk} represents an increasing trend and negative Z_{mk} represents a decreasing trend.

13.3.3 MODIFIED MANN-KENDALL TEST

The MMK test is appropriate for autocorrelated data and is based on the modified value of the variance of MK test statistics. When compared with the MK test, a superior accuracy in empirical significance level of the MMK test is observed without any loss of power (Kendall, 1975; Hamed and Rao, 1998). The MMK test is applicable for both rainfall and streamflow.

A modified variance of S, termed as Var(S)*, was used for this purpose and was expressed as follows:

$$\text{Var}(S)^* = V(S)\frac{n}{n^*} \tag{13.6}$$

where n^* denotes an effective sample size. The computation of n/n^* ratio was performed directly from the equation given by Hamed and Rao (1998) as:

$$\frac{n}{n^*} = 1 + \frac{2}{n(n-1)(n-2)}\sum_{i=1}^{n-1}(n-i)(n-i-1)(n-i-2)r_i \tag{13.7}$$

where n is the actual number of observations and r_i is the lag-i significant autocorrelation coefficient of rank I of time series. Once the computation of Var(S)* is done from eq 13.6, then further it is substituted for Var(S) in eq 13.5. Eventually, the MK and MMK Z has been tested at 95% significance level.

13.3.4 MAGNITUDE OF TREND

Along with the identification of the trend existence, the trend magnitude has also been estimated by a slope estimator, proposed by Sen (1968), commonly known as Sen slope estimator (SSE). SSE is a powerful estimation method of the trend magnitude. Also, the SSE is the median of all the feasible combinations of pairs for the entire set of data (Hirsch et al., 1982). An "upward trend" is an indication of a positive SSE value and a "downward trend" is an indication of a negative value (Xu et al., 2007; Karpouzos et al., 2010). The prediction of the trend magnitude is done by the Sen's slope estimator by using the slope (Ti) of the entire data pairs, computed as follows:

$$Ti = \frac{x_j - x_i}{j - i} \tag{13.8}$$

where x_j and x_i are assumed to be the data values at time j and i ($j > i$) correspondingly. The median of these N values of Ti is represented as Sen's estimator of slope. Sen's estimator is estimated as $Q_{\text{med}} = T(N+1)/2$ if N seems to be odd, and it is considered as $Q_{\text{med}} = [T(N/2)+T(N+2)/2]$ if N seems to be even.

An upward or increasing trend in the time series is an indication of a positive Q_i value, while a downward or decreasing trend is an indication of a negative Q_i value.

13.3.5 COEFFICIENT OF VARIATION TEST OR RAINFALL VARIABILITY TEST

The test of CV is used for the measurement of the how the individual points vary about the mean value (Landsea and Gray, 1992). The variability of the rainfall is determined by calculating the CV. A larger variability is denoted by a high value of CV while the lower CV values denote less variability. The computation is done as follows:

$$CV = \frac{\sigma}{\mu} \times 100 \tag{13.9}$$

where coefficient of variation is denoted by CV, standard deviation by μ, and mean precipitation is denoted by μ.

13.3.6 SPATIAL ANALYSIS

The spatial interpolation approach is made for determining the spatial pattern by applying Arc GIS 10.1. The IDW which is a local interpolation technique is applied for the assessment of spatial temporal precipitation trends over the basin with the help of ArcGIS (Singh and Chowdhary, 1986; Lebel et al., 1987).

The flowchart of the methodology is presented in Figure 13.2.

13.4 RESULTS AND DISCUSSION

13.4.1 PRECIPITATION TREND ANALYSIS (MONTHLY, SEASONAL AND ANNUAL)

Trend analysis for the Rampur watershed was performed by applying the MK method and the Sen slope estimator. The analysis was performed by using the monthly, average annual, and seasonal (premonsoon, monsoon, postmonsoon and winter) precipitation data.

13.4.2 MANN-KENDALL TEST OF THE PRECIPITATION SERIES

13.4.2.1 MONTHLY TRENDS

Tables 13.1 and 13.3 show the MK and MMK statistics of the monthly precipitation series at 5% significance level respectively. An increasing trend

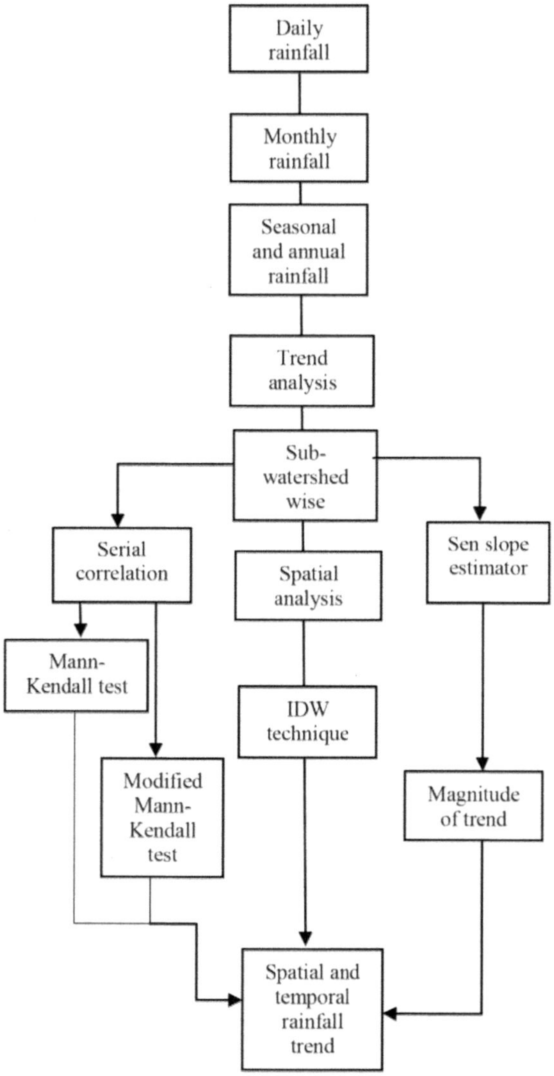

FIGURE 13.2 Flowchart of the methodology.

denotes a positive Z-statistics value whereas a decreasing trend denotes a negative Z-statistics value. Z-statistics values for nine SWS of monthly precipitation for the period 1983–2015 are presented in Table 13.1. the MK and MMK both showed that 86% of the values indicated a negative trend (either statistically significant or nonsignificant). Table 13.1 presents the MK statistics of total 108 cases (9 SW × 12 months) at 5% significance level

TABLE 13.1 Z-Statistics Values of Monthly Rainfall Using MK Test for the Nine Subbasins of Rampur Watershed during 1983–2015.

SW	Jan	Feb	Mar	Apr	May	Jun	Jul	Aug	Sep	Oct	Nov	Dec
1	−1.697	−0.466	0.416	1.056	1.408	−0.108	0.015	1.673	0.790	0.015	−1.475	−0.028
2	−0.987	−1.724	−0.266	−0.633	0.232	−0.310	0.263	0.635	0.434	−0.790	−3.565	−0.390
3	−0.435	−2.112	−0.648	0.482	−0.248	−0.976	−0.604	0.666	0.480	−0.465	−1.327	−0.123
4	−0.574	−2.336	−0.361	0.388	−0.992	−0.976	−1.038	−0.232	0.573	−0.883	−0.941	0.143
5	−0.388	−2.024	−0.533	0.558	−1.286	−1.410	−1.007	−0.139	0.480	−1.023	−1.520	−0.102
6	−0.995	−2.055	−0.675	0.620	−1.394	−1.410	−0.945	−0.108	0.573	−1.302	−1.603	−0.442
7	−1.216	−1.851	−0.512	0.864	−0.961	−1.627	−1.131	−0.170	0.480	−1.224	−1.560	−0.138
8	−0.459	−2.000	−1.566	0.202	−0.822	−0.883	−0.356	0.821	0.976	−1.255	−1.354	−2.026
9	−0.757	−2.546	−1.789	0.108	−0.419	−0.589	0.294	1.456	1.410	−1.255	1.423	−2.086

Significant at the 5% level.

whereas 77 cases showed negative trend, 9 cases showed negative significant trend, and 22 cases were obtained with positive trend. The MMK statistics of total 108 cases is presented in Table 13.3 where 79 cases showed negative trend, 22 showed positive trend, 7 showed negatively significant trend, and 1 case indicated positively significant trend at 5% significance level. The month of September more likely showed a positive trend for all SWS in both MK and MMK tests. Positive significant trend was found in only SW-1 (using MMK) in the month of August. Both MK and MMK tests showed almost similar number statistically significant trends.

13.4.2.2 ANNUAL AND SEASONAL TRENDS

The results of trend analysis at 5% significance level for seasonal and annual precipitation data for the period 1983–2015 using MK and MMK tests for the 9 SWS of Rampur watershed are presented in Tables 13.2 and 13.4. In both MK and MMK tests, only SW 1 showed positive significant trend in premonsoon season. A negative significant trend was found in SWS 8 and 9 in postmonsoon season using MMK test while all the SWS showed nonsignificant precipitation trends in both the tests. The MK and MMK statistics gave almost similar results for annual and seasonal precipitation series except for postmonsoon season, in MK test, all the 9 SWS showed negative trend, and in MMK test, 7 SWS showed negative, and the rest 2 SWS showed negative significant trend.

TABLE 13.2 Z-Statistics Values for Seasonal and Annual Rainfall Using MK Test for the Nine Subbasins of Rampur Watershed during 1983–2015.

SW	Annual	Premonsoon	Monsoon	Postmonsoon	Winter
1	0.821	2.169	0.821	*−0.418*	0.295
2	*−0.077*	*−0.015*	0.542	*−1.007*	*−1.147*
3	*−0.077*	*0.015*	0.170	*−0.418*	*−0.775*
4	*−0.697*	*−0.573*	*−0.480*	*−0.976*	*−0.558*
5	*−0.790*	*−0.589*	*−0.914*	*−1.054*	*−1.039*
6	*−0.759*	*−0.914*	*−0.852*	*−1.271*	*−1.318*
7	*−0.852*	*−0.604*	*−0.852*	*−1.007*	*−1.256*
8	0.387	*−0.325*	1.007	*1.503*	*−1.380*
9	1.100	*−0.093*	1.658	*−1.394*	*−1.302*

**Significant at the 5% level.*

TABLE 13.3 Z-Statistics Values of Monthly Rainfall Using MMK Test for the Nine Subbasins of Rampur Watershed during 1983–2015.

SW	Jan	Feb	Mar	Apr	May	Jun	Jul	Aug	Sep	Oct	Nov	Dec
1	*−0.651*	*−0.434*	0.337	0.864	1.379	*−0.226*	0.015	2.075	0.710	0.000	*−1.085*	*−0.015*
2	*−0.666*	*−1.689*	*−0.190*	*−0.620*	0.232	*−0.619*	*−0.263*	1.012	0.434	*−0.790*	*−0.790*	*−0.263*
3	*−0.356*	*−2.045*	*−0.635*	*−0.480*	*−0.248*	*−0.976*	*−0.604*	1.032	0.480	*−0.465*	*−0.744*	*−0.093*
4	*−0.387*	*−2.262*	*−0.356*	*−0.387*	*−0.992*	*−0.976*	*−1.038*	*−0.232*	0.573	*−0.883*	*−0.635*	0.108
5	*−0.294*	*−1.999*	*−0.527*	*−0.558*	*−1.286*	*−1.410*	*−1.007*	*−0.139*	0.480	*−1.023*	*−0.852*	*−0.077*
6	*−0.558*	*−2.030*	*−0.666*	*−0.620*	*−1.394*	*−1.410*	*−0.945*	*−0.108*	0.573	*−1.302*	*−0.899*	*−0.199*
7	*−0.682*	*−1.813*	*−0.496*	*−0.837*	*−0.961*	*−1.627*	*−1.131*	*−0.170*	0.480	*−1.224*	*−0.821*	*−0.061*
8	*−0.310*	*−1.937*	*−1.534*	*−0.289*	*−0.821*	*−0.883*	*−0.356*	0.821	0.976	*−2.387*	*−0.832*	*−1.859*
9	*−0.511*	*−2.495*	*−1.366*	*−0.149*	*−0.418*	*−0.589*	0.294	1.577	1.410	*−2.116*	*−0.961*	*−1.534*

Significant at the 5% level.

TABLE 13.4 Z-Statistics Values for Seasonal and Annual Rainfall Using MMK Test for the Nine Subbasins of Rampur Watershed during 1983–2015.

SW	Annual	Premonsoon	Monsoon	Postmonsoon	Winter
1	0.821	2.169	0.821	*−0.418*	0.531
2	*−0.099*	*−0.015*	1.228	*−1.007*	*−1.147*
3	*−0.089*	0.000	0.197	*−0.418*	*−0.775*
4	*−0.810*	*−0.547*	*−0.554*	*−0.976*	*−0.558*
5	*−0.928*	*−0.589*	*−1.045*	*−1.054*	*−1.038*
6	*−0.872*	*−0.914*	*−0.852*	*−1.271*	*−1.317*
7	*−0.977*	*−0.604*	*−0.852*	*−1.007*	*−1.255*
8	0.387	*−0.325*	1.007	*2.548*	*−1.379*
9	1.100	*−0.124*	1.658	*2.550*	*−1.302*

Significant at the 5% level.

13.4.3 ESTIMATION OF THE MAGNITUDE OF THE TREND SLOPE

13.4.3.1 MONTHLY PRECIPITATION DATA

A large variability can be observed in the magnitude and direction of the precipitation trends for all the 9 SWS of the Rampur watershed. Box plot of the Sen slope for the precipitation time series on a monthly basis for the period 1983–2015 was shown in Figure 13.3(a). The medians of almost 10-month slopes showed a negative value. Out of 12 whiskers or the vertical line, the lowest point was observed in the month of July (−3.59 mm/year) and the highest point was observed in the month of August (2.88 mm/year).

Total 108 values were considered for detecting the trend for the period 1983–2015 on monthly basis. Out of 108 cases, 56 values were negative and 19 values were positive (Table 13.5). The months January, November, and December showed no trend values for all the 9 SWS. The month of September had all positive values. The months February, April, May, July, and October showed negative values or decreasing trends while all the values of the June month were negative.

13.4.3.2 SEASONAL AND ANNUAL PRECIPITATION DATA

Box plots for the slopes of seasonal and annual time series for the period 1983–2015 was shown in Figure 13.3(b). The median of the slopes for all

TABLE 13.5 The Sen Slope of Monthly Rainfall Data for Rampur Watershed for the Period 1983–2015.

SW	Jan	Feb	Mar	Apr	May	Jun	Jul	Aug	Sep	Oct	Nov	Dec
1	0.000	0.000	0.000	0.065	0.256	*−0.321*	0.183	2.882	1.479	0.000	0.000	0.000
2	0.000	*−0.106*	0.000	*0.005*	0.030	*−0.454*	*−0.488*	1.276	1.015	*−0.313*	0.000	0.000
3	0.000	*−0.152*	*−0.012*	*−0.033*	*−0.042*	*−1.871*	*1.523*	0.833	1.329	*−0.227*	0.000	0.000
4	0.000	*−0.237*	*0.000*	*−0.035*	*−0.314*	*−2.443*	*−2.675*	*−0.390*	1.289	*−0.529*	0.000	0.000
5	0.000	*−0.251*	*−0.005*	*−0.069*	*−0.287*	*−3.010*	*2.481*	*−0.214*	1.404	*−0.588*	0.000	0.000
6	0.000	*−0.202*	*−0.013*	*−0.061*	*−0.326*	*−3.109*	*3.035*	*−0.156*	1.685	*−0.644*	0.000	0.000
7	0.000	*−0.224*	*0.000*	*−0.014*	*−0.160*	*−3.332*	*−3.588*	*−0.637*	1.478	*−0.632*	0.000	0.000
8	0.000	*−0.148*	*−0.114*	*−0.006*	*−0.215*	*−1.719*	*0.905*	2.034	2.058	*−0.608*	0.000	0.000
9	0.000	*−0.128*	*−0.135*	*−0.006*	*−0.097*	*−1.074*	0.973	2.623	3.228	*−0.718*	0.000	0.000

the seasons and the annual time series were negative except for the monsoon season. The lowest value the slope was observed in the annual time scale (−7.17 mm/year) and the highest value was observed in the monsoon season (7.87 mm/year).

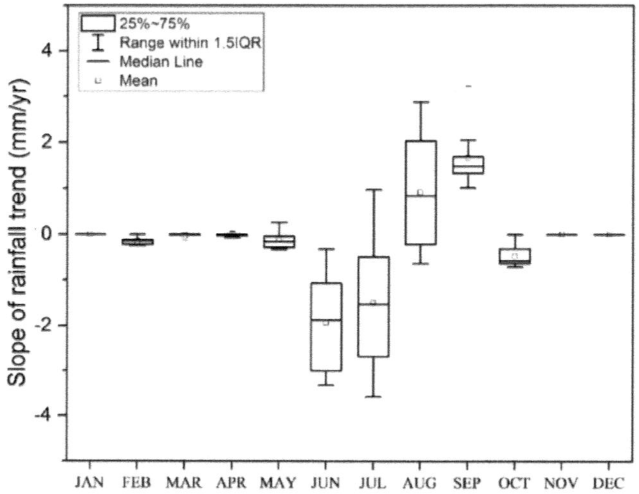

FIGURE 13.3(a) Box plot of the Sen Slopes for monthly rainfall time series of Rampur watershed of Mahanadi river basin for the period 1983–2015.

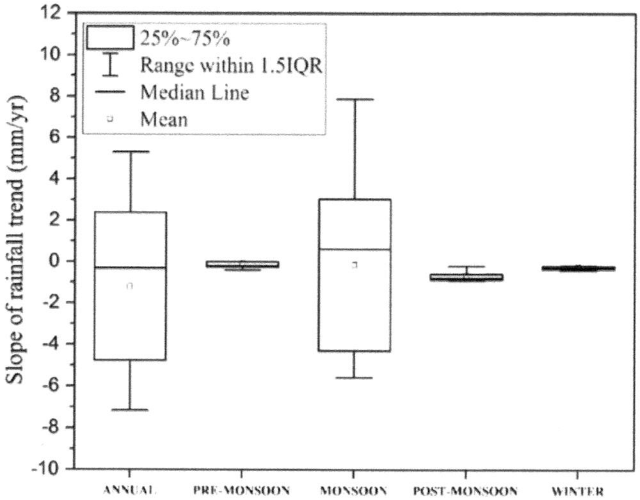

FIGURE 13.3(b) Box plot of the Sen Slopes for annual and seasonal rainfall time series of Rampur watershed of Mahanadi river basin for the period 1983–2015.

The trends magnitude for the annual and seasonal precipitation for the period 1983–2015 is presented in Table 13.6. Total 45 values are considered for the trend detection. Out of 45 values, 35 values were found to be negative and 10 values were of positive magnitude. The magnitude of the negative trend for annual precipitation series ranged from 0.13 (SW 3) to 7.18 (SW 7). The postmonsoon season gave negative values for all the 9 SW and the magnitude of the negative trend ranged from 0.19 (SW 1) to 0.91 (SW 8).

TABLE 13.6 The Sen Slope for Seasonal and Annual Rainfall Data for Rampur Watershed for the Period 1983–2015.

SW	Annual	Premonsoon	Monsoon	Postmonsoon	Winter
1	4.034	0.700	3.024	*−0.196*	0.050
2	*−0.326*	*−0.008*	2.084	*−0.591*	*−0.363*
3	*−0.126*	*−0.001*	0.609	*−0.253*	*−0.194*
4	*−4.688*	*−0.336*	*−3.064*	*−0.587*	*−0.217*
5	*−5.556*	*−0.260*	*−5.311*	*−0.806*	*−0.328*
6	*−4.774*	*−0.400*	*−4.315*	*−0.867*	*−0.412*
7	*−7.172*	*−0.269*	*−5.590*	*−0.793*	*−0.347*
8	2.384	*−0.207*	3.397	*−0.906*	*−0.311*
9	5.288	*−0.049*	7.874	*−0.902*	*−0.300*

13.4.3.3 ANALYZING ANNUAL AND SEASONAL PRECIPITATION VARIABILITY PATTERN

For acquiring a precised evaluation of an essential supplemental water, it is vital to know the seasonal variations of precipitation (Gajbhiye et al., 2015). The coefficient of variation for the annual and seasonal precipitation series of the Rampur watershed for the period 1983–2015 indicates that the maximum variability was detected in the postmonsoon season followed by the winter season and the premonsoon season (Fig. 13.4). The lowest variability was detected for the annual and monsoon precipitation series with minimum values of 20.24% and 19.93% respectively. The minimum CV values for other seasons are 83.01% (premonsoon), 103.79% (postmonsoon), and 101.59% (winter). The CV values are found to be highest in SW 7 for three seasons: 109.77% for premonsoon, 34.79% for monsoon, and 135.13% for winter. For postmonsoon season, the highest CV value was observed in SW 1 with the value of 146%. Regions with high values of coefficient of variation have greater vulnerability toward floods and droughts (Pandey and Ramasastri, 2002; Turkes, 1996).

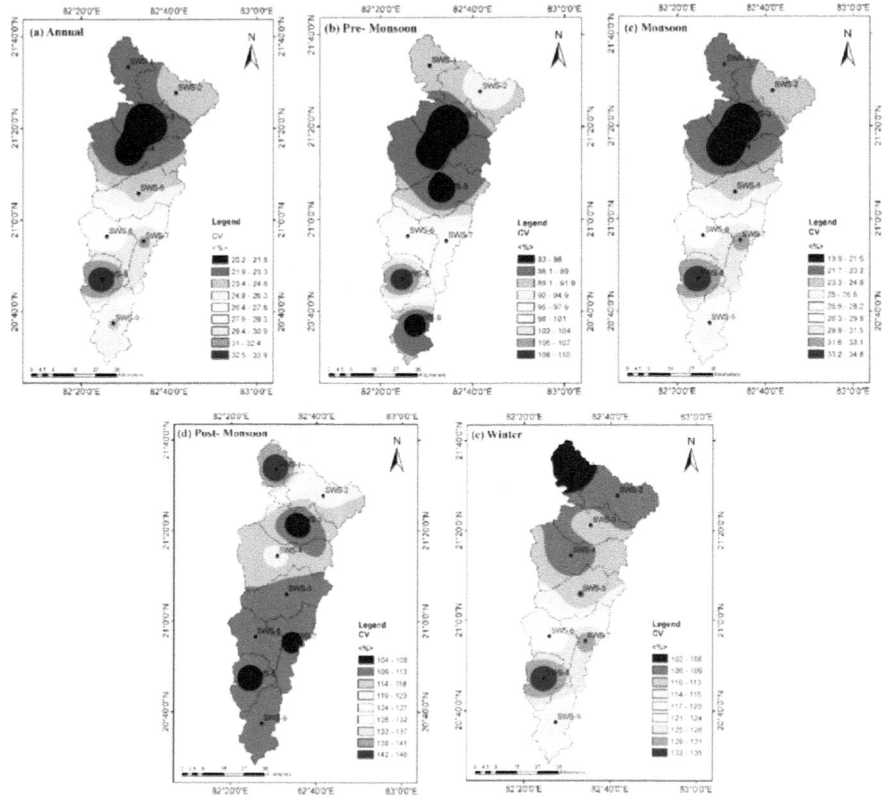

FIGURE 13.4 Interannual variability of rainfall (%CV) and its trends: (a) Annual, (b) premonsoon, (c) monsoon, (d) postMonsoon, and (e) winter seasons.

13.5 CONCLUSIONS

Trend analysis of monthly, annual, and seasonal precipitation series was performed for 9 SWS of Rampur watershed of Mahanadi river basin for the period 1983–2015. For determining the trend of the precipitation series, MK test and Sen slope estimation test were carried out. The results obtained from the study conclude that both MK and MMK tests give almost similar results for annual and seasonal precipitation series except for the postmonsoon season. The premonsoon season for both MK and MMK tests shows a positive significant trend for only SW 1. In monthly trends (using MK and

MMK), 86% of the values is indicated by a negative trend (either statistically significant or nonsignificant) and the month of September indicates a positive trend for all the SW. In MMK statistics, a positive significant trend is observed in the month of August in only SW 1.The magnitude in the postmonsoon season shows all negative values for all the SW. In monthly precipitation data, the months January, November, and December show no trend values for all the 9 SW. The months September and June show all positive and negative values respectively. Understanding the trend and variability of precipitation is necessary for determining the supplemental water required during the critical growth periods of the crops and in managing the water resources more adequately and sustainably.

KEYWORDS

- **trend analysis**
- **Mann-Kendall test**
- **modified Mann-Kendall test**
- **rainfall**
- **Rampur watershed**

REFERENCES

Bandyopadhyay, J.; Perveen, S. A Scrutiny of the Justifications for the Proposed Interlinking of Rivers in India. In *Interlinking of Rivers in India: Overview and Ken-Betwa Link*; Alagh, Y. K., Pangare, G., Gujja, B., Eds.; Academic Foundation: New Delhi, 2006; pp 23–52.

Bradley, R. S.; Diaz, H. F.; Eischeid, J. K.; Jones, P. D.; Kelly, P. M.; Goodess, C. M. Precipitation Fluctuations over Northern Hemisphere Land Areas since the Mid-19th Century. *Science* **1987,** *237*, 171–175.

Caloiero, T. Evaluation of Rainfall Trends in the South Island of New Zealand through the Innovative Trend Analysis (ITA). *Theor. Appl. Climatol.* **2019,** *139*, 493–504.

Chakraborty, S.; Pandey, R. P.; Chaube, U. C.; Mishra, S. K. Trend and Variability Analysis of Rainfall Series at Seonath River Basin, Chhattisgarh (India). *Int. J. Appl. Sci. Eng. Res.* **2013,** *2* (4), 425–434.

Chandniha, S. K.; Meshram, S. G.; Adamowski, J. F.; Meshram, C. Trend Analysis of Precipitation in Jharkhand State, India. Investigating Precipitation Variability in Jharkhand State. *Theor. Appl. Climatol.* **2017,** *130*, 261–274. doi: 10.1007/s00704-016-1875-x

Chattopadhyay, N.; Hulme, M. Evaporation and Potential Evapotranspiration in India under Conditions of Recent and Future Climate Change. *Agric. For. Meteorol.* **1997,** *87*, 55–73.

Chaudhary, A.; Abhyankar, V. P. Does Precipitation Pattern Foretell Gujarat Climate Becoming Arid. *Mausam* **1979,** *30*, 85–90.

Chowdhury, R. K.; Beecham, S. Australian Rainfall Trends and Their Relation to the Southern Oscillation Index. *Hydrol. Process* **2010,** *24* (4), 504–514.

Dash, S. K.; Jenamani, R. K.; Kalsi, S. R.; Panda, S. K. Some Evidence of Climate Change in Twentieth-Century India. *Clim. Change* **2007,** *85*, 299–321.

Dinpashoh, Y.; Mirabbasi, R.; Jhajharia, D.; Abianeh, H.; Mostafaeipour, A. Effect of Short-Term and Long-Term Persistence on Identification of Temporal Trends. *J. Hydrol. Eng.* **2014,** *19* (3), 617–625.

Fawcett, R. A Long-Term Trend in Melbourne Rainfall? *Bull. Aust. Meteor. Oceanographic Soc.* **2004,** *17*, 122–126.

Fu, T.; Jacob, D. J.; Wittrock, F.; Burrows, J. P.; Vrekoussis, M.; Henze, D. K. Global Budgets of Atmospheric Glyoxal and Methylglyoxal, and Implications for Formation of Secondary Organic Aerosols. *J. Geophys. Res.* **2008,** *113* (D15303), 1–17.

Gadagil, A. Annual and Weekly Analysis of Rainfall and Temperature for Pune: A Multiple Time Series Approach. *Inst. Indian Geogr.* **1986,** *8* (1).

Gajbhiye, S.; Meshram, C.; Mirabbasi, R.; Sharma, S. K. Trend Analysis of Rainfall Time Series for Sindh River Basin in India. *Theor. Appl. Climatol.* **2015**. doi:10.1007/s00704-015-1529-4

Gajbhiye, S.; Meshram, C.; Singh, S. K.; Srivastava, P. K.; Islam, T. Precipitation Trend Analysis of Sindh River basin, India from Long Term 102 Years of Record (1901–2002). *Atmos. Sci. Lett.* **2016,** *17* (1), 71–77.

Gallant, A. J. E.; Hennessy, K. J.; Risbey, J. Trends in Rainfall Indices for Six Australian Regions: 1910: 2005. *Aust. Meteorol. Mag.* **2007,** *56*, 223–239.

Gozzo, L.; Palma, D.; Custodio, M.; Machado, J. Climatology and Trend of Severe Drought Events in the State of Sao Paulo, Brazil, during the 20th Century. *Atmosphere* **2019,** *10*, 190.

Guhathakurta, P.; Rajeevan, M. Trends in the Rainfall Pattern over India. NCC Research Report No 2/2006. National Climate Centre. India Meteorological Department, 2006; p 23.

Hamed, K. H.; Rao, A. R. A Modified Mann-Kendall Trend Test for Auto Correlated Data. *J. Hydrol.* **1998,** *204*, 182–196.

Hennessy, K. J.; Suppiah, R.; Page, C. M. Australian Rainfall Changes, 1910–1995. *Aust. Meteor. Mag.* **1999,** *48*, 1–13.

Hirsch, R. M.; Slack, J. R.; Smith, R. A. Techniques of Trend Analysis for Monthly Water Quality Data. *Water Resourc. Res.* **1982,** *18* (1), 107–121.

Hulme, M.; Osborn, T. J.; Johns, T. C. Precipitation Sensitivity to Global Warming: Comparison of Observations with HADCM2 Simulations. *Geophys. Res. Lett.* **1998,** *25*, 3379–3382.

IPCC. Climate Change 2007: Climate Change Impacts, Adaptation and Vulnerability. Working Group II Contribution to the Intergovernmental Panel on Climate Change Fourth Assessment Report. Summary for Policymakers, 2007; p 23.

Jagannathan, P.; Parthasarathy, B. Trends and Periodicities of Rainfall over India. *Mon. Weather Rev.* **1973,** *101*, 371–375.

Kampata, J. M.; Parida, B. P.; Moalafhi, D. B. Trend Analysis of Rainfall in the Headstreams of the Zambezi River Basin in Zambia. *Phys. Chem. Earth* **2008,** *33*, 621–625.

Kendall, M. G. *Rank Correlation Methods*; Charles Graffin: London, 1975.

Kiem, A. S.; Verdon-Kidd, D. C. Towards Understanding Hydroclimatic Change in Victoria, Australia—Preliminary Insights Into the BBig Dry. *Hydrol. Earth Syst. Sci.* **2010,** *14*, 433–445.

Koteswaram, P.; Alvi, S. M. A. Secular Trends and Periodicities in Rainfall at West Coast SWS in India. *Curr. Sci.* **1969,** *38*, 229–231.

Kumar, V.; Jain, S. K. Trends in Rainfall Amount and Number of Rainy Days in River Basins of India (1951–2004). *Hydrol. Res.* **2011,** *42* (4), 290–306.

Kumar, V. P.; Bindi, M.; Crisci, A.; Maracchi, G. Detection of Variation in Air Temperature at Different Time Scale during the Period 1889–1998 at Firenze, Italy. *Clim. Change* **2005,** *72*, 123–150.

Lal, M. Climatic Change—Implications for India's Water Resources. *J. Indian Water Res. Soc.* **2001,** *21*, 101–119.

Landsea, C. W.; Gray, W. M. The Strong Association between Western Sahel Monsoon Rainfall and Intense Atlantic Hurricanes. *J. Clim.* **1992,** *5* (5), 435–453.

Lebel, T. G.; Bastin, G.; Obled, C.; Creutin, J. D. On the Accuracy of Areal Rainfall Estimation: A Case Study. *Water Resour. Res.* **1987,** *23*, 2123–2134.

Mann, H. B. Non Parametric Tests again Trend. *Econometrica* **1945,** *13*, 245–259.

Meshram, S. G.; Singh. V. P.; Meshram, C. Long-term Trend and Variability of Precipitation in Chhattisgarh State. *Theor. Appl. Climatol.* India, **2016.** doi:10.1007/s00704-016-1804-z

Mooley, D. A.; Parthasarthy, B. Fluctuations of All India Summer Monsoon Rainfall during 1871–1978. *Clim. Change* **1984,** *6*, 287–301.

Murphy, B. F.; Timbal, B. A Review of Recent Climate Variability and Climate Change in Southeastern Australia. *Int. J. Climatol.* **2008,** *28* (7), 859–879.

Praveen, B.; Talukdar, S.; Mahato, S.; Mondal, J.; Sharma, P.; Islam ARMT; Rahman, A. Analyzing Trend and Forecasting of Rainfall Changes in India Using Nonparametrical and Machine Learning Approaches. *Sci. Rep.* **2020,** *10*, 10342.

Raghavendra, V. K. Trends and Periodicities of Rainfall in Subdivisions of Maharashtra state. *Indian J. Meteorol. Geophys.* **1974,** *25*, 197–210.

Saini, A.; Sahu, N.; Kumar, P.; Nayak, S.; Duan, W.; Avtar, R.; Behera, S. Advanced Rainfall Trend Analysis of 117 Years over West Coast Plain and Hill Agro-Climatic Region of India. *Atmosphere* **2020,** *11*, 1225. doi: 10.3390/atmos11111225

Sen, P. K. Estimates of the Regression Coefficient Based on Kendall's Tau. *J. Am. Stat. Assoc.* **1968,** *63*, 1379–1389.

Shafiq, M. U.; Rasool, R.; Ahmed, P.; Dimri, A. P. Temperature and Precipitation Trends in Kashmir Valley, North Western Himalayas. *Theor. Appl. Climatol.* **2019,** *135* (1–2), 293–304. doi: 10.1007/s00704-018-2377-9

Singh, V. P.; Chowdhury, P. K. Comparing Some Methods of Estimating Mean Areal Rainfall. *Water Resour. Bull.* **1986,** *22*, 275–282.

Smith, I. An Assessment of Recent Trends in Australian Rainfall. *Aust. Meteor. Mag.* **2004,** *53*, 163–173.

Swain, S.; Verma, M.; Verma, M. K. Statistical Trend Analysis of Monthly Rainfall for Raipur District, Chhattisgarh. *Int. J. Adv. Eng. Res. Stud.* **2015,** *4* (2), 87–89.

Taschetto, A. S.; England, M. H. An Analysis of Late Twentieth Century Trends in Australian Rainfall. *Int. J. Climatol.* **2009,** *29* (6), 791–807.

Thakur, A.; Mishra, P. K.; Nema, A. K.; Sahoo, S. P. Spatio-temporal Trends and Shift Analysis of Temperature for Wainganga Sub-Basin, India. *Int. J. Environ. Stud.* **2020,** *77* (3), 464–479. doi: 10.1080/00207233.2019.1686884

Theil, H. A Rank Invariant Method of Linear and Polynomial Regression Analysis, Part 3. *Nederl. Akad. Wetensch Proc.* **1950,** *53*, 1397–1412.

CHAPTER 14

Estimation of Soil Loss and Sediment Yield Using Universal Soil Loss Equation in Jiadhal Basin of Dhemaji District, Assam

ARNAB SARMA

Department of Civil Engineering, The Assam Royal Global University, Betkuchi, Guwahati 781035, Assam, India

**Corresponding author. E-mail: arnab.sarma@rgi.edu.in, nebapghy@gmail.com*

ABSTRACT

Soil erosion can be considered to be one of the most pressing of global environmental problems currently and can be attributed to both natural and human factors. Soil erosion by water is one of the major threats to soils in most of the river basins/catchments including in the river basins/catchments of the northeast region of India. Jiadhal is one such basin in this region that contributes huge sediments to the downstream plains of Assam. Accordingly, an attempt was made to quantify the soil loss from the catchment using the Universal Soil Loss Equation (ULSE). The river Jiadhal originates at an altitude of 1247 m above the mean sea level in West Siang district of Arunachal Pradesh. It flows southward and joins the river Subansiri, which is the largest tributary of Brahmaputra. It has its extent/spread between latitude 27°15' N and 27°45' N, and longitude 94°15' E and 94°40' E. The catchment of the river can be demarcated using Survey of India Toposheets: 83 I/6, 83 I/7, 83 I/8 and 83 I/10. The catchment of Jiadhal drains an area of 437 km^2

Advances in Hydrology and Climate Change: Historical Trends and New Approaches in Water Resources Management. Surendra Kumar Chandniha, Anil Kumar Lohani, Gopal Krishan, & Ajay Krishna Prabhakar (Eds.)

considering the outlet at Bordoloni. The soil loss estimation was carried out by individually determining the parameters, viz., rainfall erosivity factor denoted by "R," soil erodibility factor denoted by "K," slope length and steepness factor denoted by "LS," crop management factor denoted by "C" and finally the support practice factor denoted by "P." All these factors were quantified based on ArcGIS10.1 using relevant data sources as well as adopting data preparation methods. As part of this study, DEM and Landsat data were used. Toward estimating the factor "R" daily rainfall data (1990–2014) were used. The value of soil erodibility factor (K) was found to vary between 0.023 and 0.0325. The LS factor for study area varies from 0 to 355.8. The C-factor was derived from normalized difference vegetation index (NDVI), obtained from Landsat images downloaded from USGS earth explorer. The value for this factor (C) is found to lie in the range of 0.10053–2.35642 and the NDVI values of the Jiadhal catchment are 0.765766 and 0.285714. Assuming that no preventive measures are taken, the P-factor was assigned a value equal to 1 throughout the basin area. The soil loss map of Jiadhal basin was also generated and depending on the soil loss intensity, the basin's area has been classified under five categories. It was found that the annual soil loss from the basin ranged from 0 to 4790.45 T/ha/year depending on the area with a mean value of 10.359 T/ha/year. Sediment yield estimate at the outlet (Kumotiya river near Samorajan) was found to be about 21.179 T/ha/year. Finally, toward planning and implementing various conservation and management practices, the subbasins (4) have been prioritized based on the soil loss rates from them.

14.1 INTRODUCTION

Soil erosion is defined as the process of removal of soil by wind or water of which water is considered to the primary factor. The process of erosion caused by rain water essentially involves dislodgement, movement/transportation and finally settlement/deposition of the transported soil particles as the velocity (kinetic energy) of the water carrying the particles diminishes. Taken together, water and wind are considered responsible to cause degradation of nearly 84% of the land on a global scale. All basins do not erode at the same rate. Even within the same basin, a certain part may erode faster than the other. This means, which part of a basin will erode more than the other parts will depend on a plethora of factors, viz., basin configuration, orientation, the soil erodibility, and the rainfall intensity. Additionally, the rate of erosion largely depends on the land use, land cover, and the land management practices adopted by people within the basin.

Evaluation of loss of soil from basin is required while assessing soil erosion and its effects on agricultural production. It also concerns workers responsible for the management of sediment in reservoirs that are part of various hydroelectric projects. Sediment yield is a measure of load passing the outlet (point of concern) of a basin. There are a number of empirical models, viz., the USLE, the modified universal soil loss equation (MUSLE) and the Revised Universal Soil Loss Equation (RUSLE) that are extensively used by researchers across the globe for estimation of soil erosion from a given unit of land. These models are used because of their simple structure, ease of calculation and also because of the fact that they provide fairly acceptable results. Research has shown that USLE provides better estimate of soil loss when applied on a field plot scale studies as compared with a catchment scale study for the simple reason that a part of the soil that is eroded and transported gets deposited within the basin itself before reaching the point of interest (outlet). This difficulty can be overcome by applying routing procedures: one of the most commonly used procedures being the concept of sediment delivery ratio. The present study was therefore undertaken in one of the major transboundary basins (Jiadhal) of Dhemaji District, Assam to estimate the soil erosion rate and also sediment yield. The study has been conducted in a GIS environment for better results.

14.2 STUDY AREA

The present study was conducted at Jiadhal river basin that lies in Arunachal Pradesh as well as in Dhemaji district, Assam (Fig. 14.1). The basin area is encompassed between 27°25′45″ N and 27°44′15″ N latitudes and 94°15′ E and 94°37′30″ E longitudes. The river Jiadhal is a transboundary river between Assam and Arunachal Pradesh and it has its origin in West Siang district of Arunachal Pradesh at an altitude of about 1247 m above the mean sea level. As can be seen from Figure 14.2, three streams (Sido, Sika, and Siri) meet at a place called Trimukh (27°37′56″ N, 94°25′15″E) in Arunachal Pradesh and then flow through a narrow gorge in Arunachal Pradesh before entering the Dhemaji district of Assam at a place called Jiadhal mukh. Thereafter, the river flows in braided channels until it meets the river Subansiri. In the past, however, the river used to debouch into Brahmaputra near a place called Selamukh in Lakhimpur district. For the current study, the outlet was considered at a place called Bordoloni Holowdunga and the basin area up to this point is about 437 sq. km. Out of this, about 79 sq. km (18%) falls in Arunachal Pradesh and about 358 sq. km (82%) falls in Assam.

FIGURE 14.1 Location of study area.

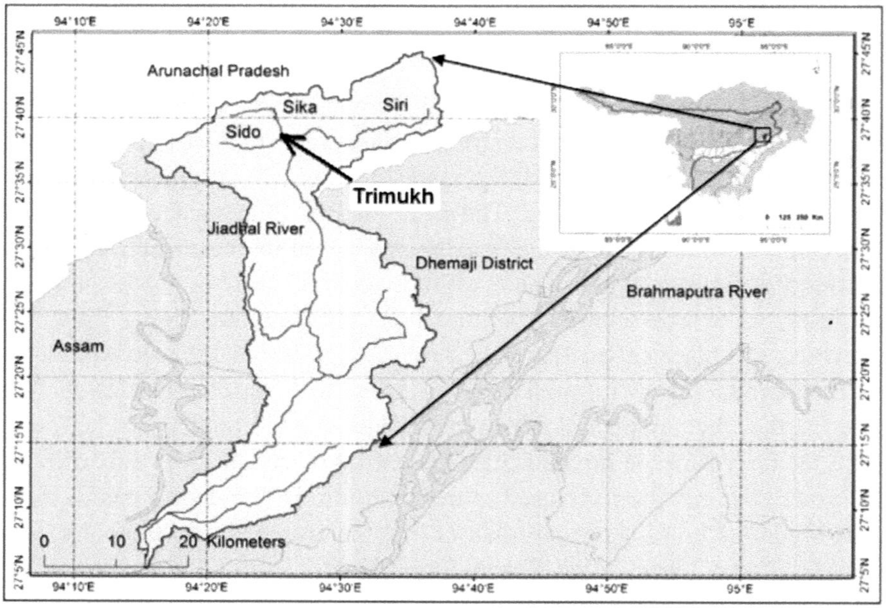

FIGURE 14.2 Sido, Sika, and Siri streams.

Being in the foothills of Arunachal Pradesh, the district of Dhemaji receives enormous quantity of rainfall (2600–3600 mm) due to orographic effect and is geographically an unstable zone. Heavy rainfall coupled with flash floods, water logging, and erosion damages lands, crops, and properties almost every year. The climate of the region is warm and temperate. The annual average temperature is 23.5°C. The minimum temperature experienced during winter is about 5.9°C. The relative humidity varies from 73% to 90%.

14.3 MATERIALS AND METHODS

14.3.1 THE SOIL LOSS ESTIMATION MODEL

As part of the present study, the USLE developed by Wischmeier and Smith (1965) was selected and used due to reasons already mentioned. The model can account for soil loss caused by sheet as well as rill erosion to a large extent. The parameters representing the model have their basis on annual rainfall distribution, soil erodibility characteristics, topographic features, land use/land cover, and erosion control practices adopted in a given area. The model can be expressed as (Fig. 14.3):

$$A = R \times K \times LS \times C \times P \qquad (14.1)$$

where

A= Estimated soil loss (T/ha/year)
R= Rainfall erosivity index
K= Soil erodibility factor
LS= Topographic factor (L: slope length & S: Slope)
C= Land cover factor
P= Conservation practice factor

14.3.2 DIGITAL ELEVATION MODEL

The digital elevation model of the Jiadhal basin is presented in Figure 14.4. This DEM of 30 m resolution was downloaded from the USGS earth explorer. For the purpose, the basin was delineated at first. The DEM of the basin was developed following the process of mask extraction on ArcGIS using the inbuilt shape file.

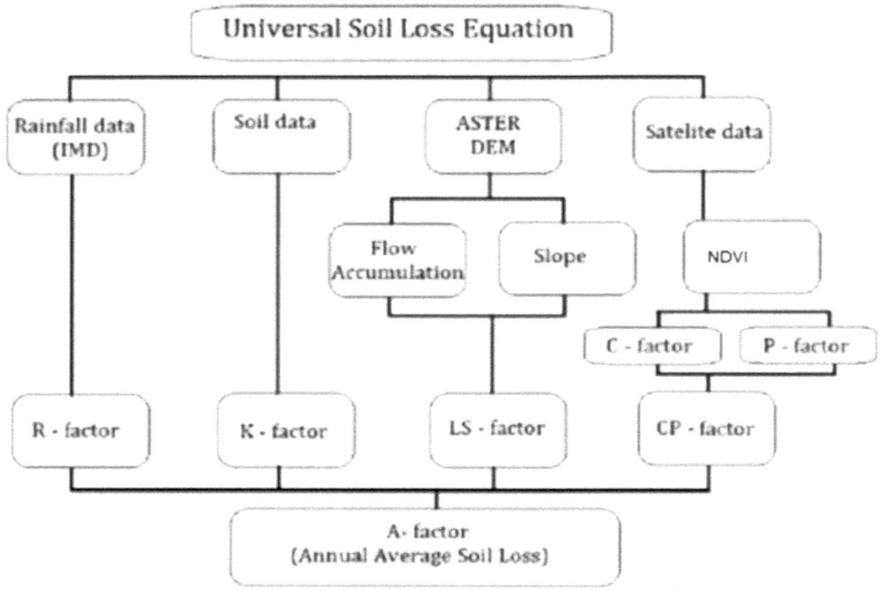

FIGURE 14.3 The component of the model.

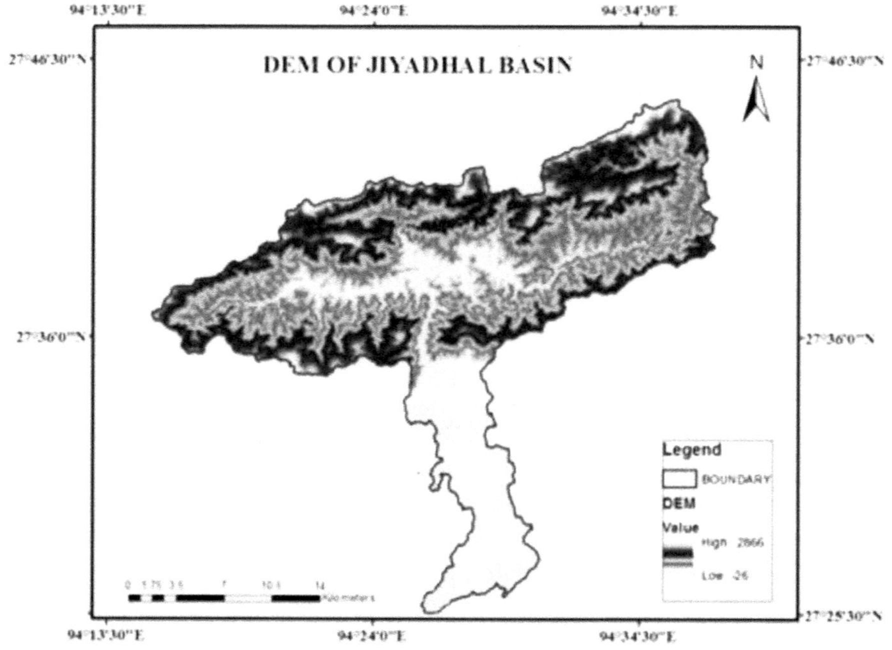

FIGURE 14.4 Digital elevation model of the Jiadhal basin.

14.4 THE USLE PARAMETERS

14.4.1 RAINFALL-RUNOFF EROSIVITY FACTOR (R)

This is a very important factor and may be more appropriately termed as erosivity index. It is calculated from a given annual summation of rainfall energy considering the maximum 30-min intensity of rainfall. It varies spatially. The factor heavily influences the erosion process that involves detachment/dislodgement, breakdown, and finally the transport of soil particles. For the purpose of the present study, precipitation data for the period from 1990 to 2014 was used to determine the "R" factor and the necessary data was collected from the Water Resources Department, GoA, Dhemaji Division. The data set used is presented in Table 14.1. It may be seen that the basin experiences heavy rainfall annually (average= 3130.15 mm). The value of the parameter "R" was calculated using the following expression.

$$R = P \times 0.5 \tag{14.2}$$

where
P = mean annual rainfall in "mm."

TABLE 14.1 Rainfall Variation in Different Years during 1990–2014.

Year	Rainfall (mm)	Year	Rainfall (mm)	Year	Rainfall (mm)	Year	Rainfall (mm)	Year	Rainfall (mm)
1990	3330.9	1995	3231.4	2000	3055.3	2005	3670.12	2010	3552.42
1991	3155.76	1996	3026.3	2001	2901.5	2006	2934.57	2011	2482.66
1992	2742	1997	3245.3	2002	2793	2007	3307.9	2012	3132.51
1993	3629.4	1998	3875.9	2003	3531	2008	2861.83	2013	1942.33
1994	3346.8	1999	2700.3	2004	5011.4	2009	2544.11	2014	2249.13

Average annual rainfall is about 3130.15 mm.

14.4.2 SOIL ERODIBILITY FACTOR (K)

This factor is used in the soil loss estimation model as it quantifies the cohesive character of a given type of soil. The factor also signifies the soil's resistance to dislodging and subsequent transport due to surface runoff. In the study area, two types of ACRISOLS, namely, Orthic Acrisol (A_o) and Eutric Cambisol (B_e) are found (Source: FAO, Digital Soil Map DSMW). The percentage of silt, clay, sand, and organic matter for each soil group is noted and the K-factors for each soil group were calculated by the following formula (Fig. 14.8; Table 14.3):

$$F_{CSand} = [0.2 + 0.3 \exp(-0.0256 \times Sand(1 - \frac{Silt}{100}))] \quad (14.3)$$

$$F_{ClSilt} = [\frac{Silt}{(Clay+Silt)}]^{0.2} \quad (14.4)$$

$$F_{org} = 1 - \frac{0.25 \times Clay}{Clay + Exp(3.72 - 2.95 \times Clay)} \quad (14.5)$$

$$F_{hisand} = [1 - \frac{0.7 \times SN1}{SN1 + \exp(-5.41 + 22.9 \times SN1)}] \quad (14.6)$$

$$K_{USLE} = F_{CSand} \times F_{ClSilt} \times F_{org} \times F_{hisand} \quad (14.7)$$

$$K_{factor} = 0.1317 \times K_{USLE} \quad (14.8)$$

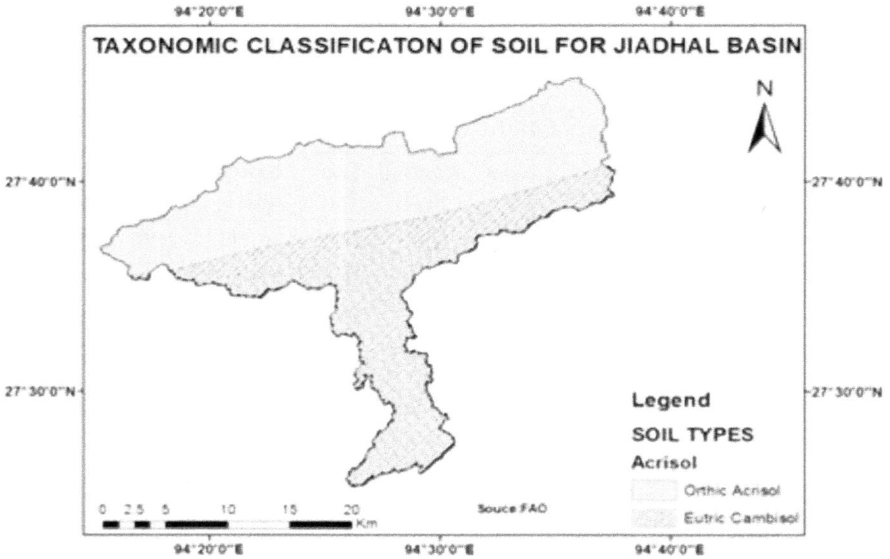

FIGURE 14.5 Soil classification for Jiadhal basin.

14.4.3 SLOPE LENGTH AND STEEPNESS FACTOR (LS)

The L and S factors represent the effect of topography on soil erosion. The factor (LS) has been considered as one of the most important parameters in USLE analysis because of its direct bearing on soil erosion by water. By integrating digital elevation model (DEM) in GIS, the slope length (L) and slope (S) can be determined accurately. Accuracy of estimation is largely dependent on the quality (resolution) of the DEM generated for the purpose. Normally, for a study of the present type, a 30 m ASTER DEM is enough and has therefore been used. To derive the required LS map Raster calculator was used. While the expression used to determine the value of "LS" for the study is given below. The slope map of the Jiadhal basin is shown in Figure 14.6.

TABLE 14.2 Percentage of Sand, Silt, Clay, and Organic Carbon for Acrisol.

Names	Sand (%)	Silt (%)	Clay (%)	Organic carbon (%)	SN1 = (1−Sand/100)
A_o	53.6	15.8	30.6	2.25	0.464
B_e	36.4	37.2	26.4	1.07	0.636

$$\begin{aligned} \text{LS} &= \text{Power}(\text{“Faccumulation”} \times 90/22.1,\ 0.4) \\ &\times \text{power}(\sin(\text{“slope”} \times 0.017450/0.09,\ 1.4) \times 1.4 \end{aligned} \tag{14.9}$$

14.4.4 LAND COVER FACTOR (C)

The land cover factor (more precisely cropping and land cover factor) is used in the model because it provides information on how efficiently the crop management methods adopted in the basin can prevent/reduce soil loss. It may be defined as the ratio of the amount of soil that is lost/eroded under a given crop cover and management system to a condition when the land is kept fallow. Therefore, it becomes incumbent on the part of the researchers to have good knowledge of the crop cover so as to determine the value of C as crop cover inhibits the loss of top soil. While estimating the value of "C" based on remote sensing index, the concept of NDVI is commonly used. This (NDVI) can be expressed with the help of the following equation (For Landsat TM):

$$\text{NDVI} = \frac{\text{NIR} - \text{RED}}{\text{NIR} + \text{RED}} \tag{14.10}$$

where

NIR: near-infrared region
RED: red (visible) region

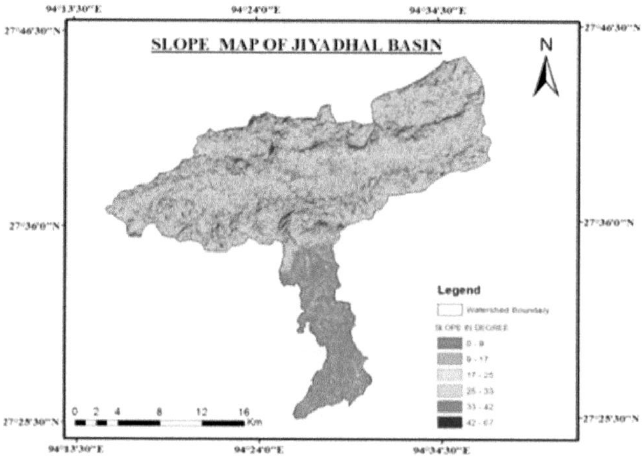

SLOPE CLASSES	SLOPE RANGE	DESCRIPTION
A	<9	Gentle sloping
B	9-17	Moderately sloping
C	17-25	Strongly sloping
D	25-33	Moderately steep
E	33-42	Steep
F	42-67	Very Steep
G	>67	Very Very Steep

FIGURE 14.6 Slope ranges of Jiadhal basin.

The value of NDVI for a given pixel ranges from −1 to +1. Increasing positive values indicate dense green cover, while increasing negative values indicate water. A value close to zero (0) indicates an urban conglomeration. For the purpose of the present study, NDVI image was generated taking recourse to Landsat data having resolution of 30 m. Following this step (production of the NDVI image), the value of factor "C" was calculated as follows:

$$C = e^{-\alpha\{(\mathrm{NDVI}/(\beta - \mathrm{NDVI})\}} \tag{14.11}$$

where

α & β are used to determine the shape of the NDVI versus C curve.

For the present study, $\alpha = 2$ and $\beta = 1$

14.4.5 SUPPORT PRACTICE FACTOR (P)

This factor is an important factor in influencing soil loss from a given area and indicates as to how various agronomic and soil conservation practices like contour and strip cropping, provision of concave slopes, various terrace types, etc. affect soil loss process. The practices adopted in a given area determine the degree of soil erosion/loss and is represented by the C-factor. The support practices adopted in an area are intended to reduce soil erosion/loss mainly by reducing the runoff velocity. As no information on strip cropping, terracing, and contour cultivation practices was available, it was assumed that no preventive measures are taken and thus, the P-factor was assigned a value of 1 throughout the basin area.

14.5 ESTIMATION OF SEDIMENT YIELD

The runoff from an area that reaches an outlet carries soil/sediment contributed by the area upstream of the outlet. Generally speaking, the total quantity of sediment that reaches the outlet is termed as sediment yield. This parameter is an important indicator of the catchment health and may be expressed either in terms of volume or in terms of weight. However, to adjust for (to compare) basins of various sizes, this parameter is expressed in terms of volume/weight per unit area of the basin. Sediment yield can also be expressed in terms of sediment delivery ratio (SDR).

$$\text{Sediment yield} = \text{SDR} \times \text{A} \tag{14.12}$$

where

SDR = sediment delivery ratio
A = USLE estimated value of soil loss

According to Williams and Berndt (1972), sediment delivery ratio is given by the following relation. The model gives acceptable results despite using only a single parameter.

$$\text{Sediment delivery ratio} = 0.627 \times SLP^{0.403} \quad (14.13)$$

where

SLP = % slope of main stream channel
Main stream length =39.14 km
Therefore

$$\text{Slope of main stream (\%)} = \frac{\text{Difference in Elevation}}{\text{Distance / Length}}$$

The main stream passes through the maximum contour of 750 m and minimum contour of 65 m (Fig. 14.7).

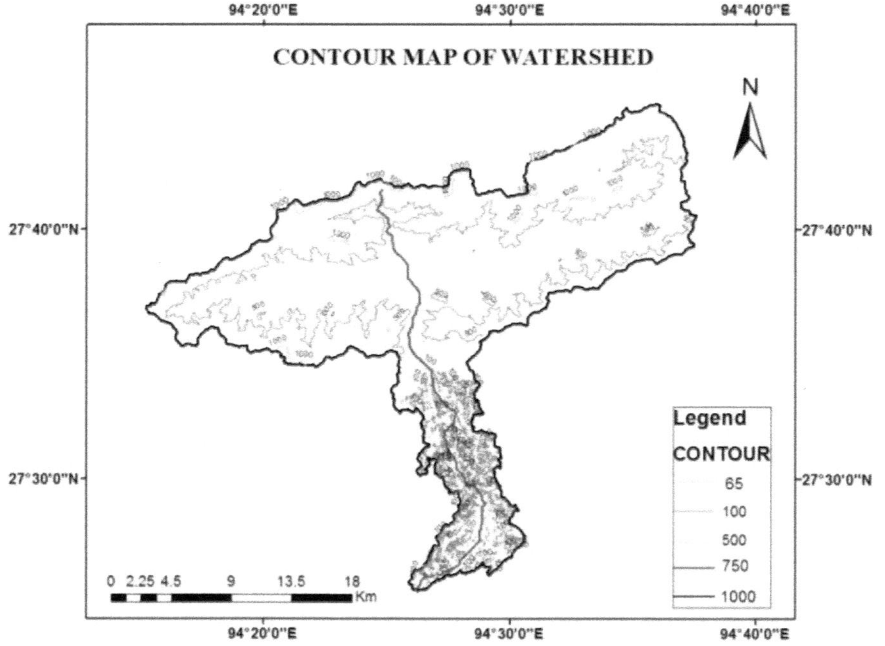

FIGURE 14.7 Contour map of Jiadhal basin.

14.6 RESULTS AND DISCUSSION

For the estimation of soil loss from the Jiadhal basin, all the five individual parameters/factors forming the USLE were calculated following standard procedures explained earlier. Calculations/estimations pertaining to the parameters are presented in the sections to follow along with necessary explanations.

14.6.1 ESTIMATION/DETERMINATION OF MODEL PARAMETERS

The following sub-sections provide the results of the estimation/computations carried out to determine the values of various input parameters used in the USLE.

14.6.1.1 RAINFALL EROSIVITY FACTOR (R)

The rainfall erosivity factor is calculated as follows:

$$R = P \times 0.5 = 3130.15 \times 0.5 = 1565.08 \text{ mm}$$

14.6.1.2 SOIL ERODIBILITY FACTOR (K)

To determine the value of soil erodibility factor (K), the equation developed by Wischmeier and Smith (1978) was used. The K-values obtained for two types of soils, that is, Orthic Acrisol (Ao) and Eutric Cambisol (Be) are 0.023 and 0.0325, respectively (Fig. 14.8; Table 14.3).

TABLE 14.3 Values of K for Orthic Acrisol (Ao) and Eutric Cambisol (Be).

Orthic Acrisol (Ao)	Eutric Cambisol (Be)
$F_{CSand} = 0.294$	$F_{CSand} = 0.367$
$F_{ClSilt} = 0.806$	$F_{ClSilt} = 0.898$
$F_{org} = 0.75$	$F_{org} = 0.75$
$F_{hisand} = 0.998$	$F_{hisand} = 0.999$
$K_{USLE} = 0.17736$	$K_{USLE} = 0.2469$
$K_{factor} = 0.023$	$K_{factor} = 0.0325$

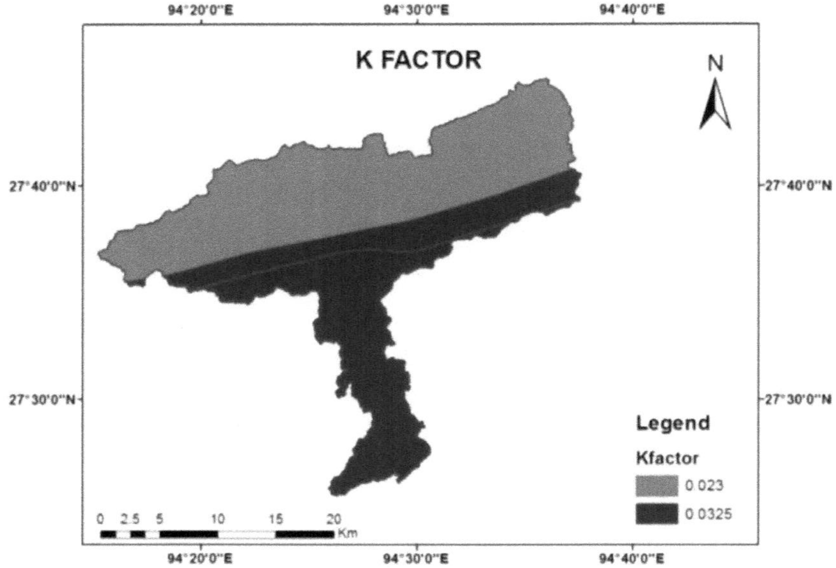

FIGURE 14.8 Spatial distribution of **K**-factor.

14.6.1.3 SLOPE LENGTH AND STEEPNESS (LS) FACTOR

The values of slope length and slope steepness were calculated using the Wischmeier and Smith equation explained earlier. Additionally, the LS factor map was created with the help of flow accumulation and slope grid derived from ASTER DEM. For the study area, the values for the LS factor were found to vary between 0 and 355.8 (Fig. 14.9).

14.6.1.4 COVER MANAGEMENT FACTOR (C)

The map showing the factor "C" was generated by using the NDVI image derived from Landsat data as mentioned earlier. For the study area, the values of the factor "C" were found to range from 0.10053 to 2.35642 (Figs. 14.10 and 14.11).

14.6.1.5 SUPPORT PRACTICE FACTOR (P)

The value of this factor ranges between 0 and 1 with 0 signifying a very good practice undertaken/available in the basin to resist soil loss and 1 signifying a condition wherein no resistance to soil loss exists. In our case, it was found that there were only negligible support practices or any other soil conservation measures per se. Therefore, the value of the conservation practice factor was assumed to be 1.

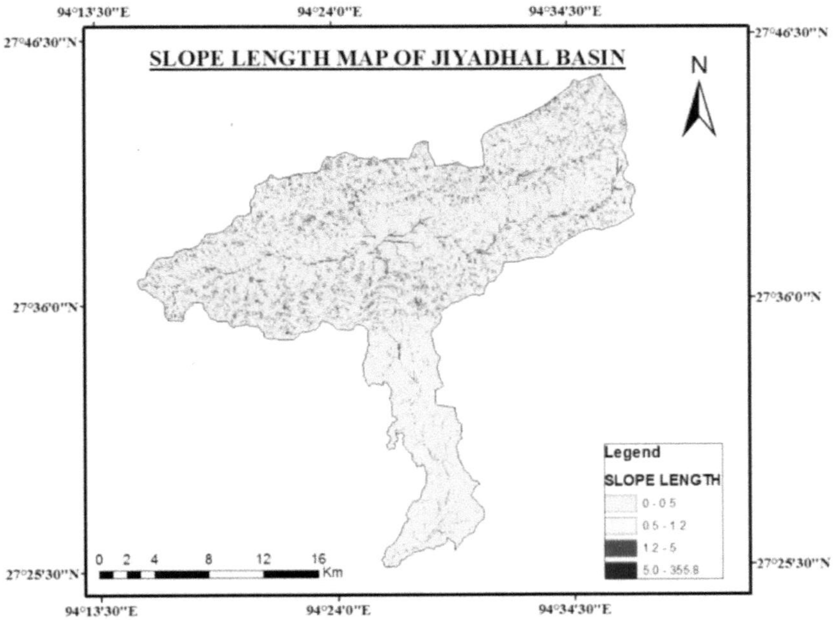

FIGURE 14.9 LS map of Jiadhal basin.

14.6.2 *SOIL LOSS IN JIADHAL BASIN*

The soil loss value calculated using USLE provides an indication about the physical health of the basin under study. It also points toward the erosion risk involved in the area. The various parameters that are part of the USLE were in spatial/graphical as well as numerical format. For understanding the severity of erosion from the basin, the entire basin area has been distributed (as % to total area) and various scales of priority fixed, viz., Slight (0–5 T/ha/year), Moderate (5–10 T/ha/year), High (10–15 T/ha/year), Very high (15–20 T/ha/year), Severe (>20 T/ha/year) depending on the severity of erosion from these areas/pockets. Information about the extent of annual erosion in the study area and soil erosion severity may be found in Figures 14.12 and 14.13, respectively. Table 14.4(a) gives information on the extent of area falling under various severity classes. It can be seen that major parts of the basin

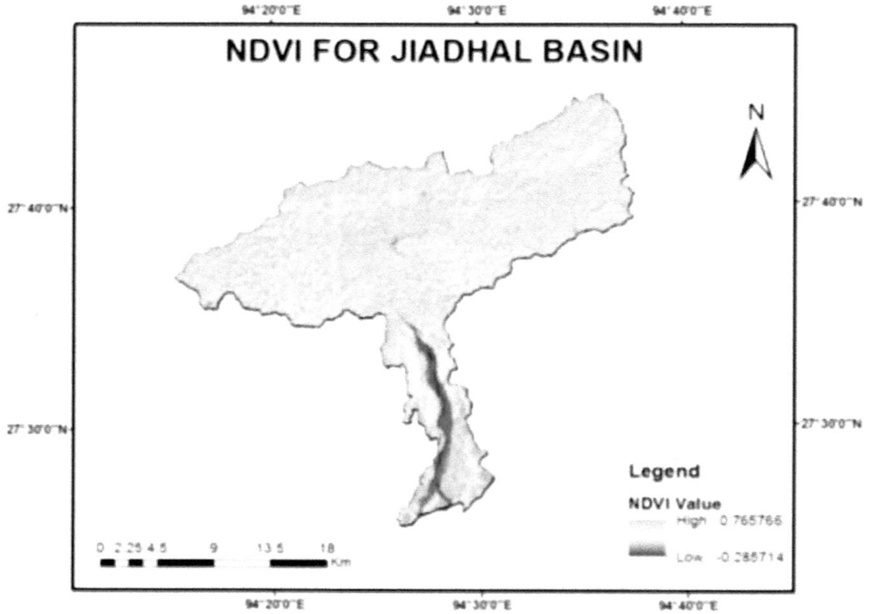

FIGURE 14.10 NDVI map of Jiadhal basin.

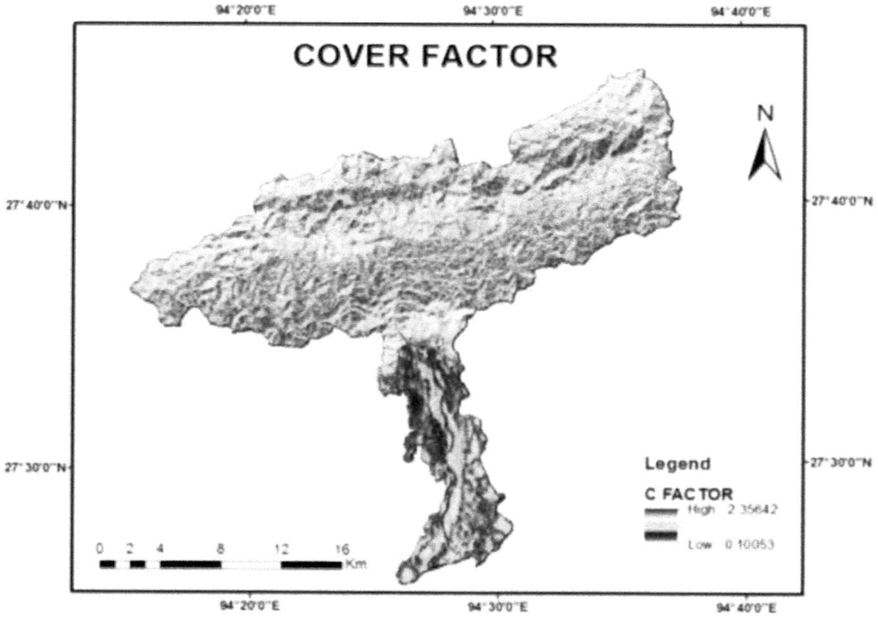

FIGURE 14.11 C-factor map of Jiadhal basin.

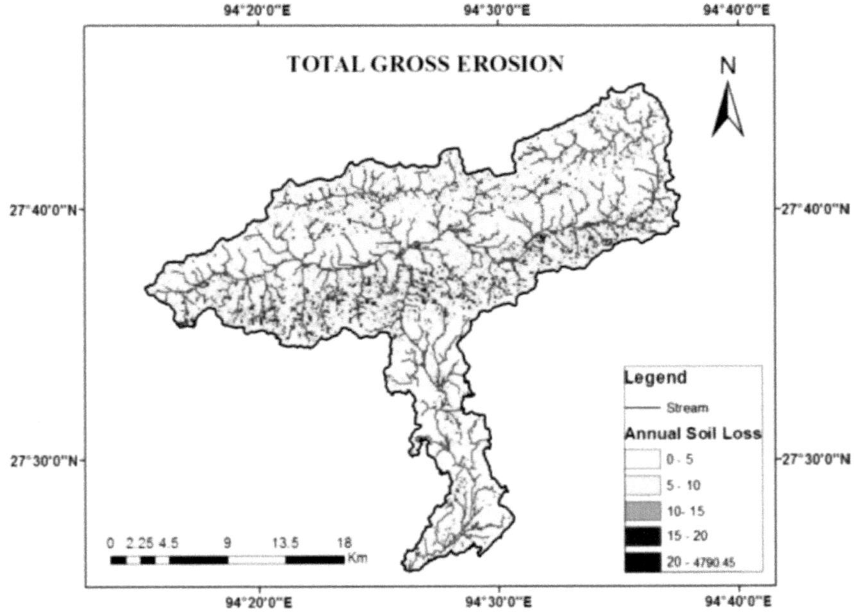

FIGURE 14.12 Erosion map of study area.

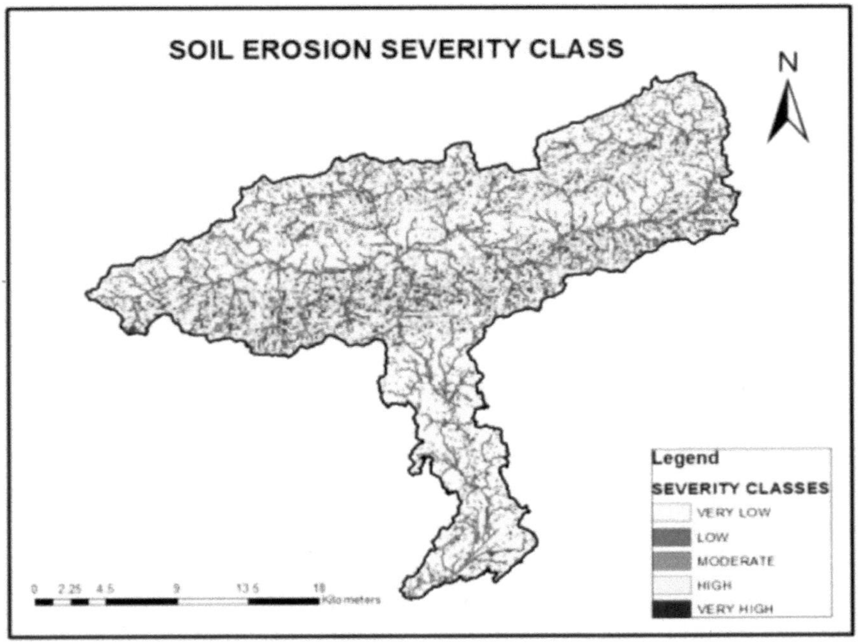

FIGURE 14.13 Soil erosion severity Class map.

(67.96%) are under the severity class SLIGHT. The annual soil loss rate of the area ranges between 0 and 4790.45 T/ha/year as calculated with the help of ArcGIS software. The mean value of soil loss rate is 10.659 T/ha/year. The total soil loss of the study watershed is 0.465 million T/year.

TABLE 14.4(a) Soil Loss Severity Classes with Rate of Loss and Area Covered.

Severity class	Soil loss (T/ha/year)	Area (km²)	Area (%)
Slight	<5	297	**67.96**
Moderate	5–10	39	**8.92**
High	10–15	26	**5.95**
Very High	15–20	29	**6.64**
Severe	>20	46	**10.53**
		437	**100.00**

14.6.3 SEDIMENT YIELD ESTIMATION

Sediment yield (SY) at an outlet of Jiadhal watershed (Kumotiya river near Bordoloni) is found to be 0.365 million T and the SDR is calculated as 0.785. The calculation is as under (Fig. 14.14):

$$\text{Sediment yield (SY)} = A \times SDR \qquad \text{eq 14.12}$$

where

A = USLE estimated value of erosion

$$SDR = 0.627\ SLP^{0.403} \qquad \text{eq 14.13}$$

Main stream length = 39136.798 m

$$\text{Slope of main stream} = \frac{\text{Difference in Elevation}}{\text{Distance / Length}} = \frac{(750-65)\times 100}{39136.798} = 1.750$$

$SDR = 0.627 \times 1.750^{0.403} = 0.786$

$SY = A \times SDR = 0.465 \times 0.786 = 0.365$ mill T/year

14.6.4 PRIORITIZATION OF THE BASIN

For the purpose of prioritization, the entire basin was divided into four (4) subbasins and soil loss from each of the subbasins was determined. Estimated values of subbasin-wise soil loss are given in Table 14.4(b). It can be seen from the table that the maximum soil loss takes place from the Subbasin I (10.98 T/ha/year) followed by Subbasin III (10.97 T/ha/year) and Subbasin II (10.35 T/ha/year), while the minimum soil loss takes place from Subbasin IV (Fig. 14.15; Tables 14.5–14.8).

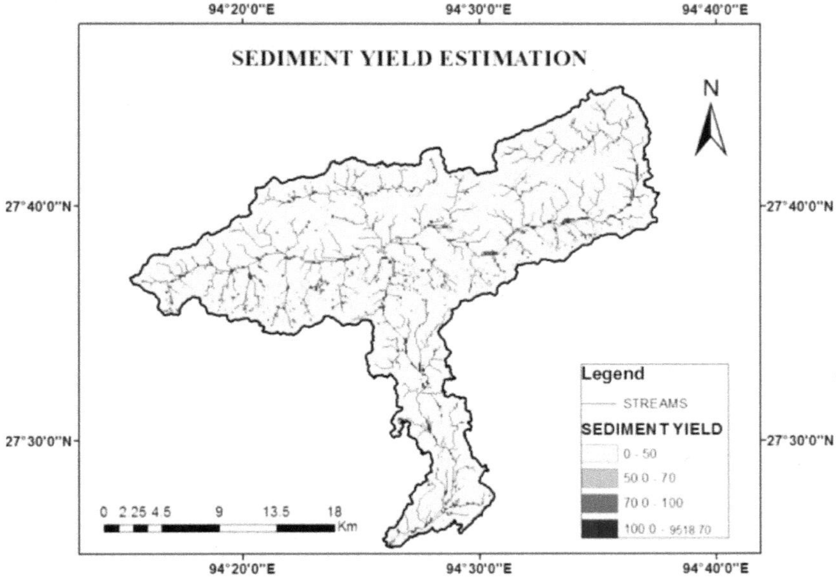

FIGURE 14.14 Sediment yield map.

TABLE 14.4(b) Priority Classes of Subbasins.

Subbasin	Soil loss (T/ha/year)	Priority class
SB-I	10.98	I
SB-II	10.35	III
SB-III	10.97	II
SB-IV	10.17	IV

TABLE 14.5 Soil Loss from Subbasin I.

Severity class	Soil loss (T/ha/year)	Area (km^2)	Soil loss (mill.T/year)
Slight	<5	70.46	0.0176
Moderate	5–10	10.66	0.0079
High	10–15	8.54	0.0131
Very high	15–20	7.53	0.0137
Severe	>20	11.20	0.0672

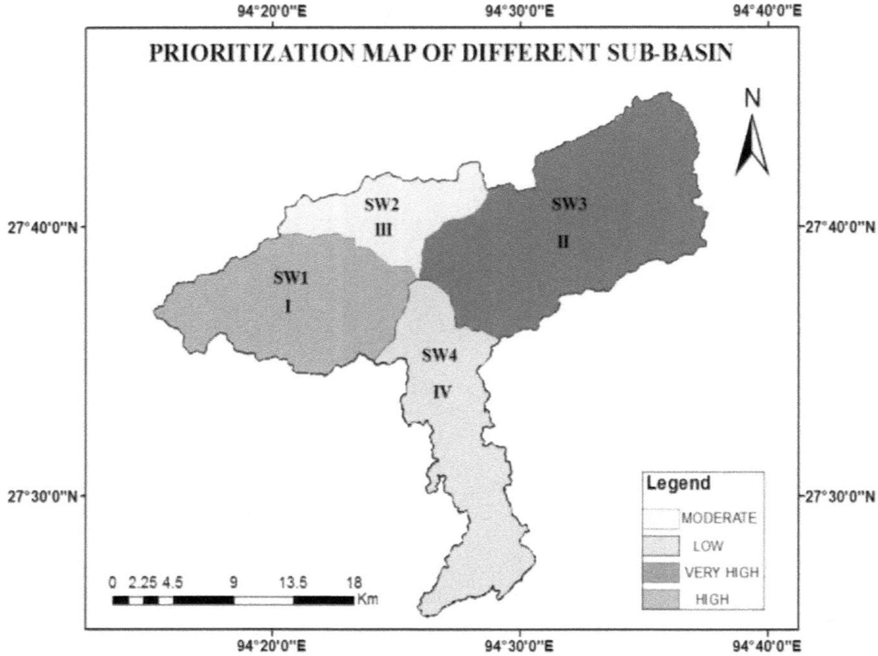

FIGURE 14.15 Prioritization map of subbasins.

TABLE 14.6 Soil Loss from Subbasin-II.

Severity class	Soil loss (T/ha/year)	Area (km^2)	Soil loss (mill.T/year)
Slight	<5	32.93	0.0082
Moderate	5–10	4.03	0.003
High	10–15	3.82	0.0047
Very high	15–20	2.61	0.0045
Severe	>20	6.85	0.0614

TABLE 14.7 Soil Loss from Subbasin-III.

Severity class	Soil loss (T/ha/year)	Area (km^2)	Soil loss (mill.T/year)
Slight	<5	120.68	0.0302
Moderate	5–10	14.59	0.0109
High	10–15	12.31	0.0153
Very high	15–20	7.79	0.0136
Severe	>20	20.52	0.1231

TABLE 14.8 Soil Loss from Subbasin-IV.

Severity Class	Soil Loss (T/ha/year)	Area (km²)	Soil Loss (mill.T/year)
Slight	<5	72.62	0.0181
Moderate	5–10	8.79	0.0065
High	10–15	5.51	0.0068
Very high	15–20	5.72	0.0100
Severe	>20	10.57	0.0634

TABLE 14.9 Total Soil Loss in Each Subbasin and Priority Class.

Subbasin	Area (km²)	Area (%)	Soil Loss (million T/year)	Priority Class
SB-I	108	24.71	0.119	II
SB-II	50	11.44	0.052	IV
SB-III	176	40.28	0.193	I
SB-IV	103	23.57	0.104	III
Total	**437**	**100.00**		

14.7 CONCLUSION

Through the present study, an attempt was made to estimate the annual soil loss and sediment yield from the Jiadhal river basin that is transboundary between Assam and Arunachal Pradesh. To facilitate estimation, the universal soil loss equation was used. Many factors contribute to the erosion process in the basin–basin's geometry, orientation, topography, land use/land cover, and hydrology. Additional factors that have bearing on the erosion process in the study area include composition of soil/rocks and their permeability, and most importantly, the very high rainfall that the basin receives annually. The estimated soil loss rate for the basin stands at 10.659 T/ha/year. This value points toward the fact that the Jiadhal basin is in a degraded state of health and it calls for immediate adoption of soil conservation measures. The total soil loss from the study area stands at about 0.465 million T/year (10.659 T/ha/year × 437 km²). The sediment yield value calculated at an outlet of the Jiadhal basin (river Kumotiya near Bordoloni) is found to be 0.365 million T/year, while the SDR is calculated as 0.786. However, more studies are needed to ascertain the SDR values in light of the factors, viz., rainfall, flood, and antecedent land surface that influence sediment delivery. Area-wise, about 297 km² (67.96%) falls under the severity class Slight (soil loss < 5 T/ha/year, about 39 km² (8.92%) falls under the severity class

Moderate (soil loss = 5–10 T/ha/year), about 26 km^2 (5.95%) falls under the severity class High (soil loss = 10–15 T/ha/year), about 29 km^2 (6.64%) falls under the severity class Very High (soil loss = 15–20 T/ha/year), and about 46 km^2 (10.53%) falls under the severity class Severe (soil loss > 20 T/ha/year). The basin is divided into four subbasins on the basis of average annual soil loss. This study confirms the effectiveness of the use of the USLE model integrated with GIS as compared with the time-consuming methods for assessing the soil loss in a basin scale.

KEYWORDS

- **digital elevation model**
- **basin**
- **soil loss**
- **sediment yield**
- **USLE model**

REFERENCES

Devatha, C. P.; Deshpande, V.; Renukaprasad, M.S. Estimation of Soil Loss Using USLE Model for Kulhan Watershed, Chattisgarh–A Case Study. *Aquatic Procedia* **2015,** *4*, 1429–1436.

Inamul, H. Soil Loss Estimation in GIS Framework: A Case Study in Champabati Watershed, 2018.

Jain, M. K.; Mishra, S. K.; Shah, R. B. Estimation of Sediment Yield and Areas Vulnerable to Soil Erosion and Deposition in a Himalayan Watershed Using GIS. *Curr. Sci.* **2010,** *98* (2, 25), 213–221.

Jain, S. K.; Varghese, J. Estimation of Soil Erosion for a Himalayan Watershed Using GIS Technique. *Water Resour. Manage* **2001,** *15*, 41–54.

Kim, H. S. *Soil Erosion Modeling Using RUSLE and GIS on the Imha Watershed, South Korea*; Colorado State University: Fort Collins, CO, USA, 2006.

Renard, K.; Foster, G.; Weesies, G.; McDool, D.; Yoder, D. Erosion by Water: A Guide to Conservation Planning with the RUSLE Model. *Agricultural Handbook* 703; USDA-ARS, 1997.

Van der Knijff, J.; Jones, R.; Montanarella, L. Soil Erosion Risk Assessment in Europe.

Vinay, M.; Ramu, Mahalingam, B. Quantification of Soil Erosion by Water Using GIS and Remote Sensing techniques: A Study of Pandavapura Taluk, Mandya District, Karnataka, India. *ARPN J. Earth Sci.* **2015,** *4* (2), 103–110.

CHAPTER 15

Climate Change Analysis of Precipitation and Mean Temperature in Jammu and Kashmir, India

DHEERAJ KUMAR*, NAYAN SHARMA, and NEERAJ KUMAR

Innovante Water Solutions Pvt. Ltd., Noida 201304, India

**Corresponding author. E-mail: ceo@inno-vante.com*

ABSTRACT

In this paper, an attempt has been made to assess and understand the hydrologic variable trends, that is, rainfall and mean temperature for the state of Jammu and Kashmir, India on monthly, annual, and seasonal basis. Past (1901–2002) and future (2016–2099) trends of hydrologic variables were studied for 14 meteorological stations, that is, for past trend and 18 CMIP5-CCSM4 grid stations, that is, for future trend in Jammu and Kashmir. Mann–Kendall (MK) test and modified MK (MMK) test were employed to identify monotonic trend magnitude as well as direction of temporal change on monthly, seasonal, and annual basis. Further, CMIP5-based CCSM4 climate system model was used for future trend detection with representative concentration pathways scenarios of RCP 4.5 for mean temperature and RCP 8.5 for precipitation. The spatial distribution of trends was explored by interpolating the value of each station using Arc-GIS toolboxes on annual, seasonal, and monthly basis. Rising trends in precipitation and mean temperature were found for annual, premonsoon, and postmonsoon (precipitation) and for annual, premonsoon, postmonsoon, and winter (mean temperature) in Jammu and Kashmir during 1901–2002. The study also concludes that the precipitation and mean

Advances in Hydrology and Climate Change: Historical Trends and New Approaches in Water Resources Management. Surendra Kumar Chandniha, Anil Kumar Lohani, Gopal Krishan, & Ajay Krishna Prabhakar (Eds.)

temperature is going to rise in future decades to come which will surely impact the hydrology, hydraulics, and the ecology of Dal Lake and its periphery. It is also estimated that the intense rainfall events induced flood scenarios is going to elevate drastically, subsequently conveying more sediment to Dal Lake which will surely impact the useful life of the Dal Lake.

15.1 INTRODUCTION

Information on hydrometeorological data as well as lake level and groundwater table are generally employed to describe the climatic condition of an area (Kothyari and Singh, 1996). Abrupt variations in the quantities of these factors are employed as parameters to measure climatological variations. Change in precipitation and rise in air temperature may also be applied as a measure to study the sensitivity of the climatic system to external forcing factors such as fluctuation in CO_2 concentration. The rise in average global air temperature is affiliated to extensive variations in weather conditions. Some research analyses specify that the extreme events for instance, large storms and heat waves are probable to become more common and intense with human-induced climate change. Persistent anomalies in climatological variables, potentially disrupt the society directly or indirectly. For instance, warm weather could raise air-conditioning cost, but it may be favorable for some crops. The anomalies in climatic variables are also a risk to community. Prevailing heat events can proliferate illnesses and deaths, particularly in susceptible population areas, damage of crop land and excessive melting of glacier in alpine region. While rise in rainfall can recharge water supplies and encourage agriculture, intense-storms can destroy property, may lead to community displacement and in extreme situation loss of life, and momentarily interrupting essential services, that is, telecommunications, electricity, water supplies, and transportation. From the past few decades, most of the climate change analysis studies using observational temperature and precipitation data have focused on changes in mean values. Precipitation in the Alpine region is to a high degree governed by the location and amplitude of the storm track, and by the interaction of approaching synoptic-scale weather systems with the Alpine topography. The impact of climatic change and its variance have been analyzed by various scientists in variety of geophysical domain.

The IPCC, 2007 stated that the spatial, interannual and interseasonal variance in rainfall trends all across Asian Continent during the last few decades (Parry et al., 2007). Kumar et al. (1992) reported the falling tendency in rainfall during monsoon in the eastern region of Madhya Pradesh, few regions of Gujarat and Kerala and northeastern region of Indian subcontinent, whereas the rising trend was detected on the western coastal region, northern region of Andhra Pradesh and northwestern region of Indian subcontinent during 1871–1984. Guhathakurta and Rajeevan (2008) reported that the three states, namely, Jharkhand, Chhattisgarh and Kerala observed significant falling tendency, whereas eight subdivisions, namely, Gangetic West Bengal, West Uttar Pradesh, Jammu and Kashmir, Konkan and Goa, central Maharashtra subdivision, Coastal Andhra Pradesh, and North Interior Karnataka observed substantial rising trend during monsoon rainfall in the period of 1901–2003. The variation in seasonal and annual land surface temperature differs widely based on the latitude, altitude, and location of an area and also hinges on its geographical features, such as water bodies (river, lake, or sea), forest, mountains, etc. Plausibly during the past century, substantial increase in the global mean temperature is one of the most cited parameter of climate change which have a crucial consequence on various sectors, including "water." As the hydrologic cycle is a thermal propelled system, increase in global air-temperature is potential to accelerate this cycle significantly. Detection of past trends in air-temperature and its forecasting has been the theme of numerous studies conducted all around the globe (Beniston, 2005). The identification of trends in various climatological variables, including those forecasted to occur from rise in greenhouse gas concentrations in atmosphere (Bradley et al., 2003), may be explored in historical or past data as long as these data are long-term and representative of the variable, that is, typically more than 100 years.

There are several statistical and nonstatistical tests available which are employed for establishing trends in hydrological and meteorological parameters time series. Moreover, current analyses indicate that nonstatistical tests are generally employed for non-normal distributed and skewed data, which are commonly found in hydrometeorological time-series. Mann–Kendall (MK) (Mann, 1945; Kendall, 1975) test is the most recognized rank-based nonparametric method for trend analysis and is recommended by several scientists because of its different benefit over parametric methods (Yue et al., 2002a). Some of the well-known advantages include (1) it does not require the presumption of homogeneity of variance or normality, (2) it equates medians instead of means, and in case the time-series data set have few biases, their influence is nullified, (3) prior conversion of the time-series data is not needed, (4) higher power is attained for the non-normal time-series, (5) data sets lying below threshold limit can be integrated without bias correction (Helsel, 1987). Regardless of the above advantages, the impact of serial correlation on the results of Mann-Kendall test have been analyzed in various literature (Yue et al., 2002a, 2002b; Von Storch, 1995; Yue and Wang 2002). Hamed and Rao (1998), introduced a modified Mann-Kendall (MMK) test for autocorrelated data by altering the variance of the MK test statistic to eradicate the effect of serial dependence. Further, a comparison using 500 samples were conducted between MMK and actual MK test, and the results suggested that the ability of MMK test is equivalent to the actual MK test. Furthermore, it was observed that when MMK is applied on autocorrelated data sets there is no-loss of power.

The "Coupled Model Inter-comparison Project Phase 5" (CMIP5) provides a revolutionary set of data which can be employed to analyze the change in climatological variables (Taylor et al., 2012). The "Fifth Assessment Report of Intergovernmental Panel on Climate Change" (IPCC AR5) was predominately based on CMIP5 modeling results (IPCC, 2013) and the "Fourth Assessment Report of Intergovernmental Panel on Climate Change" (IPCC, 2007) was based on CMIP3 project on which the CMIP5 data set was predicted. The CMIP5 models are at finer spatial resolutions than those in CMIP3 and integrated the improved or new interactive carbon cycle, aerosol processes, and land use land cover change (Meinshausen et al., 2011). "Representative Concentration Pathways" (RCPs) are four GHG concentration (not emissions) trajectories adopted by the IPCC for its fifth Assessment Report (AR5) in 2014. The RCPs are named after probable assumption of radiative forcing values adopted for the year 2100 which relates to preindustrial values, that is, +2.6, +4.5, +6.0, and +8.5 W/m^2 (Pachauri et al., 2014). These RCPs are employed for modeling climatological variables and are used to describe four probable climatic future. The "Community Climate System Model Version 4" (CCSM4) is a fully coupled atmospheric, dynamic ocean, sea ice, and land model (Gent et al., 2011). The "Community Atmosphere Model" (CAM4) is the atmospheric component which run with 26-vertical layers at 0.5° lat. resolution. The land model is a "Community Land Model V4," which also run at 0.5° resolution. The "half" degree resolution better represents extreme precipitation as they employ a narrow grid than generally applied in the CCSM4 system (Gent et al., 2010; Shieldsand Kiehl, 2016).

In the purview of potential extent and scale of climate change impacts, it is now essential to assess and understand the hydrologic variable trends on infrastructure, ecosystem, and populations. Reliable identification of these trends is really crucial in any decision-making step either small scale or at the larger scale. The core objective of this analysis was to reliably assess the trends in rainfall and mean temperature for the state of Jammu Kashmir,

India on monthly, annual, and seasonal basis (Chattopadhyay and Edwards, 2016). MK test and MMK test was employed to identify monotonic trend magnitude as well as direction of temporal change on monthly, seasonal, and annual basis. Moreover, CMIP5-based CCSM4 climate system model was used for future trend detection (i.e., 2016–2099) of mean temperature and precipitation with Representative Concentration Pathways scenarios of RCP 4.5 for mean temperature and RCP 8.5 for precipitation. The spatial distribution of trends was explored by interpolating the value of each station using Arc-GIS toolboxes on annual, seasonal, and monthly basis. The results so obtained, may serve as an essential information for designing infrastructure, policy and decision-making and planning of urban–rural development to alleviate any adverse implications of changing climate.

15.2 STUDY AREA AND DATA USED

Jammu and Kashmir is a state in northern India, often denoted by the acronym J&K. It is located within 32.17° N–37.07° N latitude and 72.49° E–80.43° E longitude with an altitude range from 220 to 8569 m above m.s.l. (Fig. 15.1). Jammu and Kashmir falls in the Himalayan range and shares borders with the states of Punjab and Himachal Pradesh. The state of Jammu and Kashmir is further divided in three regions: Jammu, the Kashmir Valley, and Ladakh.

The climate in the study area varies significantly due to its irregular topography and is generally humid subtropical. The state receives an average monthly rainfall of 53–118 mm between January and March and very heavy rainfall up to 330–350 mm in the month of July and August. The average annual precipitation received by the state is 1331.00 mm approximately, and the highest temperature recorded is 40°C whereas the lowest is −20°C.

The monthly precipitation and mean temperature data over 102 years (1901–2002) from all district of J&K (Table 15.1) were collected from the "Indian Meteorological Department" (IMD) to assess the spatiotemporal variance of the rainfall and mean temperature time-series. Trend analysis on seasonal, monthly, and annual basis of rainfall and mean temperature time-series were conducted for all districts of J&K. The seasons were divided into four categories as premonsoon, monsoon, postmonsoon, and winter for better understanding the trend associated with the time-series.

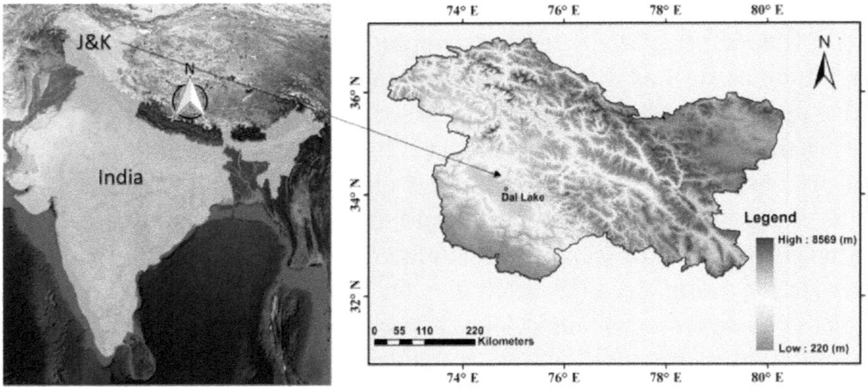

FIGURE 15.1 Index Map of the study area with digital elevation model (DEM).

TABLE 15.1 Details of Districts in J&K Employed for Trend Analysis.

S. No.	District name	Longitude	Latitude	S. No.	District name	Longitude	Latitude
1	Anantnag	75.16	33.78	8	Kupwara	74.25	34.53
2	Badgan	74.64	33.94	9	Leh	78.35	34.37
3	Baramula	74.94	34.65	10	Poonch	74.13	33.77
4	Doda	75.85	33.46	11	Pulwama	74.94	33.90
5	Jammu	74.87	32.73	12	Rajauri	74.41	33.33
6	Kargil	76.32	34.12	13	Srinagar	74.99	34.28
7	Kathua	75.55	32.59	14	Udhampur	75.07	33.08

CMIP5-CCSM4 climate model was employed for detecting the future trend in the precipitation and mean temperature from 2016 till 2099 years. 8.5 W/m^2. Representative concentration pathway in CMIP5-CCSM4 was used for precipitation trend analysis and radiative forcing of 4.5 W/m^2. Representative concentration pathway (RCP4.5) was employed for mean temperature. Model results were re-gridded to a standard resolution ($0.25° \times 0.25°$) in MATLAB which preserves the area average value. The CMIP5-CCSM4 climate model grids used for the analysis are provided in Table 15.2.

15.3 METHODOLOGY

The monthly precipitation and mean temperature data pertaining to Jammu & Kashmir, India, for the period 1901–2002 (past-IMD) and 2016–2099 (future-CMIP5-CCSM4) were used in this study. The monthly precipitation and mean temperature data were employed to evaluate the seasonal and annual time-series. The entire time-series were found autonomous and MK and MMK test were applied directly to the time-series to identify the trend using the two-sided theory (Fu et al., 2009; Xu et al., 2007) at three significant level, that is, at 1%, 5%, and 10%. The following steps which are shown in Figure 15.2 were followed to conduct the trend analysis of rainfall and mean temperature. Please see Appendix 1 for detailed description.

TABLE 15.2 Details of CMIP5-CCSM4 Climate Model Grids.

Grid No.	Latitude	Longitude	Grid No.	Latitude	Longitude
1	32.625	74.875	10	33.875	75.125
2	32.625	75.625	11	34.125	76.375
3	32.875	74.875	12	34.375	74.125
4	33.125	75.125	13	34.375	74.375
5	33.375	74.375	14	34.375	74.875
6	33.375	75.875	15	34.375	78.375
7	33.875	74.125	16	34.625	74.125
8	33.875	74.625	17	34.625	74.375
9	33.875	74.875	18	34.625	74.875

15.4 RESULTS AND DISCUSSION

15.4.1 PRELIMINARY ANALYSIS

During the period 1901 to 2011 (i.e., 102-year), the standard deviation (SD) and mean of annual rainfall datasets of various districts in the state of Jammu and Kashmir were found to be varied from 69.02 to 232.76 mm and 278.64 to 946.64 mm, respectively. Seasonal rainfall averaged values were found to be varied from 76.04 to 205.77 mm (premonsoon), 117.91 to 603.41 mm (monsoon), 15.36 to 39.23 mm (postmonsoon), and 65.32 to 177.48 mm (winter), and the standard deviation was observed as 32.43–78.20 (premonsoon), 37.69–211.22 (monsoon), 13.34–33.03 (postmonsoon), and 31.00–68.98 (winter). These statistics signify that areas with higher rainfall have less variability than areas with lower rainfall. Moreover, the mean and SD of the annual mean temperature data were found to be varied from −2.04°C to 23.55°C and 0.43°C to 0.54°C, respectively. Seasonal temperature average values varies from −3.10°C to 25.16°C (premonsoon), 7.69°C to 30.90°C (monsoon), −4.15°C to 21.76°C (postmonsoon); and −12.56°C to 13.35°C (winter), and the standard deviation was found to be 0.95°C–1.05°C (premonsoon), 0.42°C–0.53°C (monsoon), 0.68°C–0.81°C (postmonsoon), and 0.73°C–0.98°C (winter).

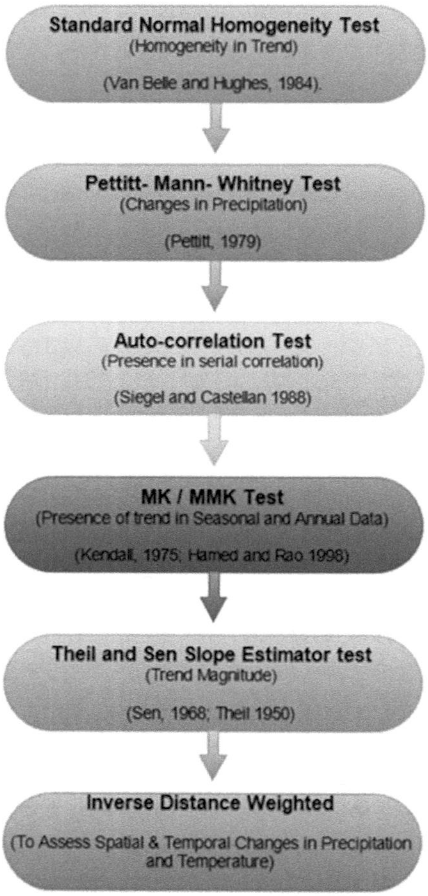

FIGURE 15.2 Flowchart of methodology employed for trend analysis.

15.4.2 PRECIPITATION TREND ANALYSIS

Precipitation trend analysis employing Mann–Kendall and Modified Mann–Kendall method were conducted for J&K state using the precipitation time-series data for seasonal (premonsoon, monsoon, postmonsoon, and winter), monthly, and average-annual.

15.4.3 PAST TREND (YEARS 1901–2002)

The statistics tests (MK and MMK) conducted at various significance levels, that is, 1%, 5%, and 10% are provided in Figures 15.3 and 15.4. A z-statistic having positive value signifies a rising trend, whereas negative value signifies a falling trend. During 1901–2002, the z-statistic of monthly rainfall of all the stations employing the MK and MMK tests are shown in Figures 15.3 and 15.4. During monthly analysis, the MK test indicated that February, March, May, July and November months demonstrated a positive trend and other months showed no trend for the period 1901–2002, whereas MMK also indicated that February, March, May and July months showed a positive trend and other months showed no trend for the same period which has a greater impact on Dal lake and its catchment. The November month showed no trend when MMK was used instead of MK. Results of the trend-analysis for annual as well as seasonal data for 1901–2002 are also presented in Figures 15.3 and 15.4 for MK and MMK, respectively. From Figures 15.3 and 15.4, it can be seen that the western part (comprising Dal Lake and its catchment) of the J&K state shows significant and nonsignificant positive trend in precipitation whereas the western part showed no trend during annual trend analysis using MK and MMK. Similarly, significant and nonsignificant positive trend in precipitation were observed for premonsoon and postmonsoon in western and central part of J&K state and other region showed no trend. During monsoon trend analysis, a nonsignificant (i.e., 10% significance level) positive trend was observed near the region of Jammu, Katra, Sunderbani and Udhampur for both MK and MMK test. In the winter season, no trend was observed in both the test conducted for the years 1901–2002.

15.4.4 FUTURE TREND (YEARS 2016–2099)

The MK and MMK tests at various significance levels (1%, 5%, and 10%) are provided in Figures 15.5 and 15.6. For the period 2016–2099, the spatially varied z-statistic values of monthly rainfall of all the stations employing the MK and MMK tests were calculated and are shown in Figures 15.5 and 15.6. The CMIP5-CCSM4 global simulation model with RCP 8.5 case scenario was employed to estimate the future trend of precipitation in J&K state. During monthly analysis, the MK test indicated that February, June, September, October and December months showed a positive trend in major part of J&K with no trend in other part and January, April, and May months

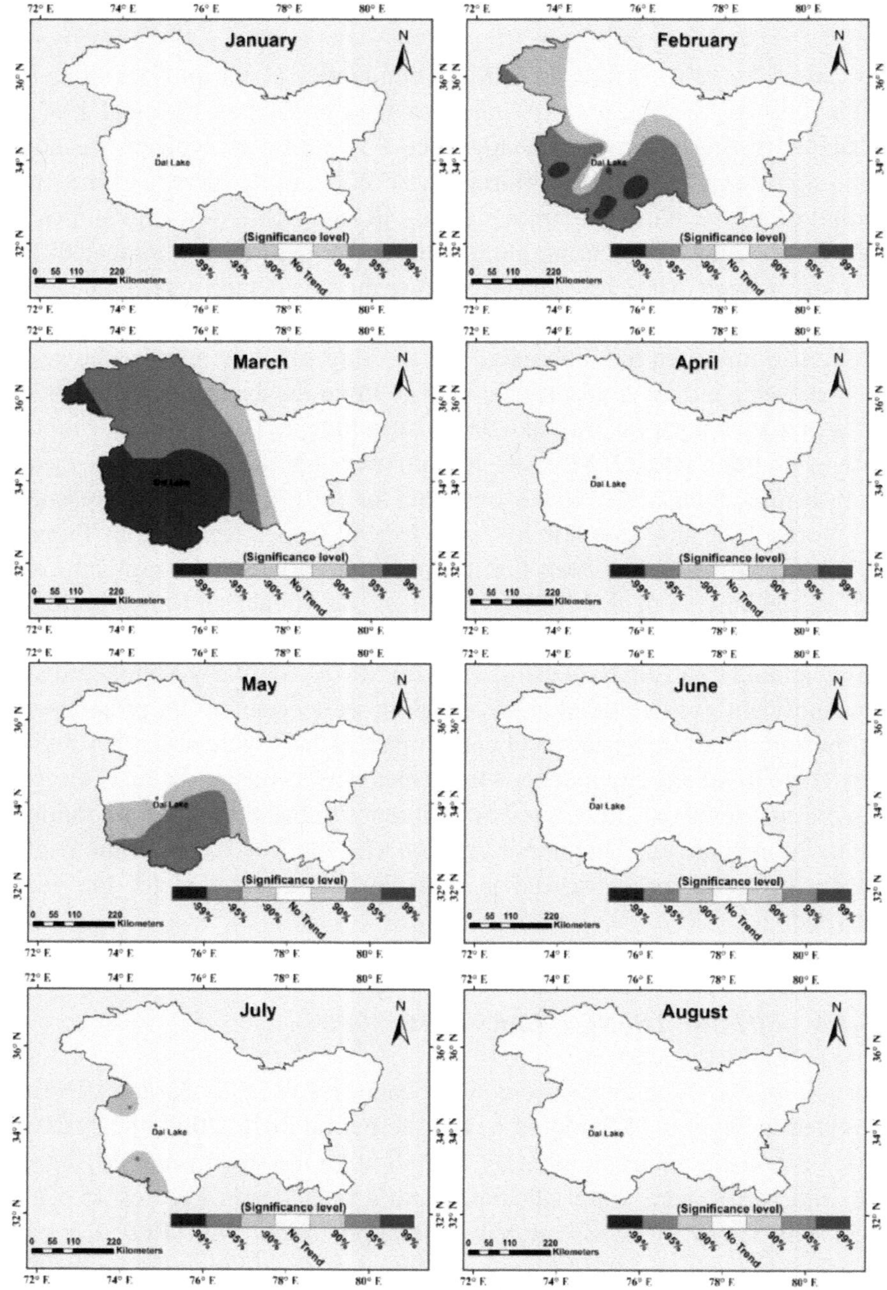

FIGURE 15.3 *(Continued)*

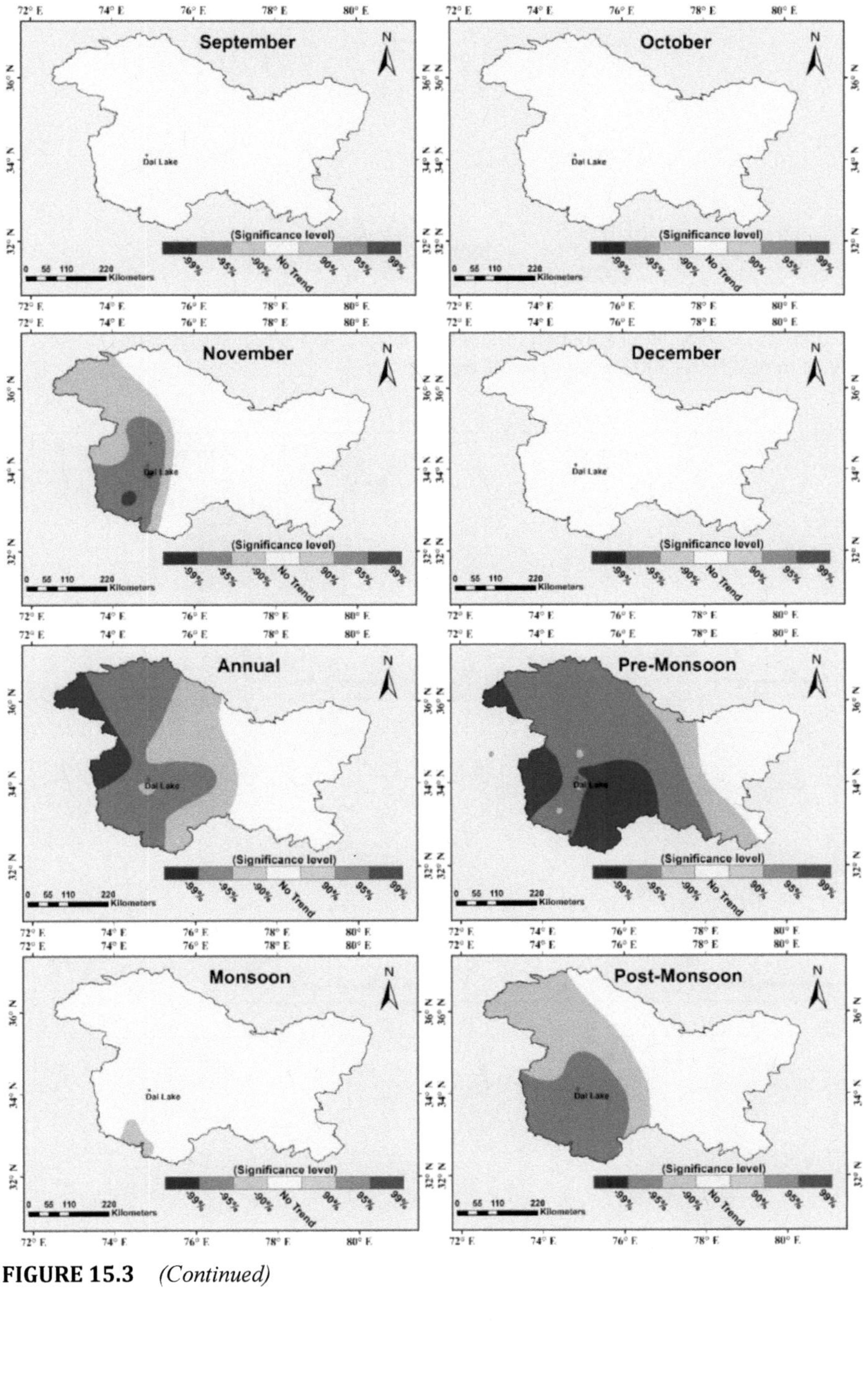

FIGURE 15.3 *(Continued)*

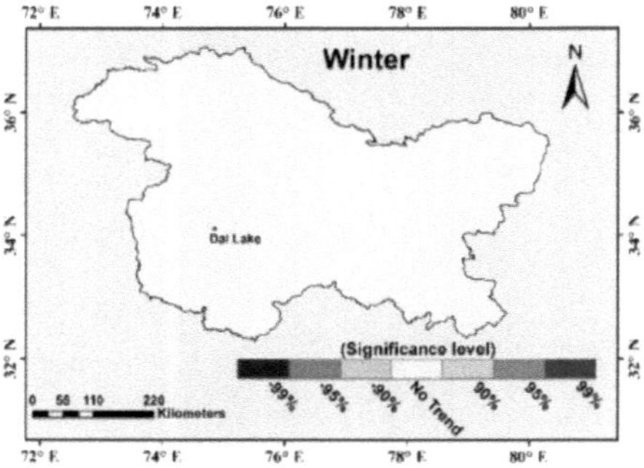

FIGURE 15.3 Spatial distribution of monthly, annual, and seasonal precipitation temporal change during 1901–2002 using Mann–Kendall.

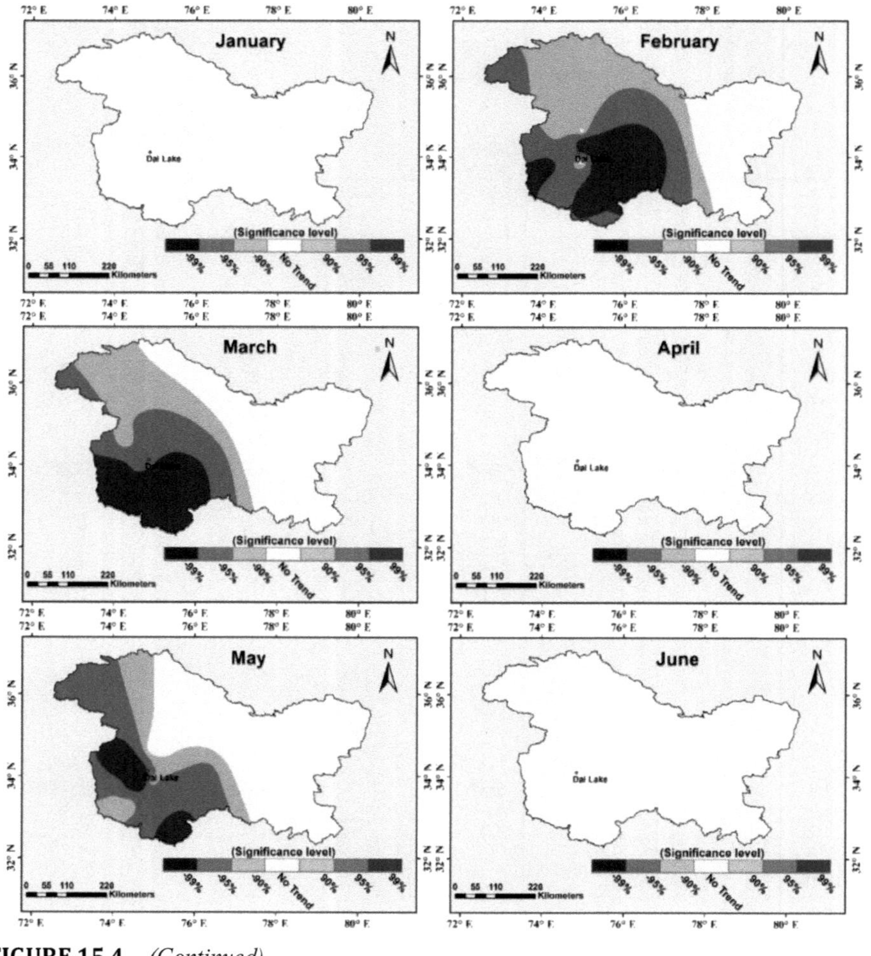

FIGURE 15.4 *(Continued)*

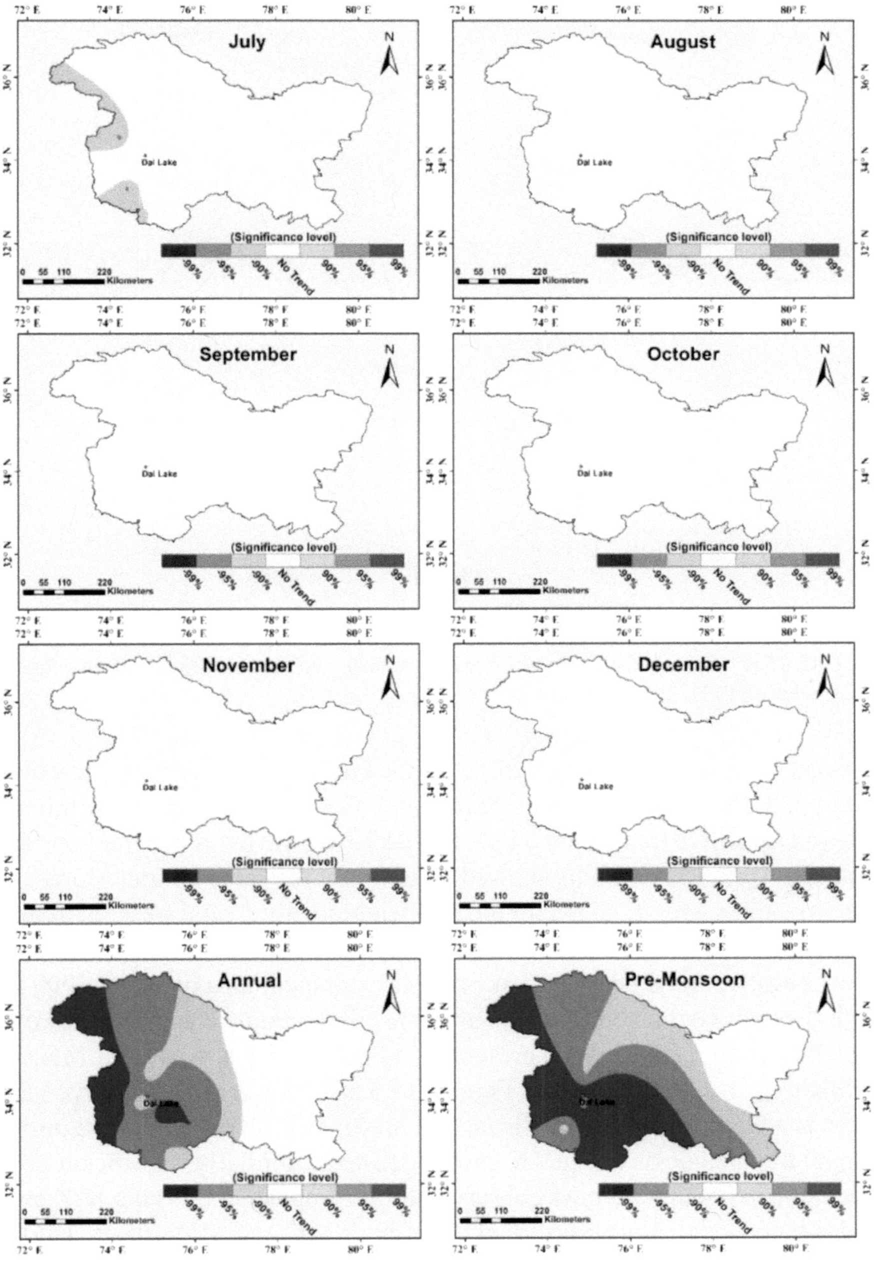

FIGURE 15.4 *(Continued)*

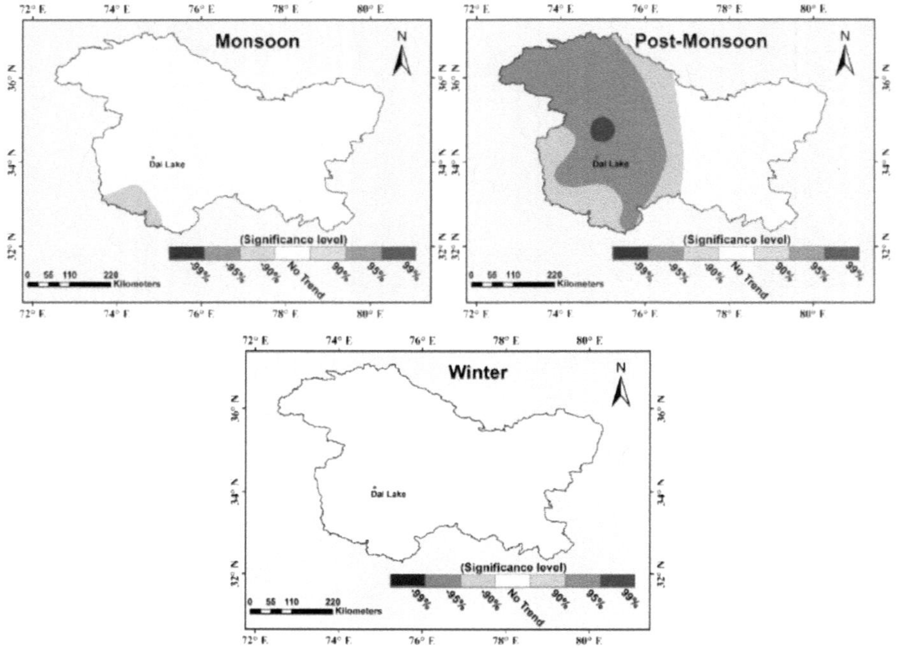

FIGURE 15.4 Spatial distribution of monthly, annual, and seasonal precipitation temporal change during 1901–2002 using MMK.

showed no trend for the period 2016–2099. Furthermore, the MMK test indicated that February, June, July, September, October and December months showed a positive trend in major parts of J&K with no trend in other parts and January, April, May months showed no trend for the same period. Moreover, the November month showed nonsignificant positive trend near the region of Poonch for both MK and MMK tests. However, July and August months show negative trend near Dal Lake and its catchment area till years 2099 for both the tests conducted. Trend analysis results for annual and seasonal data for years 2016–2099 are also presented in Figures 15.5 and 15.6 for MK and MMK tests, respectively. From Figures 15.5 and 15.6, it can be observed that there is a positive trend (significant and nonsignificant) in J&K state during annual trend analysis using MK and MMK tests. Similarly, significant positive trend in precipitation was observed for postmonsoon in whole J&K state and premonsoon showed no trend using both MK and MMK tests. During monsoon trend analysis, positive trend was observed near the region of Leh (Ladakh) and Kargil for both MK and MMK tests. In the winter season, positive trend was observed in whole J&K leaving only the Kashmir valley which shows no trend in both the tests conducted for the years 2016–2099.

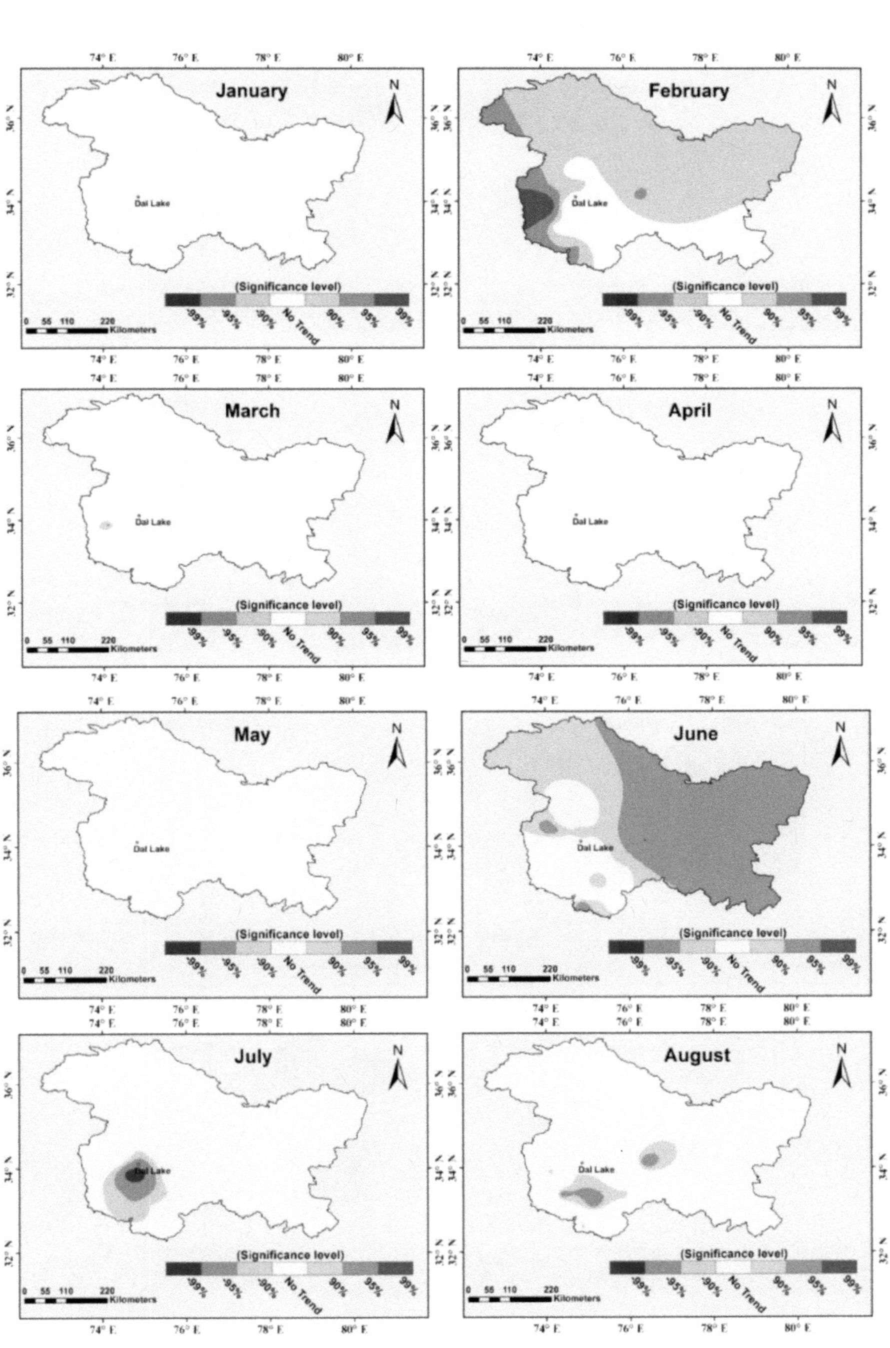

FIGURE 15.5 *(Continued)*

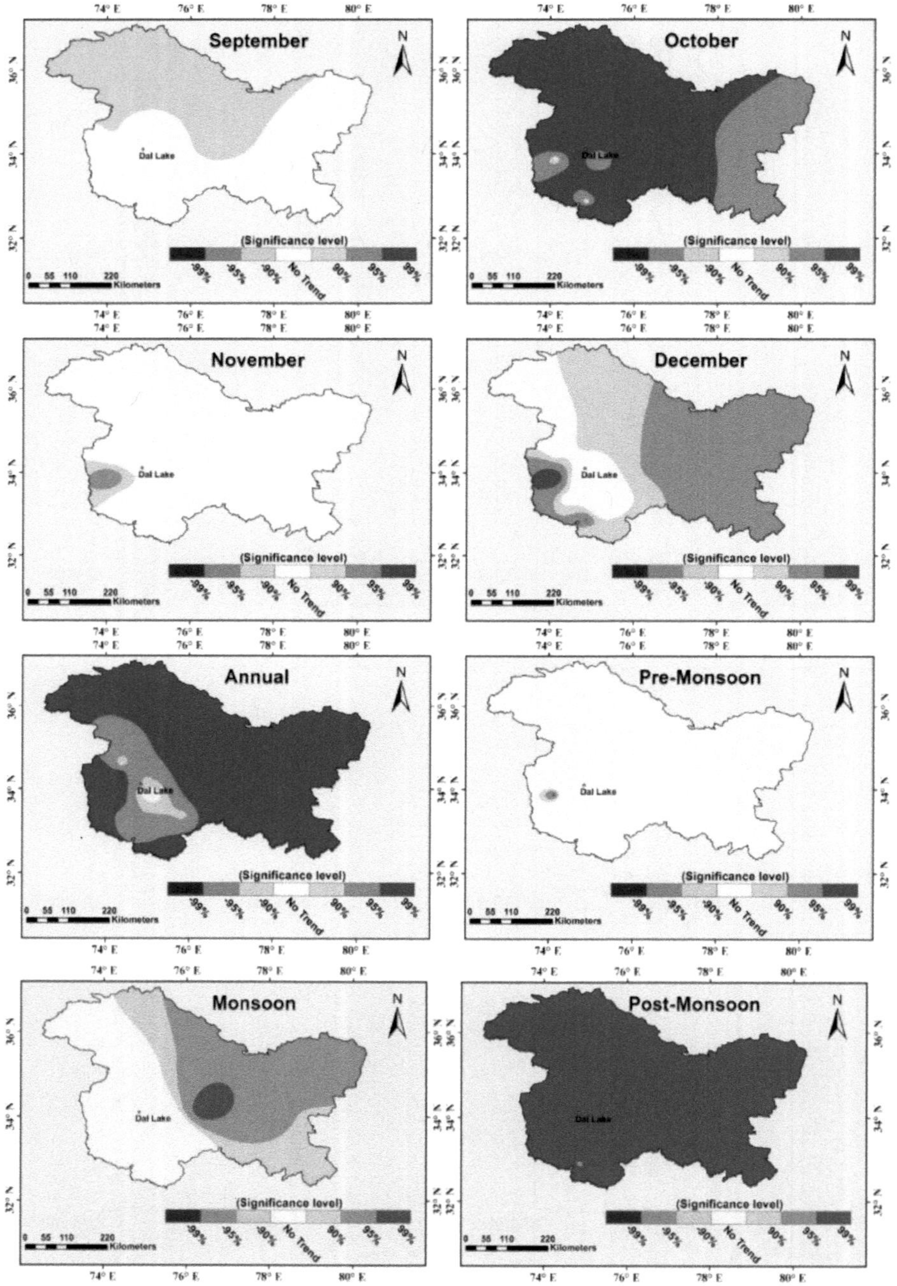

FIGURE 15.5 *(Continued)*

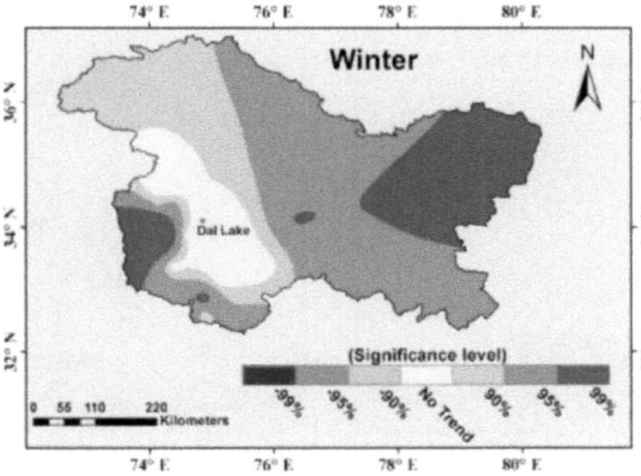

FIGURE 15.5 Spatial distribution of monthly, annual and seasonal precipitation temporal change during 2016–2099 using Mann-Kendall.

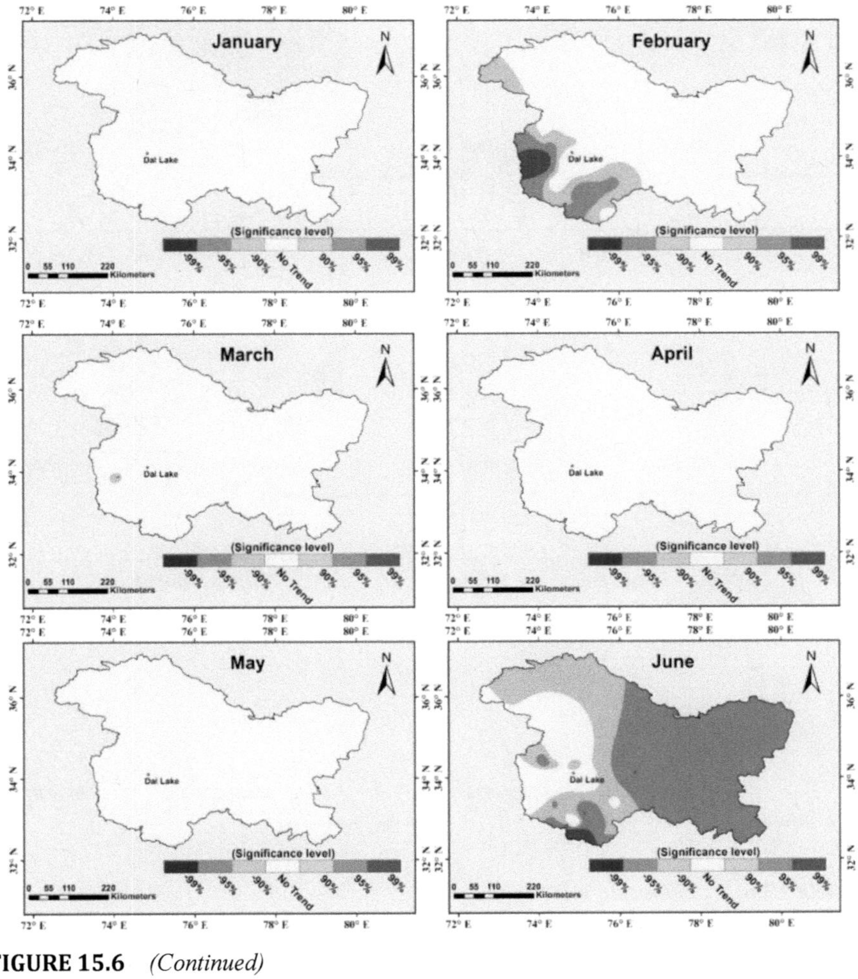

FIGURE 15.6 *(Continued)*

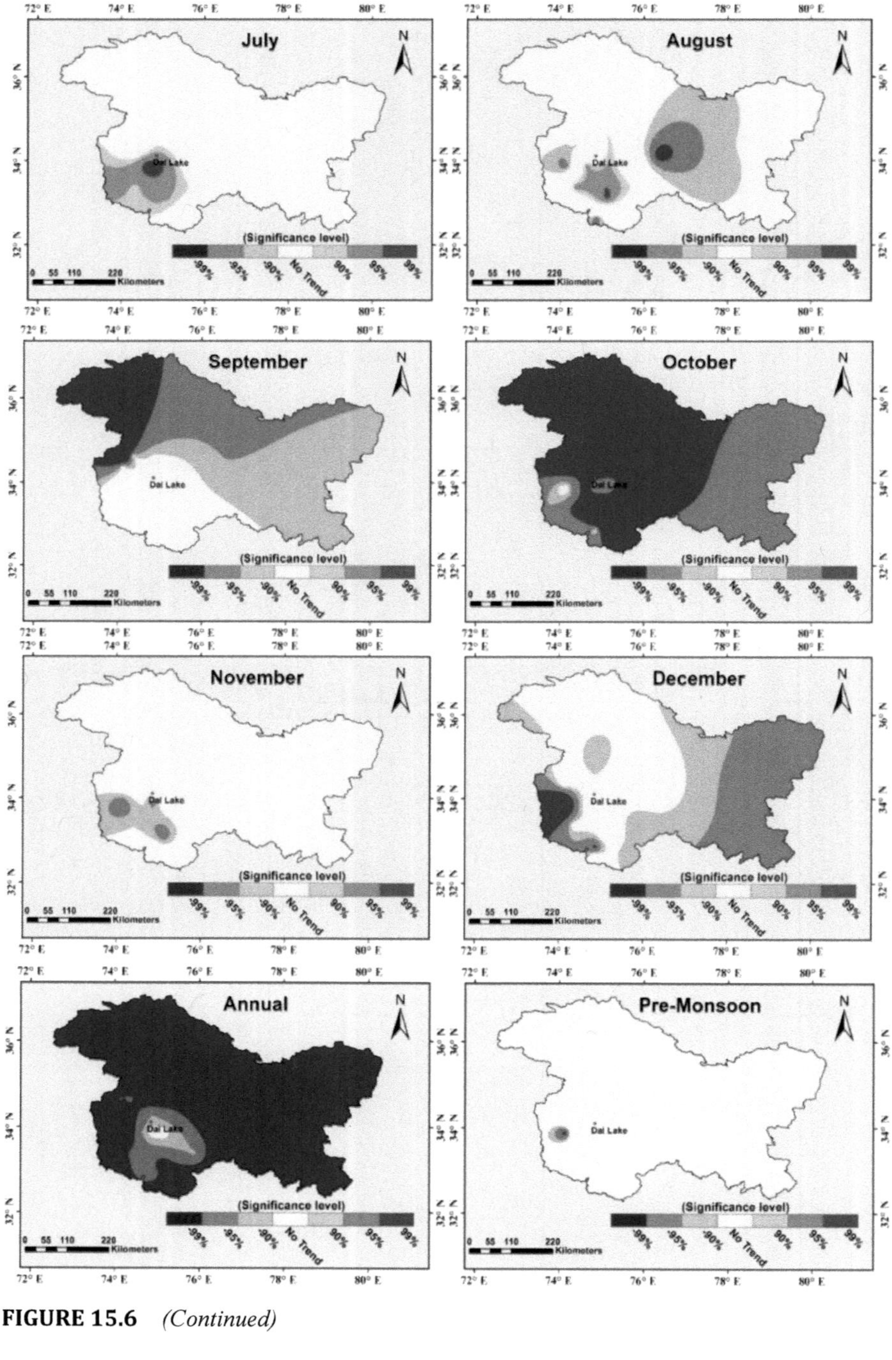

FIGURE 15.6 *(Continued)*

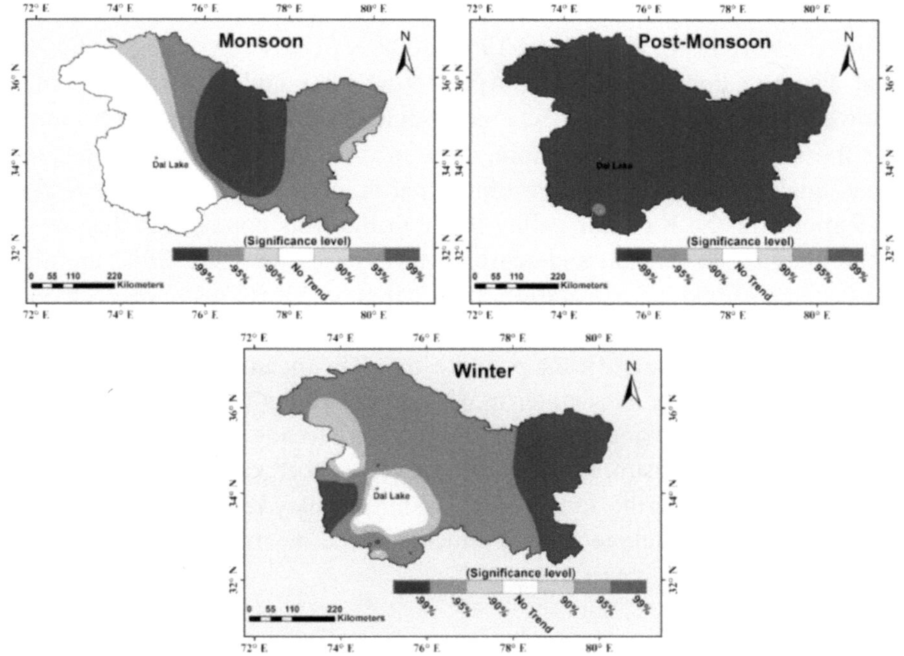

FIGURE 15.6 Spatial distribution of monthly, annual and seasonal precipitation temporal change during 2016–2099 using MMK.

15.4.5 MEAN TEMPERATURE TREND ANALYSIS

Mean temperature trend analysis employing the Mann–Kendall and modified Mann–Kendall methods was conducted for J&K State using mean temperature time-series of seasonal (premonsoon, monsoon, postmonsoon, and winter), monthly, and average annual.

15.4.6 PAST TREND (YEARS 1901–2002)

The MK and MMK tests at various significance levels of 1%, 5%, and 10% are provided in Figures 15.7 and 15.8. A positive z-statistic value signifies a rising trend in mean temperature (Hotter days), whereas a negative z-statistic value signifies a falling trend (Cooler days). For the period 1901–2002, the z-statistic values of monthly mean temperature of all the stations employing the MK and MMK tests are shown in Figures 15.7 and 15.8. During monthly analysis, MK test indicated that January, February, March, April, November and December months showed a positive trend and May and June months showed no trend for the period 1901–2002, whereas MMK test indicated that January, February, March, April, November and December months showed a positive trend and June and September months showed no trend for the same period. Furthermore, July and August months and monsoon trend analysis showed negative trend, that is, cooler days from the region of Jammu till the Kashmir valley for both the tests conducted. Results of trend analysis for annual and seasonal time-series for 1901–2002 are also presented in Figures 15.7 and 15.8 for MK and MMK tests, respectively. From Figures 15.7 and 15.8, it can be seen that entire J&K state shows significant and nonsignificant positive trend in mean temperature during annual, premonsoon, and winter trend analysis using MK and MMK tests. Similarly, significant and nonsignificant positive trends in mean temperature were observed for postmonsoon trend analysis in northern and eastern region of J&K state leaving the Jammu and Kashmir Valley region (i.e., including Dal Lake and its catchment area) which showed no trend in both the tests conducted during the years 1901–2002.

15.4.7 FUTURE TREND (YEARS 2016–2099)

The MK and MMK tests at various significance levels of 1%, 5%, and 10% are shown in Figures 15.9 and 15.10. For the period 2016–2099, the spatially varied z-statistic values of monthly mean temperature along with annual and seasonal data of all the stations using MK and MMK tests were calculated and are shown in Figures 15.9 and 15.10. The CMIP5-CCSM4 global simulation model with RCP4.5 case scenario (which assumes that global annual GHG emissions which are measured in CO_2-equivalents peaks around 2040, then decline) was employed to estimate the future trend of mean temperature in J&K state. The MK test indicated that January, February, March, April, May, September, October & November months as well as premonsoon, postmonsoon and winter showed a negative trend in the Kashmir Valley of J&K with positive trend in Jammu city and central part of J&K for the period 2016–2099. Furthermore, the MMK test indicated that January, February, March, April, May, November and December months as well as annual, premonsoon, postmonsoon, and winter seasons showed a positive trend near the Dal Lake and its catchment area of J&K for the same period. Moreover, the monsoon season analysis showed no trend for both MK and MMK tests. However, June, July, and August months show negative trend near the Dal Lake and its catchment area till year 2099 for the MK test.

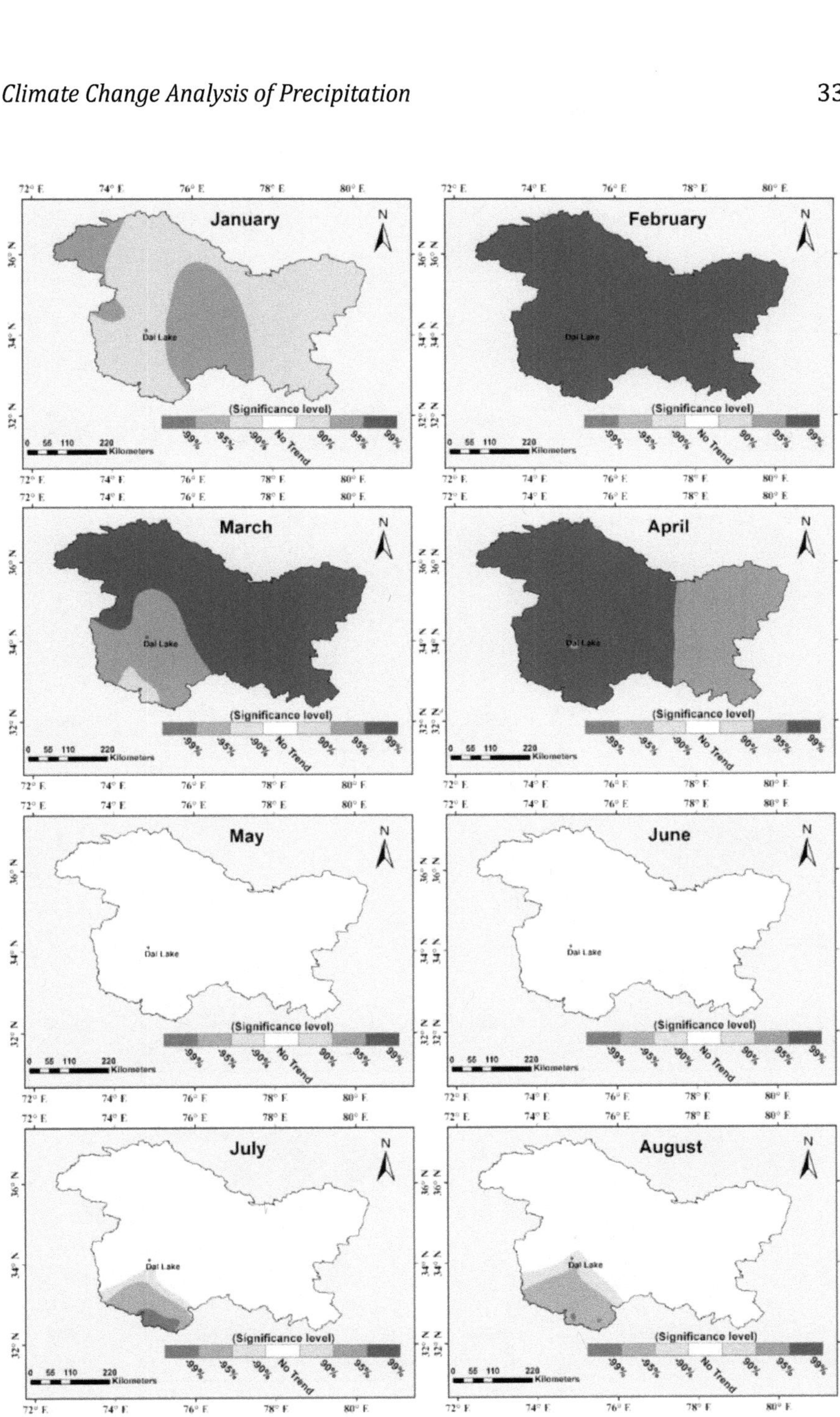

FIGURE 15.7 *(Continued)*

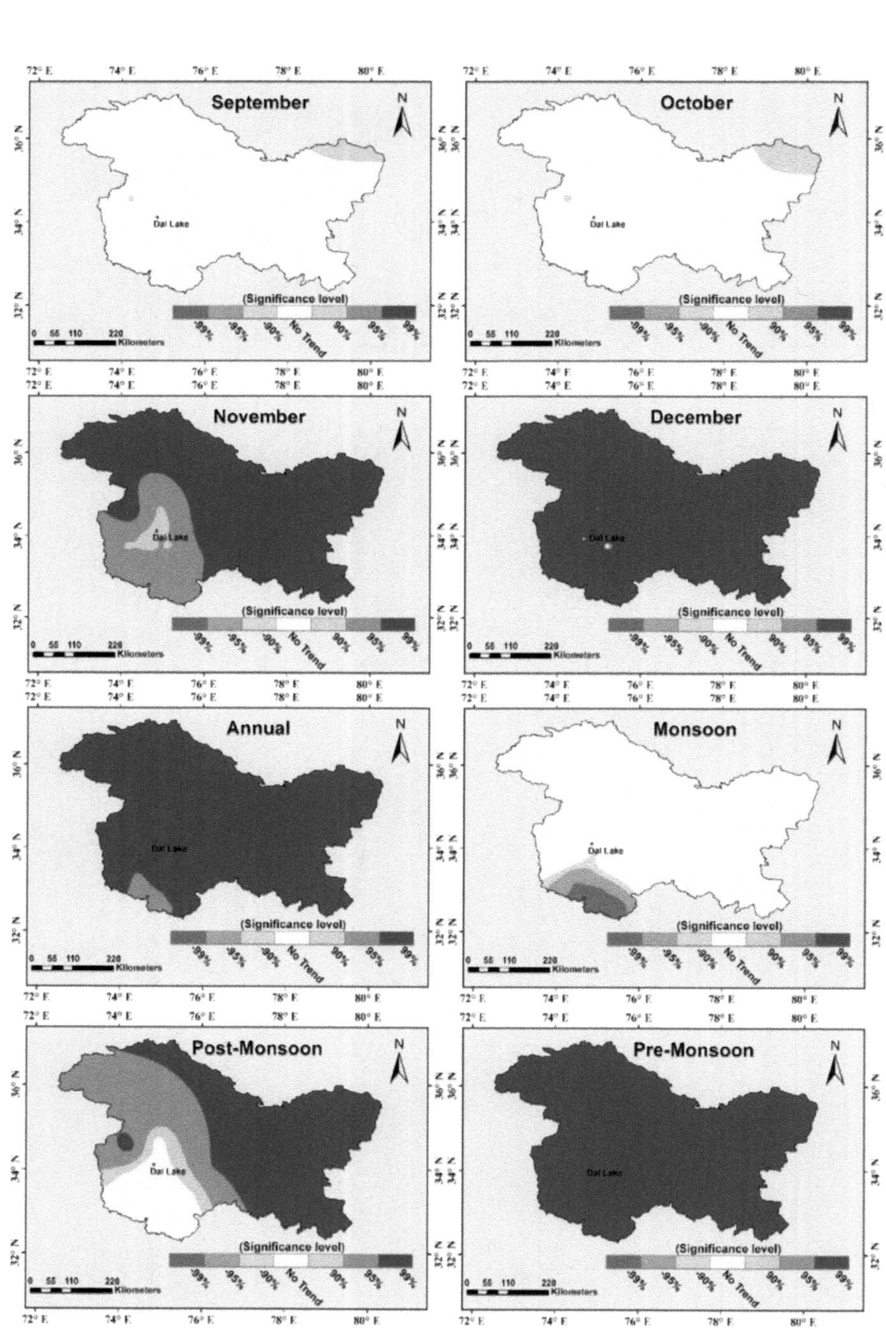

FIGURE 15.7 *(Continued)*

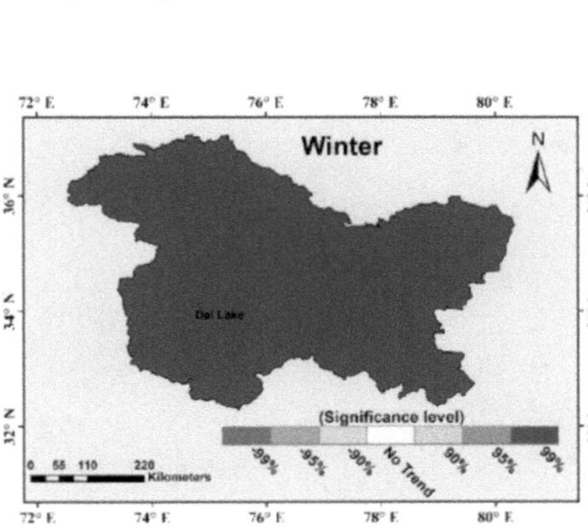

FIGURE 15.7 Spatial distribution of monthly, annual and seasonal mean temperature temporal change during 1901–2002 using Mann-Kendall.

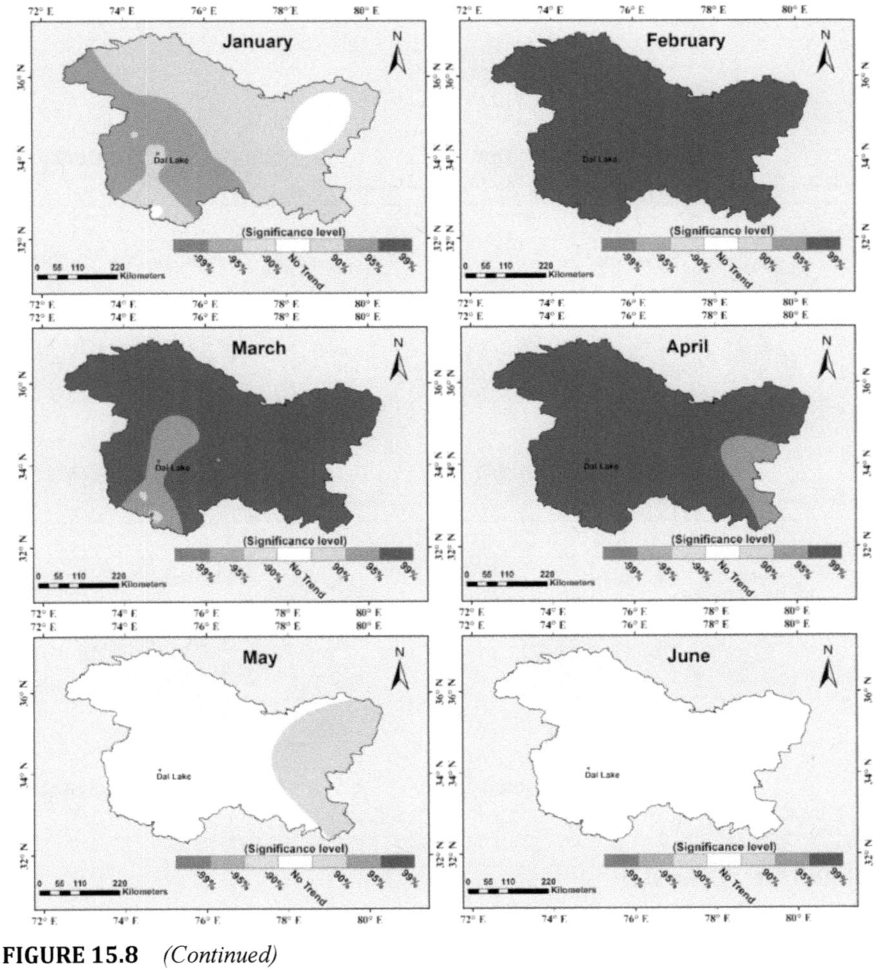

FIGURE 15.8 *(Continued)*

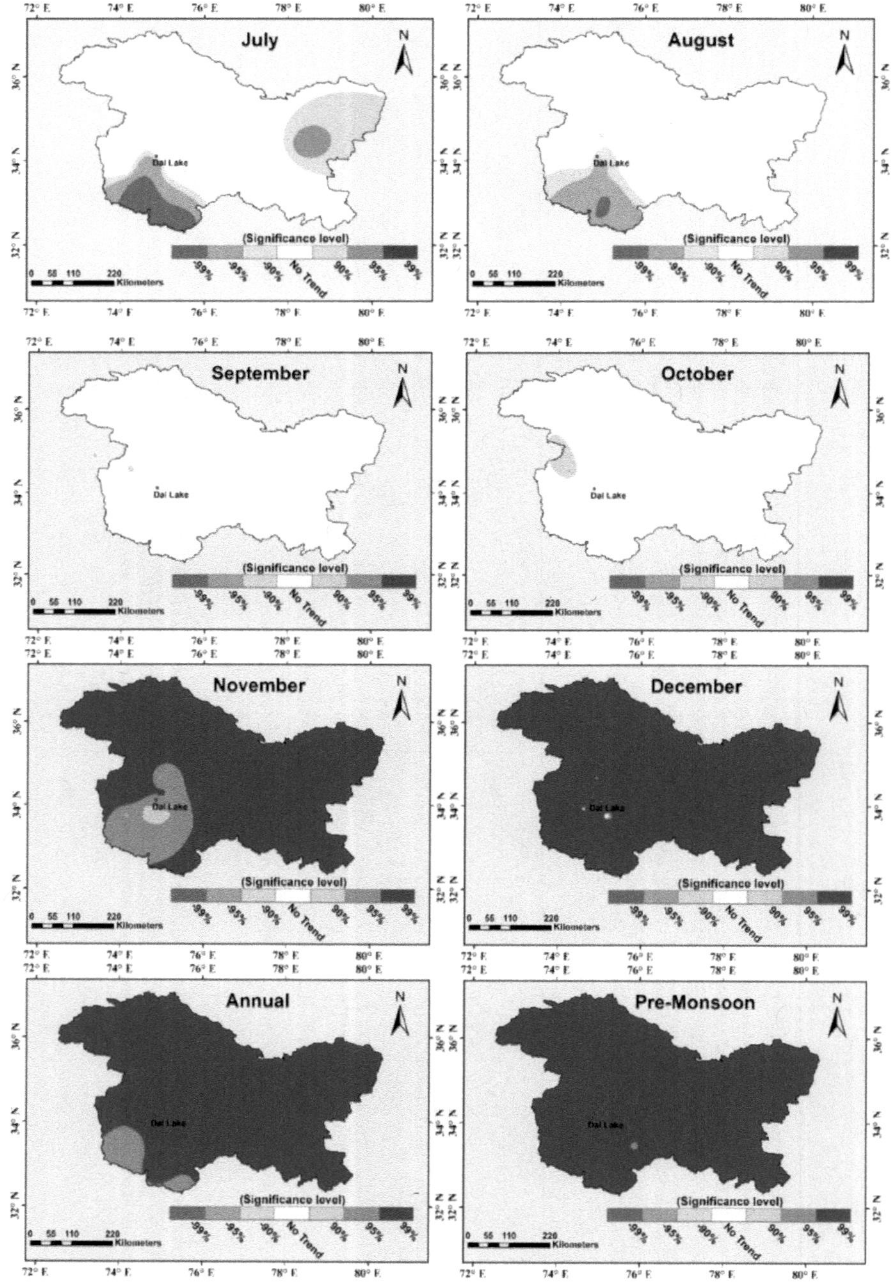

FIGURE 15.8 *(Continued)*

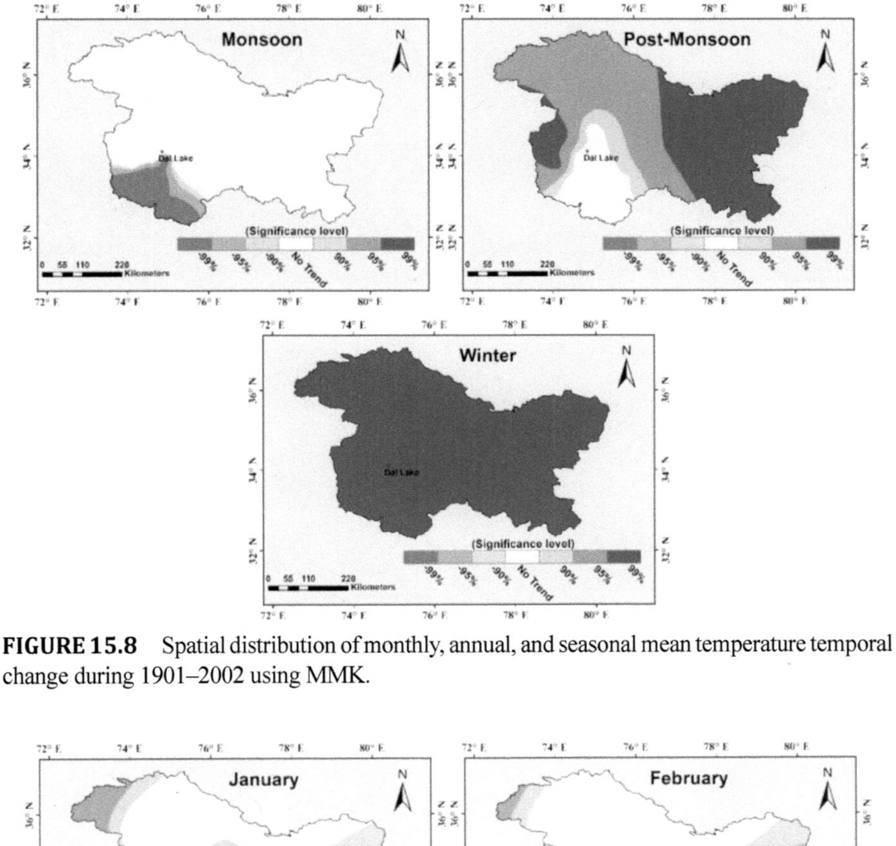

FIGURE 15.8 Spatial distribution of monthly, annual, and seasonal mean temperature temporal change during 1901–2002 using MMK.

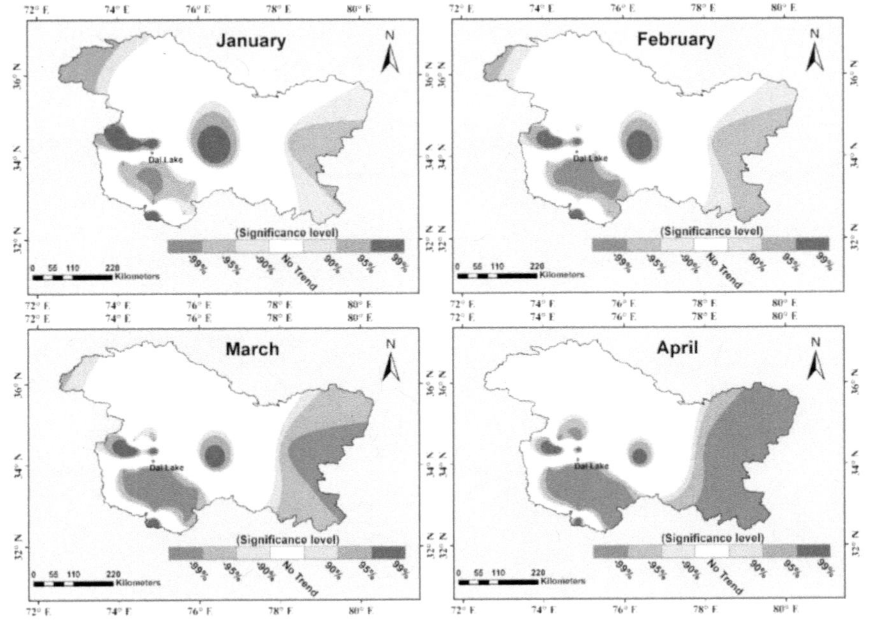

FIGURE 15.9 *(Continued)*

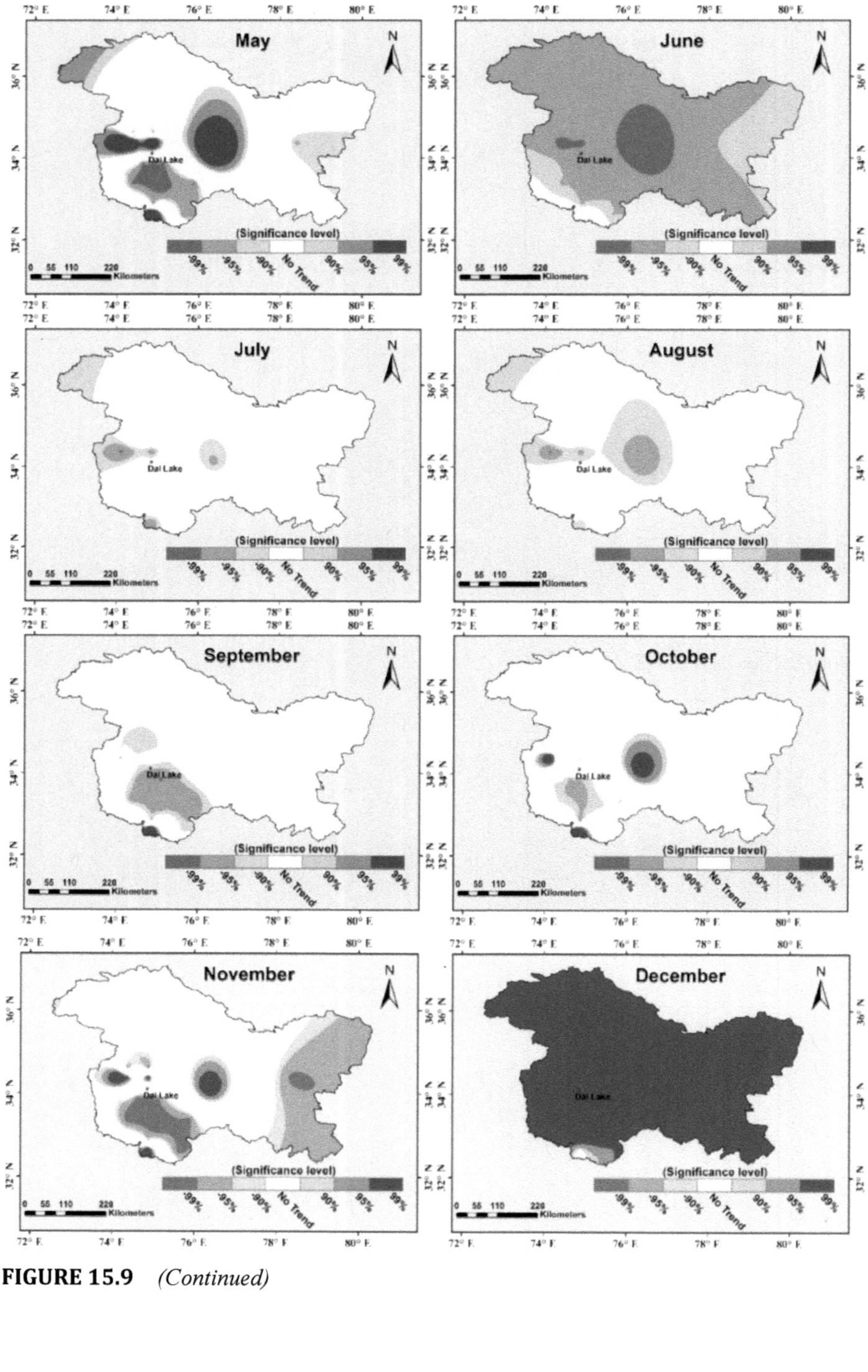

FIGURE 15.9 *(Continued)*

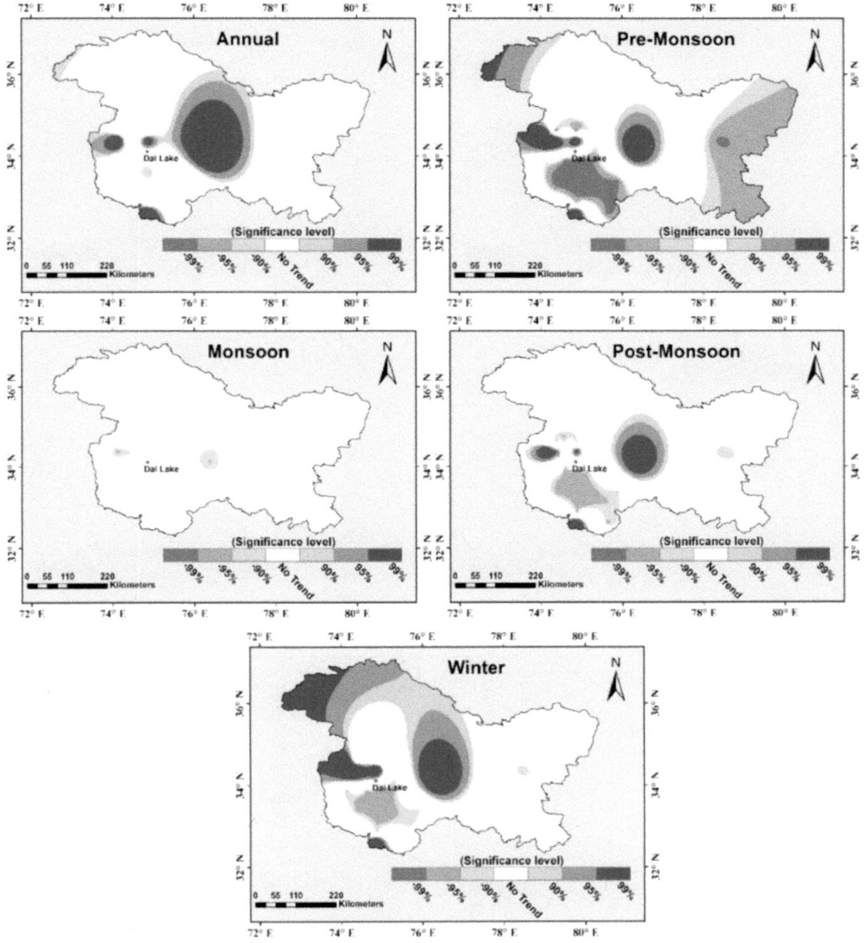

FIGURE 15.9 Spatial distribution of monthly, annual, and seasonal mean temperature temporal change during 2016–2099 using Mann–Kendall.

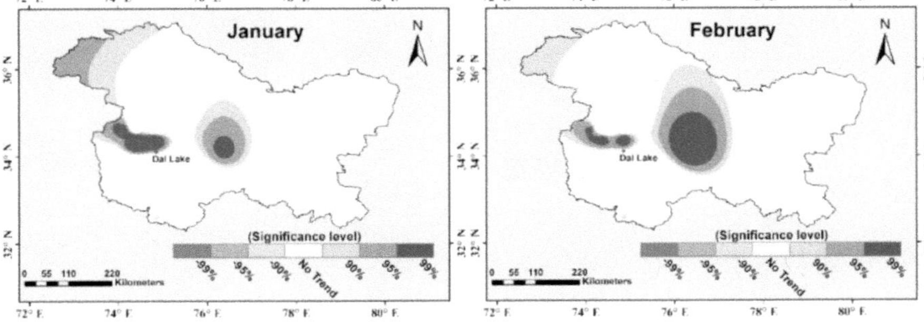

FIGURE 15.10 *(Continued)*

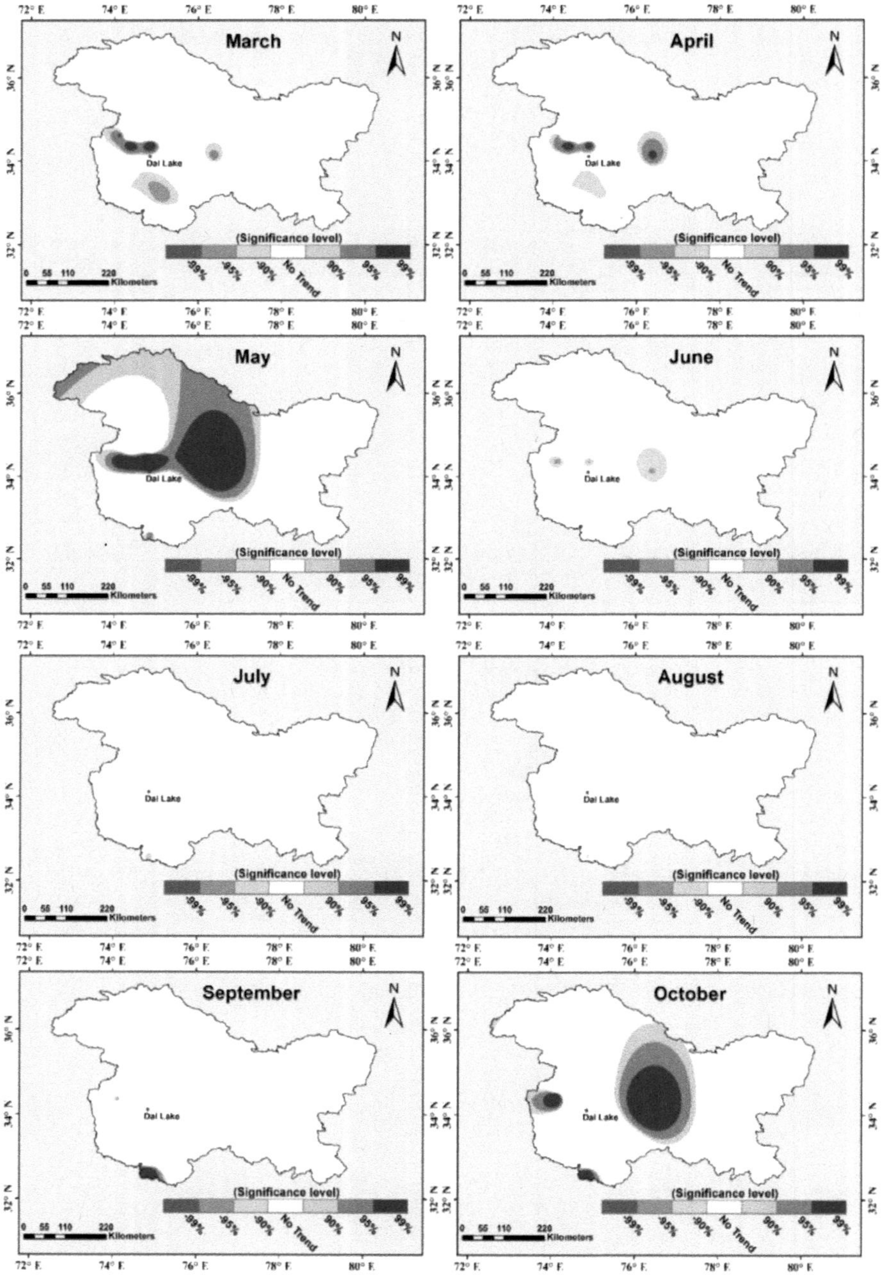

FIGURE 15.10 *(Continued)*

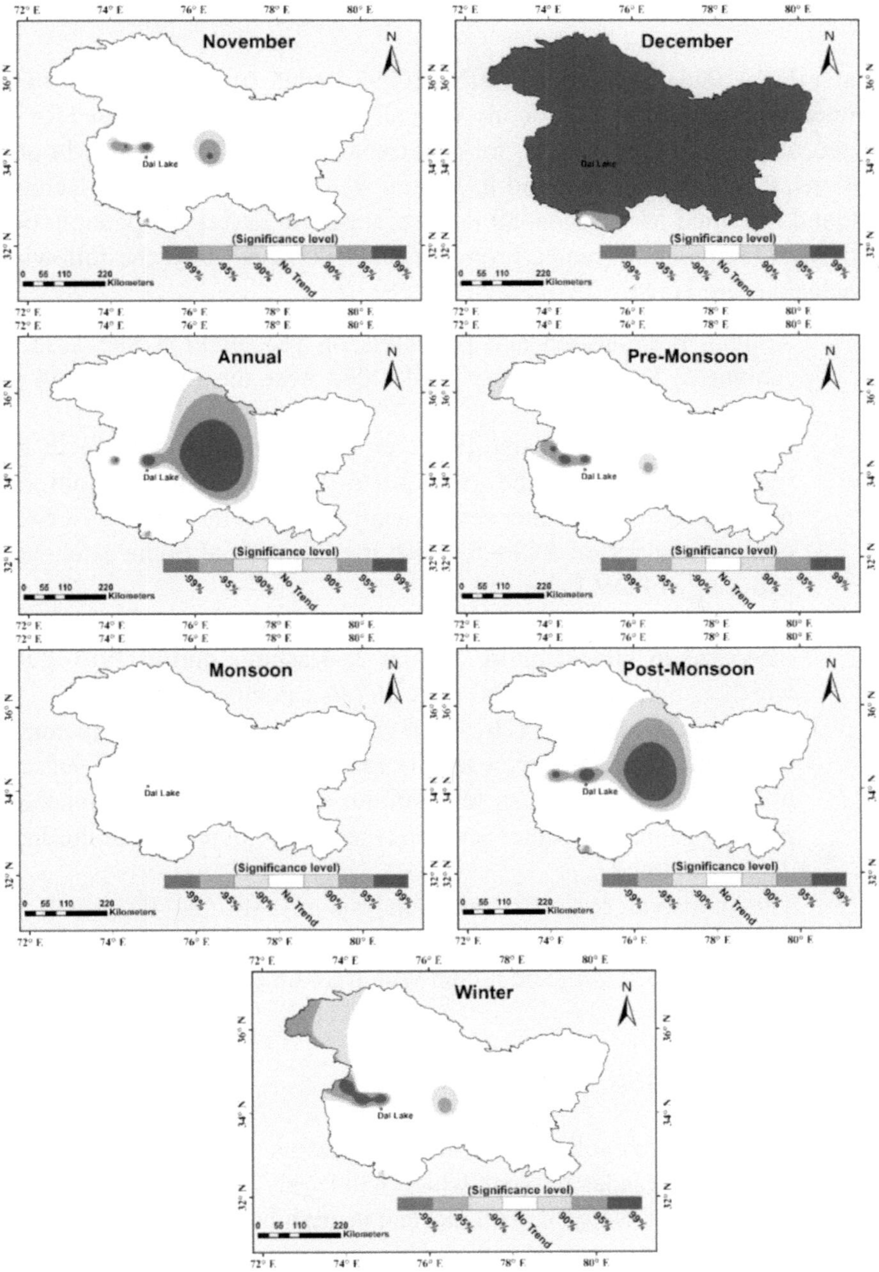

FIGURE 15.10 Spatial distribution of monthly, annual, and seasonal mean temperature temporal change during 2016–2099 using MMK.

15.5 CONCLUSIONS

Past (1901–2002) and future (2016–2099) trends of rainfall and mean temperature time-series on monthly, annual, and seasonal were studied for 14 meteorological stations, that is, for past trend, and 18 CMIP5-CCSM4 grid stations, that is, for future trend in Jammu & Kashmir. The Mann–Kendall test and modified Mann–Kendall test were used to detect the precipitation and mean temperature trends. From the results of this study, the following conclusions are drawn.

1. Annual, premonsoon and postmonsoon precipitation increased in Jammu & Kashmir during 1901–2002 near the Dal Lake and its catchment area.
2. There will be either significant or nonsignificant increase in the precipitation during the year 2016–2099 for annual, monsoon, postmonsoon, and winter season analyzed using the CMIP5-CCSM4 climate scenario of RCP8.5 which may contribute to the catchment hydrology of Dal Lake.
3. Annual, premonsoon, postmonsoon, and winter mean temperature increased in almost entire Jammu & Kashmir during 1901–2002 which also includes the Dal Lake and its catchment area.
4. Future trend analysis (2016–2099; RCP4.5) of mean temperature shows that the Dal region and its catchment area are more susceptible to increase in mean temperature during annual, premonsoon, postmonsoon, and winter seasons whereas the other part of the J&K detects no trend.
5. The monsoon seasonal trend analysis reveals that there will be no change in mean temperature for the entire J&K state using the CMIP5-CCSM4 climate model with RCP4.5 scenario.

15.6 BOTTOM LINE

The overall analysis concludes that the precipitation and mean temperature will rise in future decades to come which will surely impact the hydrology, hydraulics, and the ecology of Dal Lake and its periphery. It is also estimated that the intense rainfall events induced flood scenarios is going to elevate drastically, subsequently conveying more sediment to Dal Lake which will surely impact the useful life of the Dal Lake. Analyses of this study shall be helpful for water resource department, irrigation, agricultural practitioners, environmentalist, and in particular, the Lake and Waterway Development Authority (LDWA) of J&K and can play a vital role in the management of Dal Lake water resources more efficiently and in sustained manner in the region and for the state.

15.7 APPENDIX 1

15.7.1 AUTOCORRELATION

One of the main problems in testing and interpreting trends is the effect of serial dependence. The presence of positive or negative autocorrelation affects the detection of trend in a series (Hamed and Rao, 1998; Yue et al., 2002b). With a positive autocorrelation (persistence) in the time series, the MK test would detect a significant trend in the time series while there may be actually none or vice versa. Therefore, all the precipitation and mean temperature series are first tested for serial correlation using lag-1 autocorrelation coefficient (r1) given in eq 15.1 at 5% significant level for two-tailed test.

$$r = \frac{\sum_{i1}^{N-1}(x_t - \bar{x})(x_{t+1} - \bar{x})}{\sum_{i=1}^{N}(x_t - \bar{x})^2} \tag{15.1}$$

Where $\bar{x} = \sum_{i=1}^{N-1} x_t$ is the overall mean of N sample size.

The r_1, is tested against the null hypothesis using Anderson et al. (2000) limit for the two-tailed test.

$$r_1(95\%) = \frac{-1 \pm 1.96\sqrt{(N-2)}}{N-1} \tag{15.2}$$

If the r_1 value falls within the confidence interval given above, the data are assumed to be serially independent, otherwise, the sample data are considered to be significant serially correlated.

15.7.2 MANN-KENDALL TEST

The MK test is the rank-based nonparametric test and has been recently used by several researchers to detect the trends in data series (de la Casa and

Nasello, 2010; de la Casa and Nasello, 2012; Krishnakumar et al., 2009; Shifteh Some'e et al., 2012; Subash et al., 2011; Tabari et al., 2011). It is based on the test statics S which is defined as:

$$S = \sum_{i=0}^{n-1} \sum_{j=i+1}^{n} sgn(x_j - x_i) \tag{15.3}$$

where, x_1, x_2, … x_n represent n data points where x_j represents the data point at time j.

A very high positive value of S is an indicator of a rising trend, and a very low negative value indicates a falling trend.

$$sgn(x_j - x_i) = \begin{Bmatrix} 1 \ldots\ldots\ldots\ldots \text{if}(x_j - x_i) > 0 \\ 0 \ldots\ldots\ldots\ldots \text{if}(x_j - x_i) = 0 \\ -1 \ldots\ldots\ldots\ldots \text{if}(x_j - x_i) < 0 \end{Bmatrix} \tag{15.4}$$

It has been documented that when $n \geq 10$, the statistic S is approximately normally distributed with the mean

$$E(S) = 0$$

And its variance is

$$VAR(S) = \frac{n(n-1)(2n+5) - \sum_{i=1}^{m} t_i(t_i - 1)(2t_t + 5)}{18} \tag{15.5}$$

where n is the number of data points, m is the number of tied groups (a tied group is a set of sample data having the same value), and t_i is the number of data points in the *i*th group.

The standardized test statistic Z is computed as follows:

$$Z = \begin{cases} \dfrac{S-1}{\sqrt{VAR(S)}} \; if\ S > 0 \\ 0 \; if\ S = 0 \\ \dfrac{S+1}{\sqrt{VAR(S)}} \; if\ S < 0 \end{cases} \tag{15.6}$$

The null hypothesis, H0, meaning that no significant trend is present, is accepted if the test statistic Z is not statistically significant, that is, $-Z_{\alpha/2} < Z < Z_{\alpha/2}$, where $Z_{\alpha/2}$ is the standard normal deviate.

15.7.3 MODIFIED MANN-KENDALL TEST

The null hypothesis for the Mann–Kendall test is that the data are independent and randomly ordered, that is, there is no trend or serial correlation structure among the observations (Hamed and Rao, 1998). The existence of positive serial correlation will increase the possibility of rejecting the null hypothesis of no trend while it is actually true; while negative serial correlation will decrease the possibility of rejecting the null hypothesis (Yue and Hashino, 2003; Yue and Wang, 2004). Thus, the ESS method applied by Lettenmaier (1976) was used to correct the influence of serial correlation on the test, that is, to modify the variance of the Mann–Kendall statistic by the ESS. The modified variance $\mathrm{Var}^*(S)$ by the ESS is given by

$$Var^*(S) = Var(S) \cdot \frac{n}{n^*} \tag{15.7}$$

where ρ_1 is lag-1 serial correlation coefficient. By replacing the variance $\mathrm{Var}(S)$ with $\mathrm{Var}^*(S)$, the modified standard Mann–Kendall statistic is given by

$$Z^* = Z\sqrt{\frac{n^*}{n}} \tag{15.8}$$

15.7.4 THEIL-SEN'S ESTIMATOR

The slope of n pairs of data points was estimated using the Theil–Sen's estimator (Theil, 1950; Sen, 1968) which is given by the following relation:

$$\beta = Medium\left(\frac{x_j - x_i}{j - i}\right) \quad for\ all\ i < j \tag{15.9}$$

In which $1 < j < i < n$ and β is the robust estimate of the trend magnitude. A positive value of β indicates an "upward trend," while a negative value of β indicates a "downward trend" (Xu et al., 2007).

15.7.5 MANN-WHITNEY-PETTITT METHOD (MWP)

Consider a time series $\{X_1, X_2 \ldots, X_n\}$ with a length n. Let t be the time of the most likely change point. Two samples, $\{X_1, X_2, \ldots, X_t\}$ and $\{X_{t+1}, X_{t+2},$

..., X_n}, can then be derived by dividing the time series at time t. An index, U_t, is derived by:

$$U_t = \sum_{i=1}^{t} \sum_{j=t+1}^{n} sgn(x_i - x_j) \tag{15.10}$$

Where,

$$sgn(x_j - x_i) = \begin{Bmatrix} 1............\text{if}(x_j - x_i) > 0 \\ 0............\text{if}(x_j - x_i) = 0 \\ -1..........\text{if}(x_j - x_i) < 0 \end{Bmatrix} \tag{15.11}$$

A plot of U_t against t for a time series with no change point would result in a continually increasing value of $|U_t|$. However, if there is a change point (even a local change point), then $|U_t|$ would increase up to the change point and then begin to decrease. The most significant change point *t* can be identified as the point where the value of $|U_t|$ is maximum:

$$K_T = \max_{1 \le t £ T} |U_\tau| \tag{15.12}$$

The approximated significance probability p(t) for a change point (Pettitt, 1979) is given as:

$$p = 1 - \exp\left[\frac{-6K_T^2}{n^3 + n^2}\right] \tag{15.13}$$

The change point is statistically significant at time t with a significance level of α when probability p(t) exceeds (1 − α).

15.7.6 SPATIAL ANALYSIS OF PRECIPITATION SERIES

To explore the spatial distribution of trends on monthly, annual, and seasonal basis, the linear regression slope was interpolated using ArcGIS 10.2.2 based on each stations slope value for entire study period (1901–2002). An inverse distance weighted (IDW) interpolation method which is based on the assumption that the interpolation surface should be influenced most by nearby points and less by more distant points (Gemmer et al., 2004) was used. A variable search radius with minimum sample of 12 points was used. To control the significance of surrounding point on the interpolated value, the exponent of distance which means the fourth power was used because the higher power results in less influence from distant points.

KEYWORDS

- **Mann-Kendal**
- **modified Mann-Kendall**
- **precipitation**
- **climate change**
- **Dal lake**
- **Jammu and Kashmir**

REFERENCES

Anderson, D. R.; Burnham, K. P.; Thompson, W. L. (2000). Null Hypothesis Testing: Problems, Prevalence, and an Alternative. *J. Wildlife Management* **2000,** 912–923.

Beniston, M. Mountain Climates and Climatic Change: An Overview of Processes Focusing on the European Alps. *Pure Appl. Geophys.* **2005,** *162* (8–9), 1587–1606.

Bradley, R. S.; Briffa, K. R.; Cole, J.; Hughes, M. K.; Osborn, T. J. The Climate of the Last Millennium. In *Paleoclimate, Global Change and the Future*; Springer: Berlin, Heidelberg, 2003; pp 105–141.

Chattopadhyay, S.; Edwards, D. R. Long-Term Trend Analysis of Precipitation and Air Temperature for Kentucky, United States. *Climate* **2016,** *4* (1), 10.

De La Casa, A. C.; Nasello, O. B. Low Frequency Oscillation of Rainfall in Córdoba, Argentina, and Its Relation with Solar Cycles and Cosmic Rays. *Atmos. Res.* **2012,** *113*, 140–146.

de la Casa, A.; Nasello, O. Breakpoints in Annual Rainfall Trends in Córdoba, Argentina. *Atmos. Res.* **2010,** *95* (4), 419–427.

Fu, G.; Charles, S. P.; Yu, J.; Liu, C. Decadal Climatic Variability, Trends, and Future Scenarios for the North China Plain. *J. Clim.* **2009,** *22* (8), 2111–2123.

Gemmer, M.; Becker, S.; Jiang, T. (2004). Observed Monthly Precipitation Trends in China 1951–2002. *Theo. Appl. Climatol.* **2004,** *77* (1–2), 39–45.

Gent, P. R.; Danabasoglu, G.; Donner, L. J.; Holland, M. M.; Hunke, E. C.; Jayne, S. R.; ... & Worley, P. H. (2011). The Community Climate System Model Version 4. *J. Clim.* **2011,** *24* (19), 4973–4991.

Gent, P. R.; Yeager, S. G.; Neale, R. B.; Levis, S.; Bailey, D. A. Improvements in a Half Degree Atmosphere/Land Version of the CCSM. *Clim. Dyn.* **2010,** *34* (6), 819–833.

Guhathakurta, P.; Rajeevan, M. Trends in the Rainfall Pattern over India. *Int. J. Climatol. J. R. Meteorol. Soc.* **2008,** *28* (11), 1453–1469.

Hamed, K. H.; Rao, A. R. A Modified Mann-Kendall Trend Test for Autocorrelated Data. *J. Hydrol.* **1998,** *204* (1–4), 182–196.

IPCC. Impacts, Adaptation and Vulnerability. Contribution of Working Group II. In: Parry, M. L., Canziani, O. F., 2007.

Kendall, M. G. *Rank Correlation Methods*, 4th ed.; Charles Griffin: London, 1975; p 202.

Kothyari, U. C.; Singh, V. P. Rainfall and Temperature Trends in India. *Hydrol. Process.* **1996,** *10* (3), 357–372.

Krishnakumar, K. N.; Rao, G. P.; Gopakumar, C. S. Rainfall Trends in Twentieth Century over Kerala, India. *Atmos. Environ.* **2009,** *43* (11), 1940–1944.

Kumar, K. R.; Pant, G. B.; Parthasarathy, B.; Sontakke, N. A. Spatial and Subseasonal Patterns of the Long-Term Trends of Indian Summer Monsoon Rainfall. *Int. J. Climatol.* **1992,** *12* (3), 257–268.

Lettenmaier, D. P. Detection of Trends in Water Quality Data from Records with Dependent Observations. *Water Resour. Res.* **1976,** *12* (5), 1037–1046.

Mann, H. B. Non-Parametric Test Against Trend. *Econometrica* **1945,** *13*, 245– 259.

Meinshausen, M.; Smith, S. J.; Calvin, K.; Daniel, J. S.; Kainuma, M. L. T.; Lamarque, J. F.; ... Moss, R. The Paleoclimate Modeling Intercomparison Project Contribution to CMIP5. *WCRP Coupled Model Intercomparison Project-Phase 5-CMIP5*, 2011; p 15.

Pachauri, R. K.; Allen, M. R.; Barros, V. R.; Broome, J.; Cramer, W.; Christ, R.; Dubash, N. K. *Climate Change 2014: Synthesis Report. Contribution of Working Groups I, II and III to the Fifth Assessment Report of the Intergovernmental Panel on Climate Change*Ipcc, 2014; p 151.

Parry, M.; Parry, M. L.; Canziani, O.; Palutikof, J.; Van der Linden, P.; Hanson, C. 2007. *Climate Change 2007-Impacts, Adaptation and Vulnerability: Working Group II Contribution to the Fourth Assessment Report of the IPCC* (Vol. 4); Cambridge University Press, 2007.

Pettitt, A. N. Anon-Parametric Approach to the Change-Point Detection. *Appl. Stat.* **1979,** *28*, 126–135.

Sen, P. K. Estimates of the Regression Coefficient Based on Kendall's Tau. *J. Am. Stat. Assoc.* **1968,** *63* (324), 1379–1389.

Shields, C. A.; Kiehl, J. T. Simulating the Pineapple Express in the Half Degree Community Climate System Model, CCSM4. *Geophys. Res. Lett.* **2016,** *43* (14), 7767–7773.

Siegel, S.; Castellan, N. J. *Nonparametric Statistics for the Behavioral Sciences*; McGraw-HiU Book Company, New York, 1988.

Some'e, B. Shifeth.; Ezani, A.; Tabari, H. Spatiotemporal Trends and Change Point of Precipitation in Iran. *Atmos. Res.* **2012,** *113*, 1–12.

Subash, N.; Singh, S. S.; Priya, N. Variability of Rainfall and Effective Onset and Length of the Monsoon Season over a Sub-Humid Climatic Environment. *Atmos. Res.* **2011,** *99* (3–4), 479–487.

Tabari, H.; Somee, B. S.; Zadeh, M. R. Testing for Long-Term Trends in Climatic Variables in Iran. *Atmos. Res.* **2011,** *100* (1), 132–140.

Taylor, K. E.; Stouffer, R. J.; Meehl, G. A. An Overview of CMIP5 and the Experiment Design. *Bull. Am. Meteorol. Soc.* **2012,** *93* (4), 485–498.

Theil, H. A Rank-Invariant Method of Linear and Polynomial Regression Analysis, Part 3. In *Proc. Koninalijke Nederlandse Akademie van Weinenschatpen A*, Vol. 53; 1950; pp 1397–1412).

Van Belle, G.; Hughes, J. P. Nonparametric Tests for Trend in Water Quality. *Water Resour. Res.* **1984,** *20* (1), 127–136.

von Storch, H. (1995). Misuses of Statistical Analysis in Climate Research. In *Analysis of Climate Variability*; Springer: Berlin, Heidelberg, 1995; pp 11–26.

Xu, Z. X.; Li, J. Y.; Liu, C. M. Long-Term Trend Analysis for Major Climate Variables in the Yellow River Basin. *Hydrol. Process. Int. J.* **2007,** *21* (14), 1935–1948.

Xu, Z. X.; Li, J. Y.; Liu, C. M. Long-Term Trend Analysis for Major Climate Variables in the Yellow River Basin. *Hydrol. Process. Int. J.* **2007,** *21* (14), 1935–1948.
Yue, S.; Hashino, M. Long Term Trends of Annual and Monthly Precipitation in Japan 1. *JAWRA J. Am. Water Resour. Assoc.* **2003,** *39* (3), 587–596.
Yue, S.; Wang, C. Y. Applicability of Prewhitening to Eliminate the Influence of Serial Correlation on the Mann-Kendall Test. *Water Resour. Res.* **2002,** *38* (6), 4–1.
Yue, S.; Pilon, P.; Cavadias, G. Power of the Mann–Kendall and Spearman's RHO Tests for Detecting Monotonic Trends in Hydrological Series. *J. Hydrol.* **2002,** *259*(1–4), 254–271.
Yue, S.; Pilon, P.; Phinney, B.; Cavadias, G. The Influence of Autocorrelation on the Ability to Detect Trend in Hydrological Series. *Hydrol. Process.* **2002a,** *16* (9), 1807–1829.
Yue, S.; Pilon, P.; Cavadias, G. Power of the Mann–Kendall and Spearman's RHO Tests for Detecting Monotonic Trends in Hydrological Series. *J. Hydrol.* **2002b,** *259*(1–4), 254–271.

CHAPTER 16

Modeling 1D–2D Hydrodynamic SWMM Model to Mitigate Flooding and Prioritize Flood-Prone Areas

MANISH KUMAR SINHA[1,3*], MONIKA SHARMA[2], KLAUS BAIER[1], RAFIG AZZAM[1], and MUKESH KUMAR VERMA[3]

[1]*Department of Engineering Geology and Hydrogeology, RWTH Aachen University, Lochnestr 52064, Aachen, Germany*

[2]*Department of Environmental & Water Resources Engineering, University Teaching Department, Chhattisgarh Swami Vivekanand Technical University Bhilai, Bhilai 491107, Chhattisgarh, India*

[3]*Chhattisgarh Swami Vivekanand Technical University Bhilai, Bhilai 491107, Chhattisgarh, India*

[*]*Corresponding author. E-mail: manishsinha200389@gmail.com; sinha@lih.rwth-aachen.de*

ABSTRACT

Urban stormwater modeling is considered to be a significant tool for flood inundation analysis. Urban floods are caused by the increase of flood plains and unplanned drainage mechanisms. Due to lack of data, adoption of a simulation approach is one of the realistic ways of defining the overflooded junctions and critical subcatchments. The frequency of devastating flooding in the rainy seasons suggests that the urban drainage system in the pilot area is inadequate. This study mainly aimed to compare and simulate the calibrated model of pre- and postdevelopment scenarios to estimate surface runoff

Advances in Hydrology and Climate Change: Historical Trends and New Approaches in Water Resources Management. Surendra Kumar Chandniha, Anil Kumar Lohani, Gopal Krishan, & Ajay Krishna Prabhakar (Eds.)

depth for 10-year and 50-year return periods of rainfall. For the modeling and analysis of urban drainage, stormwater management model (SWMM) 5.1 engine was used in PCSWMM platform. For this study, 10- and 50-return periods of pre- and postdevelopment scenarios the surface runoff, infiltration, continuity error, as well as flood-prone areas were simulated. The results show the surface runoff for predevelopment scenario for 10-year and 50-year return period is 6.774 and 17.248 mm and postdevelopment for 10-year and 50-year return period is 13.588 and 20.535 mm in depth, it shows for 10-year return period there is an increase of 99.41% in surface runoff and for 50-year return period 18.86% in surface runoff has been increased. Impervious rate increased that can be seen by the decrease in infiltration rate for 10-year 1.594 mm changed to 1.503 mm and for 50-year 4.969–3.208 mm. Based on simulation B15, B2, B25, B30, B8, and WL2 are more critical subcatchments which is flood prone area, it shows that there is a need for proper drainage in the area and to provide sustainable approach to reduce flooding. This research would provide scientific support for the management of floods and disaster in the city of Raipur.

16.1 INTRODUCTION

The susceptibility to flooding is growing due to changes in the trends of land use and population growth acceleration (Miller et al., 2014). Determining and finding deficiencies in drainage systems and developing the model of surface drainage can help to reduce human-generated factors contributing to flooding in the region (Nie, 2004). Urban areas are indicative of the development's enhanced imperfection. Nevertheless, rapid urbanization contributes to rapid changes in land use and to an increasing impermeable region (Guan et al., 2015). The increased volumes and peak flow related to faster response time generate threats for urban flooding (Guan et al., 2015). It is very important to develop scientific methods for the prevention and control of urban stormwater disasters (Zhou, 2014).

In general, urbanization includes the elimination, recourse to land, and compaction with heavy machinery of natural vegetation and the topsoil. Roads are then built and services are installed, such as water drains and water supplies. The floor will then be replaced and buildings, entrances, and parking facilities will be built. Increasing the impervious area in a catchment, therefore, leads to changes to the flow regime of the stream. Stream flows could be categorized into two parts, which are considered rapid flow and a base flow portion that penetrates into the field and gradually enters the river shortly after the rainfall.

Typically, urbanization increases the component of the rapid flow, increasing the magnitude, and frequency of high flows and increasing the tempest spikes after the rain starts (Sinha et al., 2019b). This often results in the extension of the channel. At the same time, opportunities for water to penetrate into the soil are decreased and the base flow is diminished. In India, urban flooding is not new. Most Indian cities as a developing country are faced with urban floods that have been some of the last significant events in India. It is not possible to prevent these incidents, but hydrological models should help us to understand the mechanism and to develop battle strategies to tackle it. The primary element of flood forecasting is the perception and socioeconomic implications of the dynamics of floods in urban areas (Magesh and Ch, 2012). Digital elevation models (DEMs) are used to promote urban flood forecasting where fewer data are available for city-based catchments. The literature demonstrates that flood frequency analyses and maps can give clear insights into heavily affected areas and can help the authorities in developing the technical action plan. The focus of this work is on finding areas of heavy precipitation that are vulnerable to floods in a Pilot area of Raipur city. In conjunction with a 1D–2D flood modeling approach, areas susceptible to flood are identified. One-dimensional (1D) model for the pilot field is being developed using the SWMM and 2D PCSWMM. The results of the simulation run to identify overflow nodes. The overflowing nodes indicate that drainage in the area needs to be improved so that the stormwater can be removed and the floods minimized.

16.2 MATERIALS AND METHODS

Flooding has become a common phenomenon in urban areas worldwide. Considering the tremendous losses caused by such incidents, the planning and implementation of effective methods of prevention have gained due importance recently. Knowledge of the depths of flooding with its areal extent and probable junctions of drainage risk will help to take administrative decisions. In the past, the overflowing junctions in the drainage network were found using only 1D models. This approach will cause water depth to increase in the flow channel. Modeling surface flow in urban floods has equal importance that causes flooding near drains. This research uses a hybrid 1D–2D simulation approach that can understand both drain flow and surface flow to determine flood depths and areas at risk of flooding. Using the data available in the form maps and attribute values, a coupled 1D–2D PCSWMM model for Pilot Study was created. Figure 16.1 displays the PCSWMM 1D–2D concept development flowcharts.

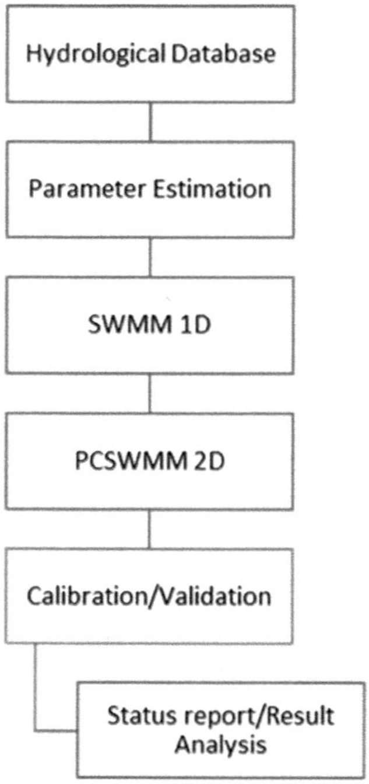

FIGURE 16.1 1D–2D concept flow chart adopted in this study.

16.3 STUDY AREA AND DATA USED

The region of research corresponds to the metropolitan catchment basin that drains the southwest portion of the city of Raipur. The urban basin has a total area of 13.58 km^2 with the extension of 3.30 km north–south and an extension of about 5.70 km in the east–west direction. Its highest height is roughly 254.9 m, the minimum height is 204.5 m above sea level. The region is flat and has a slight slope to the west-southwest (<1%).

Raipur's traditional face altered as it became a significant regional, commercial, and industrial destination following its establishment as the state's capital. Moreover, several villages were added to the town, followed by "elevated population development rates which were not matched by a respective sanitary infrastructure." Despite being a historically and archeologically significant town, with several lovely lakes in the fertile plains of the Chhattisgarh area, Raipur is unfortunately classified by the Central Pollution Control Board as a "critically polluted area." There is a strong impact on surface runoff modeling in urban catchment parameters, such as meteorology, geology, soil information, and hydrology. Knowledge of the nature of the study region can better understand and explain important catching characteristics, such as infiltration rate and precipitation. Therefore, the following sections provide a summary of key features and conditions in the urban catchment. The location of Raipur city and the area of research within the red boundary across the southwest direction are shown in Figure 16.2. Two factors contributed to the selection of the area of analysis: firstly, field survey was carried out in the city of Raipur in May 2016. Part of the research area was included in this survey. Furthermore, point of outlet of the model is projected at Kharun River. The sum of precipitation in Raipur is attributed to southwest monsoon, June to October, from June to August, heavy rainfall triggers regular flooding in the channel and river. The region receives 1282 mm of annual rainfall. The Raipur's low-lying areas would be submerged, if the rainfall exceeds 100 mm/day (Kumari et al., 2016). If the annual precipitation reaches 150 mm, 20% of the residents and 30% of roads and other services in the region are impacted. The key issue is the poor drainage network in the study area (Sinha et al., 2017).

Due to an absence of data available, modeling for the entire city of Raipur was difficult to implement. Because the city of Raipur is grouped into different wards by Raipur Municipal Corporation (RMC), a small drainage zone for the analysis of a specific ward was used. Based on accessible data, the choice of the model and its development method were made. Chhokra Nallah is the main canal that holds the water in the north. In comparison, the town of Raipur drains smaller nallahs and channel networks. Water is pumped radially through the Kharun River, as shown in Figure 16.2, on the left side of the red boundary, in its three major stream directions to the north, the south, and the west. Chingri nallah drains the study region, which goes southwest to the river Kharun. Therefore, this study's database is first presented and the method is structured. The significant information for this research is shown below.

16.4 MODEL CONCEPTUALIZATION

The capacity of SWMM to reliably represent hydrological and hydraulic processes in both urban and rural areas allows it the chosen simulation framework for this particular research project. Much of the model's input data were stored first in ArcGIS before being used in PC-SWMM. ArcGIS, or geographic information system, is a data analysis and interpretation of computer modeling software developed by ESRI. The first phase in the creation of the SWMM model was to delineate the study area using a DEM. Vaze et al. (2010) found that the quality of DEM-derived hydrological characteristics is vulnerable to both DEM accuracy and resolution and that many accurate topographical characteristics are lost as DEM resolution becomes coarser. In order to provide topographical details for the study area, a 12.5-m spatial resolution DEM, the highest resolution DEM usable for the study area, was chosen. ArcGIS was used to complete the delineation of the study area, limiting user calculation ability and calculating multiple required SWMM input parameters manually. A total of subcatchment 51 is grouped by the delineation method and according to the ward map of the study area. For each subcatchment, ArcGIS is used to measure the area, slope, duration, and elevation from the data contained in the DEM.

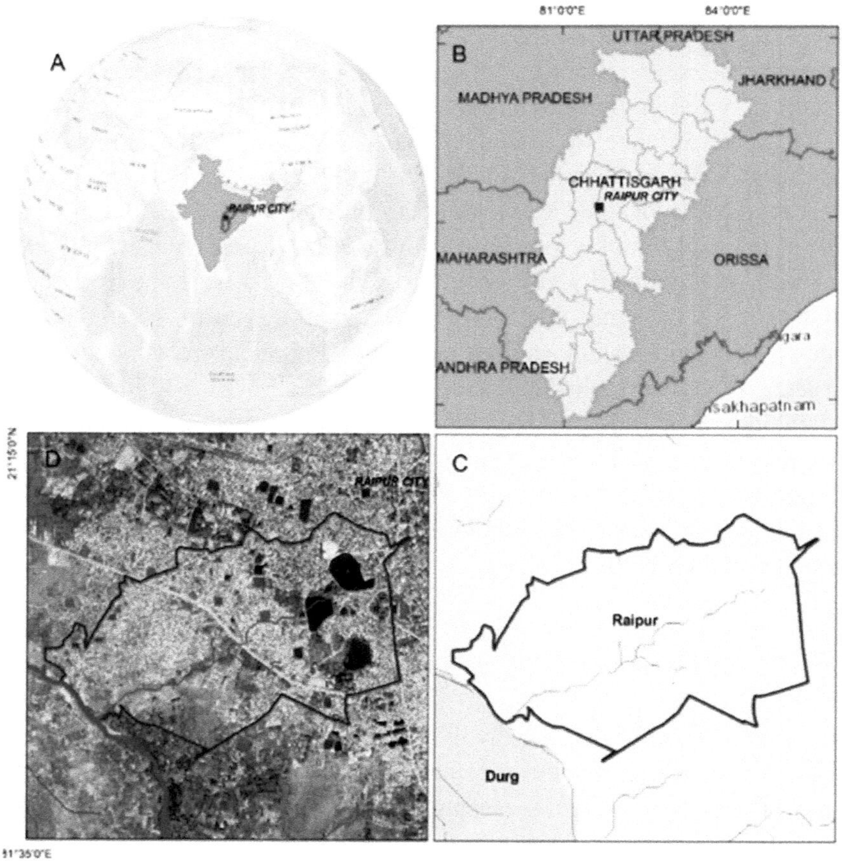

FIGURE 16.2 Location map of the study area.

TABLE 16.1 Data Used in this Analysis and SWMM Simulation.

Data	Description	Type	Source
Rainfall data	1980–2018 year	15 min interval	Indian Meteorological Department
Existing drainage network	Point line based	GIS shape file	Field surveyed
Digital elevation	12.5 × 12.5 m resolution	Raster image model	PALSAR DEM https://asf.alaska.edu/
Land use	Existing period	GIS shapefile	Used after (Sinha et al., 2019)

16.4.1 1D SWMM

The SWMM is a 1D model that calculates the flow and flow depth in the flow channel. The current drainage is used to create a 1D SWMM model for the study area. For the creation of the 1D drainage network area study region, the drainage network shapefile prepared in the ArcGIS database is used. The drainage network consists of subcatchments, canals, and junction points. In SWMM runoff, each subcatchment is called a nonlinear storage basin with a single rainfall inflow. This model is implemented by dynamic wave routing. The 1D flux equations of Saint Venant are thus solved and accurate results are obtained in SWMM. These equations consist of the continuity, momentum equations, and continuity equations at nodes in volume.

16.4.2 2D PCSWMM

The SWMM produced file is exported to PCSWMM and the 2D flow region is specified for land flow to create 2D models for the study area. Carto sat DEM applies elevation data and simulated terrain as a backdrop layer to the model. Creating 2D mesh on a 2D flow field produces 2D cells. Two-dimensional cells at the center of each 2D cell trough are generated a 2D node that takes place through a 2D flow exchange. For this study, a finer 12.5 m hexagonal mesh is adopted as Carto sat DEM is available in 12.5 m resolution.

16.4.3 MODEL DEVELOPMENT

For this research, SWMM used the EPA stormwater management model. It is a complex rainfall-runoff modeling model that can be used either for single events or long-time simulations, particularly in urban areas. The SWMM application model development is similar to that of other watershed project modeling and usually develops with the gathering and analysis of input data, followed by the simulation and calibration of the model in accordance with the local data (Gironás et al., 2010). The model can then be used to address different scenarios once the model development has been completed. Approaches for flow routing include steady, dynamic, and kinematic wave options. Infiltration rates can be measured either by using the Green-Ampt, Horton, or curve number methods (Rossman and Huber, 2016). The dynamic wave model was used for flow routing, and the Horton approach was used for infiltration in this study. The catchment was divided into 51 subcatchments, which reflects a fairly detailed discretization, taking into account the features of the study catchment. The slopes of each subcatchment are determined from the catchment's DEM. In ArcGIS, the area of the subcatchments and the percentage of impervious surfaces were determined from a land use survey. The conduit properties and junctions were derived from network data on the drainage pipe (Rossman and Supply, 2006). The width of the overland flow route in the model is an essential parameter and is very sensitive to the volume of the simulated runoff equation.

SWMM concepts a rectangular surface subcatchment with a consistent slope S and a width W, which drains to one single outlet channel, as illustrated in Figure 16.3. Overland flow is produced as a nonlinear reservoir by modeling the subcatchment, as shown in Figure 16.3.

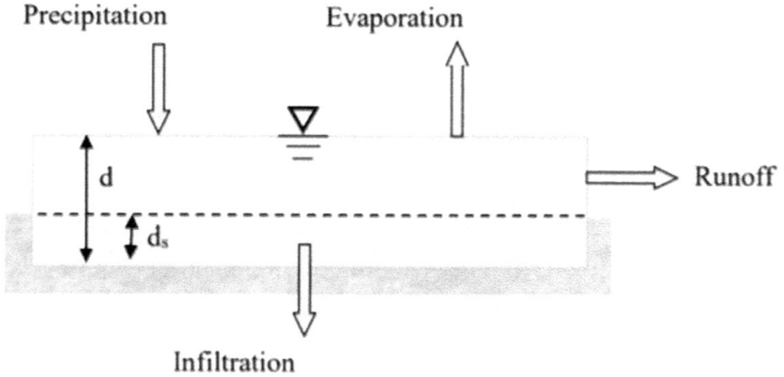

FIGURE 16.3 Conceptual nonlinear reservoir model of a subcatchment (Rossman and Huber, 2016).

In this representation, precipitation (rainfall and snowmelt) influxes and evaporation and infiltration losses occur during under catch. Net excess ponds reach a depth *d* above the under catch surface. Water contained above depression depth ds can become runoff outflow ds. Depression storage represents the first abstractions of rain, such as surface ponding, flat surfaces and vegetation interception, and surface weathering (Rossman and Huber, 2016).

The net change in depth *d* per unit of time (*t*) is the difference between input and outflow levels on the catchment from water conservation (conversation of mass):

$$\frac{\partial d}{\partial t} = i - e - f - q \tag{16.1}$$

$$Q = \frac{1.49}{n} S\frac{\frac{1}{2}}{x} R_X^{1/2} A_x \tag{16.2}$$

where:

i = rate of rainfall + snowmelt (ft/s)
e = surface evaporation rate (ft/s)
f = infiltration rate (ft/s)
q = runoff rate (ft/s)

Fluxes *i*, *e*, *f*, and *q* are indicated by unit area flow rates (cfs/ft^2 = ft/s). The method equation can be used to describe the volumetrically flow rate *Q* (cfs) of the runoff if the discharge across the surface of the subcatchment is as if it were the same discharge inside a rectangular channel of distance *W* (*f*), height *d* – *ds*, and slope *S*. Here, *n* is a coefficient of surface roughness, *S* is the noticeable or reasonable pitch (ft/ft), A_x is the region along the width of the subcatchment where the runoff (ft^2) flows, and R_x is the hydraulic distance connected with the field (ft). A_x is a rectangular field with *W* width and d–d_s height. *W* is always more than *d*, $A_x = W(d-d_s)$ and $R_x=(d-d_s)$.

$$Q = \frac{1.49}{n} Ws^{1/2}\ (d - d_s)^{5/3} \tag{16.3}$$

$$q = \frac{1.49 Ws^{\frac{1}{2}}}{An} (d - d_s)^{5/3} \tag{16.4}$$

Above is a normal nonlinear equation of differences. For known values *I*, *e*, *f*, d_s, and α, numerically it can be solved for depth *d* at each stage. Once *d* is known, eq 16.4 shows runoff rate *q* values. Note that only if *d* is larger than d_s does eq 16.5 apply. If $d \leq d_s$, ruin *q* is zero and the mass balance on *d* is simple (Rossman and Huber, 2016).

The initial values for Manning's roughness coefficient are shown in Table 16.2, that is, the width of the overland flow direction, the depression storage, the Horton equation infiltration parameters. Horton's theory is based on empirical findings that if the soil has potential for infiltration (mm h^{-1}), the infiltration decreases from an initial maximum rate to a minimal rate for an event of long rainfall, the intensity of the rain being higher than the intensity of infiltration (Rossman and Huber, 2017).

$$f_t = f_c + (f_0 - f_c)e^{kt} \qquad (16.5)$$

where f_c is the minimum infiltration rate, f_0 is the initial infiltration rate or maximum infiltration rate (at time $t = 0$), and k is the decay coefficient (Rossman and Huber, 2016).

TABLE 16.2 SWMM Hydrology and Hydraulic Module Specifications (Rossman and Huber, 2016).

Parameter	Meaning	Value range	Value used
N-Imperv	Manning's roughness coefficient for impervious area	0.011–0.015	0.01
N-perv	Manning's roughness coefficient for pervious area	0.05–0.8	0.01
Destore-imperv	Depth of depression storage on impervious area	0–3	0.05
Destore-perv	Depth of depression storage on pervious area	3–10	0.05

16.5 RAINFALL ANALYSIS

The rainfall time-series data must be presented in shorter duration rainfall format in order to operate the model in SWMM. As the rainfall data are accessible in 15 min format, it was used in 15 min time steps. Thirty-three years of precipitation data were used for creating IDF, that is, 1980–2018.

16.5.1 DEVELOPMENT OF INTENSITY FREQUENCY CURVES

The intensity frequency (IDF) curves have been developed for the rainfall measured in small time steps assess the severity of the rainfall occurrence at different times at different periods. These curves were created for 2-, 5-, 10-,

20-, 25-, 50-, 100-year return periods (Sinha et al., 2021), for the simulation of pilot area 10-year and 50-year return period is used. The fit test for the available rainfall data was performed before the curve was prepared. This test was performed for Gumbel distribution. The precipitation estimates were calculated using the severe value distribution of "Gumbel's extreme value distribution" (Type-1). In order to verify the importance thresholds for the test, values obtained from the exercise test were then correlated with the Kolmogorov–Smirnov chart.

16.5.2 DEVELOPMENT SCENARIO

The aim is to estimate and compare the stormwater discharges at the outlet of the catchment with those created before urbanization. It will use the approach typically used in stormwater drainage manuals, which is to calculate the catchment's hydrological response to a series of storms linked to the various return periods. For 15-min events with return periods 10 and 50 years, the design storms will be used here. Scenario-based simulation of hydrological response in a catchment was used. The PCSWMM software platform was carried out with the EPA SWMM 5.1 engine for simulating predevelopment and postdevelopment scenarios.

There will be two models, one representing the catchment in its current undeveloped condition (May 20, 1965) predevelopment scenario and the other representing the catchment after it has been fully developed (May 20, 2017) postdevelopment scenario. Based on the current and future conditions of this, initial estimate of discharge at the outlet of the catchment or system runoff in PCSWMM.

- Predevelopment scenario: The PCSWMM model for the undeveloped site (May 20, 1965) predevelopment scenario is shown in Figure 16.4. It comprises an R1 rain gage supplying precipitation for three subcatchments whose runoff drains into the one outfall node.
- Postdevelopment scenario: Increasing the impermeable surface area and reducing overland flow length are the main factors influencing a catchment's hydrological response when it becomes urbanized. Reducing the infiltrative surface creates additional surface runoff, as well as increased and higher peak discharges. For better accuracy, a calibrated model is used for the postdevelopment scenario, as shown in Figure 16.4.

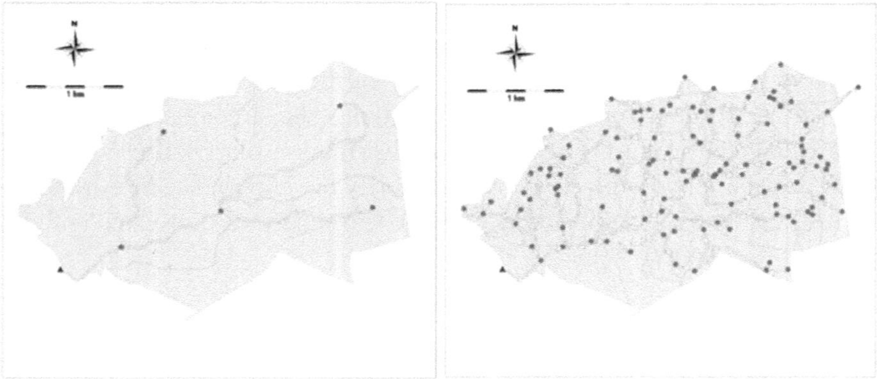

FIGURE 16.4 Conceptual model for (a) predevelopment (May 20, 1965)], (b) postdevelopment (May 20, 2017) scenarios.

Table 16.3 shows the number of entities used in two different scenarios, for predevelopment scenario on the basis of natural drainage and according to impervious and pervious area three subcatchment model is made and their parameter like, area width and slope is calculated in ArcGIS and conduits are natural open drainage, so its shape is considered as open rectangular conduits in both the scenario. One outlet is provided at the south-west part of the study area where Kharun river is located. All stormwater flows through number of natural drainage and collected at the outlet point.

TABLE 16.3 Model Assets Predevelopment and Postdevelopment.

SWMM object	Number of objects	
	Predevelopment	Postdevelopment
Subcatchments	3	51
Conduits	3	130
Junctions	5	132
Outlets	1	1

16.6 RESULT AND DISCUSSION

The study area was divided into 51 subcatchments designed along drainage line with sufficient slope at intermediate junctions and links (conduits). A combined 1D–2D model is developed and simulated for various return periods of rainfall for the study area. It was also visualized that moderately calculated surface runoff for two return periods, that is, 10- and 50-year return period of rainfall.

16.6.1 RESULT OF RAINFALL FREQUENCY ANALYSIS

Raipur rainfall information have been gathered and supplied by the Nagpur Regional Center Indian Meteorological Department. They used an automatic rain gage in the city of Raipur which continually records precipitation within 15 min. The information covers a 28-year term beginning between 1980 and 2014. The information is supplied in an excellent form. It was used to develop IDF curves and design storms for 2 years to 2 h. It was used to generate IDF curves. The input for the PCSWMM was used (Sinha et al., 2019a).

Table 16.4 shows the rainfall intensity of 10- and 50-year return period for 15, 30 min, 1, 2, 3, 6, 12, 24 h to estimate the flood prone area in the future. Figure 16.5 displays the idf curve used to calculate rainfall intensity to use for simulation and analysis of the model for different return period.

TABLE 16.4 Input IDF Data for Different Return Period.

	15 min	**30 min**	**1 h**	**2 h**	**3 h**	**6 h**	**12 h**	**24 h**
min>	**15**	**30**	**60**	**120**	**180**	**360**	**720**	**1440**
10-year	106.43	89.06	60.06	39.09	31.71	21.97	15.64	9.64
50-year	157.55	129.36	82.27	50.51	41.28	29.06	21.44	13.47

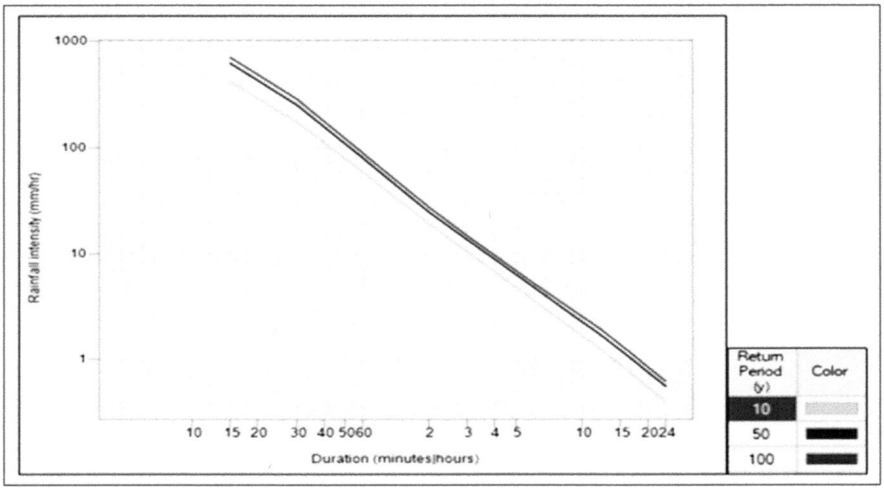

FIGURE 16.5 This figure shows IDF of different return period in PCSWMM.

16.6.2 RESULT SIMULATION OF STORMWATER MODEL

The simulation results quality is quite good, as shown in Table 16.5, continuity error of 10-year return period for pre- and post-development is −0.545 and −0.740, respectively, similarly for 50-year return period for pre- and post-development is −0.050 and −0.106, respectively. According to SWMM5.1 Manual Simulation is of poor quality when the continuity error is >10% (Rossman and Supply, 2006).

TABLE 16.5 Hydrological Behavior During Pre-development and Post-development Scenarios

	Predevelopment scenario		Postdevelopment scenario	
	10-year	**50-year**	**10-year**	**50 year**
Runoff quantity continuity	Depth (mm)			
Total precipitation	19.359	24.972	19.359	24.972
Infiltration	1.594	4.969	1.503	3.208
Surface runoff	6.774	17.248	13.588	20.535
Continuity error (%)	−0.545	−0.050	−0.740	−0.106
Maximum runoff (m^3/s):	2.056	2.649	38.66	49.73
Mean runoff (m^3/s):	1.222	0.7839	17.14	3.237
Duration of exceedances (h):	2.983	11.18	2.983	23.98
Duration of deficits (h):	0.0839	1	0	10.88

In Figure 16.6, the total rainfall (precipitation) over the subcatchment for pre- and post-development scenarios for 10-year return period is 19.359 mm. Simulation results display 34% of precipitation is converted into surface runoff for predevelopment scenario and for postdevelopment scenario 70.1% of precipitation is converted into surface runoff. Respectively for 50-year return period the total rainfall is 24.972 mm, simulation results show for pre- and post-development scenarios 71.1% and 82.2% of precipitation is converted into surface runoff. Analysis or precipitation turning into runoff shows the increment of surface runoff due to the effect of urbanization (Fig. 16.6).

The amplifying impact of urbanization on low-frequency floods is therefore evident, leading to an increase in the risks of floods in urban areas 1D–2D concept flow chart. The peak flow of after urbanization is also found to be higher than that before urbanization, which places new demands and challenges on urban infrastructure and water management. Figure 16.8 shows flood inundation map for 10- and 50-year return period, respectively. Critical subcatchments of the study area can be identified by the simulation of model by the estimation of maximum in particular subcatchment and by taking the 10- and 50-year inundation map. We can also find the flood prone area, the critical and flood-prone subcatchments of the study area that is mentioned in Table 16.6.

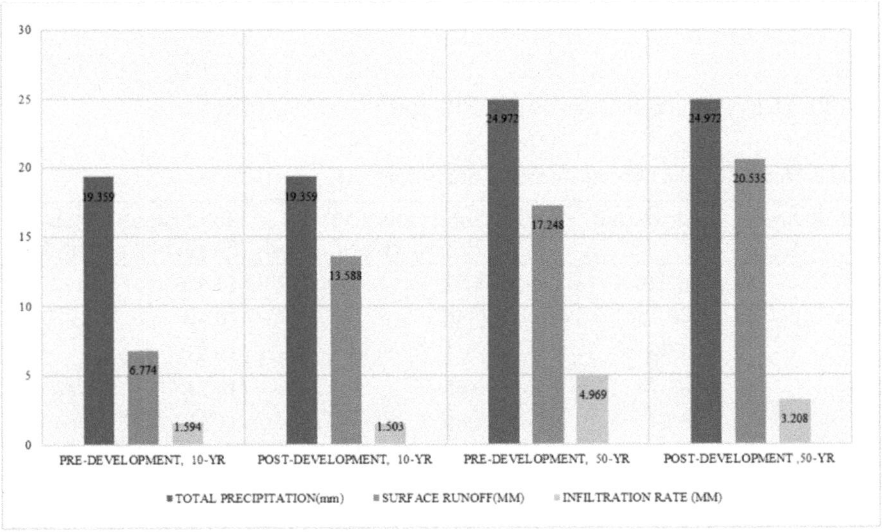

FIGURE 16.6 Comparison of surface runoff and infiltration rate of different scenario.

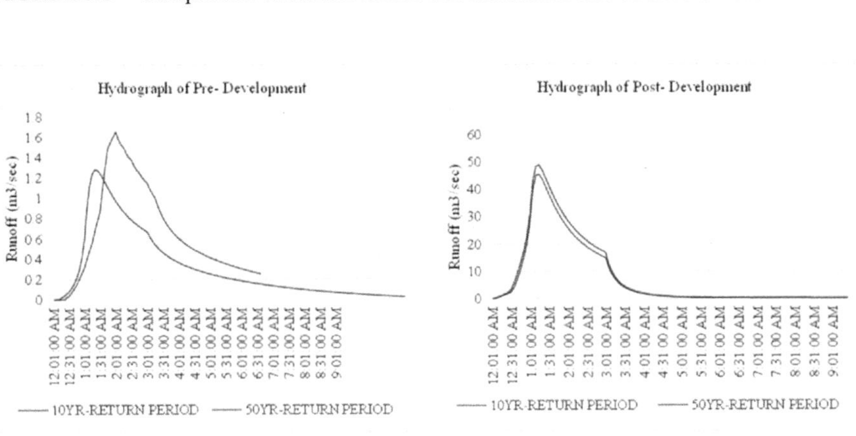

FIGURE 16.7 Changes to rainfall–runoff processes in various scenarios.

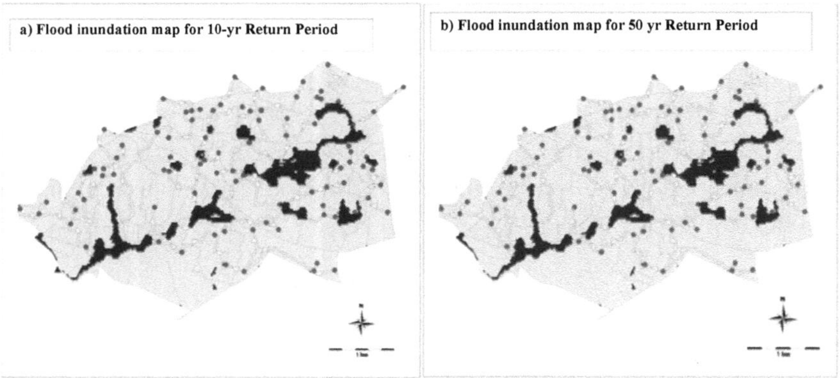

FIGURE 16.8 Flood inundation maps for 10- and 50-year design storm.

TABLE 16.6 Critical Subcatchments of Study Area.

Subcatchment	Area (ha)	Total runoff 10^6 (L) 10-year return period	Total runoff 10^6 (L) 50-year return period
B15	65.20	8.97	13.47
B2	45.15	7.46	10.49
B25	158.15	21.5	36.03
B30	43.78	6.95	10.21
B8	54.17	1.69	11.26
WL2	50.69	6.82	10.12

Study area consists B15, B2, B25, B30, B8, and WL2 are the critical and flood-prone subcatchment of area that is identified to reduce flooding in area by using some sustainable approach and supervision of drainage area. The status reports show the node that comes under the critical subcatchment are shown in Figure 16.8. Nodes j10644, j10473, j10474, and j10666 for the first profile and j7815, j7648, j7481, j7317, and j7158 for the second profile, j76856, j6697, j6696, j6995, j6854, j7012 for the third profile, and lastly fourth profile shows j14646, j14647, j14483, j14315, j14484, j14649 are critical flooded nodes (shown in Figure 16.9) in the study area. The conduits flow at various parts having maximum flow and its, velocity, duration, depths. It is used for conduit designing (Fig. 16.9).

16.7 CONCLUSION

Flood modeling plays a significant role and minimizes flood mitigation. Such equations estimate the magnitude of flood and severity of flooding.

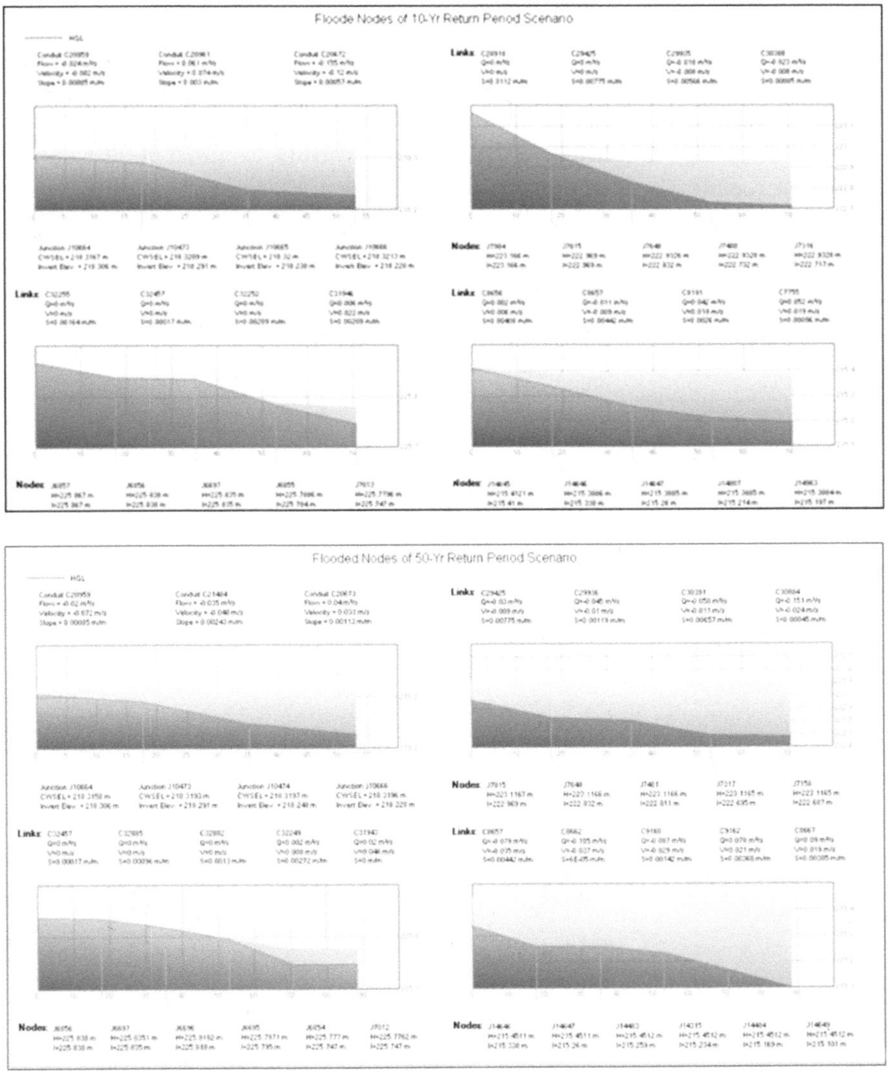

FIGURE 16.9 Flooded nodes of 10- and 50-year return period scenario.

It helps to define the geographic growth of the environment and the life and property damage may be evaluated. Such regions were associated mainly with urban colonies and were often inundated after heavy rainfall, because in places that are poor topographically, water continues to flow and accumulate. It describes a feasible application of EPA SWMM. The parameters encoded into the model were controlled by various methods and a few strategies were utilized for their assurance so as to survey their effect, moreover observed discharge and precipitation information, essential to adjust the model, were resolved either for old or current storm events. It gives a framework that comprehends the hydrological forms that occur in the urban domain and shows that a variety of indirect data can be useful for a watershed model application. It is concluded that there is need to provide a sustainable approach in the form of low impact development for stormwater management system and practices to reduce the flooding and to improve the water availability and the groundwater recharge in urban areas.

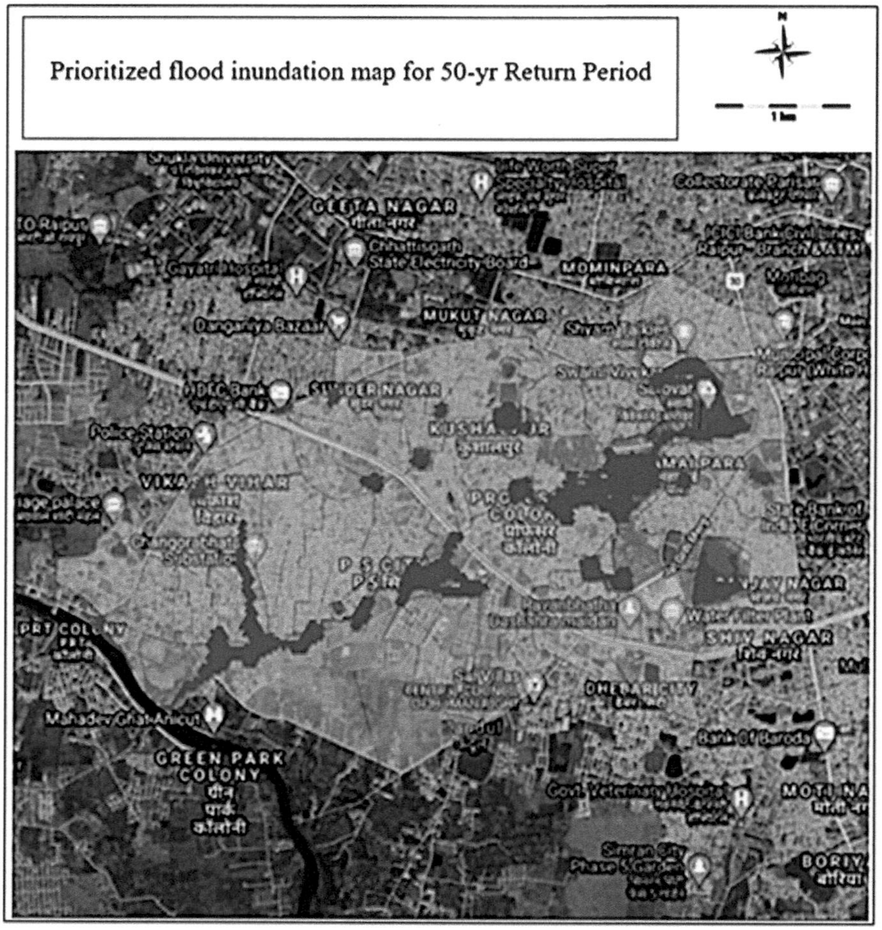

FIGURE 16.10 Prioritized flood inundation map of the study area.

16.7.1 SUSTAINABLE STORMWATER MANAGEMENT

This work can be extended to ensure the sustainable management of stormwater throughout the city. The implementation of a sustainability plan regional stormwater control management faces multiple problems, particularly in developing countries such as India, mainly due to the lack of proper results, unchecked urban growth, lack of information, legal status issues, and hydrological infiltration, runoff, evaporation, number of water sources, the consistency of the water, and eventually the discharge. The metropolitan region can be split into many tiny catchment basins on the basis of topographical details (contours). These drains may be drained into nearby ponds or low-lying areas transformed into wetlands, each pond flows to its next river or nallah, in order to create a storm drainage water drains, providing an appropriate stormwater management system for each catchment stormwater drains of a sufficient capacity can be established.

1. An adequate capability of a network of stormwater drains can ensure an efficient drainage and avoid inundation in the region.
2. The stormwater that created nuisance may be turned into a beneficial stream and solutions for water logging also can be interpreted.

ACKNOWLEDGMENT

The authors thank Deutscher Akademischer Austauschdienst- German Academic Exchange Service (DAAD), as the research material is based upon work within the program "A New Passage to India." One of the authors acknowledges the help in field data collection provided by LIH workgroup.

KEYWORDS

- **1D-2D PCSWMM model**
- **flood prone area**
- **flood inundation**
- **SWMM model**
- **urban flood**

REFERENCES

Gironás, J.; Roesner, L. A.; Rossman, L. A.; Davis, J. A New Applications Manual for the Storm Water Management Model (SWMM). *Environ. Model. Softw.* **2010,** *25*, 813–814.

Guan, M.; Sillanpää, N.; Koivusalo, H. Modelling and Assessment of Hydrological Changes in a Developing Urban Catchment. *Hydrol. Process.* **2015,** *29*, 2880–2894.

Kumari, S.; Jha, R.; Singh, V.; Baier, K.; Sinha, M. Groundwater Vulnerability Assessment Using SINTACS Model and GIS in Raipur and Naya Raipur, Chhattisgarh, India. *Indian J. Sci. Technol.* **2016,** *9*, 41.

Magesh, N.; Ch, N. A GIS Based Automated Extraction Tool for the Analysis of Basin Morphometry. *Bonfring Int. J. Indust. Eng. Manage. Sci.* **2012,** *2*, 32–35.

Miller, J. D.; Kim, H.; Kjeldsen, T. R.; Packman, J.; Grebby, S.; Dearden, R. Assessing the Impact of Urbanization on Storm Runoff in a Peri-Urban Catchment Using Historical Change in Impervious Cover. *J. Hydrol.* **2014,** *515*, 59–70.

Nie, L. Flooding Analysis of Urban Drainage Systems, 2004.

Rossman, L. A.; Huber, W. C. Storm Water Management Model Reference Manual Volume I—Hydrology (Revised). USEPA. National Risk Management Laboratory, 2016.

Rossman, L. A.; Huber, W. C. Storm Water Management Model Reference Manual Volume II—Hydraulics. US Environmental Protection Agency, Cincinnati. National Risk Management Laboratory, 2017.

Rossman, L. A.; Supply, W. Storm Water Management Model, Quality Assurance Report: Dynamic Wave Flow Routing. US Environmental Protection Agency, Office of Research and Development, 2006.

Sinha, M.; Baier, K.; Azzam, R.; Baghel, T.; Verma, M. Semi-distributed Modelling of Stormwater Drains Using Integrated Hydrodynamic EPA-SWM Model. In *Harmony Search and Nature Inspired Optimization Algorithms*; Springer, 2019a; pp 557–567.

Sinha, M. K.; Baghel, T.; Baier, K.; Verma, M. K.; Jha, R.; Azzam, R. *Impact of Urbanization on Surface Runoff Characteristics at Catchment Scale*; Springer Singapore: Singapore, 2019b; pp 31–42.

Sinha, M. K.; Baier, K.; Azzam, R.; Verma, M. K.; Jha, R. Analysis of Intensity–Duration–Frequency and Depth–Duration–Frequency Curve Projections Under Climate Variability. In *Climate Change Impacts on Water Resources: Hydraulics, Water Resources and Coastal Engineering*; Jha, R., Singh, V. P., Singh, V., Roy, L. B., Thendiyath, R., Eds.; Springer International Publishing: Cham, 2021; pp 407–421.

Sinha, M. K.; Rajput, P.; Verma, M. K.; Baier, K.; Jha, R.; Azzam, R. Response of Urban Lakes Due to Informal Urban Settlements: A Case Study in Raipur City. In *1st Asian conference on Water and Land Management for Food and Livelihood Security*; Pandey, V. K., Lakpade, R., Arora, S., Dave, A. K., Sinha, J., Eds.; IGKV Raipur and India Soil Conservation Society of India: New Delhi, India, 2017.

Vaze, J.; Teng, J.; Spencer, G. Impact of DEM Accuracy and Resolution on Topographic Indices. *Environ. Model. Softw.* **2010,** *25*, 1086–1098.

Zhou, Q. A Review of Sustainable Urban Drainage Systems Considering the Climate Change and Urbanization Impacts. *Water* **2014,** *6*, 976–992.

CHAPTER 17

Entropy-Based Measurement of Long-Term Precipitation Variability across India

PRABHASH KUMAR MISHRA[1*], HEMANT SINGH[1], SWAGATAM DAS[1], and SURENDRA KUMAR CHANDNIHA[2]

[1]*Water Resources Systems Division, National Institute of Hydrology, Roorkee, Uttarakhand, India*

[2]*Department of Soil and Water Engineering, BRSM CAET&RS, Mungeli, IGKV Raipur, India*

[*]*Corresponding author. E-mail: erprabhash@gmail.com*

ABSTRACT

Intergovernmental panel on climate change has confirmed the precipitation variability worldwide. Precipitation variability in terms of space and time is becoming one of the major concerns in India. Long-term trend analyses of climatic variables provide an estimate of increasing or decreasing trends without measuring the disorder associated with the variables. Entropy-based historical precipitation analysis measures the disorder associated with a trend apart from indicating the rising and receding trend. Accordingly, monthly precipitation data from 1871 to 2012 (142 years) from 30 subdivisions from six monsoon homogeneous regions have been investigated for variability based on entropy. Marginal entropy investigates the temporal precipitation variability in a time series. The decadal apportionment disorder index (DADI) was also analyzed to detect the decadal variability in the precipitation series. Marginal disorder

Advances in Hydrology and Climate Change: Historical Trends and New Approaches in Water Resources Management. Surendra Kumar Chandniha, Anil Kumar Lohani, Gopal Krishan, & Ajay Krishna Prabhakar (Eds.)

index (MDI) is highest in Northwest India indicating a higher variation in precipitation in the region. Rest of the regions the variability is comparatively less. The MDI values are also high during nonmonsoon months in comparison to the monsoon months.

17.1 INTRODUCTION

Climate is a complex system and is very difficult to quantify its' variables. Precipitation is an important parameter (variable) for climate change impact studies. An assessment of past precipitation events vis-à-vis its future scenarios is needed for water resources planning. Global Circulation Models are tools available to simulate the ongoing and future changes in climate at the global scale. Long-term past climatic data are investigated for hidden information and its associated variability through several techniques. Statistical approaches for measures for dispersion in a dataset indicate the data spread. Trends from long-term historical data indicate the increasing and/or decreasing nature of the data series. Mann–Kendal (MK) test and Sen's slope estimator are robust nonparametric approaches utilized to detect the trend and quantify the magnitude.

Many studies were conducted to assess the precipitation variability across India. But no particular increasing and decreasing trends were noticed in the average rainfall of whole India. There was considerable variation in seasonal rainfall in different geographical regions of India. Basistha et al. (2008, 2009) reported 1964 as the most probable year when a shift is detected in the annual and monsoon rainfall over Himalaya. Kumar and Jain (2010) analyzed rainfall trends for the Kashmir valley. The study indicated decreasing trends for the monsoon rainfall, whereas an increasing trend was found in annual, premonsoon, postmonsoon, and winter rainfall series. Kumar et al. (2010) carried out the regional trend analysis of long-term rainfall across India. No significant trends were found in the annual, seasonal, and monthly rainfall series in most of the months. Work on Indian Summer Monsoon was conducted by Duncan et al. (2012) using nonparametric MK test. The study found significant rising trends in the case of interannual variation in rainfall in major parts of the country. Patra et al. (2012) investigated rainfall trend over Orissa state for the period 1871–2006. The analysis revealed nonsignificant decreasing trend of annual and monsoon rainfall. An increasing trend was seen during postmonsoon season. The rainfall trends during the winter and summer seasons were increasing over the state. Jain and Kumar (2012) reviewed trends of rainfall, rainy days, and temperature at basin level for India. The study indicated decreasing annual rainfall trends in 16 basins. One basin showed a significant decreasing rainfall trend with 95% confidence level. The study also showed a rising trend in the mean maximum temperature series at a majority of the stations. Stations situated in the south, central, and western parts of India showed a rising trend, whereas stations located in the north and northeastern India showed a falling trend for the annual mean temperature data.

The concept of entropy dates back to 1948 when C.E. Shannon introduced in the field of communication (Shannon, 1948). Later the concept of entropy was extensively used in the field of hydrology and water resources. Singh (1997) carried out an in-depth review of entropy applications in the field of hydrology and water resources. Kawachi et al. (2001) conducted zonation of the water resources based on rainfall across Japan using the concept of entropy. Mishra et al. (2009) investigated entropy-based precipitation variability for Texas, USA. The results indicated a higher disorder in precipitation and number of rainy days which caused severe drought during the 1950s in entire Texas. Montesarchio and Napolitano (2010), using the entropy concept, disaggregated the daily rainfall into an hourly scale. Entropy-based approach to estimate natural groundwater recharge has gained momentum in recent times (Mondal and Singh, 2010; Ndatuwong and Yadav, 2012). Zhao et al. (2011) analyzed long-term precipitation variability based on entropy over Xinjiang, China. Bacanli (2012) evaluated drought in the Central Anatolian Region in Turkey utilizing entropy-based assessment of palmer drought severity index. Martino et al. (2013), using 92-year precipitation data from 400 rain gauge stations, investigated the change in precipitation in space and time using an entropy-based approach.

Geographically, India is a big country with varying climates. The climate varies from warm and humid in the south to moderate in the north. The precipitation in the region also varies spatially and temporally triggered by the monsoon system. The monsoon is dominant from June to September. The mean rainfall for the month of July during 1901–2003 was 286.5 mm, which is the highest. The month contributes about 24.2% of annual rainfall amounting to 1182.8 mm (NCC Research Report, IMD, 2006).

Thus, wide variability in the precipitation trend has triggered the growing concern of climate change and its future impact on the environment. The present study attempted to investigate the precipitation variability in the eight regional and 30 meteorological subdivisions spread over India for the period 1871–2006 using an entropy-based approach. The paper analyzed the spatio-temporal precipitation pattern with the following major objectives:

1. To investigate entropy-based precipitation variability considering both spatial and temporal scales.
2. To investigate the disorder associated with the long-term precipitation data.
3. Finally, to highlight the spatial and temporal precipitation pattern across India and its effect on the hydrological scenario.

17.2 STUDY AREA

The entropy-based precipitation variability has been analyzed across India. Indian Institute of Tropical Meteorology (IITM), Pune, divided the entire country into six homogeneous monsoon regions. These regions are classified based on the similarity in the monsoon pattern as (1) northwest, (2) central northeast, (3) northeast, (4) west-central, (5) peninsular, and (6) hilly regions. Further, these six regions are subcategorized into 36 subregions, as shown in Figure 17.1.

17.3 DATA USED

17.3.1 METEOROLOGICAL DATA

The study utilized 142 years of monthly precipitation data collected from IITM, Pune, for the periods 1871–2012 from eight regional stations and 30 subdivisions spread over India. The study did not consider the analysis of four meteorological subdivisions from the hilly regions. This is due to inadequate rain-gauge networks and their low areal representation. Two island subdivisions were also not included in the analysis. The geographical divisions of the regions and subregions are shown in Figure 17.1.

17.4 APPROACH AND METHODOLOGY

17.4.1 MEASURES OF VARIABILITY

Literally, variability refers to the quality of being subject to variation. In other words, it is the degree of being variable or changeable. Statistically, it is defined as the degree of extent a data point that differs from the mean value. Median and mean represent a distinct value that differs from the actual values. The extent to which these median and mean vary depends

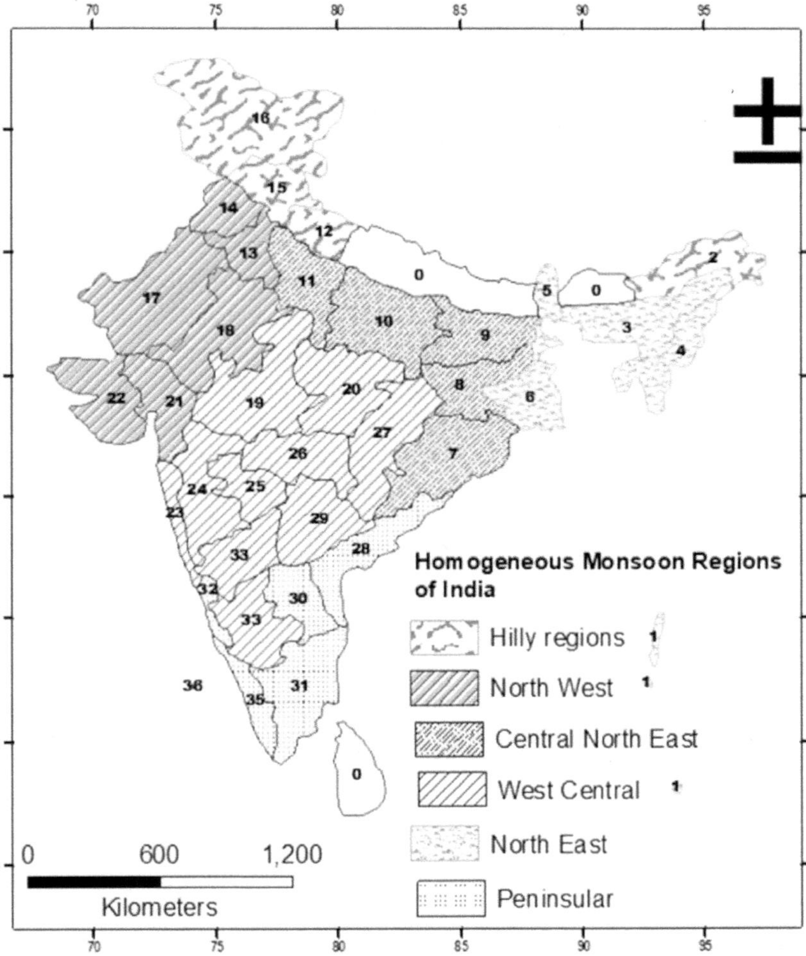

FIGURE 17.1 Homogeneous monsoon regions of India along with the subregions.

upon the variability or dispersion in the original data. Datasets become highly dispersed when they contain values considerably higher and lower than the median or the mean value. There are four commonly used measures of variability: range, mean, variance, and standard deviation. The commonly used statistical measure of variability is the variance, which measures the spread of the data points. Hence, variance is a method to capture the degree of spread of the dataset. Variability may also be measured spatially as well as temporally for a dataset. Spatial variability is represented by different values that are measured at different geographical locations. For example, spatial variability between the network of rain gauges can be investigated for the contribution of precipitation from individual station to the total amount of rainfall received in the region. Temporal variability measures randomness of a data series over different time scales.

17.4.2 ENTROPY-BASED MEASURE OF VARIABILITY

Entropy, unlike in the field of thermodynamics, measures the degree of uncertainty of a distribution. In a probability distribution function, uncertainty of the occurrence of the event is minimized by more observations. Therefore, reduction in uncertainty by increasing the information content of the variable, that is, by more observations is nothing but capturing the entropy of a series of data. Once the statistical distribution of a random variable is known, its entropy (degree of information) can be computed and expressed in specific units. The concept of entropy was introduced by C.E. Shannon in the field of communication (Shannon, 1948). Later, the concept of entropy was extensively used in a variety of areas including hydrology and water resources.

The entropy of a random variable is a measure of the information or disorder associated with it. Entropy measures the uncertainty in an outcome in a random process. Measures of information include marginal entropy (ME), joint entropy, conditional entropy, and transformation (Mishra et al., 2009).

A discrete form of entropy is given as per Shannon (1948):

$$H(X) = -\sum_{i=1}^{N} p(x_i)\log[p(x_i)] \tag{17.1}$$

where x is a discrete variable, i denotes a discrete data interval, and $p(x_i)$ is the probability of x_i. Entropy $H(X)$ is known as ME of the random variable X. Entropy is expressed in bits if the base of the logarithms is assumed to be equal to 2. H will be maximum when there is more evenness in the occurrence of the random variable, that is, the probability of chances of occurrence of one event is unbiased. H will become 0, if and only if the probability of occurrence of an event is certain, that is, we have maximum certainty of that event to occur, otherwise, we have maximum information about the event. Thus, the value of entropy varies from 0 (minimum) to $\log_2 N$ (maximum).

17.4.3 ENTROPY-BASED PRECIPITATION VARIABILITY

The precipitation variability can be measured by entropy of the data series both in terms of spatial (station-wise) and temporal variability (monthly, seasonal, and annual). Determining uncertainty or variability in a data series for a particular station or group of stations helps in decision-making of the water resources on its availability and distribution. In the present study, precipitation variability has been quantified using marginal disorder index (MDI), apportion disorder index (ADI), and decadal apportion disorder index (DADI) for the eight regional stations and 30 subregional stations spread over India. When the disorder index is calculated from the ME, it is known as MDI. Similarly, when the disorder index is calculated from apportionment entropy (AE), it is known as ADI. When the decadal analysis is carried out from decadal apportionment index, it is known as DADI. These disorder indices indicate the variability associated with a data series. The higher the disorder value (disorder index) is, the higher will be the uncertainty or variability. The ME, AE, and decadal apportionment entropy (DAE) are explained through the following equations:

$$\mathrm{ME}(X) = -\sum_{i=1}^{N} p(x_i)\log_2[p(x_i)] \tag{17.2}$$

where x is a discrete variable, i denotes a discrete data interval, and $p(x_i)$ is the probability of x_i.

$$\mathrm{AE}(X) = -\sum_{i=1}^{12} \frac{r_i}{R}\log_2[\frac{r_i}{R}] \tag{17.3}$$

where r_i is the cumulative precipitation in the month in a year and R is the annual cumulative precipitation.

$$\mathrm{DAE}(X) = -\sum_{i=1}^{12} \frac{R_i}{\mathrm{DR}}\log_2[\frac{R_i}{\mathrm{DR}}] \tag{17.4}$$

where R_i is the annual cumulative precipitation in the ith year and DR is the aggregate precipitation over 10 years.

17.5 RESULTS AND DISCUSSION

17.5.1 MARGINAL DISORDER INDEX FOR EIGHT REGIONAL STATIONS FOR THE MONTHLY SERIES (1871–2012)

MDI has been estimated monthly and seasonal for the six monsoon homogeneous region. The monthly MDI is presented in Table 17.1. It can be seen from the table that the MDI is highest in Northwest India indicating a higher variability of rainfall. Rest of the regions the variability is more or less similar

TABLE 17.1 MDI for Monthly Series (1871–2012).

Locations	Jan	Feb	Mar	Apr	May	Jun	Jul	Aug	Sep	Oct	Nov	Dec
All India	0.3502	0.3439	0.2349	0.0786	0.0657	0.0369	0.0150	0.0175	0.0340	0.0965	0.2366	0.3972
Homogeneous India	0.5466	0.5637	0.7067	0.3271	0.3474	0.0865	0.0390	0.0500	0.1034	0.3216	0.8325	1.0141
Core India	0.7248	0.7466	0.9062	0.4776	0.5370	0.0957	0.0425	0.0591	0.1312	0.4375	1.1016	1.2953
Northwest India	0.6206	0.7837	0.9461	1.0232	0.6468	0.2673	0.1004	0.1563	0.3755	1.1743	1.7950	1.1097
West-Central India	0.7686	0.6606	0.7942	0.3280	0.3893	0.0750	0.0319	0.0354	0.0769	0.2919	0.8194	1.2776
Central-East India	0.5250	0.5339	0.6523	0.3926	0.1930	0.1138	0.0276	0.0196	0.0479	0.3180	0.8051	1.0428
Northeast India	0.5357	0.3262	0.2331	0.0824	0.1404	0.0237	0.0196	0.0195	0.0298	0.1478	0.4946	0.9481
Peninsular India	0.9208	1.0125	0.6470	0.1888	0.1614	0.0290	0.0410	0.0540	0.0684	0.0796	0.2303	0.4073

in nature (Fig. 17.2). Further, the monsoon and nonmonsoon months behave differently in the rainfall variability. In general, the MDI values are high during nonmonsoon months in comparison to the monsoon months (Fig. 17.3).

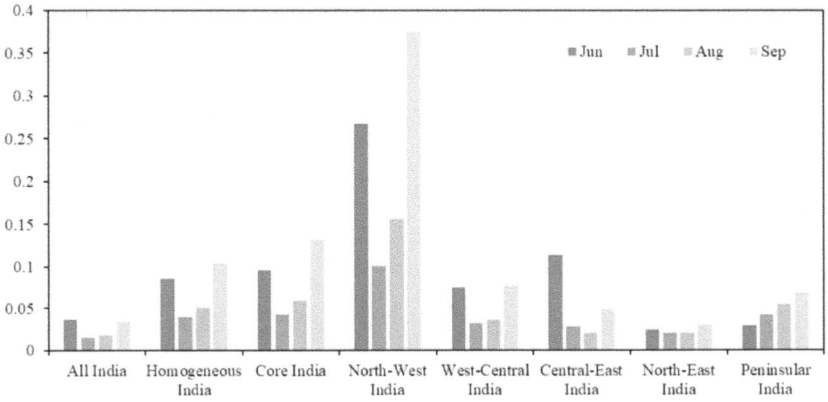

FIGURE 17.2 Marginal disorder index (MDI) during monsoon months.

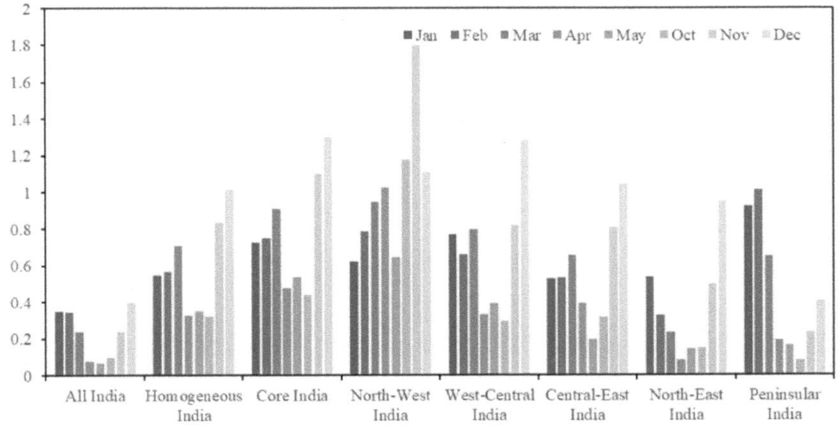

FIGURE 17.3 MDI during nonmonsoon months.

17.5.2 MARGINAL DISORDER INDEX FOR EIGHT REGIONAL STATIONS FOR THE SEASONAL SERIES (1871–2012)

The study also analyzed the MDI seasonally for the eight regional stations. The MDI values for the seasonal series are presented in Table 17.2. The seasonal analysis also confirmed that the disorder is higher in Northwest regions followed by Core India, central-east, west-central (Fig. 17.4).

TABLE 17.2 MDI for Seasonal Series (1871–2012).

Locations	Annual	Premonsoon	Monsoon	Postmonsoon	Winter
All India	0.0061	0.0313	0.0071	0.0678	0.1247
Homogeneous India	0.0160	0.1614	0.0174	0.2425	0.2390
Core India	0.0178	0.2566	0.0187	0.3288	0.3356
Northwest India	0.0461	0.3552	0.0526	0.8773	0.2722
West-central India	0.0126	0.1717	0.0130	0.2290	0.3320
Central-east India	0.0091	0.1243	0.0100	0.2578	0.2401
Northeast India	0.0071	0.0494	0.0064	0.1533	0.1885
Peninsular India	0.0096	0.0726	0.0160	0.0593	0.2229

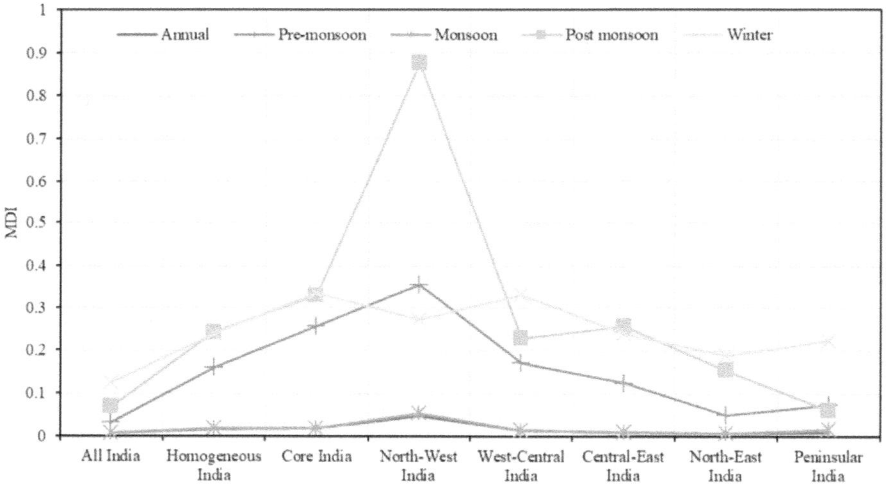

FIGURE 17.4 MDI of eight regional stations for seasonal series.

It can be seen in Figure 17.4 that the disorder for the postmonsoon period is highest. A higher disorder during postmonsoon is disadvantageous and damaging in terms of postharvest losses during Kharif harvesting. The variability is also comparatively more during the premonsoon and winter season as well. The disorder is least during annual and monsoon periods. A higher disorder in the seasonal rainfall variation has far-reaching implications in the country in the agriculture sector. The MDI is also estimated for the 30 subdivisions both for the monthly and seasonal series. Tables 17.3 and 17.4 provided the monthly and seasonal MDI values for the 30 subdivisions. It can be seen from Figure 17.5 that the MDV is less for the northeastern regions indicating a low disorder. This indicates that the rainfall received in the regions is more or less constant every year. However, in the western part of the country comprising of Gujarat, Saurashtra, and the Konkan region, the variability is more with higher MDI values.

17.5.3 MARGINAL DISORDER INDEX FOR 30 SUBREGIONAL STATIONS FOR THE MONTHLY SERIES (1871–2012)

TABLE 17.3 MDI for Seasonal Series (1871–2012).

Sl No.	Meteorological subdivisions	Jan	Feb	Mar	Apr	May	Jun	Jul	Aug	Sep	Oct	Nov	Dec
1	Assam and Meghalaya	0.4641	0.3129	0.2246	0.0824	0.0593	0.0394	0.0406	0.0386	0.0578	0.1705	0.4638	0.7889
2	Naga., Mani., Mizo. and Tripura	0.9822	0.5476	0.3905	0.1656	0.1012	0.0472	0.0410	0.0398	0.0554	0.1834	0.6143	1.5224
3	Sub-Him. W. Bengal	1.0333	0.6268	0.5181	0.2414	0.0753	0.0565	0.0479	0.0774	0.0866	0.3803	1.1527	1.7276
4	Gangetic W. Bengal	1.0567	0.7356	0.7263	0.3627	0.1793	0.1026	0.0520	0.0515	0.0885	0.2896	1.1939	2.0872
5	Odisha	1.1688	0.8179	0.7456	0.3935	0.2763	0.1246	0.0519	0.0500	0.0669	0.2734	0.7548	1.6789
6	Jharkhand	0.8778	0.7109	0.8120	0.5029	0.2668	0.1536	0.0543	0.0476	0.0787	0.4245	1.1570	1.7390
7	Bihar	0.8551	0.7765	0.8594	0.6534	0.2640	0.1563	0.0693	0.0635	0.1184	0.5897	1.8435	2.0049
8	East Uttar Pradesh	0.6754	0.7932	1.0659	1.0106	0.4942	0.2510	0.0898	0.0650	0.1508	0.9349	2.0957	1.4307
9	West Uttar Pradesh	0.5900	0.8067	0.9222	1.0770	0.5344	0.3451	0.0915	0.1096	0.2777	1.3883	1.5710	1.2277
10	Haryana, Chandigarh and Delhi	0.6632	0.7742	0.7738	1.1678	0.6346	0.3269	0.1653	0.2189	0.5309	1.5522	2.0194	1.1842
11	Punjab	0.5824	0.6105	0.6542	0.9232	0.6992	0.4636	0.1409	0.1712	0.6191	2.0194	1.7124	1.0232
12	West Rajasthan	1.0302	1.2518	1.4418	1.4284	0.7786	0.4678	0.2245	0.3283	0.6663	1.6463	2.2431	1.7059
13	East Rajasthan	0.9399	1.1332	1.4195	1.3951	0.6733	0.3143	0.1068	0.1603	0.3832	1.4184	1.9455	1.4767
14	West Madhya Pradesh	0.9251	1.1482	1.3327	0.9509	0.8448	0.2022	0.0709	0.0854	0.2574	0.9332	1.6178	1.5832
15	East Madhya Pradesh	0.8365	0.7319	0.9540	0.9390	0.7570	0.2160	0.0587	0.0624	0.1740	0.7121	1.6229	1.6093
16	Gujarat	1.9224	2.3298	3.0735	2.5406	1.9121	0.3517	0.1539	0.2158	0.4716	1.4110	1.9887	2.0493
17	Saurastra, Kutch, and Diu	1.5102	2.1099	3.2473	2.7139	2.3754	0.5012	0.3061	0.4560	0.6121	1.7275	2.3394	2.3072
18	Konkan and Goa	3.2434	2.5772	2.8668	1.9021	1.5337	0.0636	0.0687	0.1341	0.1863	0.5845	1.1435	2.8829

TABLE 17.3 *(Continued)*

Sl No.	Meteorological subdivisions	Jan	Feb	Mar	Apr	May	Jun	Jul	Aug	Sep	Oct	Nov	Dec
19	Madhya Maharashtra	2.1050	1.6769	1.2557	0.4669	0.5339	0.1282	0.0867	0.1213	0.1536	0.3588	0.8181	1.7634
20	Marathwada	1.7644	1.7209	1.5190	0.8180	0.9411	0.1377	0.1358	0.2135	0.1870	0.5617	1.0970	1.9548
21	Vidarbha	1.3393	1.0327	1.2022	0.7927	0.8961	0.1301	0.0635	0.0928	0.1782	0.6450	1.3180	1.9455
22	Chhatisgarh	1.1218	0.7848	0.9614	0.7475	0.5831	0.1757	0.0463	0.0518	0.1154	0.4760	1.2434	1.7815
23	Coastal Andhra Pradesh	1.4247	1.5657	1.8786	0.8779	0.6106	0.1529	0.0885	0.0908	0.1082	0.1960	0.3633	0.9953
24	Telangana	1.6426	1.4624	1.3012	0.4420	0.5183	0.1077	0.0936	0.1068	0.1424	0.3676	0.8474	1.7831
25	Rayalaseema	1.8524	1.9145	1.6052	0.6163	0.3468	0.2030	0.1930	0.2530	0.1537	0.2228	0.3864	0.8974
26	Tamil Nadu and Pondicherry	1.0503	1.3128	0.8413	0.3517	0.1840	0.1379	0.1710	0.1332	0.1027	0.1055	0.2503	0.4162
27	Coastal Karnataka	2.7759	3.2013	2.9655	1.0685	0.7456	0.0435	0.0609	0.1040	0.2262	0.2435	0.4717	1.7888
28	North Karnataka	2.0414	1.5890	1.3155	0.2600	0.2870	0.0931	0.0903	0.1247	0.1340	0.2408	0.6615	1.7885
29	South Karnataka	1.9173	1.6186	0.9716	0.2215	0.1146	0.0829	0.0896	0.1062	0.1396	0.1647	0.3635	1.0302
30	Kerala	1.2179	0.8630	0.4721	0.1492	0.2682	0.0585	0.0760	0.1132	0.2050	0.1010	0.2794	0.6256

17.5.4 MARGINAL DISORDER INDEX FOR 30 SUBREGIONAL STATIONS FOR THE MONTHLY SERIES (1871–2012)

TABLE 17.4 MDI for Monthly Series (1871–2012).

Sl No.	Meteorological subdivisions	Annual	Premonsoon	Monsoon	Postmonsoon	Winter
1	Assam and Meghalaya	0.0096	0.0315	0.0129	0.1455	0.1694
2	Naga., Mani., Mizo. and Tripura	0.0116	0.0532	0.0129	0.1428	0.3354
3	Sub-Him. W. Bengal	0.0126	0.0525	0.0167	0.3434	0.3907
4	Gangetic W. Bengal	0.0168	0.1260	0.0175	0.2548	0.5035
5	Odisha	0.0127	0.1619	0.0148	0.2351	0.4400
6	Jharkhand	0.0158	0.1729	0.0177	0.3630	0.4061
7	Bihar	0.0232	0.1817	0.0275	0.5101	0.4208
8	East Uttar Pradesh	0.0306	0.3300	0.0357	0.8619	0.3436
9	West Uttar Pradesh	0.0318	0.3319	0.0416	1.1439	0.3048
10	Haryana, Chandigarh, and Delhi	0.0495	0.3381	0.0679	1.1903	0.2931
11	Punjab	0.0556	0.3016	0.0832	1.4866	0.2253
12	West Rajasthan	0.0989	0.4987	0.1137	1.2754	0.5506
13	East Rajasthan	0.0456	0.4366	0.0523	1.0448	0.4677
14	West Madhya Pradesh	0.0257	0.4584	0.0282	0.6902	0.5153
15	East Madhya Pradesh	0.0200	0.4018	0.0208	0.5914	0.3922
16	Gujarat	0.0687	1.4066	0.0741	1.0421	1.0626
17	Saurastra, Kutch. and Diu	0.1252	1.6315	0.1401	1.2798	0.9907
18	Konkan and Goa	0.0264	1.2592	0.0269	0.4735	1.9733
19	Madhya Maharashtra	0.0291	0.2557	0.0334	0.2679	0.9508

TABLE 17.4 *(Continued)*

Sl No.	Meteorological subdivisions	Annual	Premonsoon	Monsoon	Postmonsoon	Winter
20	Marathwada	0.0515	0.4713	0.0554	0.4004	0.9168
21	Vidarbha	0.0259	0.4095	0.0273	0.4643	0.6093
22	Chhatisgarh	0.0210	0.3294	0.0217	0.3865	0.4774
23	Coastal Andhra Pradesh	0.0301	0.3969	0.0352	0.1417	0.5237
24	Telangana	0.0324	0.2342	0.0387	0.2781	0.7602
25	Rayalaseema	0.0375	0.1982	0.0592	0.1552	0.6022
26	Tamil Nadu and Pondicherry	0.0175	0.0982	0.0382	0.0723	0.2518
27	Coastal Karnataka	0.0192	0.5255	0.0207	0.1894	1.3712
28	North Karnataka	0.0236	0.1263	0.0300	0.1871	0.8787
29	South Karnataka	0.0198	0.0613	0.0307	0.1142	0.5627
30	Kerala	0.0151	0.1107	0.0258	0.0773	0.3043

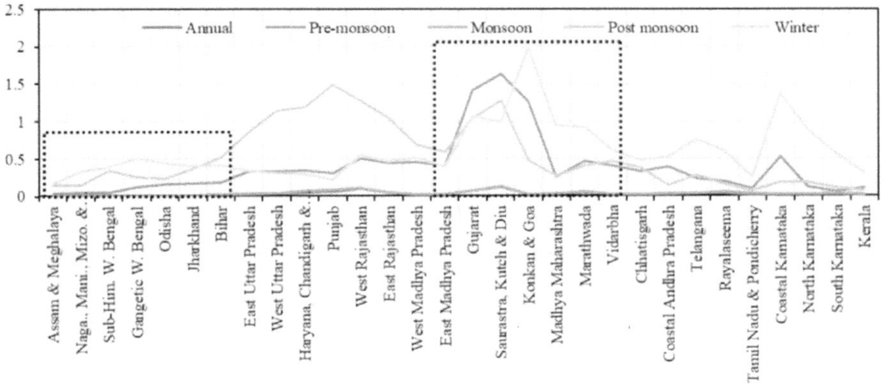

FIGURE 17.5 MDI of 30 subregional stations.

17.5.5 MEAN APPORTION DISORDER INDEX AND MEAN DECADAL APPORTION DISORDER INDEX FOR EIGHT REGIONAL STATIONS

Along with MDI, the study also analyzed the mean apportion disorder index (MADI) and mean decadal apportion disorder index (MDADI) for the eight regional centers, as given in Table 17.5. MADI is also maximum for the Northwest region followed by Core India, Homogeneous India, and central-east India (Fig. 17.6).

TABLE 17.5 MADI and MDADI for the Eight Regional Stations.

Locations	MADI	MDADI
All India	0.7606	0.0057
Homogeneous India	1.1013	0.6429
Core India	1.2068	0.1943
Northwest India	1.3570	1.1773
West-Central India	1.0722	0.0407
Central-East India	0.9785	−0.1763
Northeast India	0.6415	−1.1958
Peninsular India	0.5580	−0.0633

17.6 CONCLUSIONS

Entropy-based analysis of long-term precipitation provides an opportunity to gauge the disorder associated with a trend apart from indicating the rising and receding trend. Accordingly, monthly precipitation data from 1871 to 2012 (142 years) from 30 subdivisions were investigated for spatial and temporal variability based on entropy. ME was used to investigate the variability associated with monthly, seasonal, and annual series. Decadal apportionment disorder index was also assessed to detect the decadal variability in the precipitation series. It was seen that the MDI was highest in Northwest India indicating a higher variability of rainfall in the region. Rest of the regions the variability was more or less similar in nature. The MDI values were also high during nonmonsoon months in comparison to the monsoon months.

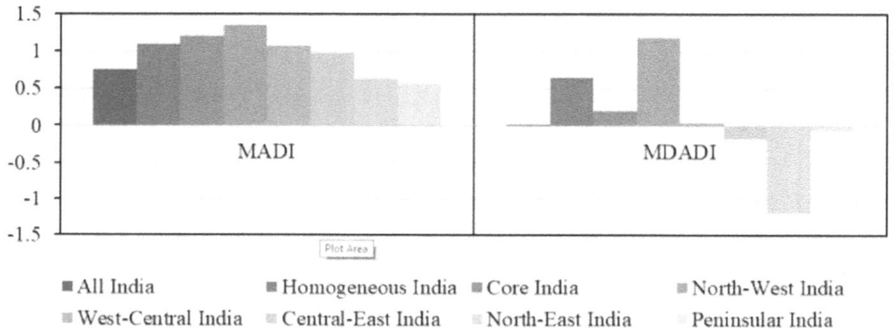

FIGURE 17.6 MADI and MDADI of eight regional stations.

KEYWORDS

- **precipitation variability**
- **entropy**
- **marginal entropy**
- **disorder index**
- **ADI**
- **DADI**

REFERENCES

Bacanli, U. G. Entropy Based Assessment and Palmer Drought Severity Index of Drought Analysis. *Sci. Res. Essays* **2012,** *7* (44), 3823–3833.

Basistha A.; Aryan, D. S.; Goel, N. K. Spatial Pattern of Trends in Indian Sub-Divisional Rainfall. *Int. J. Climatol.* **2008**. DOI: 10.1002/joc.1706

Basistha, A.; Aryan, D. S.; Goel, N. K. Analysis of Historical Changes in Rainfall in the Indian Himalayas. *Int. J. Climatol.* **2009,** *29*, 555–572.

Duncan, J. M. A.; Dash, J.; Atkinson, P. M. Analysing Temporal Trends in the Indian Summer Monsoon and Its variability at a Fine Spatial Resolution. *Clim. Change* **2012**. DOI: 10.1007/s10584-012-0537-y

Jain, S. K.; Kumar, V. Trend Analysis of Rainfall and Temperature Data for India. *Curr. Sci.* **2012,** *102* (1), 37–49.

Kawachi, T.; Maruyama, T.; Singh, V. P. Rainfall Entropy for Delineation of Water Resources Zones in Japan. *J. Hydrol.* **2001,** *246*, 36–44.

Kumar, V.; Jain, S. K. Trends in Seasonal and Annual Rainfall and Rainy Days in Kashmir Valley in the Last Century. *Quatern. Int.* **2010,** *212*, 64–69.

Kumar, V.; Jain, S. K.; Singh, Y. Analysis of Long-Term Rainfall Trends in India. *Hydrol. Sci. J.* **2010,** *55* (4), 484–496.

Martino, G. D.; Fontana, N.; Marini, G.; Singh, V. P. Variability and Trend in Seasonal Precipitation in the Continental United States. *J. Hydrol. Eng.* **2013,** *18*, 630–640.

Mishra, A. K.; Ozger, M.; Singh, V. P. An Entropy-Based Investigation Into the Variability of Precipitation. *J. Hydrol.* **2009,** *370*, 139–154.

Mondal, N. C.; Singh, V. P. Entropy-Based Approach for Estimation of Natural Recharge in Kodaganar River Basin, Tamil Nadu, India. *Curr. Sci.* **2010,** *99* (11), 1560–1569.

Montesarchio, V.; Napolitano, F. A Single-Site Rainfall Disaggregation Model Based on Entropy. In *Int. workshop Adv. in Stat. Hydrology*; Taormina, Italy, May 23–25, 2010.

Ndatuwong, L. G.; Yadav, G. S. Quantitative Estimation of Natural Recharge Due to Monsoon Rainfall Using the Principle of Information Theory in the Area of Ghorawal Block of Sonebhadra District, U.P., India. *Int. J. Environ. Sci.* **2012,** *3* (3), 976–985.

Patra, J. P.; Mishra, A.; Singh, R.; Raghuwanshi, N. S. Detecting Rainfall Trends in Twentieth Century (1871–2006) over Orissa State, India. *Clim. Changes* **2012,** *111*, 801–817.

Shannon, C. E. A Mathematical Theory of Communication. *Bell Syst. Tech. J.* **1948,** *27*, 379–423, 623–656.

Singh, V. P. The Use of Entropy in Hydrology and Water Resources. *Hydrol. Processes* **1997,** *11*, 587–626.

Zhao, C.; Ding, D.; Ye, B.; Yao, S.; Zhao, Q.; Wang, Z.; Wang, Y. An Analysis of Long-Term Precipitation Variability Based on Entropy over Xinjiang, Northwestern China. *Hydrol. Earth Syst. Sci. Discuss.* **2011,** *8*, 2975–2999.

CHAPTER 18

Evaluating NOAA and PRISM Precipitation Data in Streamflow Generation Using HAWQS Model

VIVEK VERMA[1*], MANISH KUMAR SINHA[2], and TRIAMBAK BAGHEL[3]

[1]*Water & Environmental Engineering, Florida International University, Miami, USA*

[2]*Engineering Geology and Hydrogeology, RWTH Aachen University, Germany*

[3]*Water Engineering and Management, Asian Institute of Technology, Bangkok, Thailand*

[*]*Corresponding author. E-mail: vivekverma10@gmail.com*

ABSTRACT

Hydrologic and water quality system (HAWQS) is a rainfall-runoff modeling system that utilizes the Soil and Water Assessment Tool (SWAT) model as it is core to generate flows. The HAWQS model is setup using the web interface. The models are calibrated using split-sample calibration/validation procedure. Monthly streamflow is generated for the river Penobscot at Eddington, Maine, USA, using two different weather datasets, namely, National Oceanic and Atmospheric Administration (NOAA) and Parameter-elevation Regressions on Independent Slopes model (PRISM). All the produced stream flows were compared with gaged flows using statistical parameters, such as Nash–Sutcliffe efficiency, Percentage bias, ratio of the root mean square error to the standard deviation of measured data, and coefficient of determination (R^2) to obtain most accurate flow in this region.

Advances in Hydrology and Climate Change: Historical Trends and New Approaches in Water Resources Management. Surendra Kumar Chandniha, Anil Kumar Lohani, Gopal Krishan, & Ajay Krishna Prabhakar (Eds.)

The monthly flows generated from the PRISM dataset come out to be more accurate when compared with the NOAA flows for this study area.

18.1 INTRODUCTION

The hydrologic and water quality system (HAWQS) is a web-based rainfall-runoff modeling system that produces a runoff in response to a rainfall event (Verma, 2017). At its core, it utilizes the widely accepted, public domain model Soil and Water Assessment Tool (SWAT) as its modeling engine (Arnold et al., 2012). SWAT model is a continuous and physically based model which includes processes such as movement of water, sediments, nutrient cycle, crop growth, etc. (Abraham et al., 2007). The original purpose of the SWAT model is to predict numerous land management practices on a watershed scale. Management practices include chemical yields, water, agriculture, sediments, etc., which varies over time (Arnold et al., 2013).

The model requires specific information about a watershed which includes vegetation, weather, land management practices, topography, and soil properties (Yen et al., 2016). It is computationally efficient model which employs various mathematical techniques and thus simulations of watersheds can be achieved within short interval of time (Verma, 2021). It is also designed for more advanced and specialized processes such as bacteria transport, modeling of plant growth, on-site wastewater systems, and water quality (Fant et al., 2017).

SWAT divides a watershed into various sub-basins or subwatersheds based on certain properties such as land use, soil type, and management practices. The simulation is then run for each subbasins or subwatersheds. Input information from the user is then categorized into the following five broad categories, namely, hydrologic response units (HRUs), main channel or reach, ponds/wetlands/reservoirs, climatic data, and outlet point of the subbasin.

The simulation process is divided into two major types—land phase and routing phase. Water balance equation is the soul of processes going inside the watershed. Land phase consists of water, pesticides, insecticides, sediments, nutrient cycle from the hydrologic cycle. Once land phase is over, routing phase take control to distribute abovementioned variables through defined channels in the watershed to the outlets (Neitsch et al., 2005).

Precipitation can be in the form of either rain or snow. Snow contributes in the form of rain when it gets melt. Precipitation is a part of the hydrologic cycle which falls in the entire watershed taken into consideration.

Some portion of the precipitation infiltrates inside the ground surface, some evaporates from the atmosphere, some are taken by plants for photosynthesis. Some portion of the water percolates deep inside the surface and gets collected in aquifer. Aquifer is also a storage system but inside the ground. It is of two types, namely, shallow and deep aquifer. Beside the ground surface, water starts flowing, also known as lateral flows, and sometimes meets in the streams, ponds, etc. along with pesticides transported contributing toward streamflow (Neitsch et al., 2002). When there is enough water, top layer, that is, soil surface gets saturated and thereafter no water percolates downward. At this instant, the water starts flowing in the horizontal direction which results in surface runoff. Due to this runoff, some water is collected in the ponds, reservoirs, wetland, streams, and other storage systems. The collected water can be utilized for numerous purposes such as for irrigation, industrial purposes, daily activities, cooking, and so on. The results from rainfall-runoff model can be used to perform inundation analysis using other tools and software (Kang et al., 2019).

There are losses in this entire process, for instance, evaporation loss from the soils and water bodies such as lake, streams, reservoirs, and other storage units. Evapotranspiration losses from the plants. Transmission losses due to infiltration inside the soil surface, near the bed, and in the floodplain region. After accounting the abovementioned entire process, water collected is routed to next reach or reservoir and to next until it gets off from outlet.

SWAT model employs perennial and annual plant growth model to simulate land covers. Perennial plants have their roots entire year whereas annual plants grow from their planting date until it gets harvested. This model helps to understand and quantify transpiration, nutrients removal, and biomass production. The Modified Universal Soil Loss Equation (MUSLE) is used to estimate sediment yield and soil erosion. The basic principle employed by the MUSLE method to simulate sediment yield and soil erosion is with the help of total amount of runoff (Zhang et al., 2009). Runoff volume and peak runoff rate is determined from the hydrologic cycle within subbasins or subwatershed which is then employed to calculate runoff erosive energy variable. SWAT employs various cycles such as nitrogen cycle, phosphorous cycle to track movement, and circulation of various kinds of gases within a watershed (Kemanian et al., 2011). For instance, around agricultural area, nutrients may introduce in the upstream and can be simulated to downstream and subsurface flow through surface runoff. Similarly, pesticides are simulated in the SWAT model in a stream network using surface runoff. The vertical movement of the pesticides is also estimated with the help of the soil layers and this process is known as pesticide leaching. SWAT uses flexible, user-defined management practices at every HRU level. The user can enter starting and ending period of the plant season, amount of fertilizers, pesticides, insecticides, and other chemical employed for the agriculture (Arnold et al., 2013). After sowing period, during harvest season, tillage operations can also be defined. All these inputs are mapped with the surface runoff which estimates amount of chemicals for individual HRUs.

Once the land phase is being input into the SWAT model, routing phase proceed to finish the simulation process. SWAT employs mass-balance equation to route the components through stream network in a watershed. The routing can be classified in four components, namely, water, chemicals, sediment, and nutrients (Neitsch et al., 2011). Water component deals with the flood routing. The water is routed from upstream to downstream using the Muskingum routing method or variable storage coefficient method. Losses are included in the estimation of outflow in the downstream which includes transmission loss, evaporation and transpiration loss, loss due to human consumption, agricultural use, or industrial use. Streamflow in the channel can be supplemented by direct rainfall and/or addition of water from different streams or reaches joining the channel.

Deposition and degradation are the two processes responsible for the sediment transport from one point to another and it is a function of the peak channel velocity (Bonumá et al., 2014). Sediment also results from the bed erosion of the channels. Water quality component of the SWAT model is responsible for nutrient routing. It has two components, nutrients dissolved in the streamflow and nutrients adsorbed to sediments. Thus, nutrients routing is done with streamflow and its accumulation is estimated with sediments (Douglas-Mankin et al., 2010).

All kinds of pesticides can be simulated in the SWAT model through HRUs. However, only one pesticide can be routed downstream because of the complicated simulation processes. Similar to the nutrient routing, pesticides have two components, one will route with the streamflow and other will adsorbed to sediments. Pesticide routing in the model can simulate numerous processes such as settling, transformation, resuspension, diffusion, burial, and volatilization (Neitsch et al., 2011). Routing in the reservoirs employs mass balance equation which includes rainfall, evaporation losses, outflow, seepage loss, and inflows to the reservoir. As described earlier, same mechanism is being employed in reservoir outflow, sediment routing, reservoir nutrients, and reservoir pesticides.

HAWQS provides interactive web-based maps with its own interface with preloaded input dataset as discussed above. It provides output in various forms and formats such as graphs, tables, and charts. All the execution, development, and storage of modeling projects is stored in the HAWQS database. User can retrieve, make changes, and rerun the simulation at any time (Yen et al., 2016). It employs a wide variety of management practices for the hydrology and water quality aspects. For instance, hydrology includes soils, land uses, crops, climate change scenarios, and water quality aspects include the following parameters such as pesticides, pathogens, biological oxygen demand, water temperature, sediments, and dissolved oxygen and nutrients (Hanson et al., 2017). As the software is very handy and provides estimates in relatively short interval of time, it is been employed by various federal agencies, contractors, state and local government bodies, private players, and academicians.

18.2 STUDY AREA AND DATA USED

The river Penobscot, as shown in Figure 18.1 is situated at Eddington city in the Maine state of the United States of America. It is located at latitude 44°49′36″, and longitude 68°41′48″ according to North American Datum (NAD 83). The contributing drainage area of the river Penobscot is 7764 mi^2. On an average, Penobscot receives 43 inches of rainfall every year. To run simulations, it employs three watershed scales, also known as hydrologic unit codes (HUCs), namely, large (HUC 8), medium (HUC 10), and small (HUC 12). HUC 8 covers the largest scale of 700 mi^2, then HUC 10 which covers 227 mi^2, and lastly, HUC 12 covers 40 mi^2. The temporal scale includes daily, monthly, and annual estimates of water quantity as well as water quality across entire United States as per its current database which might expand in the near future. One of the main advantages of using HAWQS is that it requires minimal computing power as it process simulations using servers located at Texas A&M University, College Station, TX, USA. It is supported by almost all the latest versions of web browsers such as Internet Explorer, Chrome, Safari, Microsoft Edge, and Firefox. Additionally, it can be accessed via mobile browsers.

18.3 MATERIAL AND METHODS

The operational framework of HAWQS can be categorized into four main categories of initialization, customization, output management, and output demonstration as illustrated in Figure 18.2. The first step is the initialization in which the study area can be selected either through latitude or longitude,

HUC ID, or with stat or end HUC ID. HUC can be further classified into large (HUC 8), medium (HUC 10), and small (HUC 12). HUC 8 covers the largest scale of 700 mi^2, then HUC 10 which covers 227 mi^2, and lastly, HUC 12 covers 40 mi^2 (Fant et al., 2017). The delineation of the watershed is done on the basis of HUC selected. Thereafter, project name is entered followed by total simulation period for which model has to run. Warm-up period can also be entered in the model which runs the simulation without being collecting any data. This is usually done to get into conditions in which the actual model has to run. Next, intervals of the simulation result are then entered which can be daily, monthly, or yearly depending upon the requirement of the user.

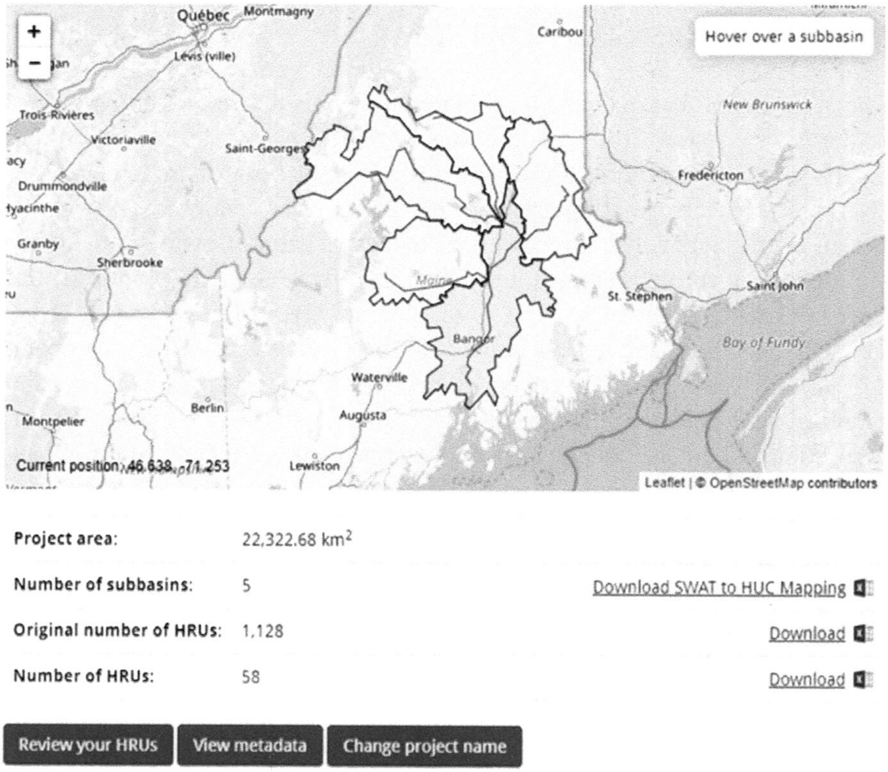

FIGURE 18.1 Location of the study area, River Penobscot (Eastern Coast USA).

After the study area is delineated the user can select threshold, in percentage, HRUs to set land use, soil, and slope class to eliminate minor land, soil, and slope in subbasins, and to reflect actual prevailing condition of the subbasins. Afterward, basin level settings such as fertilizer input data, nutrient efficiency, basin input data, and urban input data are entered in the model. Subsequently, subbasin level settings are entered which includes curve number, sediment algorithm, point source data, pothole parameters, climate inputs, and weather generator.

The third step is the output management in which the HAWQS simulation model runs based on the reach, subbasins, and HRU parameters entered by the user. The simultaneous reach output variables, subbasin output variables, and HRU output variables can be visualized. The project execution is then performed using servers located at Texas A&M University, College Station, TX, USA, and hence minimum computational power is required. The project execution includes writing of the SWAT input files or tables, running of the SWAT executable file, process of the SWAT output files, and finally creating project zip files when the simulation is over. The calibration of the model can then be performed to obtain better results by adjusting the existing model to reflect the actual condition as closely as possible.

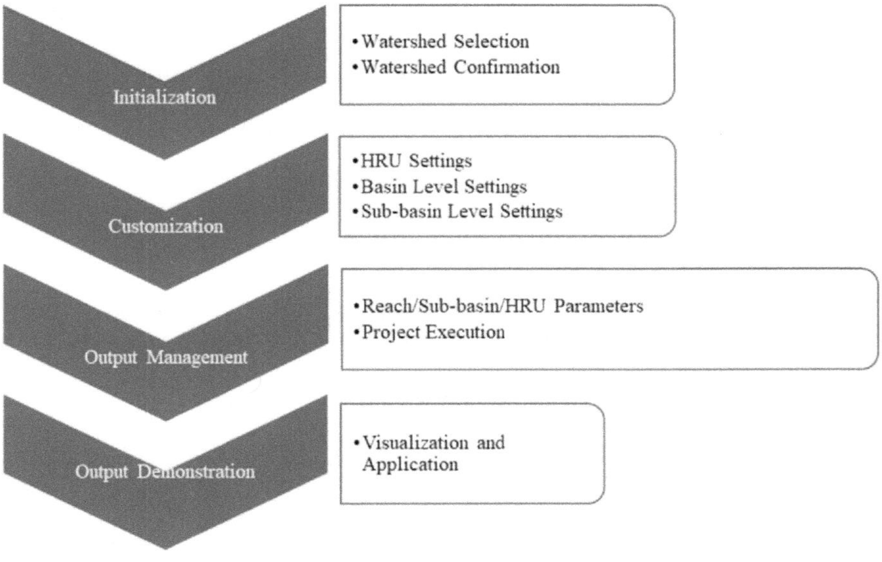

FIGURE 18.2 Operational framework of HAWQS.

The final step deals with output demonstration in the form of graphical view and applications. The output demonstration includes output reach data, output summary charts, and simulation output download for the simulated results. Also, there are certain postprocessing programs, such as SWAT check and water quality index calculation that are linked to HAWQS to evaluate simulation outputs (Yen et al., 2016). The execution of this entire framework takes relatively short interval of time as it runs on the servers with extremely high computational power located at Texas A&M University, College Station, TX, USA. The only requirement, the user must have the latest versions of web browsers, such as Internet Explorer, Chrome, Safari, Microsoft Edge, and Firefox. Moreover, it can also be performed with mobile browsers.

The model has been set up using HAWQS a web interface. The river Penobscot at Eddington, Maine is selected using HUC 8-01020005. The project area for this study is 22,322.68 km^2, with total of five subbasins and 1128 HRUs as illustrated in Figure 18.1. Thereafter, thresholds values are selected to eliminate minor slopes, soils, and land uses in each subbasin and to replicate the actual condition as closely as possible. This selection will reduce total number of HRUs for each subbasin; thus expediting the simulation process. The total HRUs reduced from 1128 to 58.

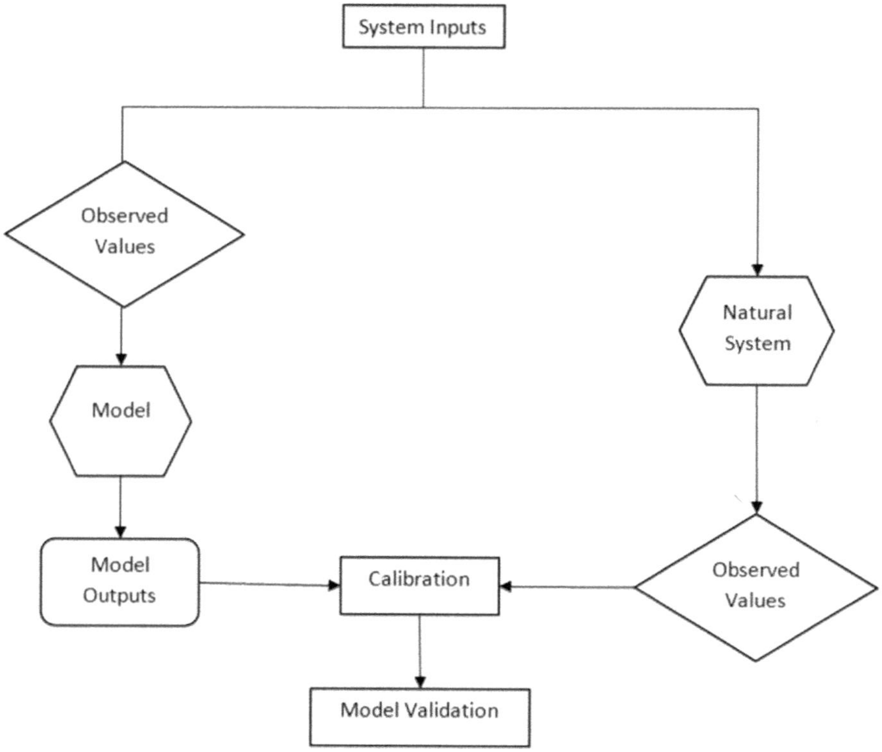

FIGURE 18.3 Calibration–validation methodology.

The flows/discharge are generated on the basis of monthly time steps using the two distinct weather sources, namely, National Oceanic and Atmospheric Administration (NOAA) and Parameter-elevation Regressions on Independent Slopes Model (PRISM). The detail about the two weather sources can be found at https://www.ncdc.noaa.gov/cdo-web/datasets for NOAA, and https://www.wcc.nrcs.usda.gov/climate/prism.html for the PRISM. The available NOAA weather dataset range from 1961 to 2018 and the PRISM dataset from 1981 to 2018. The flows generated from the NOAA and the PRISM weather datasets are called as NOAA flows and PRISM flows, respectively, in this chapter. The models are calibrated using split-sample calibration and validation procedure as depicted in Figure 18.3 (Duda et al., 2012).

The results are then compared with the available gage flows with the help of the following model evaluation guidelines—Nash–Sutcliffe efficiency (NSE), Percentage bias (PBIAS), ratio of the root mean square error to the standard deviation of measured data (RSR), and the coefficient of determination (R^2) (Moriasi et al., 2007).

18.3.1 NASH–SUTCLIFFE EFFICIENCY

In automatic model calibration, the NSE is used to quantify the quality of agreement between hydrographs (Nash and Sutcliffe, 1970). The sum or adding over the time t to 1. NSE tool the model calibration which is described as

$$NSE = 1 - \frac{\sum_{t-1}^{T}(A_m^t - A_o^t)^2}{\sum_{t=1}^{T}(A_o^t - A_o^-)^2}$$

However, there are two discharges one is A_o is observed discharges and A_m is modeled discharge. A_o^t is observed discharge at time t. An efficiency of 1 (NSE = 1) compare to modeled discharge to the observed data (Ritter and Muñoz-Carpena, 2013). A_o efficiency (NSE = 0) shows the accuracy of model predictions as middle of the observed data. However, Observed Mean is good anticipator as compare to model, when efficiency occurs lower than zero (NSE < 0).

18.3.2 PERCENTAGE BIAS

The PBIAS is a statistical tool which calculate the over prediction (PBIAS N 0) (Wagena et al., 2018) or under prediction (PBIAS b 0) of any model that is described as

$$\text{PBIAS} = 100 \times \frac{\sum_{i=1}^{n}(C_i - D_i)}{\sum_{i=1}^{n} D}$$

Meanwhile C_i is the predicted and D_i is the observed value at time *i*.

18.3.3 RATIO OF SQUARE ROOT

RSR make standard for RMSE to analyze standard deviation (Legates and McCabe Jr, 1999). RSR is measured as the ratio of the RMSE and standard deviation of calculated data

$$\text{RSR} = \frac{\text{RMSE}}{\text{STDEV}_{\text{obs}}} = \frac{[\sqrt{\sum_{i=1}^{n}(E_i^{\text{obs}} - E_i^{\text{sim}})^2}]}{[\sqrt{\sum_{i=1}^{n}(E_i^{\text{obs}} - E^{\text{mean}})^2}]}$$

RSR differ from the range of 0 optimal value to a higher positive value, For the better modeling simulation, where it reflects 0 RMSE. If there is reduction in RSR, the reduction in RMSE, meanwhile better simulation model performance.

18.3.4 RMSE

The RMSE shows good combination among observed and predicted values; however, inclining RMSE values showing a rising poor match, while it equals 0. For the better model prediction, RMSE values must be lower than half the standard deviation of the observed (measured) data which is considered low (Singh et al., 2005).

$$\text{RMSE} = \sqrt{\frac{\sum_{i=1}^{n}(C_i - D_i)^2}{n}}$$

18.3.5 COEFFICIENT OF DETERMINATION R^2

The square of the Pearson's product-moment correlation coefficient is the coefficient of determination (i.e., $R^2 = r^2$) and elaborate by the total variance portion in the observed data model shown by Legates and McCabe Jr (1999). It limits 0.0–1.0, with rising values showing good placement by the equation

$$R^2 = \{\frac{\sum_{i=1}^{N}(D_t - D)(C_t - C)}{[\sum_{i=1}^{N}(D_t - D)^2]^{0.5}[\sum_{i=1}^{N}(C_t - C)^2]^{0.5}}\}^2$$

The performance rating of the statistical parameters used for the monthly time step is shown in Table 18.1 (Moriasi et al., 2007).

TABLE 18.1 Performance Ratings of Certain Statistical Parameters for the Monthly Time Step.

Performance rating	RSR	NSE	PBIAS (%)
Very good	0.00 < RSR < 0.50	0.75 < NSE < 1.00	PBIAS < ±10
Good	0.50 < RSR < 0.60	0.65 < NSE < 0.75	±10 < PBIAS < ±15
Satisfactory	0.60 < RSR < 0.70	0.50 < NSE < 0.65	±15 < PBIAS < ±25
Unsatisfactory	RSR > 0.70	NSE < 0.50	PBIAS > ±25

18.4 RESULT AND DISCUSSION

The monthly flows from the NOAA and PRISM weather dataset are compared with the observed flows and are shown in Figures 18.4 and 18.6, respectively. The following statistical parameters, NSE, PBIAS, RSR, and R^2 have been used to evaluate the model efficiency.

The results of the statistical parameters for the NOAA model in the calibration phase come out to be 0.86 for NSE, 10.12 for PBIAS, 0.44 for RSR, and 0.85 for R^2 and for the validation, it is 0.70 for NSE, 11.35 for PBIAS, 0.58 for RSR, and 0.76 for R^2. The R^2 for the calibration and validation period of the NOAA model is shown in Figures 18.5 and 18.7.

Similarly, for the PRISM model during calibration, NSE is 0.84, PBIAS is 12.9, RSR is 0.41, and R^2 as 0.84 and for the validation, NSE is 0.67, PBIAS is 8.27, RSR is 0.60, and R^2 is 0.73. The R^2 for the calibration and validation period of the PRISM model is shown in Figures 18.6 and 18.7.

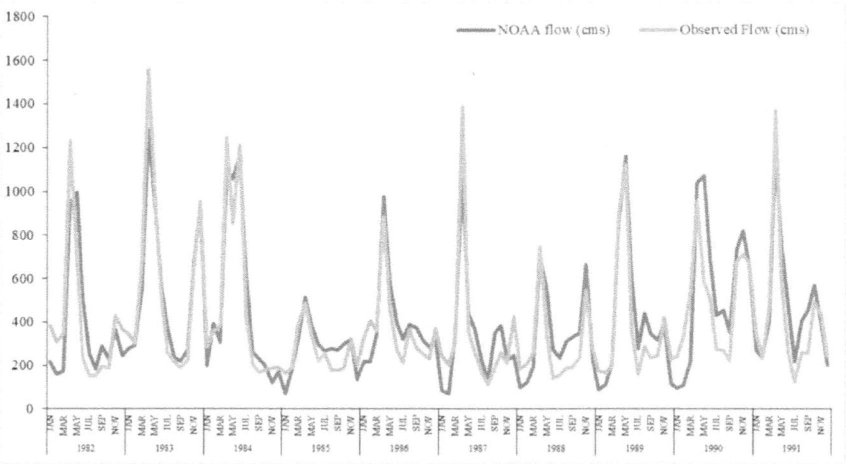

FIGURE 18.4 Observed and NOAA flow from 1982 to 1991: Calibration Period.

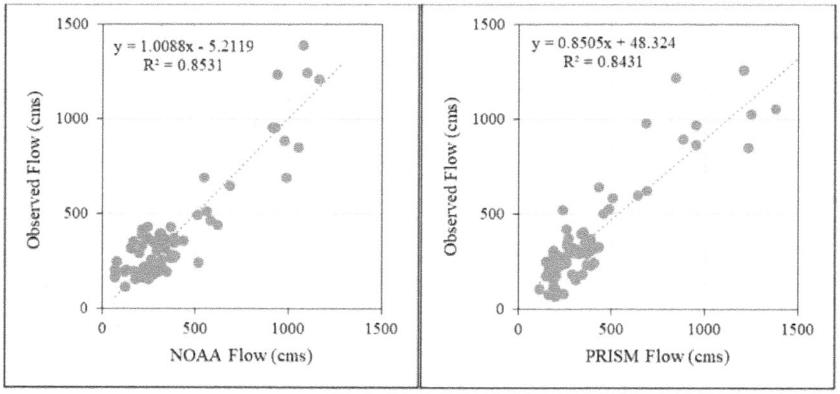

FIGURE 18.5 Linear correlation between stream flows for calibration period 1982–1987.

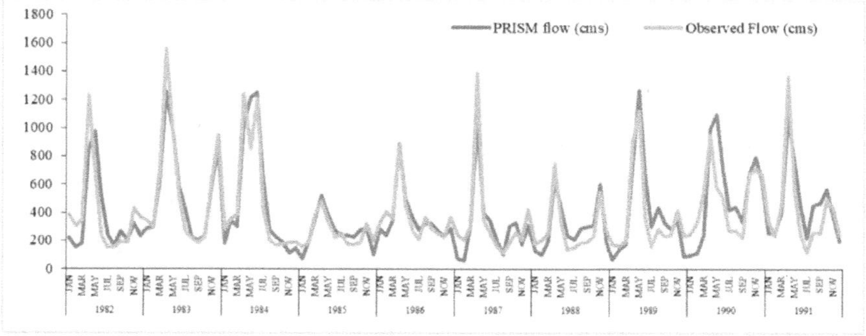

FIGURE 18.6 Observed and PRISM flow from 1982 to 1991: Validation Period.

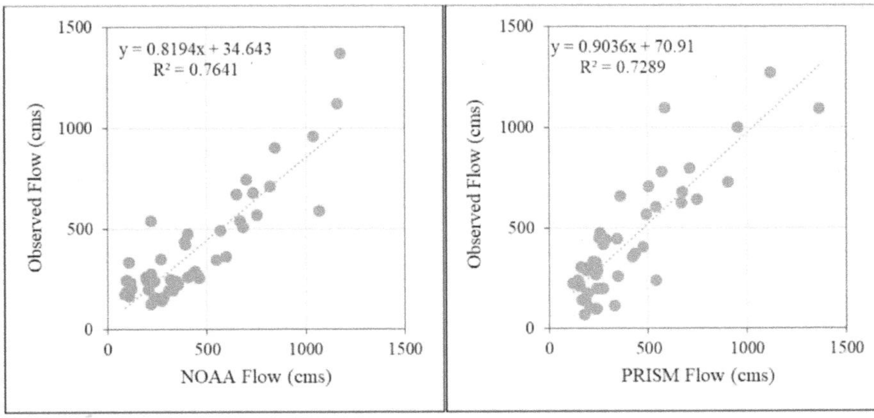

FIGURE 18.7 Linear correlation between stream flows for validation period 1988–1991.

TABLE 18.2 Statistical Parameters for the Monthly Streamflow Rate for NOAA and PRISM.

Index	**Calibration**			**Validation**		
	Observed	**Simulated**		**Observed**	**Simulated**	
		NOAA	**PRISM**		**NOAA**	**PRISM**
Count	72	72	72	48	48	48
Mean	754	722	742	735	724	736
Maximum	1554	1434	1515	1520	1540	1555
SD	5.42	5.30	5.24	5.11	3.4	3.7
r^2		0.85	0.76		0.73	0.84
E_{NS}		0.86	0.70		0.67	0.84
RSR		0.44	0.58		0.60	0.41
PBIAS		10.12	11.35		8.27	12.9

The NSE results for the calibration of NOAA and PRISM model are very good and for the validation they are good. Similarly, for the PBIAS the results are very good for the calibration of NOAA model and during calibration and validation of the PRISM model. It is good for the validation of the NOAA model. The RSR results are very good for the calibration of NOAA and PRISM model, and good for the validation of NOAA and PRISM model. Overall, the model results show acceptable results which signifies that the calibration technique performed well for the study area.

18.5 CONCLUSION

The rainfall-runoff model, HAWQS, has been employed to generate monthly simulated flows for the years 1982–1991. The HAWQS model utilizes SWAT model algorithm to perform simulation. The model has been calibrated using split-sample calibration/validation procedure. The monthly flows were generated for two different weather datasets—NOAA and PRISM. The simulated results were calibrated and validated and compared with the observed flows using NSE, PBIAS, and RSR statistical parameters; the performance rating reflects very good and good results. Based on the results, the monthly flows generated from the PRISM dataset come out to be more accurate when compared with the NOAA flow for this study area.

KEYWORDS

- **stream flows**
- **HAWQS**
- **SWAT**
- **NOAA precipitation**
- **PRISM precipitation**

REFERENCES

Abraham, L. Z.; Roehrig, J.; Chekol, D. A. Calibration and Validation of SWAT Hydrologic Model for Meki Watershed, Ethiopia. In *Conference on International Agricultural Research for Development*; University of Kassel-Witzenhausen and University of Gottingen, 2007.

Arnold, J.; Kiniry, J.; Srinivasan, R.; Williams, J.; Haney, E.; Neitsch, S. *SWAT 2012 Input/Output Documentation*; Texas Water Resources Institute, 2013.

Arnold, J. G.; Moriasi, D. N.; Gassman, P. W.; Abbaspour, K. C.; White, M. J.; Srinivasan, R.; Santhi, C.; Harmel, R.; Van Griensven, A.; Van Liew, M. W. SWAT: Model Use, Calibration, and Validation. *Trans. ASABE* **2012,** *55* (4), 1491–1508.

Bonumá, N. B.; Rossi, C. G.; Arnold, J. G.; Reichert, J. M.; Minella, J. P.; Allen, P. M.; Volk, M. Simulating Landscape Sediment Transport Capacity by Using a Modified SWAT Model. *J. Environ. Qual.* **2014,** *43* (1), 55–66.

Douglas-Mankin, K.; Srinivasan, R.; Arnold, J. Soil and Water Assessment Tool (SWAT) Model: Current Developments and Applications. *Trans. ASABE* **2010,** *53* (5), 1423–1431.

Fant, C.; Srinivasan, R.; Boehlert, B.; Rennels, L.; Chapra, S. C.; Strzepek, K. M.; Corona, J.; Allen, A.; Martinich, J. Climate Change Impacts on US Water Quality Using Two Models: HAWQS and US Basins. *Water* **2017,** *9* (2), 118.

Hanson, L.; Habicht, S.; Daggupati, P.; Srinivasan, R.; Faeth, P. Modeling Changes to Streamflow, Sediment, and Nutrient Loading from Land Use Changes Due to Potential Natural Gas Development. *JAWRA J. Am. Water Resour. Assoc.* **2017,** *53* (6), 1293–1312.

Kang, D.-H. D. K.; Kim, J.-G. J. G.; Verma, V.; Leon, A. S.; Kang, B. Evaluation of Urban Inundation under Changing Landcover— Application of EPA SWMM-LID to Andong City in South Korea. In *World Environmental and Water Resources Congress*, 2019; pp 83–88.

Kemanian, A. R.; Julich, S.; Manoranjan, V. S.; Arnold, J. R. Integrating Soil Carbon Cycling with That of Nitrogen and Phosphorus in the Watershed Model SWAT: Theory and Model Testing. *Ecol. Model.* **2011,** *222* (12), 1913–1921.

Legates, D. R.; McCabe Jr, G. J. Evaluating the Use of "goodness-of-fit" Measures in Hydrologic and Hydroclimatic Model Validation. *Water Resour. Res.* **1999,** *35* (1), 233–241.

Moriasi, D. N.; Arnold, J. G.; Van Liew, M. W.; Bingner, R. L.; Harmel, R. D.; Veith, T. L. Model Evaluation Guidelines for Systematic Quantification of Accuracy in Watershed Simulations. *Trans. ASABE* **2007,** *50* (3), 885–900.

Nash, J. E.; Sutcliffe, J. V. River Flow Forecasting through Conceptual Models Part I—A Discussion of Principles. *J. Hydrol.* **1970,** *10* (3), 282–290.

Neitsch, S. L.; Arnold, J. G.; Kiniry, J. R.; Williams, J. R. *Soil and Water Assessment Tool Theoretical Documentation Version 2009*; Texas Water Resources Institute, 2011.

Neitsch, S.; Arnold, J.; Kiniry, J.; Williams, J.; King, K. *SWAT Theoretical Documentation Version 2005*; Soil and Water Research Laboratory, ARS: Temple Texas, USA, 2005.

Neitsch, S.; Arnold, J.; Srinivasan, R. Pesticides Fate and Transport Predicted by the Soil and Water Assessment Tool (SWAT). Atrazine, Metolachlor and Trifluralin in the Sugar Creek Watershed: BRC Report 3, 2002.

Ritter, A.; Muñoz-Carpena, R. Performance Evaluation of Hydrological Models: Statistical Significance for Reducing Subjectivity in Goodness-of-fit Assessments. *J. Hydrol.* **2013,** *480*, 33–45.

Singh, J.; Knapp, H. V.; Arnold, J.; Demissie, M. Hydrological Modeling of the Iroquois River Watershed Using HSPF and SWAT 1. *JAWRA J. Am. Water Resour. Assoc.* **2005,** *41* (2), 343–360.

Verma, V. Comparative Analysis of Daily Flow Pattern Hydrographs Used to Disaggregate Monthly Naturalized Flows to Daily, 2017.

Verma, V. *Multisite Monthly to Daily Naturalised Streamflow Disaggregation Using Daily Flow Pattern Hydrograph. Climate Change Impacts on Water Resources: Hydraulics, Water Resources and Coastal Engineering*; Springer International Publishing: Cham, 2021; pp 519–527.

Wagena, M. B.; Collick, A. S.; Ross, A. C.; Najjar, R. G.; Rau, B.; Sommerlot, A. R.; Fuka, D. R.; Kleinman, P. J.; Easton, Z. M. Impact of Climate Change and Climate Anomalies on Hydrologic and Biogeochemical Processes in an Agricultural Catchment of the Chesapeake Bay Watershed, USA. *Sci. Total Environ.* **2018,** *637*, 1443–1454.

Zhang, Y.; Degroote, J.; Wolter, C.; Sugumaran, R. Integration of Modified Universal Soil Loss Equation (MUSLE) Into a GIS Framework to Assess Soil Erosion Risk. *Land Degrad. Develop.* **2009,** *20* (1), 84–91.

CHAPTER 19

Hydrogeology, Groundwater Regimes, and Quality Issues in Punjab

GOPAL KRISHAN[1*], ANOOP NAGAR[1], M. L. ANGURALA[2], RAKESH RANA[2], and N. SUDARSAN[1]

[1]*National Institute of Hydrology, Roorkee, Uttarakhand, India*

[2]*Central Ground Water Board, North Western Region, Chandigarh, India*

[*]*Corresponding author. E-mail: drgopal.krishan@gmail.com*

ABSTRACT

Punjab is an agrarian state, which relies heavily on groundwater resources for irrigation in spite of a well-organized canal irrigation system distributing water from three perennial rivers namely Sutlej, Beas, and Ravi rivers. High dependence on groundwater for meeting out the ever increasing demands of water putting pressure on these resources and resulting in its over-exploitation as the number of tube wells increased from 0.5 to 14.76 lakhs in about last six decades (1960–2017). In the last three to four decades (since 1980s), the number of overexploited blocks increased from 48% to 93%. Overdevelopment of groundwater and anthropogenic activities have also brought in various groundwater issues of quantity and quality due to decline in groundwater levels, increased salinity, increasing incidences of fluoride occurrences, and presence of arsenic, nitrate, sulfate, and iron contamination in groundwater. However, there are issues of rise in water levels and waterlogging in areas of southwestern Punjab. Keeping this in view, there is an urgent need to carry out a detailed study to assess the

Advances in Hydrology and Climate Change: Historical Trends and New Approaches in Water Resources Management. Surendra Kumar Chandniha, Anil Kumar Lohani, Gopal Krishan, & Ajay Krishna Prabhakar (Eds.)

groundwater conditions of Punjab. In this chapter, an in-depth analysis is carried out to assess effect of groundwater development on groundwater regime and groundwater resources of Punjab, which will be helpful in the management of this precious resource in the state.

19.1 INTRODUCTION

Punjab state with geographical area of 50,362 km^2 is predominantly an agrarian state contributing around two/third of the food grains procured annually in the country and is devoid of any other mineral or natural resource except water. Agriculture in the state is highly intensive in terms of agricultural inputs, water, and energy with cropping intensity of 186% and having cultivable area 83.4% of total geographical area of Punjab (http://farmech.dac.gov.in). The hard work of the farmers has proudly earned the state with the name of "Food Basket of Country" and "Granary of India" contributing 40% of rice, 50–70% of wheat, and around 60% of food grains to the central pool (http://farmech.dac.gov.in).

In Punjab state, for irrigating such a large area there is a heavy requirement of water, which sustains well-being of the farmers and economy of the state. Although a well-organized canal irrigation system spanning over 14,500 km of length has been developed distributing water from Satluj, Beas, and Ravi rivers, the available surface water resources of the state are not being able to fulfill the irrigation demands thereby putting pressure on groundwater resources and resulting in its overexploitation. This is evident from the statistics that the number of tube wells increased from 0.5 in 1960s to 0.7 lakhs in 1980s, to about 10.70 in year 2001, 11.80 lakhs in the year 2005–2006 to 14.06 in the year 2014–2015, same has increased to 14.76 Lakhs (2017–2018) as per statistical abstract of Punjab (2018).

Overdevelopment of groundwater has also brought in various groundwater issues of quantity and quality as decline in groundwater levels (Krishan et al., 2014a), increased salinity (Krishan et al., 2013, 2014b), increasing incidences of Fluoride occurrences (Krishan et al., 2017), nitrate and iron (Krishan et al., 2015; Lapworth et al., 2015), and arsenic contamination in groundwater (Lapworth et al., 2015, 2017). Nevertheless, there are waterlogged areas in southwestern Punjab (Chopra and Krishan, 2014). An analysis is carried out to assess effect of groundwater development on the groundwater regime and groundwater resources of Punjab.

19.2 SALIENT FEATURES

Punjab state lies between latitude 2930′ N to 32°32 N and longitude 73°55′ E to 76°50′E shares an international boundary with Pakistan on west and national boundary with states Jammu & Kashmir, Himachal Pradesh, Haryana, and Rajasthan on north-east, south-east, south, and west, respectively. Administratively, state has been divided into 22 districts and 147 blocks.

The Punjab state has major part flat alluvial plains in Indo-Gangetic basin (MacDonald et al., 2016; Bonsor et al., 2017) mountainous in northeastern border rising to a height from 100 to 400 m mid Miocene to Pleistocene (Wadia, 1979) and stable sand dunes in the southwestern parts. The slope of the plain is toward south and south west, which seldom exceeds 0.4 m/km. State has Sutlej, Beas, and Ravi (perennial) and Ghaggar (nonperennial) rivers feeding a vast network of canal system in the state and even provide water to Haryana, Rajasthan, and Jammu & Kashmir.

19.3 HYDRO-METEOROLOGY

In conformity with its inland position and subtropical location, Punjab is characterized by a continental, semiarid to subhumid climate. It varies from tropical steppe to subtropical steppe from northeast to southwest, respectively. Due to the semiarid and hot climate, it is mainly dry except in rainy months and characterized by intensely hot summer and cold winter. Its climatic rhythm, as of other parts of India, is defined by the change of seasons. The contrast between summer and winter is well marked, and these two primary seasons are associated with two main crop seasons of Kharif (June–November) and Rabi (December–April), respectively.

The rainy season or also known as south-west monsoon is defined in the months of July–September and accounts for 80% of the total rainfall. Another additional period of rainfall is the winter rain, which is defined in the months of December to March. This rainfall is mostly absorbed by the soil does not help in groundwater recharge. The distribution of rainfall in Punjab is inconsistent of time and space, that is, varies from ~1000 mm in northeastern part to <300 mm in southwestern parts and 400–600 mm in central parts, so it is always better to consider rainfall in annual average or seasonal average of decades. The areas to the north of Pathankot and near the Shivalik hills receive maximum amount of rainfall while the areas situated in the southwestern side of Punjab (Fazilka) receive minimum amount of rainfall. Long dry spells are often experienced necessitating irrigation from man-made systems for agriculture.

19.4 HYDROGEOLOGY AND AQUIFERS

The rivers deposited sediments from the hills through Kandi belts followed by Sirowal in the alluvial plains and there are piedmont zones (upper and lower) between hills and plains called transition zones. Saturated sand, gravel, or boulder beds constitute the aquifers. In major part of the state, multiple aquifer system exists, the aquifers are formed by granular horizons present consisting of Sand/Gravel. These are separated by finer sediments in the form of thick clays. In general, three broad aquifers have been identified in the state, which diversify from single aquifer system near foothills (Fig. 19.1).

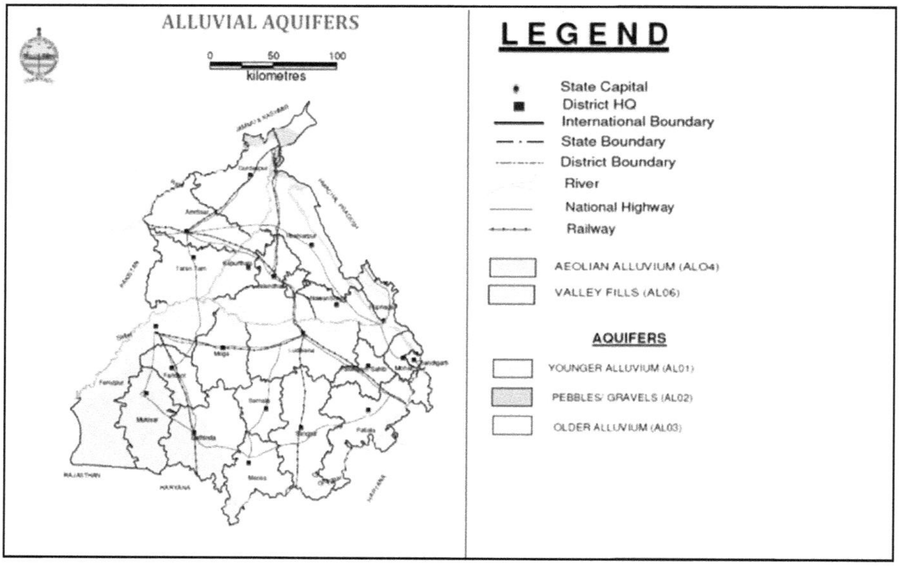

FIGURE 19.1 Major aquifer system in Punjab.

Source: Reprinted from CGWB, 20218.

The state Punjab has both fresh water and saline water aquifers. Aquifer systems in Punjab have been summarized in Table 19.1.

19.5 GROUNDWATER REGIME MONITORING

For groundwater monitoring in Punjab, 964 observation wells have been established by NWR (North Western Region), CGWB (Central Ground Water Board). About 770 wells are in canal command areas and the remaining wells are in districts adjoining the hills or Kandi areas SAS Nagar, Rupnagar, SBS

Nagar, Hoshiarpur, and Pathankot. In addition to these, some are in parts of Ludhiana, Jalandhar, and Gurdaspur.

TABLE 19.1 Details of Fresh and Saline Aquifers in Punjab.

Fresh water aquifers	Saline water aquifers	Description
Single Aquifer System Kandi Formation	Single Aquifer System	Single Aquifer with limited freshwater at shallow depth
Pathankot, Parts of- Ropar, SAS Nagar and Hoshiarpur districts.	Muktsar, Parts of Faridkot, Ferozepur, Sangrur, Bathinda, Moga, Mansa	NIL
Multiple Aquifer System (2–3)	Multiple Aquifer System (2–3)	Water Logged Areas-
Amritsar, Barnala, Fatehgarh Sahib, Gurdaspur, Hoshiarpur, Jalandhar, Kapurthala, Ludhiana, SBS Nagar, Patiala, Taran Taran, Parts of- Faridkot, Ferozpur, Sangrur, Bathinda, Moga, Mansa, Ropar	NIL	Muktsar, Parts of- Bathinda, Fazilka and Mansa.

With reference to the above state summary of aquifers in Punjab, aquifer lithologs are shown in Figure 19.2.

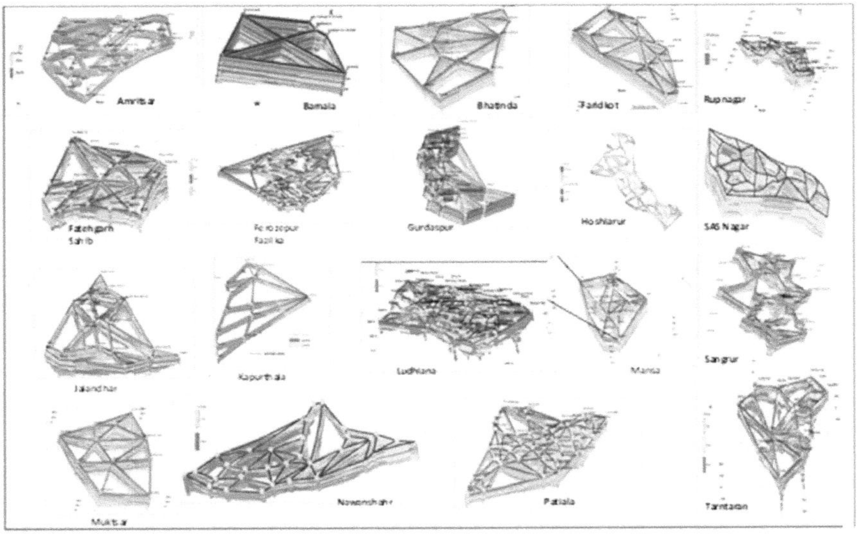

FIGURE 19.2 Aquifer disposition in Punjab.

Source: Adapted from CGWB, 20218.

19.6 GROUNDWATER LEVEL BEHAVIORS

Groundwater levels are fluctuating due to the recharge and extraction processes in different seasons. To find out and quantify the variations, water levels were measured during year 2017–2018 for starting premonsoon of year 2017 (May), monsoon (August), postmonsoon (November), and winter (January) shown the highest and lowest values in Table 19.2 and are discussed.

TABLE 19.2 Lowest and Highest Values of Water Levels during 2017–2018.

Particular	Premonsoon	Monsoon	Postmonsoon	Winter
Lowest (mbgl)	1.31	0.39	0.70	0.45
Highest (mbgl)	59.18	41.68	50.20	50.40

Table 19.2 reveals that the lowest water level is found in southwestern parts of Punjab comprising of Fazilka and Mansa districts in all the seasons as 1.31 mbgl in premonsoon in Kondal (Fazilka district); in monsoon season 0.39 mbgl Burj Bhalai ke in Mansa district; in postmonsoon 0.70 mbgl at Aryan wala (Fazilka district) and in winter 0.45 mbgl at Mothian wala (Fazilka district) while the highest water levels are in central and northeastern parts comprising of Mansa, Rupnagar, and SAS Nagar districts. Water level 59.18 mbgl in premonsoon in Sangha (Mansa district); in monsoon season 41.68 mbgl Chatarnali in Roopnagar district; in postmonsoon and winter 50.20, 50.40 mbgl, respectively, at Pedol (SAS Nagar district). Seasonal changes in water level are discussed below:

19.6.1 PREMONSOON

The water level difference map of the whole state for the premonsoon (May 2017) season is shown in Figure 19.3a. The groundwater level varied from 0.31 to 59.18 mbgl, where the shallowest level was found in Kondal village of Fazilka district and the deepest level was found in Sangha village of Mansa district.

The waterlogging problems are most probably caused if the water level is less than or equal to 2 m below the ground level. This water level (0–2 m) was found in more than 2% of observatory wells and it covers nearly 1% of the state and is found in Fazilka and Muktsar districts.

The water level range from 2 to 5 m below ground level is considered as shallow groundwater level, this was mostly found in southwestern part Muktsar, Fazilka, and Faridkot districts of Punjab and also in the northern part of the state in Gurdaspur and Pathankot districts. This level was found in 12% of the groundwater observatory wells and covering nearly 11% of the total area of Punjab.

The observatory wells monitored with 5–10 m of water level, was found widely in many districts. From the total wells monitored, 23% of the observatory wells were observed with this water level and it envelops an area of about 20% from the whole Punjab. In northern Punjab, it was observed in the districts of Pathankot, Gurdaspur, and Hoshiarpur, and in the Malwa region of Punjab, it was observed in districts of Fazilka, Ferozepur, Faridkot, Muktsar, Bathinda, and Mansa.

The water level from 10 to 20 m was considered as moderately deep water level. This water level was observed mainly in the central region of Punjab due to more exploitation of groundwater. This moderately deep water level was observed in 28% of the wells from the total wells examined and it wraps an area of 31% of Punjab.

The groundwater level with 20–40 m of depth is defined as deep water level, which was observed in 35% of the observatory wells installed in the state. This water level is collectively found in most of the districts of Punjab which covers about 37% area of Punjab. In the Malwa region, it was found in Bathinda, Fatehgarh Sahib, Barnala, Sangrur, Patiala, Ludhiana, and Moga districts; in Majha it was observed in Amritsar and Tarntaran districts; in the Bist-Doab region, it was observed in Jalandhar, Kapurthala, and SBS Nagar districts.

Groundwater level at very deep levels is a rare case in Punjab as it has three perennial rivers flowing across the state. Even though, water level above 40 m was examined in 1% of the observatory wells and 1% area of the state in the Sangrur and SBS Nagar districts in small patches.

19.6.2 MONSOON

The water level difference map of the whole state for the monsoon (August 2017) season is shown in Figure 19.3b. The groundwater level varied from 0.39 to 41.68 mbgl, where the shallow level was found in Burj Bhalaike village of Mansa district and the deep level was found in Chatamali village of Rupnagar district. Water level (0–2 m) was found in more than 5% of observatory wells and it covers nearly 2% of the state in Fazilka and Muktsar districts only. Water level range from 2 to 5 m below ground level was mostly found in southwestern part of Punjab in Muktsar, Fazilka, and Faridkot districts, and also in the northern part of the state in districts Hoshiarpur, Gurdaspur, and Pathankot. This was found in 18% of the groundwater observatory wells and by area it covers nearly 12% of the total area of Punjab. The observatory wells monitored with 5–10 m of water level found widely in many districts. From the total wells monitored, 22% of the observatory wells were observed with this water level and it envelops an area of about 16% from the whole Punjab. In northern Punjab, it was observed in Amritsar, Gurdaspur, and Hoshiarpur districts, and in the Malwa region of Punjab, it was observed in Bathinda, Mansa, Faridkot, Fazilka, Ferozepur, Rupnagar, and SAS Nagar districts.

The water level from 10 to 20 m was observed in 24% of the wells from the total wells examined and covering an area of 29% of Punjab, spread over Mansa, Faridkot, Bathinda, Patiala, Ferozepur, Rupnagar, Fatehgarph Sahib, SBS Nagar, Jalandhar, Taran Taran, Ludhiana, Hoshiarpur, Kapurthala, and Amritsar districts. The groundwater level with 20–40 m of depth was observed in 27% of the observatory wells which covers totally 39% area of Punjab. In the Malwa region, it was found in Fatehgarh sahib, Sangrur, Patiala, Ludhiana, and Moga dsitricts; in Majha it was observed in Amristar and Tarntaran districts; in the Bist-Doab region, it was observed in Jalandhar and SBS Nagar districts. Water level above 40 m was examined in 3% of the observatory wells and covering 2% area of the state in the Patiala and Sangrur Nagar districts in small patches.

19.6.3 POSTMONSOON

The water level variation in postmonsoon (November 2017) along with depth to water level map is shown in Figure 19.4a. The groundwater level varied from 0.70 to 50.20 mbgl, where the shallow level was found in Aryanwala village of Fazilka district and the deep level was found in Pedol village of SAS Nagar district. Very shallow water level (0–2 m) was found in more than 3% of observatory wells and it covers nearly 2% area of the state and found in Fazilka, Faridkot, and Muktsar districts. The water level range from 2 to 5 m was mostly found in southwestern part of Punjab in Muktsar, Fazilka, Faridkot, and Ferozepur districts and also in the northern part of the state in districts Gurdaspur, Hoshiarpur, and Pathankot and also few places at the foothills of Shivaliks. This was found in 13% of the groundwater observatory wells and by area it covers nearly 10% of the total area of Punjab. The observatory wells monitored with 5–10 m of water level, was found 19% of the observatory wells covering an area of about 18% of Punjab. In northern Punjab, it was observed in the Amristar, Pathankot, Gurdaspur, and Hoshiarpur districts and in the Malwa region of Punjab it was observed in Fazilka,

Ferozepur, Faridkot, Muktsar, Bathinda, SAS Nagar, Rupnagar, and Mansa districts. The water level from 10 to 20 m was observed in 27% of the wells covering an area of 30% of Punjab, spread over the districts Mansa, Barnala, Bathinda, Ferozepur, Rupnagar, Fatehgarph Sahib, SBS Nagar, Jalandhar, Taran Taran, Ludhiana, Hoshiarpur, Gurdaspur, Kapurthala, and Amritsar. The groundwater level with 20–40 m was observed in 37% of the observatory wells installed in the State covering an area of 39%. In Malwa region, it was found in Bathinda, Fatehgarh Sahib, Barnala, Sangrur, Patiala, Ludhiana, and Moga districts; in the Majha region, it was observed in Amristar and Tarntaran districts; in the Bist-Doab region, it was observed in Jalandhar, Kapurthala, and SBS Nagar districts. Groundwater level above 40 m was examined in 1% of the observatory wells covering area of 1% of the state in Sangrur, Shaheed Bhagat Singh Nagar, and Patiala districts.

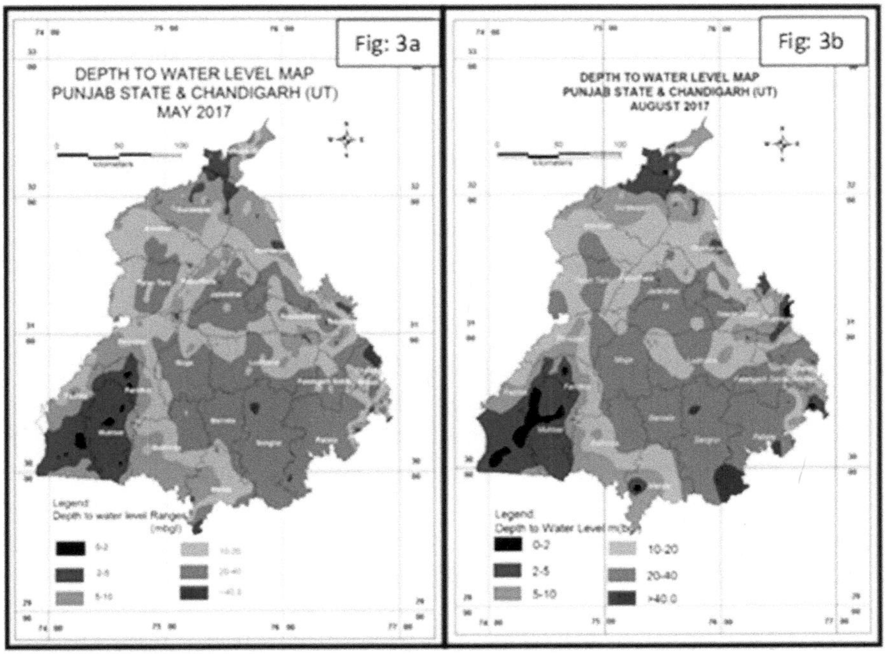

FIGURE 19.3(a and b) Water level variation in premonsoon and monsoon of 2017.
Source: Reprinted from CGWB, 20218.

19.6.4 WINTER

The water level difference map of the entire state for the winter season (January 2018) is shown in Figure 19.4b. The groundwater level varied from

0.45 to 50.40 mbgl, where the shallow level was found in Mothian wala village of Fazilka district and the deep level was found in Pedol village of SAS Nagar district. The water level (0–2 m) was found in more than 3% of observatory wells and it covers nearly 4% area of the Punjab, in Fazilka, Faridkot, and Muktsar districts. The water level in the range from 2 to 5 m was found in 13% of the groundwater observatory wells and by area it covers nearly 10% of the total area of Punjab. The above-mentioned districts are mainly canal command areas and use canal or river water for their agricultural needs. Water level 10–15 m was found in 20% of the observatory wells covering an area of about 19% of the state in Amristar, Pathankot, Gurdaspur, Hoshiarpur, Fazilka, Ferozepur, Faridkot, Muktsar, Bathinda, SAS Nagar, Rupnagar, and Mansa districts.

The water level from 10 to 20 m was observed in 27% of the wells from the total wells examined and covering area of 30% of Punjab, spread over Mansa, Barnala, Bathinda, Ferozepur, Rupnagar, Fatehgarph Sahib, SBS Nagar, Jalandhar, Taran Taran, Ludhiana, Hoshiarpur, Gurdaspur, Kapurthala, and Amritsar districts. The groundwater level with 20–40 m of depth was observed in 37% of the observatory wells covers 38% area of Punjab in Fatehgarh sahib, Barnala, Sangrur, Patiala, Ludhiana, Moga, Jalandhar, and SBS Nagar districts.

Groundwater level above 40 m was examined in 2% of the observatory wells covering 1% of the area in SAS Nagar and Hosiarpur districts.

19.7 GROUNDWATER ELEVATION AND FLOW DIRECTION

Generally, groundwater flows from higher hydraulic gradient to the areas having lower hydraulic gradient due to lateral and vertical variations in hydraulic heads in a ground-water system. Groundwater movements generally have a vertical as well as a horizontal component. The horizontal flow of groundwater was determined by analyzing the water level contour maps.

The Groundwater flow in the aquifers can be calculated from the water level. The water level contour maps have been prepared and flow direction has been interpreted from that (Messene, 2017). The recharge zones and discharge zones of the study area in different study seasons can be identified from the water level and flow direction maps. Water-table maps are prepared from water levels measured in groundwater observation wells at the elevation of the water-table surface.

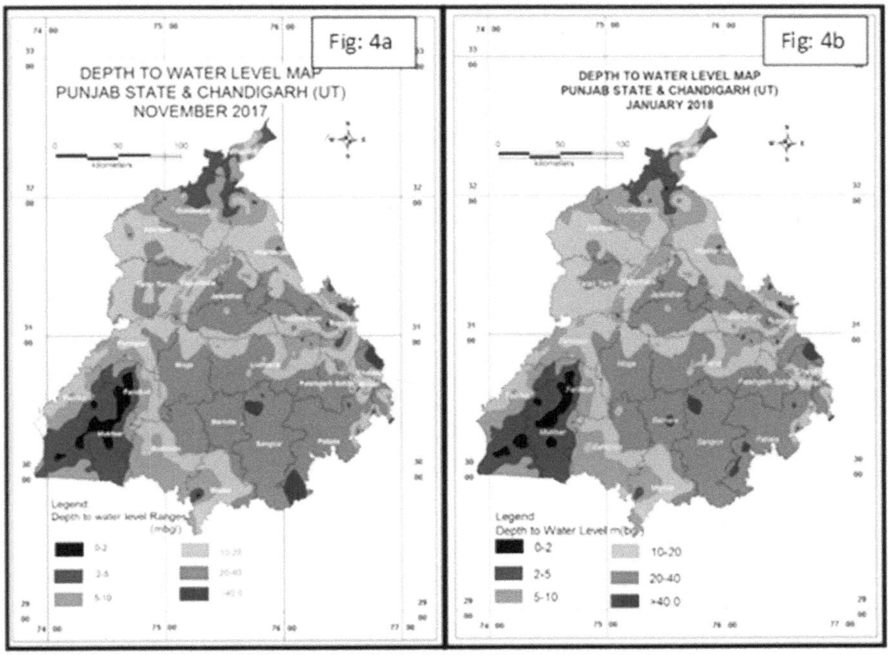

FIGURE 19.4(a and b) Water level variation in postmonsoon (2017) and winter (2018).

Source: Reprinted from CGWB, 20218.

19.7.1 PREMONSOON

From the groundwater level analysis of May-2017, it is observed that the Dhar Kalan Block of Pathankot district has the highest depth of 636.78 m amsl and the lowest depth was 147.82 m amsl from Sardulgarh Block of Mansa district. The difference in hydraulic gradient and the groundwater flow direction can be determined by the map. Generally, groundwater flows in accordance with the natural slope, in the same way the flow direction of Punjab was from the northeastern direction to southwestern part. There were also groundwater troughs created in the districts of Bathinda and Jalandhar. In the districts of Muktsar and Fazilka, the gradient of contour varies from 195 to 165 m, with a very gentle slope slowing down the movement of groundwater and finally resulting in waterlogging. In the map, it is also clear that the waterlogged area spreads to even adjoining other districts. The water table elevation map for premonsoon (May 2017) is shown in Figure 19.5a.

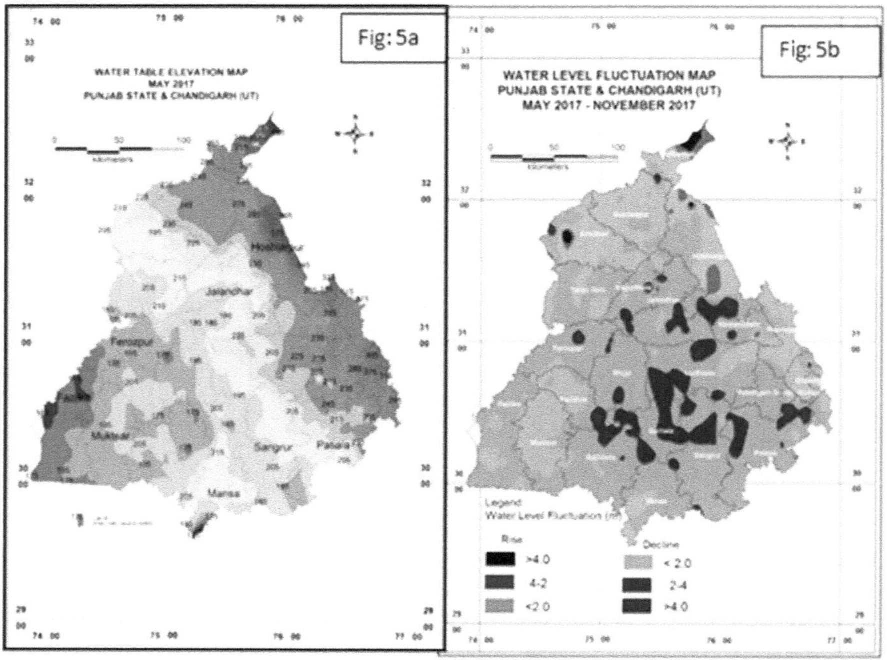

FIGURE 19.5 (a and b) Groundwater elevation map premonsoon (2017) and Groundwater level fluctuation between premonsoon and postmonsoon.

Source: Reprinted from CGWB, 20218.

19.7.2 SEASONAL FLUCTUATIONS

The groundwater level fluctuations can be measured by comparing any two water level data. The seasonal variations can be computed by comparing two different seasons of the same year and simultaneously, yearly difference can be derived by comparing two same seasons of different years.

19.7.3 PREMONSOON—POSTMONSOON

Water level seasonal difference is computed between the data of premonsoon (May 2017) and postmonsoon (November 2017). The fluctuation map (Fig. 19.5b) is used to discuss the behavioral pattern of this seasonal fluctuation. General decline in groundwater level was observed mostly in all districts, 65% of the observed wells were found to be declining in water table, which envelops an area of about 59% of total Punjab. The clear picture of quantity of declination can be derived by dividing the total number of declining wells into three range groups such as 0–2 m, 2–4 m, and >4 m. Slight decline in groundwater level of about 0–2 m was found in 54% of the wells observed, which covers 53% of the area. Moderate declination of water level of about 2–4 m was observed in 9% of the wells, covering 5% of the total area that lies in Bathinda, Moga, Ludhiana, Jalandhar, Barnala, Sangrur, and Fatehgarh Sahib districts. Most declination of water level of about more than 4 m was found in 2% of the wells and this covers 1% of the total area.

The groundwater level rise was observed in few parts of Punjab, 35% of the observed wells were found to rise in the water table, which envelopes an area of 41% of the state. The water table rise was observed in Hoshiarpur, Gurdaspur, Muktsar, and Amritsar districts. The clear picture of quantity of augmentation can be derived by dividing the total number of rising wells into three range groups such as 0–2 m, 2–4 m, and greater than 4 m. Slight rise in groundwater level of about 0–2 m was found in 32% of the wells observed, which covers 38% of the area. Moderate rising of water level of about 2–4 m was observed in 2% of the wells, covering 2% of the total area. Most rising of water level of about more than 4 m was found in 1% of the wells and this covers 1% of the total area.

19.8 GROUNDWATER QUALITY

19.8.1 SALINITY VARIATION

Groundwater is extensively used for drinking, irrigation, and industrial activities in Punjab. Suitability of groundwater for these uses depends on its ambient quality that is reflected in the concentrations of chemical constituents present in it at that time. There is a wide variation in concentration of various chemical constituents in groundwater and 70% of groundwater is fresh and potable in the north and the northeastern part of the state (Fig. 19.6). Issues of groundwater salinity, fluoride, and nitrate contaminations were found in the districts Bhatinda, Faridkot, Ferozepur, Ludhiana, Muktsar, and Sangrur of southern and southwestern parts of Punjab.

19.8.2 ION CONCENTRATION

Majority of the groundwater samples collected from, Bhatinda, Faridkot, Fatehgarh Sahib, Ferozepur, Mansa, Muktsar, and SAS Nagar districts in southern and southwestern parts of Punjab have nitrate concentrations above 100 exceeding 45 mg/L (>BIS Limit). A total of 10.7% groundwater samples collected from agriculture--dominated areas of Bathinda, Faridkot, Fatehgarh Sahib, Ferozepur, Mansa, Muktsar, Patiala, Sangrur, SAS Nagar, and Tarantaran districts have fluoride concentrations above 1.5 mg/L indicating the possibility of phosphatic fertilizers as source fluoride contamination.

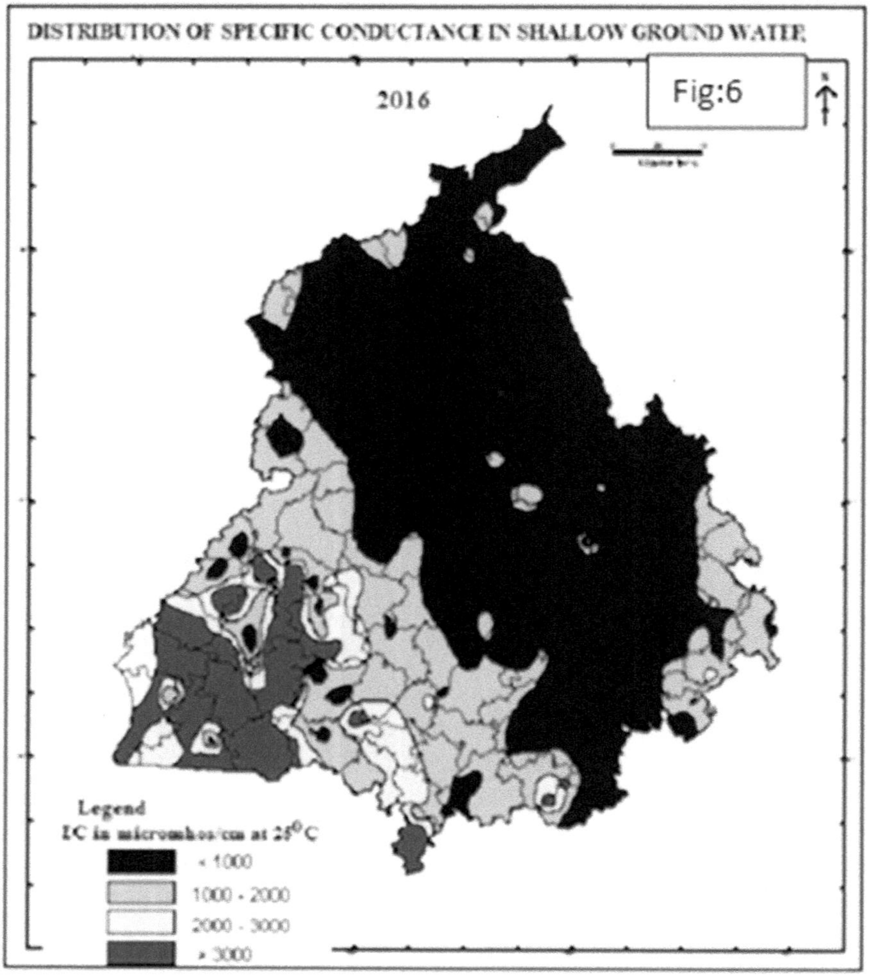

FIGURE 19.6 Salinity variation map of Punjab for the year 2016.

Source: Reprinted from CGWB, 20218.

There is wide variability in irrigation rating of groundwater samples collected from Bhatinda, Faridkot, Ferozepur, Mansa, Muktsar, Patiala, and Sangrur districts. Majority of groundwater samples collected from Amritsar,

Fatehgarh Sahib, Gurdaspur, Hoshiarpur, Jalandhar, Kapurthala, and Ropar districts are suitable for irrigation for semi-salt tolerant crops.

19.8.3 TRACE METALS

Occurrence of heavy metals in groundwater is also being monitored and it was found that in 16.1% groundwater samples collected from Bathinda, Faridkot, Firozepur, Ludhiana, Mansa, Muktsar, Sangrur, and SAS Nagar cadmium concentrations were above permissible limits of drinking water (IS 10500, 2012); 6.3% samples from Bhatinda, Gurdaspur, Hoshiarpur Jalandhar, and Pathankot districts have manganese concentrations above 0.3 mg/L. Sporadic cases of high manganese have been reported from Hoshiarpur, Jalandhar, Kapurtahala, Patiala, and Sangrur districts. Copper concentration is generally below 1.5 mg/L in groundwater of the state. Low concentration of Lead has been reported in Faridkot Gurdaspur Kapurthala, Nawanshar districts but 23.8% samples of shallow aquifers from rest of the districts have Lead concentrations above 0.01 mg/L.

Groundwater samples collected by NWR, CGWB in the year 2015 under quality-monitoring program found Arsenic concentrations >0.01 mg/L desirable limit in isolated cases from Amritsar, Faridkot, Gurdaspur, Hoshiarpur, Nawanshahr, Ropar, Sangrur, and Taran Taran districts. Highest concentration was recorded in water drawn from handpump at Harike (0.200 mg/L). During 2017, the highest concentration of arsenic has been recorded in groundwater of Bara chaunta (0.0375 mg/L) village in Rupnagar district. 16% of the groundwater samples have high Iron with the highest value recorded of 25.825 mg/L at village Issarpur in SAS Nagar.

19.9 GROUNDWATER RESOURCES

The fact is depicted in groundwater resources that are computed during different years. As per the groundwater resources estimation in 1984, which was the first year of groundwater resource estimation, 54 blocks were over-exploited out of 118 blocks, whereas during 2013 groundwater resources 113 blocks were found to be overexploited out of 138 blocks. As per the various previous studies conducted by the state of Punjab for estimation of the Dynamic groundwater scenario of the Punjab depicting the categorization of various blocks is given in Table 19.3. Also, the graphical representation of the data was shown in Figure 19.7.

TABLE 19.3 Showing Categorization of Blocks during Different Years as per Groundwater Resources Assessments.

Category of blocks/ study year	1984	1986	1989	1992	1999	2004	2009	2011	2013	2017
Overexploited	53	55	62	63	73	103	110	110	105	109
Critical	7	9	7	7	11	5	3	4	4	2
Semicritical	22	18	20	15	16	4	2	2	3	5
Safe	36	36	29	33	38	25	23	22	26	22
Total	118	118	118	118	138	138	138	138	138	138

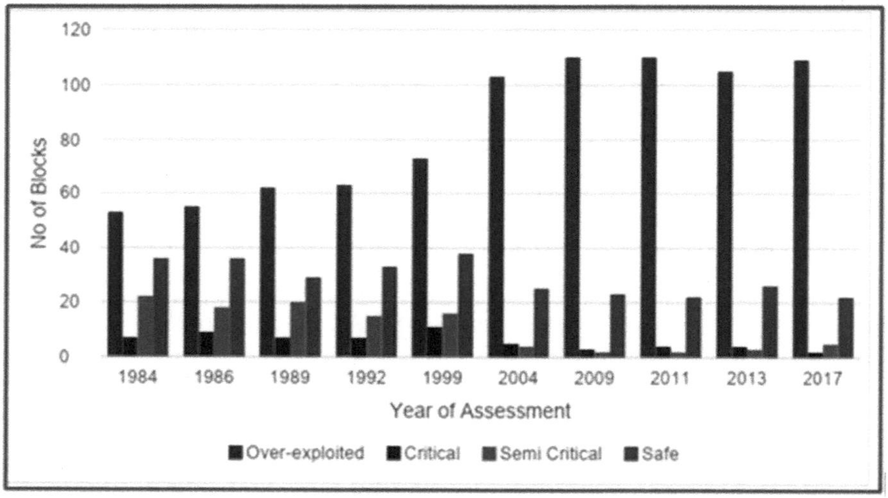

FIGURE 19.7 Categorization of blocks as per groundwater resources estimations.

Source: Reprinted from CGWB, 20218.

The recent Groundwater Resource assessment following Groundwater Resources Estimation Committee Report, 2015 has revealed that for the year 2017 the net annual groundwater availability (net annual replenishable groundwater resources) is 21,650 MCM/year against which the groundwater draft for all uses is 35,812 MCM/year which is 165% of ground. Out of 144 blocks, 109 blocks are overexploited and two are critical, whereas only 22 blocks are safe and 5 are semicritical (Fig. 19.8).

The comparison of groundwater resource estimates including, net water availability, gross draft, and net future availability was estimated for 5 years and listed in Table 19.4.

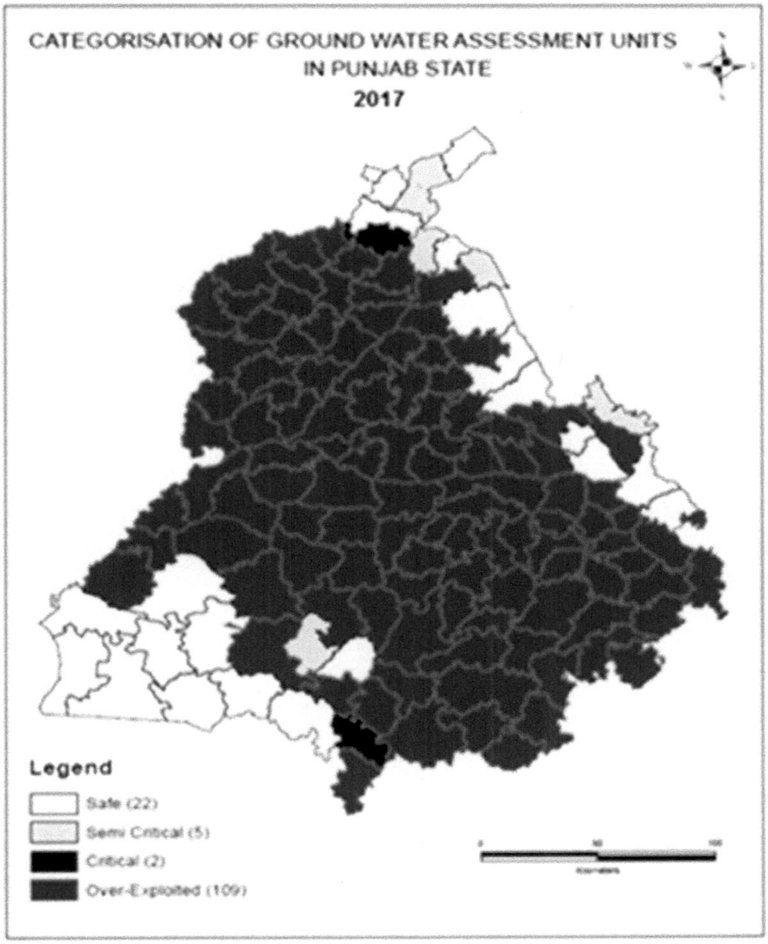

FIGURE 19.8 Categorization of blocks as per groundwater resources estimations, 2017.
Source: Reprinted from CGWB, 20218.

19.10 CONCLUSIONS

This chapter describes the groundwater resources of Punjab in detail. Rise in water demand with the help of government cooperation, there was 30 times increase in groundwater structures in last six decades. Higher rates of groundwater extraction resulted in overexploitation of groundwater and there was a net increase in blocks in overexploited category by 45%. Three zones are clearly separated out in the entire state on the basis of groundwater use. The area in Shiwalik zone falling in districts of Roopnagar, SBS Nagar,

Hoshiarpur, Pathankot, Gurdaspur, and areas in southwest Punjab comprising districts Fazilka, Muktsar are categorized in safe zone, central portion of Punjab comprising of districts Amritsar, Tarntaran, Jalandhar, Kapurthala, parts of Hoshiarpur and SBS Nagar, SAS Nagar, Patiala, Sangrur, Fatehgarh Sahib, Ludhiana, Moga, Bhatinda, Mansa, Firozepur is overexploited in terms of groundwater usage. Water levels are found rising in south-west Punjab.

TABLE 19.4 Comparison of Groundwater (GW) Resource Estimates.

Item	Year of GWR estimation				
	2004	**2009**	**2011**	**2013**	**2017**
Net GW availability (MCM)	21,443	20,349	20,321	23,390	21,584
Gross GW draft (MCM)	31,162	34,661	34,881	34,810	35,782
Net GW availability for future Irrigation development (MCM)	–	–	–	–	–
Stage of GW development in %	145	170	172	149	166
No. of blocks	137	138	138	138	138
No. of OE blocks	103	110	110	105	109
No. of critical blocks	05	03	04	04	02
No. of semicritical blocks	04	02	02	03	05
No. of safe blocks	25	23	22	26	22

Groundwater quality is also a concern in Punjab, where salinity is rising in southwestern parts of the state, arsenic is reported higher in Amritsar and Tarntaran districts, and heavy metal contamination is reported in industrial areas of Ludhiana. Considering all these, there is a urgent need to formulate area-specific plans in the entire state to save the groundwater resources.

KEYWORDS

- **hydrological regime**
- **groundwater**
- **overexploitation**
- **contamination**
- **water logging**
- **Punjab**

REFERENCES

Bhatti, S. S.; Kumar, V.; Kumar, A. et al. Potential Carcinogenic and Non-Carcinogenic Health Hazards of Metal(loid)s in Food Grains. *Environ. Sci. Pollut. Res.* **2020,** *27*, 17032–17042. https://doi.org/10.1007/s11356-020-08238-w

Bonsor, H. C.; MacDonald, A. M.; Ahmed, K. M.; Burgess, W. G.; Basharat, M.; Calow, R. C.; Dixit, A.; Foster, S. S. D.; Gopal, K.; Lapworth, D.; Lark, R. M.; Moench, M.; Mukherjee, A.; Rao, M. S.; Shamsudduha, M.; Smith, L.; Taylor, R.; Tucker, J.; van Steenbergen, F.; Yadav, S. K. Hydrogeological Typologies of the Indo-Gangetic Basin Alluvial Aquifer, South Asia. *Hydrogeol. J.* **2017,** *25* (5), 1377–1406. doi: 10.1007/s10040-017-1550-z

CGWB. *Aquifer Mapping and Management Plan, North Western Region*; Central Groundwater Board, Ministry of Water Resources Government of India Chandigarh, 2017.

CGWB, Central Ground Water Board Ministry of Water Resources, River Development and Ganga Rejuvination Government of India Ground Water Year Book of Punjab and Chandigarh (UT) 2017-2018 North Western Region Chandigarh, September 2018. http://cgwb.gov.in/Regions/NWR/Reports/2017-18%20PUNJAB%20GWYB.pdf

Chopra, R. P. S.; Krishan, G. Analysis of Aquifer Characteristics and Groundwater Quality in Southwest Punjab, India. *J. Earth Sci. Eng.* **2014,** *4* (10), 597–604. http://doi: 10.17265/2159-581X/2014. 10. 001

https://farmech.dac.gov.in/FarmerGuide/PB/index1.html#:~:text=The%20total%20 geographical%20area%20of,95.7%25%20of%20cultivable%20area) (assessed Aug 8, 2020).

Krishan, G. et al. Expansion of Salinization in Aquifers of Punjab, India. Report submitted to PSFC, Govt. of Punjab, 2021a.

Krishan, G.; Rao, M. S.; Kumar, C. P.; Kumar, S.; Loyal, R. S.; Gill, G. S.; Semwal, P. Assessment of Salinity and Fluoride in Groundwater of Semi-Arid Region of Punjab, India. *Curr. World Environ.* **2017,** *12* (1), 34–41.

Krishan, G.; Rao, M. S.; Kumar, C. P.; Garg, P.; Semwal, P. Assessment of Salinity and Groundwater Quality with Special Emphasis to Fluoride in a Semi-Arid Region of India. *J. Earth Sci. Clim. Change* **2014b,** *5* (6), 149. http://dx.doi.org/10.4172/2157-7617.S1.016

Krishan, G.; Rao, M. S.; Kumar, C. P.; Semwal, P. G. S. Identifying Salinization Using Isotopes and Ionchemistry in Semi-Arid Region of Punjab, India. *J. Geol. Geosci.* **2013,** *2*, 129. http://dx.doi.org/10.4172/jgg.1000129

Krishan, G.; Rao, M. S.; Loyal, R. S.; Lohani, A. K.; Tuli, N. K.; Takshi, K. S.; Kumar, C. P.; Semwal, P.; Kumar, S. Groundwater Level Analyses of Punjab, India: A Quantitative Approach. *Octa J. Environ. Res.* **2014a,** *2* (3), 221–226.

Krishan, G.; Singh, R. P.; Rao, M. S.; Gupta, S.; Tiwari, P. K. Fluoride, Iron and Nitrate Affected Areas of Punjab. *Suresh Gyan Vihar Int. J. Water Res.* **2015,** *1* (1). http://www.gyanvihar.org/researchjournals/c3w_vol1_2.pdf

Krishan, G.; Sudersan, N.; Sidhu, B. S.; Vashisth, R. Impact of Lockdown due to COVID 19 Pandemic on Groundwater Salinity in Punjab, India: Some Hydrogeoethics Issues, Sustainable Groundwater Resources Management (in press), 2021c.

Krishan, G.; Vashisth, R.; Sudersan, N.; Rao, M. S. Groundwater Salinity and Isotope Characterization: A Case Study from south-west Punjab, India. *Environ. Earth Sci.* **2021b,** *80*, 169. https://doi.org/10.1007/s12665-021-09419-7

Lapworth, D. J.; Krishan, G.; MacDonald, A. M.; Rao, M. S. Groundwater Quality in the Alluvial Aquifer System of Northwest India: New Evidence of the Extent of Anthropogenic and Geogenic Contamination. *Sci. Total Environ.* **2017,** *599–600*, 1433–1444.

Lapworth, D. J.; MacDonald, A. M.; Krishan, G.; Rao, M. S.; Gooddy, D. C.; Darling, W. G. Groundwater Recharge and Age-Depth Profiles of Intensively Exploited Groundwater Resources in Northwest India. *Geophys. Res. Lett.* **2015,** *42* (18), 7554–7562. doi:10.1002/2015GL065798.

MacDonald, A.; Bonsor, H.; Ahmed, K.; Burgess, W.; Basharat, M.; Calow, R.; Dixit, A.; Foster, S.; Krishan, G.; Lapworth, D.; Lark, M.; Moench, M.; Mukherjee, A.; Rao, M. S.; Shamsudduha, M.; Smith, L.; Taylor, R.; Tucker, J.; Steenbergen Frank van; Yadav, S. Groundwater Depletion and Quality in the Indo-Gangetic Basin Mapped from in situ Observations. *Nat. Geosci.* **2016,** *9*, 762–766.

Statistical Abstract of Punjab. Economic & Statistical Organisation, Government of Punjab, 2018.

Wadia, D. N. *Geology of India*; Tata McGraw Hill Publishing Co., Ltd.: New Delhi, 1979; p 508.

CHAPTER 20

Data Assimilation in Groundwater Modeling

NITESH PATIDAR* and ANJALI

National Institute of Hydrology, Roorkee 247667, Uttarakhand, India

Corresponding author. E-mail: npatidar.nihr@gov.in

ABSTRACT

Numerical groundwater models have become an important tool for assessing groundwater availability under different scenarios of natural and anthropogenic changes. With the recent developments in modeling, various groundwater models are available that can handle simple to complex groundwater problems in an effective manner. However, despite these developments, uncertainty associated with the model results has been a prime concern of modelers and policy makers. Data assimilation has played a key role for quantification and reduction of parameter uncertainty in hydrological modeling in a cohesive and systematic manner. In this chapter, various data assimilation approaches and observations that can be assimilated in groundwater models are described. The data assimilation can help quantifying and reducing the parameter uncertainty in groundwater modeling and thereby increase reliability of models for better decision-making.

20.1 INTRODUCTION

Groundwater resources have paramount importance to meet the development and societal needs of water. With the increased groundwater exploitation

Advances in Hydrology and Climate Change: Historical Trends and New Approaches in Water Resources Management. Surendra Kumar Chandniha, Anil Kumar Lohani, Gopal Krishan, & Ajay Krishna Prabhakar (Eds.)

and pollution, groundwater predictions under different natural and anthropogenic scenarios have become important. To assess the groundwater resource availability and to investigate the impacts of various alterations in natural system, the numerical groundwater models have been utilized effectively. Nowadays, models play a key role in decision-making on issues, such as optimal groundwater usage, artificial recharge planning, and groundwater pollution (Refsgaard et al., 2012).

A variety of groundwater models of varying complexity have been developed during the past decades. Simple groundwater models consider the process in a simplified manner while the complex models elaborate the groundwater system in a more detailed fashion considering various hydrogeological properties and process. With the increased computational power, improved understanding of groundwater systems and ever-increasing availability of distributed hydrologic observations, the models have become more complex. Paradoxically, while these developments help understanding the complex process and interactions in the groundwater systems, these have also increased the need for understanding, quantifying, and handling the uncertainty associated with the models themselves, and with the observations required for driving and evaluating the models (Rojas et al., 2010). With the increased understanding of uncertainties and their sources in modeling, it is being broadly recognized that knowledge of uncertainty associated with model outputs could help effective planning and management. One critical issue to modelers is how to handle the uncertainties arising in groundwater modeling in a cohesive and systematic way. In this context, the data assimilation has played a key role for quantification and reduction of parameter uncertainty in hydrological modeling.

Data assimilation is a scientific method to introduce observations into a process model. The key aim of data assimilation is to provide the best estimate of the state of a physical system by combining the modeling and observations in such a way that the combined estimates are more certain. It is assumed that both the model predictions and observations are imperfect and a more accurate estimate of state variable can be obtained through synergy of model predictions and observations.

The application of data assimilation was first started with numerical weather forecasting models to improve short-term predictions of meteorological conditions. Its application in groundwater modeling is relatively new, although model predictions by combining groundwater level observations, and parameter estimation through inverse groundwater modeling have become reasonably mature (Montzka et al., 2012; Panzeri et al., 2013, 2015; Crestani et al., 2015).

20.2 GROUNDWATER MODELING

Groundwater models are set of mathematical equations that mimic the natural groundwater system. Models help water managers in multiple ways to help understand the natural processes and enabling impact assessments. Models are useful to investigate the response of groundwater system under certain conditions, or to predict/hindcast the system behavior under anthropogenic or natural changes in groundwater recharge/abstraction, land scape, climate and hydrogeology. A verity of groundwater models have been developed in the past. Among these models, the physically based process-driven models remain the most widely used. These models use simplified laws of fluid dynamics and mimic the natural groundwater system utilizing various properties and parameters of the aquifer system. These models allow to handle the heterogeneity and anisotropy by discretizing the study domain into several spatial grids. Similarly, the temporal variations of aquifer stress can be considered by discretizing the temporal domain into several stress periods. The complex multidimensional flow equations are typically solved in these models utilizing numerical solution techniques, such as finite-difference method (FDM) and finite-element method (FEM).

MODFLOW, modular finite-difference flow model, is a numerical groundwater model developed by the United States Geological Survey (USGS) and is the most widely used model for groundwater modeling. MODFLOW is a FORTRAN-based program that was first launched in 1983 and had seen several modifications since then to expand its capabilities. The latest version of MODFLOW is MODFLOW-6. The groundwater flow equation in MODFLOW-6 is discretized using a control-volume finite-difference (CVFD) method. The finite-difference equations are formulated in terms of inter-cell conductance terms, boundary condition conductance, and specified flows. MODFLOW solves a three-dimensional partial differential equation which is derived by combining Darcy's Law with water balance within a control volume. The 3D movement of groundwater of constant density through porous earth material is described by Darcy's Law:

$$\mathbf{q} = -\mathbf{K}\nabla h = -\begin{pmatrix} K_{xx} & \mathrm{EQ0} & 0 \\ 0 & K_{yy} & 0 \\ 0 & 0 & K_{zz} \end{pmatrix} \nabla h \qquad (20.1)$$

where, $\mathbf{q}$ is the specific discharge vector (LT^{-1}), $\mathbf{K}$ is the hydraulic conductivity tensor (LT^{-1}), K_{xx}, K_{yy}, K_{zz} indicate the hydraulic conductivity in x, y, and z directions, h is represented as the potentiometric head (L) and ∇h

is head-gradient vector. Darcy's Law when combined with water balance, it results in a partial differential equation of groundwater flow:

$$\frac{\partial}{\partial x}\left(K_{xx}\frac{\partial h}{\partial x}\right)+\frac{\partial}{\partial x}\left(K_{yy}\frac{\partial h}{\partial y}\right)+\frac{\partial}{\partial z}\left(K_{zz}\frac{\partial h}{\partial z}\right)+Q_s=SS\frac{\partial h}{\partial t} \quad (20.2)$$

where Q_s is incoming/outgoing volumetric flux to/from the control volume. It is negative when water is going out, and positive when water is added to the control volume. *SS* is the specific storage of the porous material (L^{-1}), and *t* represents the time (*T*). To obtain the solution of eq 20.2, various numerical methods are available. One such method available in MODFLOW is CVFD in which the partial differential equation is replaced with finite set of discrete points in space and time. The partial derivatives are replaced with the finite-difference terms calculated at these points lead to a system of simultaneous linear algebraic difference equations. The head values at each node is then calculated by solving the linear equations. The MOFLOW program is built in separate modules which allow to customize the model code for specific groundwater systems.

An alternative modeling approach is to obtain an input–output relationship by disregarding physical theories of groundwater flow. In such approaches, an empirical relationship between groundwater levels and one or a more predictor variables is established. Typically, these models use statistical transfer functions and sometimes are referred to as "black box" models (Mackay et al., 2014). A number of empirical models have been developed, for example, the Boxe-Jenkins autoregressive integrated moving average (ARIMA) model has been used in a number of studies to simulate the meteorological impacts on groundwater levels (Aflatooni and Mardaneh, 2011). These models have gained popularity recently as these can handle complex seasonal and random components of the system. Moreover, these models can be combined easily to develop ensemble model to deal with uncertainty (Yadav et al., 2020).

20.3 DATA ASSIMILATION

Data assimilation aims at improving the reliability of model predictions by exploiting the information embedded in in situ observations and combining them with the model state sequentially. It is well understood that both the field observations and model predictions are inherently associated with uncertainties. Data assimilation utilizes the information of both the imperfect model and limited observations knowing the uncertainty of both to provide more accurate prediction. This way, the data assimilation could also be useful to fill the data gaps, combine multiple models and observations, and quantify uncertainties.

Data assimilation approaches can be divided into sequential and nonsequential approaches. The former assimilates observations from the past until the time of analysis and is typically used in real-time assimilation systems. The latter allows to incorporate observations from the future as well and is generally employed in re-analysis exercises. The data can be assimilated in either intermittent or continuous manner. In an intermittent method, observations can be processed in small batches, which is usually technically convenient. In a continuous method, observation batches over longer periods are considered, and the correction to the analyzed state is smooth in time, which is physically more realistic.

Figure 20.1 shows a typical way of data assimilation in groundwater models. Groundwater flow models require inputs, such as aquifer properties (hydraulic conductivity, storage coefficient), system states (hydraulic head, head gradient, and fluxes) at initial time step, and boundary conditions with source/sink terms (recharge/discharge and extractions). Inaccuracy in such data sets leads to uncertain model outputs, which can be reduced by conditioning the model on measured values of parameters describing system properties and of system states at various locations in space-time (Panzeri et al., 2013). For instance, if the hydraulic head observations are available at different time and locations, the model-derived hydraulic head can be corrected considering the errors in both modeled and observed hydraulic heads through a data assimilation algorithm. The combined heads can be referred to as analysis and is added to the model state for the next iteration. Utilizing the updated state, the model parameters can also be updated accordingly. These corrected states and parameters are then utilized in model for obtaining the outputs at next time step. In this manner, the model trajectory is corrected each time the head observations are available.

20.4 GROUNDWATER OBSERVATIONS FOR ASSIMILATION

Prevailing practice in groundwater modeling has been to assimilate hydraulic heads to improve model simulations. The key objective of such assimilation is to improve the hydraulic conductivity values in the model by conditioning the model-simulated heads using data assimilation algorithms (Li et al., 2012, 2018). Ensemble of hydraulic heads is derived considering the errors in simulated and observed heads and the hydraulic conductivity is then

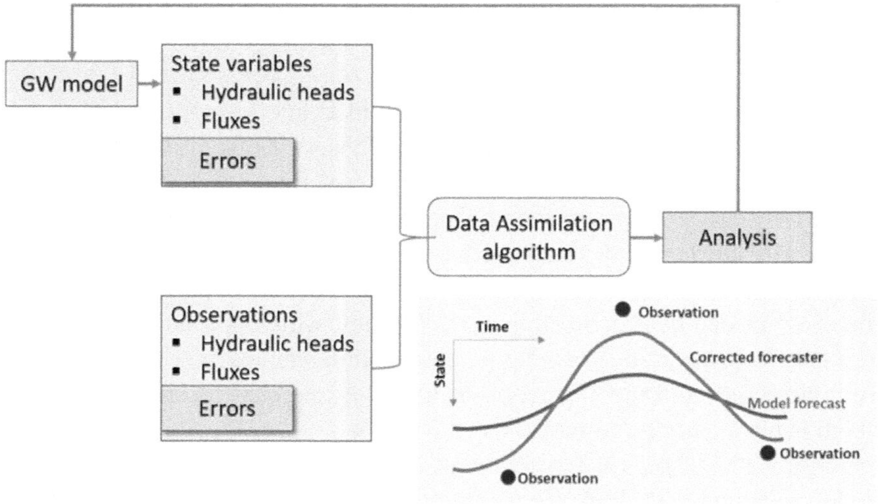

FIGURE 20.1 Schematic of data assimilation in groundwater models.

calibrated such that the simulated heads fall within the limits of ensemble heads. This leads to a considerable reduction in parameter uncertainty and makes the results more reliable (Li et al., 2018). Another potential application of data assimilation is to improve the flow and transport simulations by assimilating concertation data. The plume concentration is measured at multiple locations and is utilized to condition the simulated concentrations. In such applications, porosity, hydraulic conductivity, and heads are updated using an assimilation algorithm. Some researchers have also demonstrated the assimilation of electrical resistivity data with groundwater models. The tracer tests are combined with electrical resistivity tomography (ERT) to estimate solute travel time. The travel time is then assimilated to improve hydraulic conductivity distribution in horizontal and vertical planes. The key advantage of assimilating travel time directly instead of concentration values is that it does not require the calibration of a petrophysical law (Crestani et al., 2015). The observations in vadose zone, such as soil moisture and chloride concentrations, could be useful to reduce uncertainty in groundwater recharge estimation. Observations are assimilated in unsaturated zone models, such as soil–water–atmosphere–plant (SWAP) model, that accounts for moisture and solute transport (Ng et al., 2009). Using such assimilation, values of soil parameters (such as unsaturated hydraulic conductivity) and vegetation parameters (such as rooting depth and evapotranspiration parameters) can be improved.

Recently, multivariate data assimilation has gained popularity. Multiple data, such as soil moisture, hydraulic heads, concentration, are assimilated together in a model to improve the simulations. Multivariate data assimilation allows data assimilation with integrated surface–subsurface models and aims at improving simulation of all hydrological variables simultaneously (Zhang et al., 2016). With the advancement of assimilation algorithms, it is also possible to assimilate both state variable and model parameters simultaneously. For example, in integrated hydrological models, Leaf area index (LAI) is used as a parameter which is easily available at suitable spatial and temporal resolution through remote sensing, and thus can be assimilated in the models (Montzka et al., 2012). Additionally, state variables obtained from other models (single or multiple) can also be assimilated and could be advantageous in areas where observations are not available. Assimilation of sate variables from multiple models could also help reducing model structural uncertainty to some extent (Refsgaard et al., 2012). A few selected studies are presented in Table 20.1 elaborating types of groundwater observations and assimilation techniques used in these studies.

TABLE 20.1 Selected Studies on Data Assimilation in Groundwater.

Sr. No.	Reference	Assimilation algorithm	Observations	Groundwater model
1.	Li et al. (2012)	EnKF	Hydraulic head and concentration	MODFLOW and MT3D
2.	Panzeri et al. (2013)	EnKF	hydraulic heads and fluxes	Stochastic moment equations of transient groundwater flow
3.	Kurtz et al. (2014)	EnKF	Hydraulic heads and groundwater temperatures	SPRING
4.	Crestani et al. (2015)	EnKF	Electrical resistivity tomography (ERT)-based travel time	Developed a transport model
5.	Zhang et al. (2016)	Ensemble transform Kalman filter (ETKF)	Soil moisture and piezometric head	MIKE-SHE
6.	Zovi et al. (2017)	Normal score ensemble Kalman filter (NS-EnKF)	Electrical resistivity tomography (ERT) and piezometric head	Developed a two-dimensional groundwater flow model
7.	Chen et al. (2018)	EnKF	Solute concentration	MODFLOW and MT3D
8.	Li et al. (2018)	Ensemble smoother (ES) and EnKF	Piezometric head	MODFLOW

EnKF, Ensemble Kalman Filter.

20.5 DATA ASSIMILATION ALGORITHMS

A variety of data assimilation algorithms have been developed and most of these algorithms are based on Bayesian filter. The use of ensemble Kalman filter, ensemble transform Kalman filter, ensemble smoother, and 4-DVAR can be found in groundwater literature, among which the ensemble Kalman filter (EnKF) has been the most popular in groundwater modeling. Theoretical description of EnKF and ensemble smoother (SM) is provided in the following subsections.

20.5.1 ENSEMBLE KALMAN FILTER

The EnKF is a variant of standard Kalman filter (KF), named after Rudolf E. Kálmán, which uses a series of observations with associated errors and estimates the ensemble of variables that tend to be more accurate than the single observations alone by estimating a joint probability distribution over the variables for each time frame. A key limitation of KF is that it assumes the distribution of errors to be Gaussian which may not hold true in the case of groundwater where the input–output relationship is often nonlinear. Therefore, a variant of KF, called extended Kalman filter (EKF), was developed which can handle nonlinearity. However, as neither of the KF and EKF is suitable for large complex models, such as groundwater flow and transport models, the EnKF was developed and has been the most widely used in hydrology. EnKF employs Monte Carlo (MC) approach to generate the ensemble (a collection of random model representations). The ensembles are then utilized to estimate first and second statistical moments, also referred to as ensemble moments of model parameters and system states. Both set of variables are then updated using Kalman gain matrix obtained from ensemble moments. A detailed description of EnKF can be found in Evensen (2003). Procedure to update both parameters (e.g., hydraulic conductivity and porosity) and states (piezometric head h and concentration c) using EnKF is described here. Implementation of EnKF includes forecast and analysis steps.

In the forecast step, system states are forecasted using a forward model simulation by solving groundwater flow and transport equations.

$$\mathbf{Y}_k = f\left(\mathbf{X}_{k-1}, \mathbf{Y}_{k-1}\right) \tag{20.3}$$

Equation 20.3 represents a forward model where the system state at time k ($\mathbf{Y}_k$), including hydraulic heads and concentration data, is forecasted by groundwater flow and transport model with predefined boundary conditions and aquifer stresses indicated by *f*. The model parameters (hydraulic conductivity and porosity) are denoted by $\mathbf{X}_{k-1}$, where the subscript (k–1) indicates the latest update. There could be multiple cases depending on the availability of observations:

$$\begin{cases} \text{Case A} : \mathbf{X} = \left[\ln K\right]^T & \mathbf{Y} = \left[h\right]^T, \text{ if only h is available} \\ \text{Case B} : \mathbf{X} = \left[\ln K, \phi\right]^T & \mathbf{Y} = \left[c\right]^T, \text{ if only c is available} \\ \text{Case C} : \mathbf{X} = \left[\ln K, \phi\right]^T & \mathbf{Y} = \left[h, c\right]^T, \text{ if h and c both are available} \end{cases} \tag{20.4}$$

In the analysis step, the model parameters are updated using the observations as follows:

- A joint vector (ψ_k) of model parameters (**X**) and system states (**Y**) is created as:

$$\psi_{k,j} = \begin{bmatrix} \mathbf{X} \\ \mathbf{Y} \end{bmatrix}_{k,j} \tag{20.5}$$

The subscript *k* and *j* indicate time step and ensemble member, respectively. This joint vector can be split into as many members as there are realizations in the ensemble. For instance, if both *h* and *c* observations are available and both parameters *K* and ϕ are being updated for *N* discretization blocks, dimension of ψ_k would be $4 \times N$.

Analysis is created to update the ψ_k using observations $\mathbf{Y}_k^{obs}$ as follows:

$$\psi_{k,j}^{a} = \psi_{k,j}^{f} + \mathbf{G}_k \left(\mathbf{Y}_{k,j}^{obs} + \mathring{\boldsymbol{a}} - \mathbf{H}\psi_{k,j}^{f} \right) \tag{20.6}$$

where superscripts *a* and *f* denotes analysis and forecast, respectively. The error in observations is denoted by the vector ε, operator **H** is introduced here to linearly interpolate the forecasted values at the point where observations are available. For the cases where the observations and forecasts belong to the same grid point, *H* is taken as 1. The Kalman gain $\mathbf{G}_k$ can be estimated as:

$$\mathbf{G}_k = \mathbf{P}_k^f \mathbf{H}^T \left(\mathbf{H}\mathbf{P}_k^f \mathbf{H}^T + \mathbf{R}_k \right)^{-1} \tag{20.7}$$

where $\mathbf{R}_k$ is the measurement error covariance matrix, $\mathbf{P}_k^f$ contains the covariance between the different components of the state vector and can be estimated using forecasted ensemble members as:

$$\mathbf{P}_k^f \approx E\left[\left(\psi_{k,j}^{f} - \overline{\psi}_{k,j}^{f} \right) \left(\psi_{k,j}^{f} - \overline{\psi}_{k,j}^{f} \right)^T \right]$$

$$\approx \sum_{j=1}^{N_e} \frac{\left(\psi_{k,j}^f - \overline{\psi}_{k,j}^f\right)\left(\psi_{k,j}^f - \overline{\psi}_{k,j}^f\right)^T}{N_e} \tag{20.8}$$

where the overbar indicates the mean value over the ensemble members, N_e is the number of ensemble members.

Once the states are updated, these become the current states and forecast-analysis loop is started again each time the observation is available.

20.5.2 ENSEMBLE SMOOTHER

The ensemble smoother (ES) is a viable alternative to EnKF when recursive updates are not required. It is first proposed by Van Leeuwen and Evensen (1996). Like EnKF, ES also includes two steps: forecast and update. The key difference between EnKF and ES is that ES starts with time zero ($t = 0$) unlike EnKF which starts with the previous update step ($t = k-1$). This way ES considerably reduces computation time as no recursive update is performed, and the parameters and the states are updated in one global forecast-update step. The forecast step is similar to EnKF eq 20.3. Similar to EnKF, a joint vector is created but here the states at all times ($t = 1, 2, 3,\ldots k$) are included as:

$$\psi_j = \begin{bmatrix} \mathbf{X} \\ \mathbf{Y}_1 \\ \mathbf{Y}_1 \\ \ldots \\ \mathbf{Y}_k \end{bmatrix}_j \tag{20.9}$$

Where **X** indicates parameter vector to be updated and **Y** indicates the system states from $t = 0$ to $t = k$. Analysis is created to update the ψ_j using observations $\mathbf{Y}^{obs}$at all times as follows:

$$\psi^a = \psi^f + \mathbf{G}\left(\mathbf{Y}^{obs} + \mathring{a} - \mathbf{H}\psi^f\right) \tag{20.10}$$

where superscripts a and f denote analysis and forecast, respectively. The error in observations is denoted by the vector ε; **H** is an operator for linearly spatial interpolation. The Kalman gain **G** can be estimated as:

$$G = \mathbf{P}^f\mathbf{H}^T\left(\mathbf{H}\mathbf{P}^f\mathbf{H}^T + R\right)^{-1} \tag{20.11}$$

The ES has an exact solution of the Bayesian formulation with increasing ensemble size, if parameters follow a multi-Gaussian distribution and the transfer function is linear.

20.6 TOOLS AND SOFTWARE PROGRAMS

Data assimilation algorithms described in the above sections typically involve multiple iteration of groundwater models and heavy computation which requires efficient computer program that can couple the model with assimilation algorithms. It is also important to have groundwater model source code/executable file in order to achieve a closed coupling of the model and assimilation algorithm for an efficient recursive forecast-update operation. The USGS's groundwater model, MODFLOW, has been the most widely used in previous groundwater modeling studies for a variety of applications.

Several tools for implementing data assimilation algorithms are also available which can be utilized with groundwater models. OpenDA is a tool that provides various assimilation algorithms and can be coupled with numerical models. It includes almost all the commonly used algorithms, such as EnKF, ensemble square root Kalman filter (EnSR), steady state Kalman filter, particle filter and 3DVAR. An alternative is to use a programming language, like R, python or MATLAB to perform data assimilation. Most of these include well-developed packages that include common assimilation algorithms.

20.7 CONCLUSION

The groundwater models have always been recognized as import supporting tools for groundwater water management. Recently, with the improved understanding of hydrogeological processes and computational abilities, groundwater models have become more complex. This has necessitated uncertainty estimation for the models. The knowledge on uncertainties associated with a model help better decision-making considering various best and worst scenarios possible. Recently, in order to improve the reliability of model outputs, researchers developed various techniques to handle model uncertainties. Data assimilation is one of those techniques which has widely been tested with a variety of numerical models. In this chapter, commonly used data assimilation techniques and groundwater observations are discussed. Commonly used groundwater observations are piezometric heads, solute concentrations and travel time. Among various available assimilation techniques, EnKF is the most widely used with groundwater flow and transport models and has recognized as an effective technique to reduce parameter uncertainty to a considerable extent. This is also found that updated system states using assimilation can reduce model structural uncertainty to some extent. In addition, like EnKF, other variations of the Kalman filter, such as extended Kalman filter, have also been found to be effective in groundwater modeling. Application of several other assimilation techniques, including particle filtering, ensemble smoother, 3DVAR, etc., have also been tested with groundwater models. With improved observation network and advanced sensor technology, the groundwater observations are increasingly available which could help improve model predictions to a significant level using data assimilation.

KEYWORDS

- **data assimilation**
- **groundwater**
- **uncertainty**
- **policy**
- **decision-making**

REFERENCES

Aflatooni, M.; Mardaneh, M. Time Series Analysis of Ground Water Table Fluctuations Due to Temperature and Rainfall Change in Shiraz Plain. *Int. J. Water Resour. Environ. Eng.* **2011,** *3*, 176–188. doi: https://doi.org/10.5897/IJWREE.9000061

Chen, Z. et al. Joint Identification of Contaminant Source and Aquifer Geometry in a Sandbox Experiment with the Restart Ensemble Kalman Filter. *J. Hydrol.* **2018,** *564*, 1074–1084. doi: 10.1016/j.jhydrol.2018.07.073

Crestani, E.; Camporese, M.; Salandin, P. Assessment of Hydraulic Conductivity Distributions through Assimilation of Travel Time Data from ERT-Monitored Tracer Tests. *Adv. Water Resour.* **2015,** *84*, 23–36. doi: 10.1016/j.advwatres.2015.07.022

Evensen, G. The Ensemble Kalman Filter: Theoretical Formulation and Practical Implementation. *Ocean Dyn.* **2003,** *53*, 343–367. doi: https://doi.org/10.1007/s10236-003-0036-9

Kurtz, W. et al. Joint Assimilation of Piezometric Heads and Groundwater Temperatures for Improved Modeling of River-Aquifer Interactions. *Water Resour. Res.* **2014,** 1665–1688. doi: 10.1002/ 2013WR014823

Van Leeuwen, P. J.; Evensen, G. Data Assimilation and Inverse Methods in Terms of a Probabilistic Formulation. *Mon. Weather Rev.* **1996,** *124* (12), 2898–2913.

Li, L. et al. Jointly Mapping Hydraulic Conductivity and Porosity by Assimilating Concentration Data via Ensemble Kalman filter. *J. Hydrol.* **2012,** *428–429*, 152–169. doi: 10.1016/j.jhydrol.2012.01.037

Li, L.; Puzel, R.; Davis, A. Data Assimilation in Groundwater Modelling: Ensemble Kalman Filter versus Ensemble Smoothers. *Hydrol. Process.* **2018,** *32* (13), 2020–2029. doi: 10.1002/hyp.13127.

Mackay, J. D.; Jackson, C. R.; Wang, L. A Lumped Conceptual Model to Simulate Groundwater Level Time-Series. *Environ. Model. Softw.* **2014,** *61*, 229–245. doi: 10.1016/j.envsoft.2014.06.003

Montzka, C. et al. Multivariate and Multiscale Data Assimilation in Terrestrial Systems: A Review. *Sensors* **2012**. doi: 10.3390/s121216291

Ng, G. H. C. et al. Using Data Assimilation to Identify Diffuse Recharge Mechanisms from Chemical and Physical Data in the Unsaturated Zone. *Water Resour. Res.* **2009,** *45* (9). doi: 10.1029/2009WR007831

Panzeri, M. et al. Data Assimilation and Parameter Estimation via Ensemble Kalman Filter Coupled with Stochastic Moment Equations of Transient Groundwater Flow. *Water Resour. Res.* **2013,** *49* (3), 1334–1344. doi: 10.1002/wrcr.20113

Panzeri, M. et al. EnKF Coupled with Groundwater Flow Moment Equations Applied to Lauswiesen aquifer, Germany. *J. Hydrol.* **2015,** *521*, 205–216. doi: 10.1016/j.jhydrol.2014.11.057.

Refsgaard, J. C. et al. Review of Strategies for Handling Geological Uncertainty in Groundwater Flow and Transport Modeling. *Adv. Water Resour.* **2012,** *36*, 36–50. doi: 10.1016/j.advwatres.2011.04.006.

Rojas, R. et al. Application of a Multimodel Approach to Account for Conceptual Model and Scenario Uncertainties in Groundwater Modelling. *J. Hydrol.* **2010,** *394* (3–4), 416–435. doi: 10.1016/j.jhydrol.2010.09.016

Yadav, B. et al. Ensemble Modelling Framework for Groundwater Level Prediction in Urban Areas of India. *Sci. Total Environ.* **2020,** *712*, 135539. doi: 10.1016/j.scitotenv.2019.135539

Zhang, D. et al. Multivariate Hydrological Data Assimilation of Soil Moisture and Groundwater Head. *Hydrol. Earth Syst. Sci. Discus.* **2016,** *20* (10), 4341. doi: 10.5194/hess-2016-126

Zovi, F. et al. Identification of High-Permeability Subsurface Structures with Multiple Point Geostatistics and Normal Score Ensemble Kalman filter. *J. Hydrol.* **2017,** *548*, 208–224. doi: 10.1016/j.jhydrol.2017.02.056

CHAPTER 21

Decision Support System for Surface Water Planning

ANIL KUMAR LOHANI[1*], RAHUL KUMAR JAISWAL[2], SURENDRA KUMAR CHANDNIHA[3], and SUKANT JAIN[1]

[1]*Surface Water Hydrology Division, National Institute of Hydrology, Roorkee, Uttarakhand, India*

[2]*Central India Hydrology, Regional Centre, WALMI Campus, Near Kaliasot Dam, Post Ravishankar Nagar, Bhopal, India*

[3]*Department of Soil and Water Engineering, BRSM CAET & RS, Mungeli, IGKV Raipur, Chhattisgarh, India*

[*]*Corresponding author. E-mail: aklnih@gmail.com*

ABSTRACT

A decision support system (DSS) is well-structured and user-friendly computer program which provides an alternative solution to water management-related issues to the water managers. Thus, it supports in proper decision-making by comparing different strategies under different scenarios for managing the water resources issues in a judicious manner. Moreover, modern visualization techniques enable the managers to get a good understanding about various choices and trade-offs. In that respect, a DSS can also be quite useful for priority ranking in master planning studies. In this paper, a conceptual frame work of surface water planning DSS is presented. Further, key tasks carried out for the development and implementation of DSS (Planning) activities are elaborated in detail. A number of issues related to the use of hydrological data for DSS (Planning) are also presented.

Advances in Hydrology and Climate Change: Historical Trends and New Approaches in Water Resources Management. Surendra Kumar Chandniha, Anil Kumar Lohani, Gopal Krishan, & Ajay Krishna Prabhakar (Eds.)

21.1 INTRODUCTION

Water is the liquid that lubricates the development and it is as much an economic input as are infrastructure and energy (Pate et al., 2007). Society demands a safe supply of water for drinking and other socioeconomic activities as well as protection against floods and droughts. Cheng and Hu (2012) suggested effective and efficient strategies for development, control, protection, regulation, and beneficial use of water resources are required for the water supply for irrigation, industries, hydropower generation, and domestic use, recreation/water sports, navigation, etc.

About 80% of Indian rainfall occur in monsoon period (Lal, 2005). India receives an average annual rainfall of 1170 mm. However, spatial and temporal variability in rainfall results in drought and floods at the same time in different parts of the country. Added to this large diversity in space and time, there are large variations from 1 year to another resulting in floods and droughts years. In India, besides a good amount of average annual rainfall, some basins are categorized as *water scarce* and *water stressed*, and some as *absolute scarcity* (Sinha et al., 2019).

The planning and judicious management of water resources thus appears as a challenging task, as it requires not only an in-depth understanding of hydrological processes, human activities, socioeconomic conditions, societal, political, and cultural contexts, but also the understanding about their correlation with each other (Di Baldassarre et al., 2019). It was only very recently the river basins became not only the focus of hydrological investigations, but also for the management of water resources. Envisaged benefits can only be achieved through a well-planned integrated river basin development schemes based on the reliable data/information with due regard to improved methodologies available for the purpose (Caratti et al., 2004). Generally, in water resources planning, development and management consist of formulation of policies and management strategies for local, national, or regional scale, planning, design of the water resources structures, planning the optimum operation policies; and training and research (Airoldi et al., 2005).

Decision-making in water resources is a multifaceted and challenging job (Guswa et al., 2014). This is because of complexity associated with the management of water supply and demand, and economic inequalities. Further, there is spatial and temporal variability in the rainfall pattern and thus in water availability (Raziei et al., 2009). This intensifies the intricacy of the system. Thus, decision-making in water resources is a very complex task (Nicklow et al., 2010). A decision support system (DSS) on geographical information system platform with a repository of useful models and tools can handle large amount of spatial and temporal data for modeling and decision-making.

21.2 NEED FOR A DECISION SUPPORT SYSTEM

The planning of water resources for sustainable development and management is an important task in view of the increasing demands. As water system is a very complex system due to spatial and temporal variation in the availability of the water resources and various conflicting demands in time and space. Further various environmental issues also play an important role in the water resources development process. Decision-making should be based on water policy guidelines. DSS provides various scenarios envisaged by the planner so that an optimum and reliable decision can be taken for water resources management (Kumar et al., 2005; Jain et al., 2009; Nyimbili and Erden, 2018). It supports in appropriate decision-making for the planning and management of water resources considering future climate change scenarios. The water resource decision-making requires:

- Data monitoring and information acquisition from different sources.
- Knowledge of stakeholders in appropriate areas.
- System for presenting the information to focus important issues.
- Provide various scenarios for different development aspects and climate change to support the decision-making process.

In a decision-making process, people, groups, or institutions play an important role (Cubillo et al., 2006). The decision maker selects and implement a solution for a specific problem. A decision taken by the decision maker has direct or indirect impact on affected people. Figure 21.1 illustrates the various steps involved in the decision-making.

21.3 DECISION SUPPORT SYSTEM APPROACH

Problems for decision-making are generally unstructured or semistructured. Now for making decisions for such type of problems, decision theory plays an important role. Decision alternatives help the decision maker to select the appropriate combination for decision-making. Decision alternatives also help in organizing and synthesizing the data and facts on a multifaceted and opposing issues.

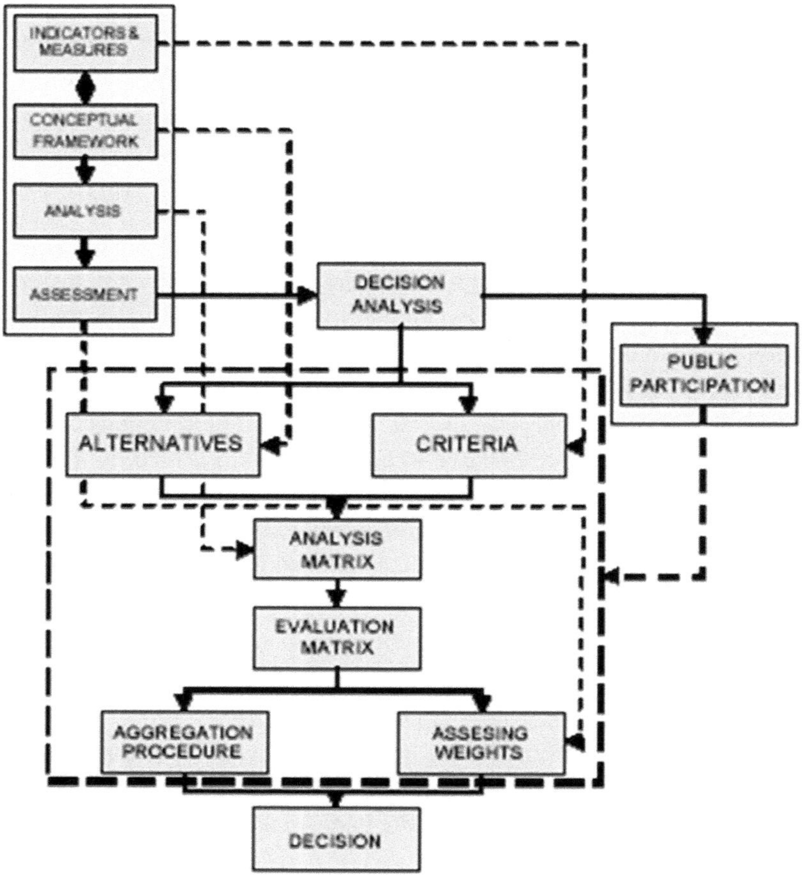

FIGURE 21.1 Steps for decision-making process.

Sprague and Carlson (1982) defined DSS as "computer software that supports decision makers in utilizing data and models for solving unstructured or underspecified problems." A DSS is a user-friendly software useful for supporting individuals or the institutions in decision-making for real-time operation and/or future planning. DSS can also permit man–machine interaction by integration of modelers judgment into the analysis process.

A well-defined water resources problem with different management scenarios has great potential for conservation and management of water. A DSS can answer the relevant questions related to availability of water and its optimal use in various sectors, such as agriculture, industries, and municipal sector. A number of decision methods available in the literature are developed so far with different thoughts and have different theoretical basis. Selection of a specific method depends on the type of problem, preferences, interaction approach, etc. Different types of decision methods may provide different answers, therefore, a combination of different methods helps in choosing the most suitable one. Alternatives obtained from different decision-making methods show different ranks. A number of DSSs have been hypothesized in the scientific literature. These DSS tools generally deals with drought management, flood management, water resources planning, pollution prevention and control, etc. WATERWARE (Fedra, 1994; Jamieson and Fedra, 1996a; Jamieson and Fedra, 1996b), AQUATOOL (Andreu et al., 1996), NELUP (O'Callaghan, 1995; Dunn et al., 1996), FLOODSS (Catelli et al., 1998), DSSIPM (da Silva et al., 2001) are the DSS tools developed to deal with different water resources problems.

DSS design basically depends on the main purpose for which it is required. Therefore, DSS developers must have the knowledge about the information needed in the suitable format. Various tools used for the analysis in a DSS depend on the spatial and temporal resolutions required. Modeling of the hydrological processes constitutes as one of the important tools in the DSS. Surface water planning and management-based DSS is shown in Figure 21.2.

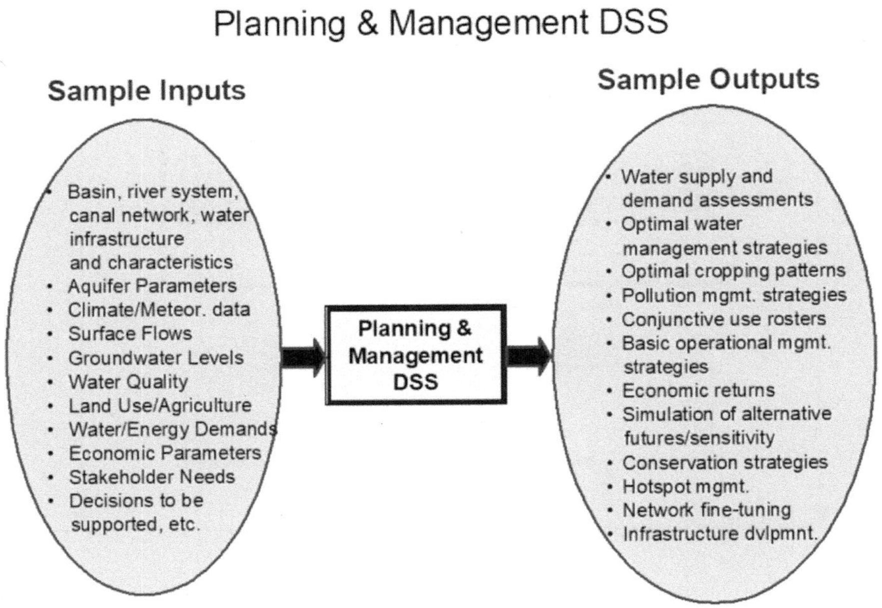

FIGURE 21.2 Surface water planning and management DSS.

The other tools may include models of environment, economic, social processes, etc. A DSS generally has the following components:

1. Data base management system.
2. Software for hydrological and other processes to be considered in the DSS.
3. Knowledge-based expert system.
4. Integrated software for database management system along with the software for the different processes as well as knowledge based expert system.
5. Software for retrieval and display of DSS output in the desired format.

A typical architect of the surface water planning DSS is shown in Figure 21.3.

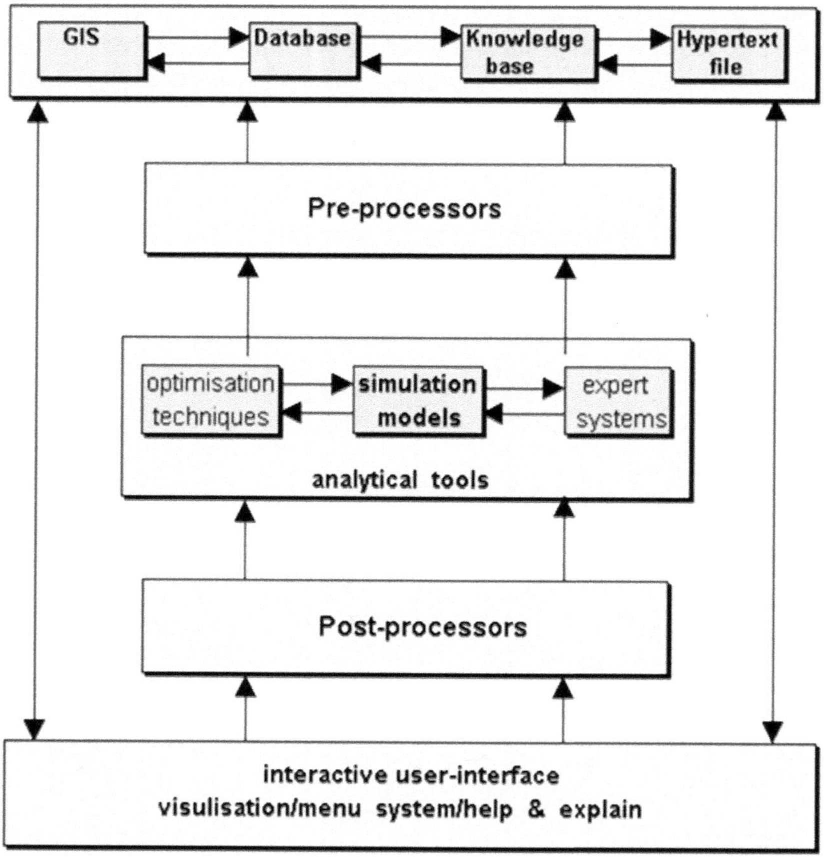

FIGURE 21.3 Architect of the surface water planning DSS.

21.4 DSS STRUCTURE AND FUNCTIONS

The structure and the functionalities of a water resources planning and management DSS reflect some conceptual steps in order to generate various planning scenarios. These steps are the following:

1. Estimation of water resources availability in a basin/subbasin.
2. Hydrological simulation for estimating surface and groundwater.
3. Generation of future water availability and demand scenarios.
4. Evaluation of the alternative scenarios.

The major objective of a surface water planning DSS is to develop a policy framework for managing water resources based on a sustainable water use policy spanning the next 5, 10, 20, 25 years, etc. The policy framework should identify the interrelationships and ensure integration between water use and the following:

- Surface water availability.
- Water supply from surface water, groundwater, rain, wastewater reuse, and desalination.
- Storm drainage.
- Project water demand considering various development activities.
- Water quality.
- Coastal zone management.
- Land/agriculture.
- Industry.
- Tourism.

Suitable indicators should be provided in a DSS to provide a picture of existing and future scenarios for selecting a suitable alternative. For water quality, quantity, environmental, social and development aspects, different indicators may be provided in the DSS. Furthermore, the policy framework should spell out the role of different stakeholder groups and make each group accountable for its implementation.

21.5 DECISION SUPPORT SYSTEM CONCEPTUAL DESIGN

Increasing demand of water due to increasing population, industrialization and irrigation, management of water is becoming increasingly complicated. Keeping in view the demand of water by various sectors, multidisciplinary and integrated approach of water management must be considered. A DSS should analyze scenarios and provide alternative solutions to satisfy the water resource demand by considering the future development activities and/or climatic scenarios. In the recent past, the development in GIS technology has facilitated the DSS developers to design a DSS on GIS platform. The DSS can systematically analyze and evaluate various alternative solutions. The merging of geographical information system (GIS) with the DSS adds a spatial dimension to the analysis process and provides a more effective tool for a fast and transparent decision-making. This is the basis for the development of spatial decision support system (SDSS). A DSS is an integrated software tool which aggregates several models, procedural and heuristic together to enhance the modeling process and to aid in providing a more comprehensive understanding of the physical system being studied. DSS can handle uncertainty and risk in the development of decision problems

21.6 DECISION SUPPORT SYSTEM COMPONENTS

21.6.1 GEOGRAPHIC INFORMATION SYSTEM

User friendliness and applicability of a DSS is increased by incorporating GIS in the aspects of database, interface, and model connection (Keenan, 1995). A GIS provides spatial dimensions into the database and also integrates environmental, social, and economic aspects. GIS serves both as spatial database as well as a toolbox. A GIS module in the DSS also provides data to other modules. GIS is the process of gathering information associated with physical or geographical space, converting it into digital form that a computer can recognize, and then manipulating the information to make evaluation and decision-making more cost-effective. GIS technology replaces standard uses of text files, spreadsheets, and databases, allowing users to analyze information in new and dynamic way. GIS enables linking of spatial data with nonspatial data (such as tabular data) available from different sources and helps in displaying required information graphically and in the form of map layers. GIS contains all spatial data comprising remote sensing imagery, DEM, geology, soil types, rivers, roads, etc. (Fig. 21.3).

21.6.2 DATABASES

It consists of a combination of static, slow-moving and dynamic data, all of which are generally linked to the GIS through georeferencing. A DSS provides a choice to perform statistical analysis and present results in a variety of ways (Fig. 21.4).

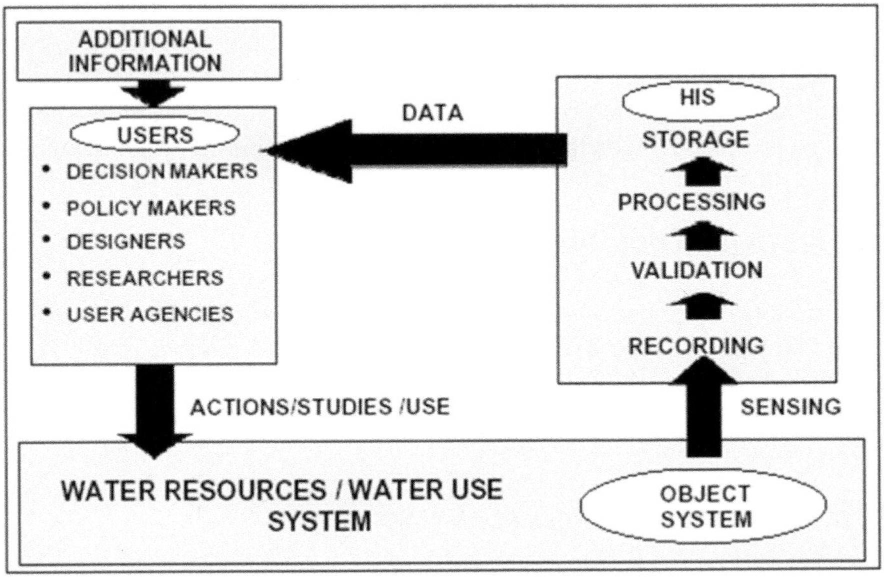

FIGURE 21.4 Components of hydrological information system.

21.6.3 GROUNDWATER POLLUTION CONTROL

A 2-D groundwater flow and contaminant transport model can be used to model the pollutant movement. Depending on the availability of required data and specific need, a 3-D model can be used. A typical DSS can deliver a procedure in three steps (1) locating potential sites for scavenger boreholes, (2) evaluating the performance of different arrangements of boreholes, and (3) choosing a cost-effective solution for prescribed contamination level.

21.6.4 SURFACE WATER POLLUTION CONTROL

A DSS generally has a 1-D model for river water-quality. The DSS should also have a waste-load allocation module. The modules should be able to ascertain a cost-effective solution.

21.6.5 DEMAND FORECASTING

CROPWAT model can be irrigation scheduling and demand forecasting. Data required for the CROPWAT model can be accessed from the GIS module. The module provides the alternate options of selecting different cropping patterns for optimal use of water. Furthermore, it should be able to provide various scenarios for improving irrigation efficiency. This module should also incorporate methods for estimating municipal/domestic water demands.

21.6.6 HYDROLOGIC PROCESSES

This module comprises of a rainfall-runoff model for simulating flow series. Depending on the type of rainfall-runoff model incorporated in the DSS, spatial and temporal data requirements vary. Various spatial and/or temporal data required by the model can be retrieved from the GIS module.

21.6.7 WATER RESOURCES PLANNING

This module generally runs on daily, weekly, 10-daily or monthly time-steps depending on the planner's requirement. This module generates different scenarios for water resources planning in accordance with available of water and projected demands.

21.7 HYDROLOGICAL INFORMATION SYSTEMS AND DATA PROCESSING

Information related to available water resources is most important for any water resources planning DSS. As the water demand by various sectors is increasing with growing population. Therefore, advances in water resources planning are dependent on the quality of the measured hydrological variables. This can be achieved through a well-developed hydrological information system (HIS). An HIS has various physical interface for the collection, processing, storing, and disseminating the hydrological and hydrometeorological data. In HIS, the physical infrastructure contains metrological and hydrological networks, data processing centers, standard laboratories, data storage and retrieval system and data communication system. It also has well-trained staff in the field of data collection activities, data processing, and analysis.

The HIS activities are categorized as:

- Users need assessment.
- Optimal hydrological network.
- Time series of historical hydrological data.
- Collection and management of spatial and temporal data.
- Data processing, analysis, and reporting.
- Data transfer, exchange, storage, and dissemination.
- Creating connections between HIS and DSS.

21.7.1 HYDROLOGICAL DATA PROCESSING

All the collected hydrologic data are to be necessarily processed before using in hydrologic analysis. It is important to note that the processing of the collected hydrological is to be performed before taking up any hydrological study. Data processing is classified as (1) preliminary processing, and (2) secondary data processing. Before using the hydrological data for any analysis, its processing is must. The objectives of hydrological data processing are (1) to evaluate the hydrological data accuracy, (2) to prepare the hydrological data in a useful form. Computerized data processing provides an opportunity to handle large amount of data and to process it timely. Most of the hydrometeorological data are observed from manual instruments. However, a few of the data are observed using automatic instruments. Hydrological data processing can be carried out using data processing software. Various data stored in the HIS are classified as: Time-oriented data: time series data of meteorology, hydrology, climatic, and water quantity.

- Space-oriented data: catchment boundary and other features and water resources infrastructure.
- Location-oriented data: Data related to the location of observation stations and hydraulic structures.

Now more accurate data processing is possible with the use of data processing software. HIS has the following advantages:

- Promotes the standardization procedures for hydrological data processing, validation and reporting.
- Storage, processing, and analysis of very large amounts of time series data.
- Large data processing, storage, and dissemination.
- Dissemination of data in tables and graph.

21.7.2 HYDROLOGICAL DATA PROCESSING

Data processing is a recent terminology, although this process is not new. The activity of data processing includes collection and recording of facts/figures related to various events and present the collected data in a meaningful information.

Hydrological data processing has various steps, that is, recording, validation, process, storage, and finally in the form of hydrological information. The processed data serve as the input to various hydrological modeling and decision support systems. The processed data improve the modeling results and thus the decisions are taken related to planning and management of water resources.

21.7.3 DATA AND INFORMATION

The term "data" denotes raw facts and figures that are generated by and could record an event or an activity or a situation. Sources of data may be personal, business, trading, technical, scientific, and legislative. Sources may be internal as well as external to the organization. Such data when procured provide different classes, namely, operating data, reference data, control data, and management data. These are isolated facts and figures, which may not mean anything, in particular, but may be transferred into meaningful information.

Data are defined as "things known" and information as "items of knowledge". Data are used to generate information. Information consists of selected data. Data are selected and organized with respect to user, problem, time, place, and the function. The conversion of data to information is the primary function of data processing.

21.7.4 HYDROLOGICAL DATA PROCESSING CYCLE

The hydrological data processing has a cyclic form. In the hydrological data processing cycle, the first step is to collect hydrological data processing. The hydrological data collected from various instruments/gauges serve as the input for the second step, that is, preliminary hydrological data processing. In the third step, secondary data processing is carried out. In the next step, the processed data are stored for analysis and retrieval. After that information is generated from the processed data. Now on the basis of the feedback appropriate correction/modification are applied. Again, the revised system is operated for generating accurate and reliable information. The data processing life cycle activities are illustrated in Figure 21.5.

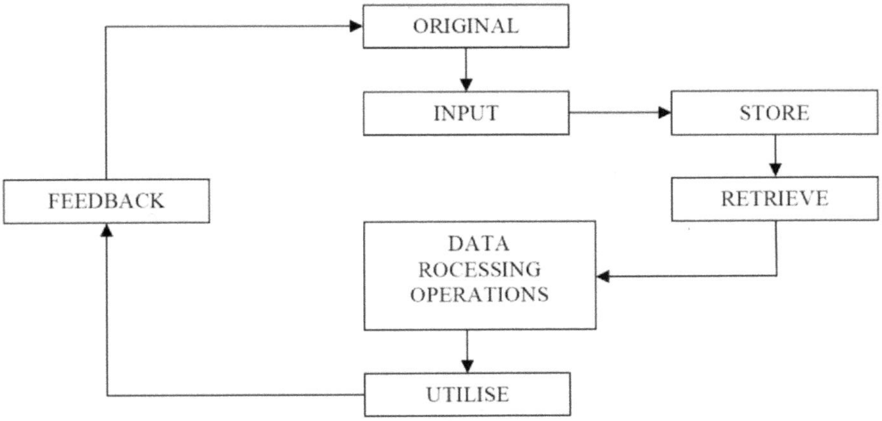

FIGURE 21.5 An illustration of data processing life cycle activities.

21.8 DATA ACQUISITION SYSTEM

The data acquisition system provides a facility to collect generic data. Generally, the time series data are collected by manual or automatic sensors and spatial as well as metrological data by remote sensors. Some of the data are also collected by surveys, interviews, and reports. Data acquisition system consists of standard gauges and optimal network for the collection of the hydrometrological and spatial data. Illustrative example has been shown in Figure 21.6.

21.9 WATER RESOURCES PLANNING DECISION-MAKING

A DSS provides the water management authorities a user-friendly system so that a decision maker can take appropriate decision with its support. Various components/models described in the previous section become a part of the planning for DSS. A planning for DSS provides a comparison of different strategies and scenarios and presents the results in both in tabular/graphical form and in GIS platform so that a quick and reliable decision can be taken.

DSS is an arrangement of various tools through which one can carry out hydrologic data analysis, simulate runoff, generate water allocation plan using allocation models, and analyze various strategies and scenarios for making the appropriate decisions. A DSS can help in planning and managing the water resources in a river basin so that future demand can be fulfilled through supply/or demand management. The DSS has capability to simulate various scenarios for operation alternatives, cropping pattern, water use efficiency, and surface water–groundwater conjunctive use.

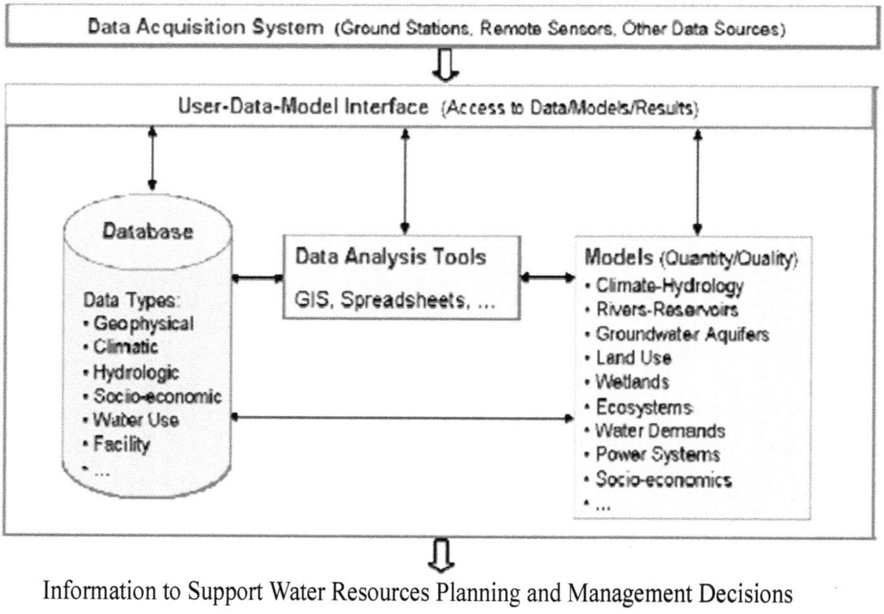

FIGURE 21.6 Data acquisition and data analysis module.

The DSS for surface water planning should be flexible and user-friendly. The following sections highlight the characteristics of a DSS (Simonovic, 1996).

21.9.1 WATER MANAGEMENT PROBLEM IDENTIFICATION

The issue of sustainable management of water resource is defined by semi and nonstructured problems. A well formulated management problem is defined as structured problem. Decisions in this case are straightforward because alternative solutions are known. A semistructured problem consists of inadequate data, nonquantifiable variables, and complexity. Problem identification is one of the basic step for developing an effective and useful DSS. Water management problem identification consists of decision-making domain for solving the anticipated problems in water resources supply and management in future.

21.9.2 PROBLEM FORMULATION (LEARNING)

The term "problem" related to sustainable consists of problem as (1) objective reality or (2) mental construct (Landry et al., 1985). The problem for decision-making is to be defined objectively by the stakeholder. Further, everyone involved in the decision-making perceives it in the same manner. In the case of problem as a mental construct, it must be considered as subjective. Problem formulation is a very important task in decision-making, and it is not a technical process but a social process.

21.9.3 ADAPTABILITY

A DSS allows a decision maker to generate different scenarios. It helps in establishing the areas of uncertainty and also helps in analyzing the sensitivity of the results. Thus, it supports a decision maker in selecting a most optimal solution for high-quality decision-making.

21.9.4 FACILITATION

DSS for water-resources planning should be developed on GIS platform with plugging to a number of modeling techniques and of evaluation tools. It should have a user-friendly interface for allowing scenario analysis by multidisciplinary stakeholders.

21.9.5 INTERACTION

A DSS user can interact with it in its own way. This functionality is very important as the DSS is used by multidisciplinary stakeholders

21.10 DATA MANAGEMENT AND PLANNING MODELS

A typical surface water planning DSS includes various modules, which require wide range of data related to river network, hydrology, hydropower facilities, water use, and reservoir regulation. Data management models provide facility to view, add/modify the data. It has a user-friendly interface which represents river network by reaches, junctions, nodes, and reservoirs with associated attributes. In the river network local inflow, return and/or water withdrawals locations are represented by river nodes. River segments of the natural rivers are represented by river reaches and are associated with attributes related to their hydrological/hydraulic characteristics. Man-made or natural lakes are represented as reservoir associated with attributes, such as water supply for various uses, environmental flow, and flood control, etc.

21.10.1 MODELS

Data management and planning model has a combination of models for rainfall-runoff modeling, flow forecasting, simulation at the basin nodes and reservoir operation and management at the reservoir nodes. After analyzing the seasonal and long-term variability in streamflow data, future flow series are generated by hydrologic or stochastic models.

21.10.1.1 HYDROLOGIC MODELING

Hydrologic models may be applied for simulating runoff by considering precipitation and other meteorological variables as input. Flow series are generated and then aggregated to weekly, ten daily or monthly, depending upon the time-steps required for planning purpose. These models may be applied for small basins to large basins. Various features of this module are given below:

- Input data for hydrological models are classified as rainfall, temperature, and/or evapotranspiration. G and D data are required at basin outlet for model calibration/validation.
- Output of the model is defined as flow series at the basin outlet.
- After dividing a large basin in various subbasins, a network model may be developed for the simulation of stream flow.

21.10.1.2 CHANNEL AND RESERVOIR ROUTING MODEL

In DSS, a channel and reservoir routing module is used with rainfall-runoff network module to route water through the channels and reservoirs so as to simulate runoff at outlet. Hydrologic or hydraulic routing methods are used depending on the available information.

21.10.1.3 RESERVOIR MANAGEMENT MODEL

This model is used to determine the reservoir release sequences in order to fulfil the various demand, for example, irrigation, municipal, industrial, hydropower, etc. Besides meeting the demand targets, the release sequence should be able to manage water during the hydrological extremes, for example, floods and droughts.

21.10.1.4 SEASONAL PLANNING

A major portion of the water is being utilized by the agricultural sector. Keeping in view the importance of agriculture, seasonal planning for irrigation in command area is very much needed. This model should be capable to develop irrigation strategies for drought periods, enhancing cropped areas and considering change in cropping pattern.

21.10.1.5 GIS AND REMOTE SENSING

Besides spatial information of the basins, such as DEM, LULC, soil etc., remote sensing can also provide rainfall information. Generally, the collection of the rainfall using rain gauges is limited to certain extent in the basin. However, the satellite-based rainfall information can provide data uniformly for the whole basin, and is now becoming more popular due to its increasing accuracy. The DSS model can have a connectivity with the satellite data available online and field measured data for reliable estimates.

21.11 SURFACE WATER PLANNING DECISION SUPPORT SYSTEM DATA REQUIREMENTS

21.11.1 RIVER NETWORK

- River network configuration (principal river reaches their tributaries, and water resources projects).

- System node locations: System nodes include major tributary confluences; river locations where reliable flow records are available; sites of existing and planned reservoirs, barrages, or other hydraulic works (such as diversions); locations of water supply withdrawals (for agricultural, domestic, or industrial use); and flood prone or environmentally sensitive areas.

21.11.2 RIVER HYDROLOGY

- Streamflow records should be done at weekly, 10 day, or monthly time resolution at system nodes; streamflow records should preferably be unimpaired, indicating streamflows that would have been realized in the absence of upstream withdrawals and returns; as a minimum, streamflow records should be available for at least 20 years.
- Estimated streamflow losses between system nodes.

21.11.3 RESERVOIRS

- Year of construction and reservoir filling.
- Reservoir storage–area–elevation curves.
- Reservoir storage ranges.
- Rate of siltation, hydrographic survey details carried out time to time.
- Spillway details.
- Operating rules.
- Evaporation rate and other metrological data.
- Reservoir inflows series.

21.11.4 HYDROPOWER FACILITIES

- Penstock hydraulic loss functions.
- Penstock and outlet leakage.
- Number and type turbines and operational ranges.

21.11.5 WATER USE

- Water requirements.

- Historical water use.
- Return flows as percentage of water withdrawals.
- Future water demand.
- In-stream flow requirements (for pollution abatement and ecological purposes) at system nodes, if any.
- Flood and drought prone areas.

21.12 CONCLUSION

The planning and management of water resources is an important task due to increasing water demands by various sectors. In order to develop a proper water management plan, it is important to split the problem in supply side and demand side management. Further, for this, a proper assessment of available water resources and their changes in future due to land use and climate change must be analyzed. After analyzing the spatial and temporal data and developing various hydrological models for the simulation of the future runoff data series and water resources management, it is important to develop a decision support system (DSS) for the planning and management of water resources. Such DSS may help decision makers in taking quick and reliable decisions. In view of the climate change, more urbanization changes in the water demand patterns, and the use of the DSS should be encouraged and more research is required in this area for suggesting methodologies which utilize the advanced techniques for decision-making.

KEYWORDS

- **decision support system**
- **surface water**
- **planning**
- **modeling**
- **GIS**

REFERENCES

Airoldi, L.; Abbiati, M.; Beck, M. W.; Hawkins, S. J.; Jonsson, P. R.; Martin, D., Åberg, P. An Ecological Perspective on the Deployment and Design of Low-Crested and Other Hard Coastal Defence Structures. *Coastal Eng.* **2005,** *52* (10–11), 1073–1087.

Andreu, J.; Capilla, J.; Sanchís, E. AQUATOOL, a Generalized Decision-Support System for Water-Resources Planning and Operational Management. *J. Hydrol.* **1996,** *177* (3–4), 269–291.

Caratti, P.; Dalkmann, H.; Jiliberto, R. (Eds.). *Analysing Strategic Environmental Assessment: Towards Better Decision-Making*; Edward Elgar Publishing, 2004.

Catelli, C.; Pani, G.; Todini, E. FLOODSS: FLOod Operational Decision Support System. In *Hydroinformatics' 98. Proceedings of the Third International Conference on Hydroinformatics*; Copenhagen, Denmark, August 24–26, 1998; Volume 1, pp 255–262.

Cheng, H.; Hu, Y. Improving China's Water Resources Management for Better Adaptation to Climate Change. *Clim. Change* **2012,** *112* (2), 253–282.

Cubillo, J. M.; Sánchez, J.; Cerviño, J. International Students' Decision-Making Process. *Int. J. Edu. Manag.* **2006.**

da Silva, L. M.; Park, J. R.; Keatinge, J. D. H.; Pinto, P. A. II. The Use of the DSSIPM in the Alentejo Region of Southern Portugal. *Agric. Water Manag.* **2001,** *51* (3), 203–215.

Di Baldassarre, G.; Sivapalan, M.; Rusca, M.; Cudennec, C.; Garcia, M.; Kreibich, H.; Blöschl, G. Sociohydrology: Scientific Challenges in Addressing the Sustainable Development Goals. *Water Resour. Res.* **2019,** *55* (8), 6327–6355.

Dunn, S. M.; Mackay, R.; Adams, R.; Oglethorpe, D. R. The Hydrological Component of the NELUP Decision-Support System: An Appraisal. *J. Hydrol.* **1996,** *177* (3–4), 213–235.

Fedra, K. Models, GIS, and Expert Systems: Integrated Water Resources Models. In *Applications of Geographic Information Systems in Hydrology and Water Resources Management. Proc. International Conference*, Vienna, 1993; No. 211, pp 297–308.

Fedra, K.; Jamieson, D. G. The 'WaterWare' Decision-Support System for River-Basin Planning. 2. Planning Capability. *J. Hydrol.* **1996,** *177* (3–4), 177–198.

Guswa, A. J.; Brauman, K. A.; Brown, C.; Hamel, P.; Keeler, B. L.; Sayre, S. S. Ecosystem Services: Challenges and Opportunities for Hydrologic Modeling to Support Decision Making. *Water Res. Res.* **2014,** *50* (5), 4535–4544.

Jain, S. K.; Chalisgaonkar, D.; Bhunya, P. K.; Kumar, V. 118-Decision Support System for Integrated Reservoir Management in India—Need, Concept and Framework, 2009.

Jamieson, D. G.; Fedra, K. The 'WaterWare' Decision-Support System for River-Basin Planning. 1. Conceptual Design. *J. Hydrol.* **1996,** *177* (3–4), 163–175.

Keenan, P. Using a GIS as a DSS Generator. University College Dublin, Department of Management Information Systems, 1995.

Kumar, R.; Singh, R. D.; Sharma, K. D. Water Resources of India. *Curr. Sci.* **2005,** 794–811.

Lal, M. *Climatic Change–Implications for India's Water Resources. Climate Change and Water Resources in South Asia*; Monirul Qader Mizra, M., Ahmad, Q.K., Ed.; Taylor & Francis: Leiden, the Netherlands, 2005; pp 155–196.

Landry, M.; Pascot, D.; Briolat, D. Can DSS Evolve without Changing Our View of the Concept of 'problem'?. *Dec. Support Syst.* **1985,** *1* (1), 25–36.

Nicklow, J.; Reed, P.; Savic, D.; Dessalegne, T.; Harrell, L.; Chan-Hilton, A.; ASCE Task Committee on Evolutionary Computation in Environmental and Water Resources

Engineering. State of the Art for Genetic Algorithms and beyond in Water Resources Planning and Management. *J. Water Resour. Plann. Manag.* **2010,** *136* (4), 412–432.

Nyimbili, P. H.; Erden, T. Spatial Decision Support Systems (SDSS) and Software Applications for Earthquake Disaster Management with Special Reference to Turkey. *Nat. Haz.* **2018,** *90* (3), 1485–1507.

O'Callaghan, J. R. NELUP: An Introduction, 1995.

Pate, R. C.; Hightower, M. M.; Cameron, C. P.; Einfeld, W. Overview of Energy-Water Interdependencies and the Emerging Energy Demands on Water Resources (No. SAND2007-1349C). Sandia National Lab. (SNL-NM), Albuquerque, NM (United States), 2007.

Raziei, T.; Saghafian, B.; Paulo, A. A.; Pereira, L. S.; Bordi, I. Spatial Patterns and Temporal Variability of Drought in Western Iran. *Water Resour. Manag.* **2009,** *23* (3), 439–455.

Simonovic, S. I. Decision Support Systems for Sustainable Management of Water Resources: 1. General Principles. *Water Int.* **1996,** *21* (4), 223–232.

Sinha, D.; Syed, T. H.; Reager, J. T. Utilizing Combined Deviations of Precipitation and GRACE-based Terrestrial Water Storage as a Metric for Drought Characterization: A Case Study over Major Indian River Basins. *J. Hydrol.* **2019,** *572*, 294–307.

Sprague Jr, R. H.; Carlson, E. D. *Building Effective Decision Support Systems*; Prentice Hall Professional Technical Reference, 1982.

CHAPTER 22

Optimal Multiobjective Multireservoir Operation Using Interactive Decision Making by Parametric Approach

R. U. KAMODKAR[1*], J. B. GURAV[2], and D. G. REGULWAR[3]

[1]*Department of Civil Engineering, Government Polytechnic, Jalgaon 425001, Maharashtra, India*

[2]*Department of Civil Engineering, Amrutvahini College of Engineering, Sangamner 422608, Maharashtra (Savtiribai Phule Pune University, Pune) India*

[3]*Department of Civil Engineering, Government College of Engineering, Aurangabad 431005, India*

[*]*Corresponding author. E-mail: ravikamodkar@gmail.com*

ABSTRACT

In this study, the multipurpose multireservoir operating model is built, and its applicability is demonstrated through the case study of multireservoir network in a subbasin of the Godavari river, Maharashtra state, India. The present technique offers the practical solution approach to incorporating fuzzy parametric programming (FPP) and fuzzy linear programming (FLP). The model considered all reservoir operation parameters to be fuzzified simultaneously. Exponential membership functions define the objective function coefficients and resources, and the technological coefficients are defined by a linear membership function, which is simultaneously fuzzified over the tolerance range. The interactive concept of the procedure is carried

Advances in Hydrology and Climate Change: Historical Trends and New Approaches in Water Resources Management. Surendra Kumar Chandniha, Anil Kumar Lohani, Gopal Krishan, & Ajay Krishna Prabhakar (Eds.)

out according to decision makers choice to find the optimum solutions for both the objective functions from various grades of precision (μ). The degree of satisfaction (λ) obtained corresponding to ($\mu = 0$) which is equal to 0.893; however, the corresponding ($\mu = 1$) to is equal to 0.65. The study included the knowledge, information and preferences of the decision makers acquired through interactions. Hence, all solutions are optimal and effective at any level of precision.

22.1 INTRODUCTION

To optimize benefits, an efficient management of the available water resources in multiobjective reservoir operations is needed. An optimal reservoir release policy should be obtained for a year to accomplish this, and the water released should be allocated optimally among all the objectives. Management of the water resources system is an important work in framing various important policy decisions regarding the planning, design, operation and maintenance of such systems. To identify, define, verify, review, and execute efficient and effective operational plans, the data collection and workflow strategies for each water supply network have their own structure. In order to satisfy the future water requirements for various beneficial uses, the projected gaps between total demand of water and supply of water must be significantly reduced.

It is not only essential to deal with the uncertainty in the hydrological variables but also to tackle the complicated social and economic systems with different interest groups having multiple and conflicting objectives, since it makes the implementation of policies of water resources management difficult. The reservoir operating policy model needs to consider variation in hydrological variables, such as rainfall, inflows in the reservoir, reservoir storage capacity, evaporation, and soil moisture. Some kinds of uncertainties occur not only because of randomness but also because of inaccuracy, subjectivity, vagueness, and lack of adequate data. The random nature of inflow to reservoirs, inaccurate stream flow data, and insufficient storage and release data when assessing the reservoir's water balance and operational planning and imprecise water demand values are important factors in simulating estimates of water usage value in a river basin.

The reservoir operation model for irrigation is associated with the various hydrologic uncertainties with reference to the rainfall, inflow, crop water requirement, and evaporation which make the problem formulation more complicated. This may be due to the climate change, land use pattern, technology changes, and the living standard of the people. Further, the multiple and conflicting nature of objectives of variable demands and supply makes problem more complicated. The fuzzy sets can be used to represent the uncertainty associated with these types of variables. The principle of fuzzy sets gives a systematic aid in dealing with any imprecision.

Fuzzy logic is also the most suitable methodological tool to model reservoir operations as the methodology is more versatile and enables expert viewpoints to be integrated, which would make the methodology more reasonable for operators. The traditional optimizing techniques cannot match the field conditions. Fuzzy system modeling takes into account the fuzziness in the variables and the model parameters that can be applied in the field.

22.2 LITERATURE SURVEY

After Zadeh (1964) Fuzzy set was introduced and a number of authors have further developed the applications of fuzzy set theory. Fuzzy set theory can be efficiently used to deal vagueness while planning multiple objectives involving imprecision in the various parameter values. The history of the Fuzzy set theory and Fuzzy number is available in several books like the books byDubois and Prade (1980), Klir and Yuan (1995), Zimmermann (1996), Fuller (1998), Ibrahim (2003), and Sivanandam et al. (2007). A two-phase method for solving multiobjective fuzzy linear programming (FLP) problems is introduced by Feyzan and Zulal (2007). The technique offers a realistic platform to a solution, which is an application of fuzzy parametric programming (FPP) and FLP to solve multiobjective programming models with all real-life coefficients fuzzy. The all-objective functions are optimized simultaneously by using the procedure of interactive concept for different grades of precision according to the decision maker (DM) preferences.

Moorthi et al. (2018) discussed water resource regulation based on fuzzy "if-then" rules that link concepts of interpolative reasoning, logical implications, and some inference tools to infer knowledge about a water resource system using linguistic descriptions. Moeini et al. (2011) implemented a fuzzy rule-based model for hydropower reservoir operations. The proposed fuzzy rule-based model provides a set of appropriate operating rules based on optimal or target storage levels for release from the reservoir. The infeasible conditions and curse of dimensionality are two significant challenges in the case of stochastic dynamic programming method. Saadat and Asghari (2017) have suggested a new approach for avoiding unfeasible conditions and increasing the performance of the solution with modern discretization technique. To tackle with, an optimization module is integrated into the standard structure of stochastic dynamic programming (SDP), so that the near optimum values of the state variables are calculated based on the constraints available. Sangeeta and Mujumdar (2015) developed a fuzzy stochastic dynamic programming (FSDP) model which considered the inflow in the reservoir as a stochastic variable while reservoir storage and soil moisture are treated as a fuzzy numbers. The objective of the study is to minimize the crop yield deficits which result in optimal water allocations to the crops while maintaining the soil moisture balance and storage continuity. To find the optimum surface and groundwater withdrawal, a linear fuzzy optimization model is developed by Milan et al. (2018). The results of this model are used to develop the fuzzy inference system (FIS) to assess the withdrawal of groundwater automatically.

Safavi et al. (2013) developed operating rules for the Zayandehrood reservoir, Iran, using the fuzzy inference method and adaptive neuro-fuzzy inference method from a combined surface and ground water resources perspective. The aim is to decide the proportion of reservoir Zayandehrood in order to meet downstream water demands. The uncertainties in the objectives and the various parameters of the reservoir optimization model are addressed by the Fuzzy Set Theory using the membership function by Kamodkar and Regulwar (2010). Practical swarm optimization (PSO) has been introduced by Saberchenari et al. (2016) to solve the multipurpose Mahabad Dam Reservoir operation problem in the northwest Iran with the objective of minimization of the difference between the monthly downstream demand and releases.

Choong et al. (2017) developed the Artificial bee colony (ABC) model to solve the problem of optimizing the Chenderoh Reservoir project, Perak State, Malaysia. The objective of the algorithm was to minimize the water scarcity in the reservoir system and to observe its performance obtained from month wise and week wise input data. An improvement in the reservoir operation was achieved by using the evolutionary algorithm named Borg MOEA by AI-Jawad and Tanyimboh (2017). The algorithm is implemented to the real-world reservoir to test its performance. The algorithm is aimed to obtain the operating policy of the reservoir to satisfy the downstream water demands in water scarcity while maintaining the sufficient storage in it. A monthly operation release policy is obtained by Moreira and Celeste (2016) by using Implicit Stochastic Optimization (ISO). This model uses the prediction of the mean inflow instead of the current monthly inflows to the reservoir. The optimization problem can be solved by using the fuzzy multi-objective goal programming (FMGP). The FMGP model is developed by

Srivastava and Singh (2017) to optimize the cropping pattern by optimizing the land allocated for cultivation.

Gaur et al. (2015) has developed a multiobjective fuzzy optimization model. The particle swarm optimization (PSO) is used to achieve upper and lower bounds of each objective function. The sustainable groundwater management policy is obtained by using the PSO and the analytical element method (AEM) by considering three conflicting objectives. Kamodkar and Regulwar (2014) developed the multipurpose single reservoir operation model by treating model parameters as a triangular fuzzy number distribution. The Jayakwadi Project Stage-2 in the State of Maharashtra, India is used to observe the model performance.

Based on the aforementioned literature, the present study aims to establish the reservoir operation model for optimum reservoir operation under a linear programming system under a fuzzy environment. The model is applied to the case study of multireservoir system in the subbasin of the Godavari River, Maharashtra state, India. A two-phase method to solve multiple objective, fuzzy linear programming models is demonstrated in the next paragraph. The methodology provides a realistic way that integrates the FPP and FLP to solve multiobjective programming models with all fuzzy coefficients in real-life.

22.3 THE FUZZY LINEAR PROGRAMMING PROBLEM WITH FUZZY COEFFICIENTS OF OBJECTIVE FUNCTIONS, FUZZY TECHNOLOGICAL COEFFICIENTS AND FUZZY RESOURCES ($\tilde{C}_j$, $\tilde{a}_{ij}$ and $\tilde{b}_i$)

In this section, the methodology for the simultaneously optimizing the all-objective functions according to the DM preferences for the different grades of precisions is demonstrated. The steps of the methodology are stated below. All the steps of the suggested methodology are summed up in main two stages. The FPP approach proposed in stage-I to solve single-objective mathematical model is applied to solve multiobjective mathematical model. Then in stage-II, the FLP approach is used to solve the FPP multiobjective model for varying level of precisions.

The FPP approach is used to express the model of the vector optimization in each precision level and it contains the fuzziness for various executable levels. FLP is used at each level of precision for simultaneous objective functions optimization which assures an efficient solutions. Also, in the present methodology, the DM's preferences are considered before the optimization which implies that the DM and analyst are different persons. The FLP model is formulated by using the maximum and minimum values (upper and lower bounds) obtained from the solution of the vector optimization model, so the solutions are not depending on the DM. However, the relative flexibility in the model is achieved by using the FPP method.

22.4 THE STEPS OF THE METHODOLOGY PROPOSED ARE EXPLAINED BELOW

22.4.1 FORMULATION OF THE MULTIOBJECTIVE MATHEMATICAL MODEL

The mathematical expression is given by eq 22.1 in which $f_k(x)$ is the kth objective function, the objective function value is z_k, however, c_k, x, b and A represent $m \times n$-sized matrix.

$$\begin{array}{lll} Max(Min) & \{z_k = f_k(x) = c_k x\}, & p = 1,2,\ldots,P \\ s.t. & A_x \leq, \geq, = b, \; x \geq 0. & \end{array} \quad (22.1)$$

22.4.2 FUZZY PARAMETERS DETERMINATION

The imprecise parameters are decided by the DM and/or analyst. The present methodology is worth considering when all the parameters or only A and c or A and b of the model are treated as a fuzzy, and it is beneficial. It may be noted that when all the parameters or A and b, A and c or only A are fuzzy in the model, the present method gives realistic solutions.

22.4.3 DEVELOPMENT OF MEMBERSHIP FUNCTION

As per the FPP methodology, the completely trade-off membership functions are formulated in linear, nonlinear, and/or piece-wise linear form as required by the imprecise parameters.

22.4.4 FORMULATION OF MULTIOBJECTIVE DECISION PROBLEMS FOR EACH MEMBERSHIP PRECISION LEVEL

By using developed membership functions, the FPP models are formulated for each precision level for completely trade-off membership (μ = (μ_{ck} *or*

$\mu_{zk}) = \mu_A = \mu_b)$ Each model represents the problem of vector maximization. Each fuzzy parameter in the model is represented by membership function. It is assumed that the linear membership function is associated with a_{ij}, the concave membership function is associated with b_i and convex membership function is associated with c_k parameters, that is, technological coefficients, resources and objective functions coefficients. These membership functions are represented graphically by Figures 22.1–22.3.

where

$$A = [a_{ij}], b = [b_i], c_k = [c_{kj}], i = 1, 2, ...m \ \ and\ j = 1, 2, n\ and\ k = 1, 2.$$

$$\begin{aligned} \mu_{aij} &= (a_{ij} - a_{ij}^1) / (a_{ij}^0 - a_{ij}^1), \\ \mu_{bi} &= 1/(1 - \exp(-0.8))[1 - \exp(-0.8(b_i - b_i^1)/(b_i^0 - b_i^1))] \\ \mu_{ckj} &= 1/(1 - \exp(3))[1 - \exp(3(c_{kj} - c_{kj}^1)/(c_{kj}^0 - c_{kj}^1))] \end{aligned} \tag{22.2}$$

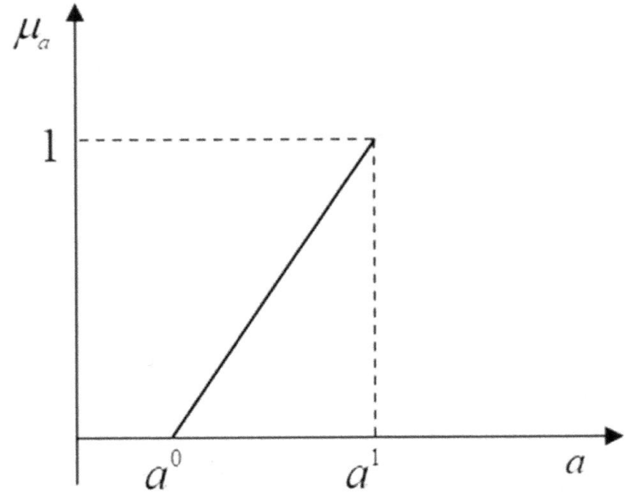

FIGURE 22.1 Linearly increasing membership function for technological constraint coefficients "Risk-free" value "Unimplementable" value.

The FLP model is developed from the membership function as follows:

$$\begin{aligned} &\text{Max } Z_k = [c_k^1 + (1/3)\text{In}[1 - \mu_{ck}(1 - \exp(3))](c_k^0 - c_k^1)]x,\ k = 1, 2 \\ &[A^1 + \mu_a(A^0 - A^1)]x \le b^1 + (1/(-0.8)In[1 - \mu_b(1 - \exp(-0.8))](b^0 - b^1), \\ &\forall i, j \text{ and } \mu \text{ level} \end{aligned} \tag{22.3}$$

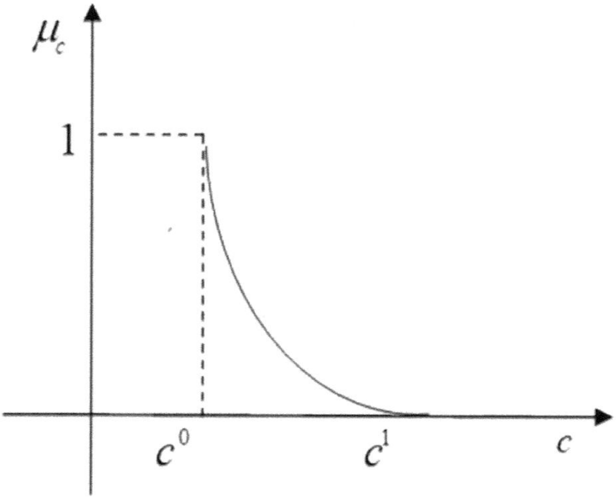

FIGURE 22.2 Membership function for objective function coefficients "Unimplementable" value "Risk-free" value.

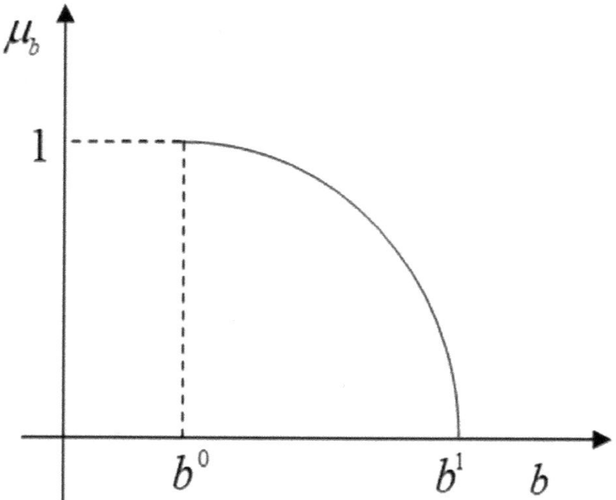

FIGURE 22.3 Membership function for right hand sides "unimplementable" value "risk-free" value.

22.4.5 USE IDEAL SOLUTIONS TO OBTAIN FUZZY INTERVAL

For each precision level or some preferred levels, the ideal solutions of the vector maximization problems are obtained. Then, the corresponding value of each objective obtained from each solution is determined, and the payoff matrix is formulated. Finally, the maximum (u) and the minimum value (l) for each objective corresponding to the solutions set are determined.

22.4.6 FUZZY LINEAR PROGRAMMING SOLUTIONS

The FLP models are constructed to simultaneously optimize the objectives by using the $[u,l]$ values of each objective for each vector maximization problem (of which constructed for each precision levels) given by eq 22.4.

$$\begin{aligned}
&\text{Max}\,\lambda,\\
&\lambda \le (z_k - l_k)/(u_k - l_k), \qquad k = 1,2,\ldots,k\\
&[A^1 + \mu_a(A^0 - A^1)]x \le b^1 + (1/(-0.8)\text{In}[1 - \mu_b(1 - \exp(-0.8))](b^0 - b^1), \quad (22.4)\\
&0 \le \lambda \le 1,\\
&x \ge 0.
\end{aligned}$$

22.4.7 SOLUTION PRESENTATION TO THE DECISION MAKER

Models represented by eq 22.4 are solved by using the FLP model and solutions are presented to the DM. The alternative decision plan is obtained for each precision level. For DM, if the alternative decision plans are not satisfactory, then model must be solved again for different precision levels from step 2 or 3.

22.5 RESERVOIR SYSTEM AND RESERVOIR OPERATION MODEL

The methodology explained under Section 22.3 is implemented to the physical system under consideration comprises of four reservoirs, R_1 (i.e., Jayakwadi stage-I) on Godavari river, R_2 (i.e., Jayakwadi stage-II) on Sindaphana river, R_3 and R_4 (i.e., Yeldari, and Siddheshwar) on Purna river, in the Maharashtra State, India. The schematic representation of this basin is shown in Figure 22.4.

22.6 DEVELOPMENT OF RESERVOIR OPERATION MODEL FOR MULTIOBJECTIVE MULTIRESERVOIR SYSTEMS

In the present section, a fuzzy multiobjective multireservoir operation model is formulated to obtain the optimal release policy for optimum use of available water in the subbasin of river Godavari.

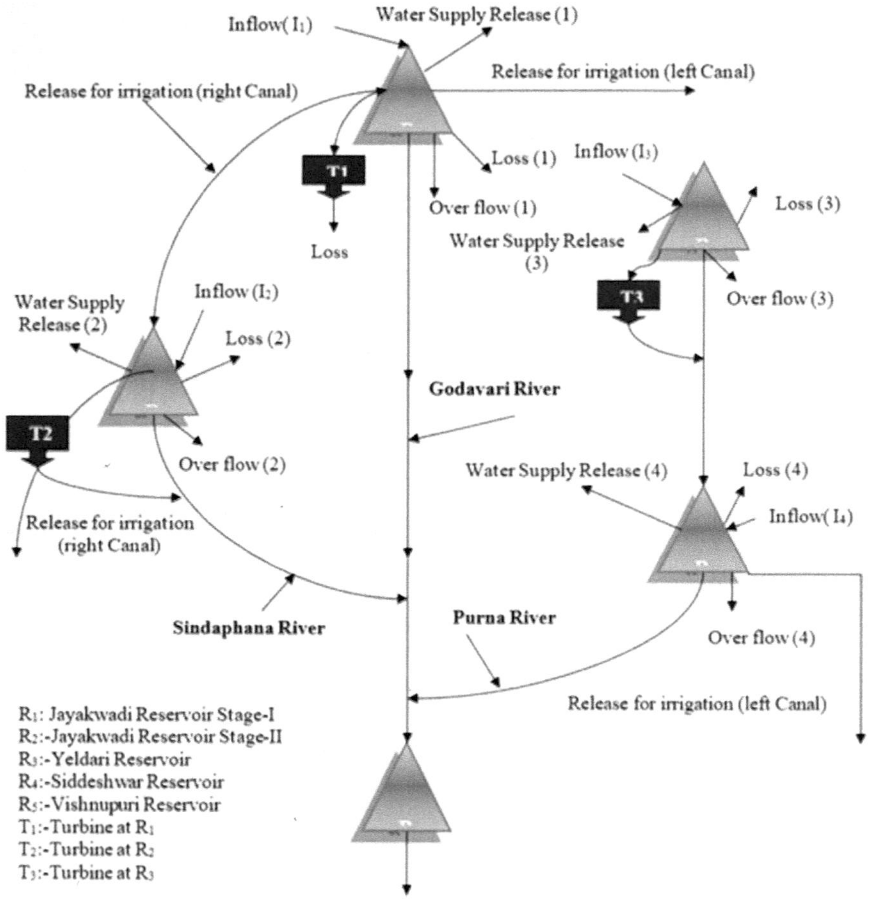

FIGURE 22.4 Schematic representation of the physical system.

22.6.1 OBJECTIVE FUNCTIONS

In the present study, following objectives are considered

1. Maximization of releases for irrigation (i.e., RIRN).
2. Maximization of releases for hydropower generation (i.e., RHPG).

$$\text{Maximize } Z_1 = \sum_{m=1}^{4}\sum_{n=1}^{12}(\text{RIRN})_{mn} \tag{22.5}$$

$$\text{Maximize } Z_2 = \sum_{m=1}^{4}\sum_{n=1}^{12}(\text{RHPG})_{mn} \tag{22.6}$$

where

$m = 1, 2, 3, 4$ (number of reservoirs, i.e., 4 reservoirs) and n = time interval (i.e., 12 months).

22.6.2 CONSTRAINTS

22.6.2.1 RELEASE FOR HYDROPOWER GENERATION–TURBINE CAPACITY CONSTRAINTS

For all the months, the water released for the hydropower generation through the turbine must be less than or equal to the maximum turbine capacities (TCMax). Similarly, the water released for the hydropower generation through turbine should be more than or equal to the water needed to generate the firm power (FPR) through turbine. This constrains are given as follows:

$$\mathrm{RHPG}(m,n) \le \mathrm{TCMAX}(m) \qquad \forall m = 1,2,\ldots\ldots,4. \tag{22.7}$$

$$\mathrm{RHPG}(m,n) \ge \mathrm{FPR}(m) \qquad \forall n = 1,2,3,\ldots\ldots\ldots,12. \tag{22.8}$$

22.6.2.2 RELEASE FOR IRRIGATION-DEMAND CONSTRAINTS

The water released in to the canal for the irrigation (RIRN) must be less than or equal to the maximum demand for irrigation (MAXID) for all reservoirs in all time period. Similarly, the water released in to the canal for the irrigation (RIRN) must be more than or equal to the lowest demand for irrigation (MINID). The irrigation release–demand constraint, can, therefore be written as:

$$\mathrm{RIRN}(m,n) \le \mathrm{MAXID}(m,n) \qquad \forall m = 1,2,\ldots\ldots\ldots,4. \tag{22.9}$$

$$\mathrm{RIRN}(m,n) \ge \mathrm{MINID}(m,n) \qquad \forall n = 1,2,3,\ldots\ldots\ldots,12. \tag{22.10}$$

22.6.2.3 STORAGE CAPACITY CONSTRAINTS

The reservoir storage (RS) in all the time periods must be less than or equal to the maximum storage capacity (SCMAX) and more than or equal to the minimum storage capacity (SMIN) of the reservoir. The storage available at the beginning of the month is taken as initial storage in the reservoir. The reservoir storage capacity constraints are given as follows:

$$\text{RS}(m,n) \leq \text{MAXSC}(m) \qquad \forall m = 1,2,......,4. \qquad (22.11)$$

$$\text{RS}(m,n) \geq \text{MINS}(m) \qquad \forall n = 1,2,3,........,12. \qquad (22.12)$$

22.6.2.4 HYDROLOGIC CONTINUITY CONSTRAINTS

These constraints are related to the releases for hydropower generation (RHPG), releases for irrigation (RIRN), constant domestic water supply (DWS) releases, storage in the reservoir (RS), inflows in to the reservoir (INF), and reservoir losses for all months. The reservoir losses are the function of the storage as given by Loucks et al. (1981).

Let A_o is water surface area of the reservoir with respect to the dead storage in the reservoir and e_n is rate of evaporation with respect to the time period n. A_a is the water spread area of the reservoir per unit active storage volume.

Then the actual rate of evaporation during the time period n is given by

$$\text{Loss due to Evaporation} = A_0 e_n + A_a e_n \frac{(S_n + S_{n+1})}{2}$$

$$\text{Put } a_n = \frac{A_a e_n}{2} = 0.5 A_a e_n \quad \text{then,}$$

$$\text{Loss due to Evaporation} = A_0 e_n + a_n (S_n + S_{n+1})$$

Then, the hydrologic continuity constraints can be written for all the reservoirs as

(i) Reservoir (R_1)

$$(1+a_n(1,n))\ \text{RS}(1,n+1) = (1-a_n(1,n))\ \text{RS}(1,n) + \text{INF}(1,n) - \text{RHPG}(1,n) - \text{RIRN}(1,n) - \text{OVF}(1,n) - \text{DWS}(1,n) - \text{FCR}(1,n) + \alpha_1 \text{RHPG}(1,n) - A_0 e_n(1,n)\ n = 1,2,3,..........,12. \qquad (22.13)$$

(ii) Reservoir (R_2)

$$(1+a_n(2,n))\ \text{RS}(2,n+1) = (1-a_n(2,n))\ \text{RS}(2,n) + \text{INF}(2,n) + \alpha_2\ \text{FCR}(1,n) - \text{RHPG}(2,n) - \text{RIRN}(2,n) - \text{OVF}(2,n) - \text{DWS}(2,n) - A_0 e_j(2,n) \qquad (22.14)$$

$n = 1,2,3,..........,12.$

(iii) Reservoir (R_3)

$$(1+a_j(3,n))\ \text{RS}(3,n+1) = (1-a_n(3,n))\ \text{RS}(3,n) + \text{INF}(3,n) - \text{RHPG}(3,n) - \text{OVF}(3,n) - \text{DWS}(3,n) - A_0 e_n(3,n) \qquad (22.15)$$

$\forall\, n = 1,2,3,..............,12.$

(iv) Reservoir (R_4)

$$(1+a_j(4,n))\ \text{RS}(4,n+1) = (1-a_n(4,n))\ \text{RS}(4,n) + \text{INF}(4,n) + \alpha_3 \text{OVF}(3,n) + \alpha_4 \text{RHPG}(3,n) - \text{RIRN}(4,n) - \text{DWS}(4,n) - \text{OVF}(4,n) - A_0 e_j(4,n) \qquad (22.16)$$

$n = 1,2,3,..........,12.$

$$\text{RS}(m,1) = \text{RS}(m,13). \tag{22.17}$$

Equation 22.17 is essential to bring the state of the reservoir at the end of the year to the initial storage at the beginning of the next year.

In the above model, the Reservoir R_1 is the pumped storage project. Hence, the loss in the conveyance of the turbine water back into the reservoir is considered as 10% of the turbine water. So the constant α_1 is taken 0.9 in the constraint of the Reservoir R_1. The constant domestic water supply (DWS) considered for reservoirs R_1, R_2, R_3, and R_4 are 31.63, 3.55, 2.0, and 2.0 Mm^3, respectively. For feeder canal release (FCR) from R_1 to R_2, the conveyance losses are considered 10% of the FCR. Hence, the constant α_2 is taken as 0.9 in the constraint of the reservoir R_2. The 10% losses are considered in the conveyance of overflow (OVF) from R_3 to reach to R_4. Hence, the constant α_3 is taken as 0.9 in the constraint of the reservoir R_4. Similarly, the loss in the conveyance of turbine water (RHPG) from reservoir R_3 to reach R_4 is considered as 10% of RHPG. Therefore, the constant α_4 is considered as 0.9 in the constraint of reservoir R_4.

22.7 RESULTS AND DISCUSSIONS

In this study, the methodology explained in Section 22.3 is applied to the reservoir operation model as presented in Section 22.4.1, where fuzzy sets represent all the parameters (C_j, A_{ij}, b_i) of the reservoir optimization model. The coefficients of the objective function ($\tilde{c}_j$), the technological coefficients ($\tilde{a}_{ij}$), and the resources ($\tilde{b}_j$) are simultaneously treated as a fuzzy set. The model is solved in two phases which is the integration of FPP and FLP approaches. To solve the multiobjective reservoir operation problem, it is assumed that the parameters of the reservoir operation model are known with only some degree of precision, and are not fully known. It is also shown that the model is parameterized so that the optimal solutions are governed by the degree of precision (μ). The imprecision is quantified by the membership functions. For the objective functions and the resources, the nonlinear membership function is used whereas for the technological coefficients, the linear membership function is used. For the objective functions, technological coefficients, and the resources, the lower limit is treated as risk-free value and the upper limit is treated as the unimplementable value of the fuzzy set. The developed model is solved for different values of the precision level (μ) ranging from 0 to 1 with an interval of 0.1. Results obtained by solving the model for the precision level ($\mu = 0$) and ($\mu = 1$) are presented here as the results give the best and worst values of the objective functions. The results obtained by solving these models are given in Tables 22.1–22.4. Table 22.1 represents the operating policy for the maximization of the releases for the irrigation when (μ = 0). Table 22.2 represents the operating policy for the maximization of the releases for irrigation when (μ = 1). From these results, it is observed that the total releases for irrigation are varied from 2173.6 to 2077.6 Mm^3. The releases obtained for the reservoir R_1 vary from 1096.7 to 1090 Mm^3. For the reservoir R_2, the releases vary from 471.6 to 468.5 Mm^3 and for the reservoir R_4, the releases vary from 605.3 to 518.8 Mm^3 with (μ = 0) and (μ = 1), respectively. In this case, the releases obtained for hydropower generation are equal to 824 Mm^3.

Next, the model is solved for the second objective function of the maximization of the releases for hydropower generation. In this case, the objective of the maximization of the releases for irrigation has been given less priority. The release policy corresponding to the value (μ = 0) is given in Table 22.3 and (μ = 1) is given in Table 22.4. From the results, it is observed that the values of the releases obtained for the hydropower generation vary from the 1540.1 to 1434.6 Mm^3 corresponding to the (μ = 0) and (μ = 1) respectively. The releases obtained for the reservoir R_1 vary from 407.5 to 351.8 Mm^3. For the reservoir R_2, releases vary from 348 to 300 Mm^3 and for the reservoir R_3, the releases vary from 784.69 to 782.3 Mm^3.

From these results, the best and worst values of each objective are determined, which are given in Table 22.5. These values are used to decide the lower and upper bounds of both the objectives. The objectives are fuzzified with lower and upper bounds using linear membership function. The membership function for the objective of the maximization of the releases for the irrigation (Z_1) and hydropower generation (Z_2) is given by eqs 22.18 and 22.19. The graphical representation of these membership functions is given by Figures 22.5 and 22.6.

Phase II of the FLP model has performed to achieve the maximum degree of satisfaction (λ) for each precision level. For Phase II, the information obtained from eqs 22.18 and 22.19 is used to develop the compromised model by introducing the new variable (λ) given by the eq 22.20 and all other original constants. The compromised model is solved for the maximization of the degree of satisfaction (λ) for (μ) ranging from 0 to 1. The results obtained corresponding to the precision level (μ = 0) and (μ = 1) are presented here because the results represent the best and worst optimal operating policies. The results given in Tables 22.6 and 22.7 represent the results corresponding to (μ = 0) and (μ = 1), respectively, for the maximized degree of satisfaction

TABLE 22.1 Operating Policy for the Maximization of the Releases for Irrigation (Precision Level $\mu = 0$).

Months	Irrigation releases (Mm^3)				Hydropower releases (Mm^3)				Feeder canal releases (Mm^3)			
	R_1	R_2	R_4	Total	R_1	R_2	R_3	Total	R_1	R_2	R_3	R_4
June	18.6	2.4	33.1	54.0	22.7	24.0	22.0	68.7	0.0	0.0	0.0	0.0
July	26.7	20.8	35.2	82.8	22.7	24.0	22.0	68.7	0.0	0.0	0.0	0.0
August	25.4	37.6	35.2	98.3	22.7	24.0	22.0	68.7	0.0	0.0	0.0	0.0
September	85.8	46.0	93.5	225.3	22.7	24.0	22.0	68.7	0.0	0.0	0.0	0.0
October	267.9	132.0	77.6	477.5	22.7	24.0	22.0	68.7	50.0	50.0	0.0	0.0
November	228.7	127.1	74.7	430.5	22.7	24.0	22.0	68.7	80.0	80.0	0.0	0.0
December	210.9	39.3	65.1	315.3	22.7	24.0	22.0	68.7	70.0	70.0	0.0	0.0
January	143.2	30.2	65.1	238.5	22.7	24.0	22.0	68.7	90.0	90.0	0.0	0.0
February	25.5	9.0	35.5	70.0	22.7	24.0	22.0	68.7	60.0	60.0	0.0	0.0
March	21.0	8.7	37.4	67.1	22.7	24.0	22.0	68.7	0.0	0.0	0.0	0.0
April	25.6	10.7	30.5	66.8	22.7	24.0	22.0	68.7	0.0	0.0	0.0	0.0
May	17.4	7.8	22.3	47.5	22.7	24.0	22.0	68.7	0.0	0.0	0.0	0.0
Total	1096.7	471.6	605.3	2173.6	272.0	288.0	264.0	824.0				

Total irrigation releases = 2173.6 Mm^3; Total hydropower releases = 824 Mm^3.

TABLE 22.2 Operating Policy for the Maximization of the Irrigation Releases (Precision Level $\mu = 1$).

Months	Irrigation releases (Mm^3)				Hydropower releases (Mm^3)				Feeder canal releases (Mm^3)			
	R_1	R_2	R_4	Total	R_1	R_2	R_3	Total	R_1	R_2	R_3	R_4
June	15.9	2.4	28.4	46.7	22.7	24	22	68.7	0	0	0	0
July	22.9	17.9	30.2	70.9	22.7	24	22	68.7	0	0	0	0
August	21.8	32.3	30.2	84.2	22.7	24	22	68.7	0	0	0	0
September	73.5	39.4	80.1	193.1	22.7	24	22	68.7	0	0	0	0
October	229.6	113.2	66.5	409.3	22.7	24	22	68.7	50	50	0	0
November	196.1	108.9	64	369	22.7	24	22	68.7	80	80	0	0
December	180.8	76.7	55.8	313.2	22.7	24	22	68.7	70	70	0	0
January	197.4	41.7	55.8	295	22.7	24	22	68.7	90	90	0	0
February	73.1	9	30.4	112.5	22.7	24	22	68.7	60	60	0	0
March	36.3	8.7	32.1	77.1	22.7	24	22	68.7	0	0	0	0
April	25.6	10.7	26.1	62.4	22.7	24	22	68.7	0	0	0	0
May	17.4	7.8	19.1	44.3	22.7	24	22	68.7	0	0	0	0
Total	1090.4	468.5	518.8	2077.6	272	288	264	824				

Total irrigation releases = 2077.6 Mm^3; Total hydropower releases = 824 Mm^3.

TABLE 22.3 Operating Policy for the Maximization of the Hydropower Releases (Precision level $\mu = 0$).

Months	Irrigation releases (Mm^3)				Hydropower releases (Mm^3)				Feeder canal releases (Mm^3)			
	R_1	R_2	R_4	Total	R_1	R_2	R_3	Total	R_1	R_2	R_3	R_4
June	5.53	2.14	9.93	17.60	29.32	25.0	63.8	118.2	0.0	0.0	0.0	0.0
July	8.01	6.25	10.56	24.82	29.32	25.0	107.5	161.8	0.0	0.0	0.0	0.0
August	7.64	11.29	10.56	29.49	29.32	25.0	142.0	196.4	0.0	0.0	0.0	0.0
September	25.68	13.81	28.04	67.53	29.32	25.0	142.0	196.4	0.0	0.0	0.0	0.0
October	80.28	39.60	23.82	143.70	29.32	25.0	107.5	161.8	50.0	50.0	0.0	0.0
November	68.42	38.12	22.40	128.94	29.32	25.0	82.8	137.1	80.0	80.0	0.0	0.0
December	63.23	26.83	19.54	109.60	29.32	25.0	26.7	81.1	70.0	70.0	0.0	0.0
January	69.55	30.20	19.54	119.29	29.32	25.0	22.0	76.4	90.0	90.0	0.0	0.0
February	25.51	9.01	10.65	45.17	29.32	25.0	22.0	76.4	60.0	60.0	0.0	0.0
March	21.02	8.69	11.22	40.93	29.32	25.0	22.0	76.4	0.0	0.0	0.0	0.0
April	25.63	10.67	9.15	45.45	29.32	25.0	22.0	76.4	0.0	0.0	0.0	0.0
May	17.44	7.76	6.69	31.89	29.32	25.0	22.0	76.4	0.0	0.0	0.0	0.0
Total	417.9	204.4	182.1	804.4	351.8	300.5	782.3	1434.6				

Total irrigation releases = 804.4 Mm^3*; Total hydropower releases = 1434.6* Mm^3*.*

TABLE 22.4 Operating Policy for the Maximization of the Hydropower Releases (Precision level $\mu = 1$).

Months	Irrigation releases (Mm^3)				Hydropower releases (Mm^3)				Feeder canal releases (Mm^3)			
	R_1	R_2	R_4	Total	R_1	R_2	R_3	Total	R_1	R_2	R_3	R_4
June	5.53	2.14	9.93	17.60	29.32	25.0	63.8	118.2	0.0	0.0	0.0	0.0
July	8.01	6.25	10.56	24.82	29.32	25.0	107.5	161.8	0.0	0.0	0.0	0.0
August	7.64	11.29	10.56	29.49	29.32	25.0	142.0	196.4	0.0	0.0	0.0	0.0
September	25.68	13.81	28.04	67.53	29.32	25.0	142.0	196.4	0.0	0.0	0.0	0.0
October	80.28	39.60	23.82	143.70	29.32	25.0	107.5	161.8	50.0	50.0	0.0	0.0
November	68.42	38.12	22.40	128.94	29.32	25.0	82.8	137.1	80.0	80.0	0.0	0.0
December	63.23	26.83	19.54	109.60	29.32	25.0	26.7	81.1	70.0	70.0	0.0	0.0
January	69.55	30.20	19.54	119.29	29.32	25.0	22.0	76.4	90.0	90.0	0.0	0.0
February	25.51	9.01	10.65	45.17	29.32	25.0	22.0	76.4	60.0	60.0	0.0	0.0
March	21.02	8.69	11.22	40.93	29.32	25.0	22.0	76.4	0.0	0.0	0.0	0.0
April	25.63	10.67	9.15	45.45	29.32	25.0	22.0	76.4	0.0	0.0	0.0	0.0
May	17.44	7.76	6.69	31.89	29.32	25.0	22.0	76.4	0.0	0.0	0.0	0.0
Total	417.9	204.4	182.1	804.4	351.8	300.5	782.3	1434.6				

Total irrigation releases = 804.4 Mm^3; Total hydropower releases = 1434.6 Mm^3.

(λ). It is observed that the degree of satisfaction (λ) corresponding to ($\mu = 0$) is maximum and is being equal to 0.893, however, corresponding to ($\mu = 0$), it is equal to 0.65. The optimal irrigation releases are ranging from 2001.5 to 1977.4 Mm^3. The releases for hydropower generation are varying from 1437.54 to 1326.19 Mm^3. There is a marginal change in the releases obtained for the irrigation for the reservoir R_1 vary from 1083.4 to 1082.9 Mm^3, but the distribution of the releases pattern is different. For the reservoir R_2, the releases vary from 323.8 to 375.7 Mm^3. For the reservoir R_4, the releases are ranging from 594.28 to 518.75 Mm^3. In the case of the optimal releases obtained for hydropower generation, for the reservoir R_1, it varies from 407.5 to 351.9 Mm^3. For the reservoir R_2, the releases range from 253.8 to 201.5 Mm^3 and for the reservoir R_3, the releases range from 776.18 to 772.7 Mm^3 corresponding to the precision level (μ).

TABLE 22.5 Upper and Lower Bounds for Irrigation and Hydropower Generation MODEL V: (C_j, A_{ij}, b_i Fuzzification).

Objective function (maximization)	Maximum (upper bound)	Minimum (lower bound)
Irrigation releases (Z_1)	2173.6 Mm^3	804.4 Mm^3
Hydropower releases (Z_2)	1540.10 Mm^3	824.0 Mm^3

$$\mu_{Z_1}(x) = \begin{cases} 0 & \text{if } Z_1 \le 804.4 \\ \dfrac{Z_1 - 804.4}{2173.6 - 804.4} & \text{if } 804.4 \le Z_1 \le 2173.6 \\ 1 & \text{if } Z_1 \ge 2173.6 \end{cases} \tag{22.18}$$

$$\mu_{Z_2}(x) = \begin{cases} 0 & \text{if } Z_1 \le 824.4 \\ \dfrac{Z_2 - 824.4}{1540.1 - 824.4} & \text{if } 824.4 \le Z_1 \le 1540.1 \\ 1 & \text{if } Z_1 \ge 1540.1 \end{cases} \tag{22.19}$$

$$\left.\begin{array}{l} \text{Max} = \lambda, \\ \text{Subject } to, \\ (Z_1 - 804.4)\,/\,(2173.6 - 804.4) \ge \lambda, \\ (Z_2 - 824.4)\,/\,(1540.4 - 824.4) \ge \lambda \end{array}\right\} \tag{22.20}$$

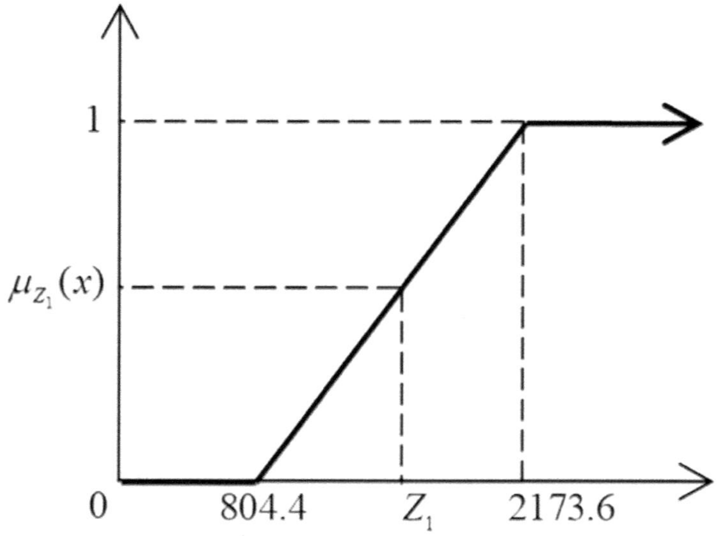

FIGURE 22.5 Linear membership function for irrigation.

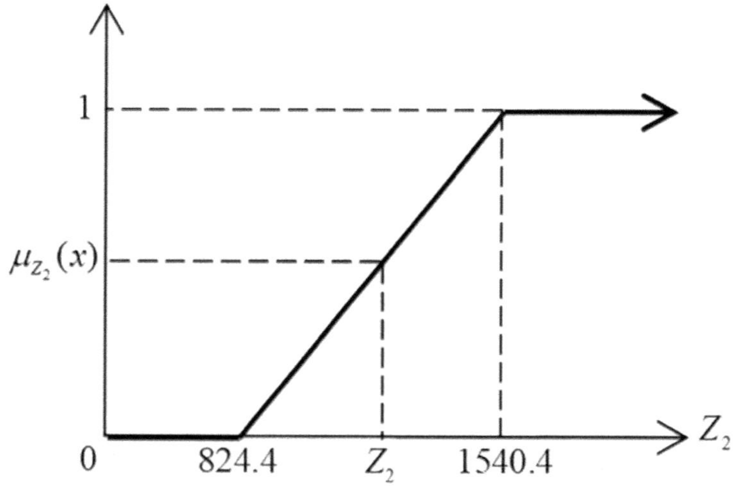

FIGURE 22.6 Linear membership function for hydropower.

22.8 CONCLUSIONS

In this study, the multipurpose multireservoir optimization model is developed, and the model's applicability is demonstrated through the case study of

TABLE 22.6 Optimal Operating Policy (Precision level $\mu = 0$).

Months	Irrigation releases (Mm³)				Hydropower releases (Mm³)				Feeder canal releases (Mm³)			
	R_1	R_2	R_4	Total	R_1	R_2	R_3	Total	R_1	R_2	R_3	R_4
May	18.6	2.1	33.1	53.8	33.96	9.0	63.80	106.71	0.0	0.0	0.0	0.0
June	26.7	6.3	35.23	68.2	33.96	27.3	124.21	185.49	0.0	0.0	0.0	0.0
July	25.4	37.6	35.23	98.3	33.96	29.0	36.28	99.24	0.0	0.0	0.0	0.0
August	85.8	46.0	93.46	225.3	33.96	29.0	104.09	167.05	0.0	0.0	0.0	0.0
September	267.9	100.5	77.6	445.9	33.96	29.0	93.59	156.55	50.0	50.0	0.0	0.0
October	228.7	38.1	74.68	341.5	33.96	29.0	90.07	153.03	80.0	80.0	0.0	0.0
November	210.9	26.8	65.14	302.9	33.96	29.0	79.11	142.07	70.0	70.0	0.0	0.0
December	129.8	30.2	54.14	214.2	33.96	29.0	78.99	141.95	90.0	90.0	0.0	0.0
January	25.5	9.0	35.5	70.0	33.96	17.5	40.05	91.48	60.0	60.0	0.0	0.0
February	21.0	8.7	37.4	67.1	33.96	8.7	22.00	64.66	0.0	0.0	0.0	0.0
March	25.6	10.7	30.5	66.8	33.96	8.7	22.00	64.66	0.0	0.0	0.0	0.0
April	17.4	7.8	22.3	47.5	33.96	8.7	22.00	64.66	0.0	0.0	0.0	0.0
Total	1083.4	323.8	594.28	2001.5	407.52	253.8	776.18	1437.54				

Total irrigation releases = 2001.5 Mm^3; Total hydropower releases = 1437.54 Mm^3.

TABLE 22.7 Optimal Operating Policy (Precision level $\mu = 0$).

Months	Irrigation releases (Mm³)				Hydropower releases (Mm³)				Feeder canal releases (Mm³)			
	R_1	R_2	R_4	**Total**	R_1	R_2	R_3	**Total**	R_1	R_2	R_3	R_4
April	15.9	2.1	28.37	46.4	29.329	9	63.8	102.08	0.0	0.0	0.0	0.0
May	22.9	8.5	30.19	61.6	29.329	25	107.5	161.85	0.0	0.0	0.0	0.0
June	21.8	32.3	30.19	84.2	29.329	25	60.1	114.44	0.0	0.0	0.0	0.0
July	73.5	39.4	80.1	193.1	29.329	25	88.2	142.56	0.0	0.0	0.0	0.0
August	229.6	113.2	66.51	409.3	29.329	25	80.4	134.79	50.0	50.0	0.0	0.0
September	196.1	87.1	64.011	347.2	29.329	25	77.5	131.87	80.0	80.0	0.0	0.0
October	180.8	26.8	55.83	263.4	29.329	23.8	68.1	121.22	70.0	70.0	0.0	0.0
November	197.4	30.2	55.83	283.5	29.329	8.7	67.9	105.96	90.0	90.0	0.0	0.0
December	73.1	9	30.42	112.5	29.329	8.7	43.247	81.28	60.0	60.0	0.0	0.0
January	28.8	8.7	32.05	69.5	29.329	8.7	49.6	87.64	0.0	0.0	0.0	0.0
February	25.6	10.7	26.14	62.4	29.329	8.7	44.4	82.45	0.0	0.0	0.0	0.0
March	17.4	7.8	19.11	44.3	29.329	8.7	22	60.03	0.0	0.0	0.0	0.0
Total	1082.9	375.7	518.75	1977.4	351.95	201.5	772.7	1326.19				

Total irrigation releases = 1977.4 Mm³; Total hydropower releases = 1326.19 Mm³.

multireservoir system in a subbasin of the Godavari river in Maharashtra state, India. The model deals with the simultaneous fuzzification of all reservoir operation parameters. Over the tolerance range, the objective function coefficients, technological coefficients, and resources are simultaneously fuzzified. The model uses the exponential membership function for objective function coefficients and resources. However, the linear membership function is used for technological coefficients.

The degree of satisfaction (λ) corresponding to ($\mu = 0$) is equal to 0.893, however, corresponding to ($\mu = 1$), it is equal to 0.65. The optimal irrigation releases are varying from 2001.5to 1977.4 Mm^3. The releases for hydropower generation are varying from 1437.54 to 1326.19 Mm^3. The interactive concept of the procedure is carried out according to the decision maker's choice to find the optimum solutions for both the objective functions from various grades of precision. The model assumes that parameters involved in the problem of reservoir operation are not completely defined, only with some degree of precision, and shows that the model can be parameterized such that the optimal solutions are the function of the degree of precision.

This study includes the experience of the decision makers, information; expectations obtained through interactions. Therefore, all the solutions are optimal and effective at any level of precision.

KEYWORDS

- **parametric programming**
- **exponential membership function**
- **decision making**
- **precision level**
- **reservoir operation**

REFERENCES

Al-Jawad, J. Y.; Tanyimboh, T. T. Reservoir Operation Using a Robust Evolutionary Optimization Algorithm. *J. Environ. Manag.* **2017,** *197,* 275–286. doi:10.1016/j.jenvman.2017.03.081

Choong, S.; El-Shafie, A.; Mohtar, W. H. Optimisation of Multiple Hydropower Reservoir Operation Using Artificial Bee Colony Algorithm. *Water Resour. Manag.* **2017,** *31* (4), 1397–1411. doi:10.1007/s11269-017-1585-x

Dubois, D.; Prade, H. *Fuzzy Sets and Systems: Theory and Application*; Academic Press, Inc.: New York, 1980.
Fayzan A.; Zulal, G. A Two-Phase Approach for Multi-Objective Programming Problems with Fuzzy Coefficients. *Inform. Sci.* **2007,** *177* (23), 5191–5202.
Fullér, R. *Fuzzy Reasoning and Fuzzy Optimization*; Turku Centre For Computer Science, 1998.
Gaur, S.; Srinivasa Raju, K.; Nagesh Kumar, D.; Graillot, D. Multiobjective Fuzzy Optimization for Sustainable Groundwater Management Using Particle Swarm Optimization and Analytic Element Method. *Hydrol. Processes* **2015,** *29* (19), 4175–4187. doi. 10.1002/hyp.10441
Ibrahim, A. M. *Fuzzy Logic for Embedded Systems Applications. Newnes*; Elsevier: New York, 2003.
Kamodkar, R. U.; Regulwar, D. G. Optimal Multiobjective Reservoir Operation with Fuzzy Decision Variables and Resources: A Compromise Approach. *J. Hydro-Environment Res.* **2014,** *8* (4), 428–440. doi:10.1016/j.jher.2014.09.001
Klir, G. J.; Yuan, B. *Fuzzy Sets and Fuzzy Logic: Theory and Applications*; Prentice-Hall of India Private Limited: New Delhi, Indian Reprint, 1995.
Loucks, D. P.; Stedinger, J. R.; Haith, D. A. *Water Resources Systems Planning and Analysis*; Prentice-Hall: Englewood Cliffs, 1981.
Milan, S. G.; Roozbahani, A.; Banihabib, M. E. Fuzzy Optimization Model and Fuzzy Inference System for Conjunctive Use of Surface and Groundwater Resources. *J. Hydrol.* **2018,** *566*, 421–434. doi:10.1016/j.jhydrol.2018.08.078
Moeini, R.; Afshar, A.; Afshar, M. H. Fuzzy Rule-Based Model for Hydropower Reservoirs Operation. *Int. J. Electr. Power Energy Syst.* **2011,** *33*(2), 171–178. doi:10.1016/j.ijepes.2010.08.012
Moorthi, P. V.; Singh, A. P.; Agnivesh, P. Regulation of Water Resources Systems Using Fuzzy Logic: A Case Study of Amaravathi Dam. *Appl. Water Sci.* **2018,** *8* (5), 1–11. doi:10.1007/s13201-018-0777-8
Moreira, R. M.; Celeste, A. B. Performance Evaluation of Implicit Stochastic Reservoir Operation Optimization Supported by Long-Term Mean Inflow Forecast. *Stoch. Environ. Res. Risk Assess.* **2016,** *31* (9), 2357–2364. doi:10.1007/s00477-016-1341-4
Regulwar, D. G.; Kamodkar, R. U. Derivation of Multipurpose Single Reservoir Release Policies with Fuzzy Constraints. *J. Water Resour. Protect.* **2010,** *2* (12), 1030–1041. doi:10.4236/jwarp.2010.212123
Saadat, M.; Asghari, K. Reliability Improved Stochastic Dynamic Programming for Reservoir Operation Optimization. *Water Resour. Manag.* **2017,** *31* (6), 1795–1807. doi:10.1007/s11269-017-1612-y
Saberchenari, K.; Abghari, H.; Tabari, H. Application of PSO Algorithm in Short-Term Optimization of Reservoir Operation. *Environ. Monit. Assess.* **2016,** *188* (12). doi:10.1007/s10661-016-5689-1
Safavi, H. R.; Chakraei, I.; Kabiri-Samani, A.; Golmohammadi, M. H. Optimal Reservoir Operation Based on Conjunctive Use of Surface Water and Groundwater Using Neuro-Fuzzy Systems. *Water Resour. Manag.* **2013,** *27* (12), 4259–4275. doi:10.1007/s11269-013-0405-1
Sangeeta, K.; Mujumdar, P. P. Reservoir Operation with Fuzzy State Variables for Irrigation of Multiple Crops. *J. Irrig. Drain. Eng.* **2015,** *141* (11), 04015015. doi:10.1061/(asce)ir.1943-4774.0000893

Sivanandam, S. N.; Sumathi, S.; Deepa, S. N. *Introduction to Fuzzy Logic Using MATLAB*; Springer, TLFebook, 2007.

Srivastava, P.; Singh, R. M. Agricultural Land Allocation for Crop Planning in a Canal Command Area Using Fuzzy Multiobjective Goal Programming. *J. Irrig. Drain. Eng.* **2017,** *143* (6), 04017007. doi:10.1061/(asce)ir.1943-4774.0001175

Zadeh, L. A. Fuzzy Sets, 1964. doi:10.21236/ad0608981

Zimmermann, H. J. *Fuzzy Set Theory- and Its Applications*; Allied Publishers Limited: New Delhi, IInd Revised Edition, 1996.

CHAPTER 23

Projection of Future Precipitation over Gangotri Glacier at Himalayan Belt Using CMIP5 Climate Model

SURENDRA KUMAR CHANDNIHA[1*], MANOHAR ARORA[2], LALIT PAL[3], AMIT KUMAR[4], PRAGYA DAS[5], and G. K. DAS[6]

[1]*Department of Soil and Water Engineering, BRSM CAET & RS, IGKV, Mungeli, Chhattisgarh, India*

[2]*National Institute of Hydrology, Roorkee, Uttarakhand, India*

[3]*Department of Civil Engineering, Indian Institute of Technology, Roorkee, India*

[4]*School of Geography, University of Nottingham, United Kingdom*

[5]*Department of Soil and Water Engineering, SV CAET&RS, IGKV, Mungeli, Chhattisgarh, India*

[6]*Department of Agrometeorology, IGKV, Raipur, India*

[*]*Corresponding author. E-mail: chandniha.surendra@gmail.com*

ABSTRACT

In the twenty-first century, climate change is considered to be one of the most important environmental threats for the survival of living beings. Climatic variability and its behavior is a complex phenomenon that is directly associated with uncertainties. Changes in climate extremes adversely impact the sustainability in terms of social, environment, and economy consideration. In the climate change study, particularly in hydrological aspects, it is necessary to

Advances in Hydrology and Climate Change: Historical Trends and New Approaches in Water Resources Management. Surendra Kumar Chandniha, Anil Kumar Lohani, Gopal Krishan, & Ajay Krishna Prabhakar (Eds.)

identify the parameters (predictors) that are directly or indirectly associated with predictands. The forecasted results are directly associated with the selection of predictors. In the Fifth Assessment Report of the Intergovernmental Panel on Climate Change (IPCC), representative concentration pathways (RCPs) are used for climate projections under the Coupled Model Intercomparison Project 5 (CMIP5). This study represents the multiscenario precipitation projections for the Gangotri Glacier region in the Himalayan belt during the long-term period 1961–2100. From the study, it was noticed that CMIP5 ensemble mean climate is closer with observed precipitation dataset. For the purpose of study, CanESM3 model has been utilized to estimate the future projections using three climate scenarios, that is, RCP-2.6, RCP-4.5, and RCP-8.5.

23.1 INTRODUCTION

In 2000, based on different socioeconomic development assumptions, the IPCC developed four families of emission pathways, viz., A1, B1, A2, and B2 in its emission scenario (SRES) special report (Nakicenovic et al., 2000). Because of these emission scenario, future climate is projected using global climate models, also known as IPCC climate models under the Coupled Model Intercomparison Project 3 (CMIP3) (http://www. ipcc-data.org/). In order to characterize the impact of future climate change on society and ecosystem, climate projections based on SRES emission scenario from CMIP3 are used in the Fourth Assessment Report of IPCC (Solomon et al., 2007; Parry et al., 2007). In the past, various future climate change impact studies in India have relied solely on the CIMP3 model climate projections. For instance, Krishnakumar et al. (2010) projected surface temperature and monsoon rainfall using CMIP3 multimodel data outputs for the period 1901–2098 over entire India. The Hadley Centre's Coupled Model (HadCM3) generated climate data are downscaled to fine resolution data and are readily available for Indian regions which have been developed under Providing Regional Climate for Impact Studies (PRECIS) project (Rupakumar et al., 2006); Krishnakumar et al., 2011). Regional climate of India was simulated under SRES scenarios, that is, A2 and B2 for the baseline period as well as long-term projections using PRECIS data sets (Rupakumar et al., 2006). Krishnakumar et al. (2011) also used PRECIS model to simulate SRES scenario A1B-based regional climate of India for the period 1961–2098. Three simulations for three different future projections segments, that is, short (2020s), medium (2050s), and long term (2080s) were employed to drive PRECIS using 17-member perturbed physics-based ensembles from

HadCM3 for quantifying uncertainty in model predictions (QUMP) project. An increase of 4–5% by 2030s and 6–14% toward the end of the century (2080s) in all India precipitation had been projected by under the business-as-usual scenario as compared with the 1961–1990 baseline. Despite the temperature projection being considered, more reliable than precipitation projection, an increase in agreement in precipitation projection had been observed from RCP2.6 to RCP8.5 and from short-to long-term projections. The long-term precipitation projections are presented to be more robust than their short-term counterparts (Chaturvedi et al., 2012).

Several improvements have been made in the new generation climate models used within Coupled Model Intercomparison Project Phase 5 (CMIP5) in terms of physics and resolution (Taylor et al., 2012). However, there exist a lack of accuracy in the simulations made using CMIP5 outputs for regional scale projections, as the outputs are available at relatively coarser resolution. Various statistical downscaling approaches have been developed in the past to obtain regional/local scale projections (Fowler et al., 2007). Ghosh and Mujumdar (2006) applied a modified linear regression model to examine future rainfall scenarios from outputs of Coupled global Climate model (CGCM2) over Odisha state in India. The linear regression model was modified by incorporating principle component analysis (PCA) and fuzzy clustering technique. Ghosh (2010) established a relationship between predictors and predictand using support vector machine (SVM) to predict monsoon rainfall for SRES A2, B2, and 20C3M scenarios in Assam and Meghalaya meteorological subdivision from GCM outputs. The methodology developed is highlighted to be computationally intensive especially for downscaling at daily time steps. Anandhi et al. (2008) used SVM to downscale monthly precipitation to river basin in India using third-generation Canadian general circulation model (CGCM3) for SRES emission scenarios A1B, A2, B1, and COMMIT. The study results projected an increase in the future precipitation for almost all emission scenarios. Ojha et al. (2010) generated the projections of mean monthly precipitation in the Pichila lake catchment in India using a linear multiple regression (LMR) and artificial neural networks (ANNs). CGCM3 simulations were used to project for emission scenario A1B, A2, B1, and COMMIT. An increase in projected precipitation was stated for A2 and A1B scenarios. The present study aims to project precipitation over Gangotri glacier in the Himalayan region by incorporating CMIP5 climate model outputs using multiple linear regression (MLR) technique. It also aims to analyze for any temporal pattern, if exists. The scenarios which are studied in this paper are relevant to IPCC's Fifth Assessment Report (AR5) which was released in 2014 (IPCC, 2014).

23.2 STUDY AREA

This study area is located under Uttarakhand State of India. The areal extension of the glacier lies between 30°43′00″ N and 31°0103″ N latitude and between 79°00′08″ E to 79°17′41″ E longitude and elevation varying from 4000–7000 m (msl). The Gangotri glacier system consists of various small glaciers having dimension of main Gangotri glacier as length: 30.20 km; width: 0.20–2.35 km; and area about 86.32 km². Raktvarn, Chaturangi, Swachand, and Maiandi glaciers contribute to the Gangotri glacier system from the northeast and Meru, Kirti, and Ghanohim glaciers join the trunk glacier from the southwest (Arora et al., 2014). Location map of the study area is shown in Figure 23.1.

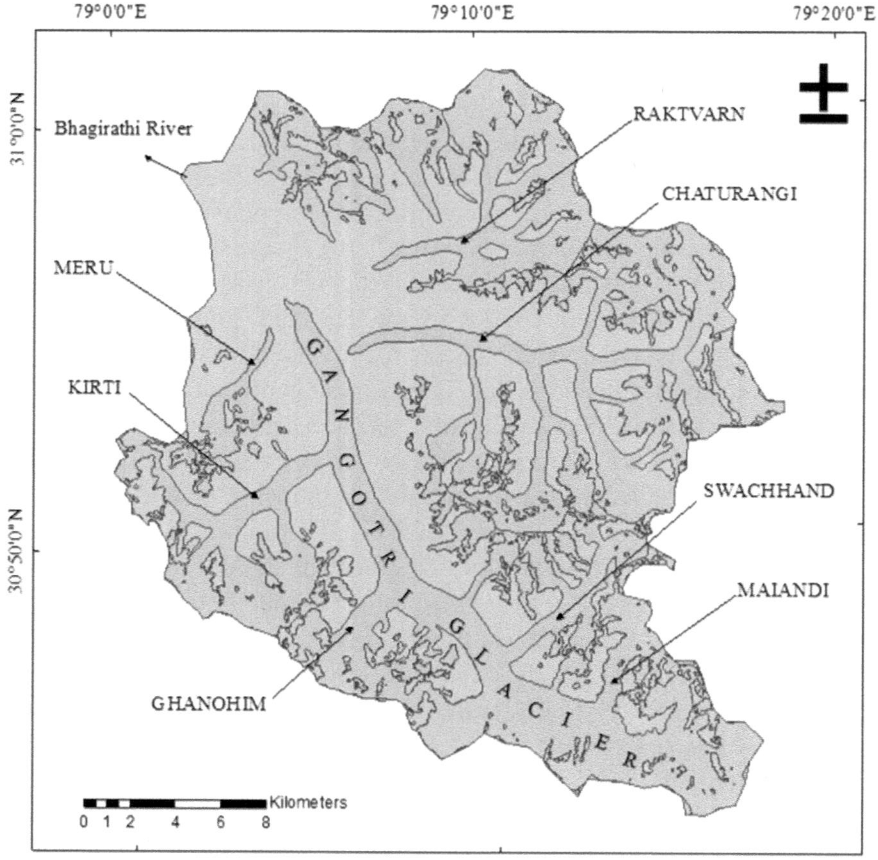

FIGURE 23.1 Map of Gangotri glacier in the Himalayas.

23.3 MATERIAL AND METHODS

The Statistical Downscaling Model (SDSM) is a multiple regression-based tool for generating future scenarios to assess the impact of climate change. It has the ability to capture the interannual variability better than other statistical downscaling approaches (Chandniha and Kansal, 2016). This approach involves three subclasses, such as weather typing, weather generator, and regression/transform function. The model requires two types of daily data, that is, (1) the local data known as "Predictand" (rainfall) and (2) the different atmospheric variables known as "Predictors." Formulating an empirical relationship between predictand and predictor is central to the downscaling technique. This can be derived by various methods such as parametric (Multiple Linear Regression) and nonparametric (artificial neural network; support vector machine). This study uses the multiple linear regression method which falls under parametric methods. The downscaling has been carried out using SDSM tool version 4.2.9. The proposed methodology has been shown in the form of a flow chart as shown in Figure 23.2.

23.3.1 SELECTION OF PREDICTORS

The selection of appropriate predictors is considered to be one of the most important task during downscaling of predictand. The appropriate predictor variables are selected through scatter plots, positive and negative correlation, and partial correlation analysis between predictand (rainfall) and predictors (most appropriate out of 26 possible parameters). The observed daily data set of National Center of Environmental Prediction (NCEP) reanalysis for the period 1961–2001 was used for the selection of predictors. The list of predictors and details of variables used in the present study are given in Tables 23.1 and 23.2.

23.3.2 MODEL CALIBRATION AND VALIDATION

Model calibration is carried out to development of an empirical relationship between the predictand and the predictors using multiple linear regression. NCEP reanalysis data for the period 1961–1990 was used for model calibration, and rest of the data from 1991 to 2001 is used for validation purpose.

Validation process enables to produce synthetic daily data based on inputs of the data considered during the model calibration. The model performance was evaluated based on the coefficient of correlation (R) between the observed values during the validation period and the modeled values, which can be calculated as

$$R = \frac{\sum\left(X_{obs} - \bar{X}_{obs}\right)\left(X_{\text{mod}} = \bar{X}_{\text{mod}}\right)}{\sqrt{\sum\left(X_{obs} - \bar{X}_{obs}\right)^2 \sum\left(X_{obs} - \bar{X}_{obs}\right)^2}}$$

where

X_{obs} = observed value, $\bar{X}_{obs}$ = mean observed value, $\bar{X}_{\text{mod}}$ = Modeled value, $\bar{X}_{\text{mod}}$ = Mean modeled value.

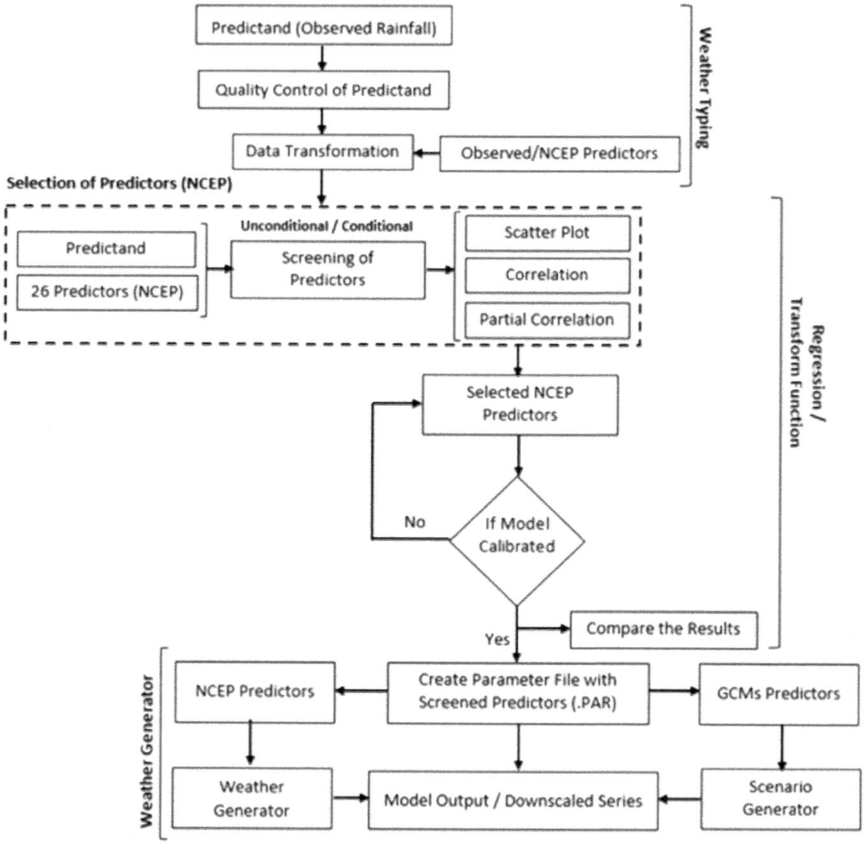

FIGURE 23.2 Methodology of rainfall downscaling.

TABLE 23.1 List of Predictor Variables for Model Calibration, Validation, and Projection.

Predictors	Origin	Duration
Mean sea level pressure	CanESM2	1961–1990 (baseline climate)
1000 hPa wind speed		2000–2099
1000 hPa zonal velocity		
1000 hPa meridional velocity		
1000 hPa vorticity		
1000 hPa wind direction		
1000 hPa divergence		
500 hPa wind speed	CanESM2	1961–1990 (baseline climate)
500 hPa zonal velocity		2000–2099
500 hPa meridional velocity		
500 hPa vorticity		
500 hPa geopotential		
500 hPa wind direction		
500 hPa divergence		
850 hPa wind speed	CanESM2	1961–1990 (baseline climate)
850 hPa zonal velocity		2000–2099
850 hPa meridional velocity		
850 hPa vorticity		
850 hPa geopotential		
850 hPa wind direction		
850 hPa divergence		
Specific humidity at 500 hPa	CanESM2	1961–1990 (baseline climate)
Specific humidity at 850 hPa		2000–2099
Specific near surface humidity		
Mean temperature at 2 m		

Source: climate-scenarios.canada.ca

23.3.3 SCENARIO GENERATION

The validated regression model is applied to generate future scenario for the watershed utilizing the simulated CanESM2 climate model with RCP2.6, RCP4.5, and RCP4.8 scenarios. The study assumes that the relationship between predictor and predictand remains valid under the future climate conditions.

TABLE 23.2 Description of Predictors (Variables).

Variable	Description	Additional notes
Temp	Mean temperature at 2 m	
Mslp	Mean sea level pressure	
p500	500 hPa Geopotential height	The height of this surface will vary depending on the temperature of the atmospheric column: warmer = higher; cooler = lower
p850	850 hPa Geopotential height	The height of this surface will vary depending on the temperature of the atmospheric column: warmer = higher; cooler = lower
Rhum	Near surface relative humidity	The vapor content of air as a percentage of the vapor content needed to saturate air at the same temperature
r500	Relative humidity at 500 hPa height	
r850	Relative humidity at 850 hPa height	
Shum	Near surface specific humidity	The mass of water vapor as a proportion of the total mass of moist air of which it is a part; can be used for tracking air masses
s500	Specific humidity at 500 hPa height	
s850	Specific humidity at 850 hPa height	
Derived variables	The following variables have been derived using the Geostrophic Approximation	
**_f	Geostrophic air flow velocity	
**_z	Vorticity	A measure of the rotation of the air
**_u	Zonal velocity component	Velocity component along a line of latitude (i.e., east-west)
**_v	Meridional velocity component	Velocity component along a line of longitude (i.e., north–south)
**zh	Divergence1	Relates to the stretching of a fluid, and usually refers to the outflow of air from the base of an anticyclone in meteorology
**th	Wind direction	This is the only variable which is NOT normalized

*******Refers to different atmospheric levels*: the surface (p_), 850 hPa height (p8) and 500 hPa height (p5).

Source: climate-scenarios.canada.ca

23.4 DATA USED

23.4.1 METEOROLOGICAL DATA

The daily rainfall data (0.25° × 0.25° grid) were collected from the India Meteorological Department (IMD), Pune, for the period 1961–2005. Statistical downscaling has been performed using the daily rainfall time series as input predictand in SDSM software. Further, daily observed and estimated time series converted into monthly and annual time series.

23.4.2 REANALYSIS DATA (NCEP)

The daily observed predictor data (re-analysis data) of atmospheric variables, derived from the NCEP on 1° latitude × 1° longitude grid-scale for 45 years (1961–2005) are obtained from the Canadian Climate Impacts Scenarios (CCIS) website (http://www.cccsn.ec.gc.ca/index.php?page=gridded-data).

23.4.3 CMIP5 CLIMATIC SCENARIO (CANESM2)

The CMIP-CanESM2 climate scenarios on 1° latitude × 1° longitude grid-scale are also obtained from the CCIS website (http://www.cccsn.ec.gc.ca/index.php?page=gridded-data). Canadian Earth System Model (CanESM2) is a CMIP5 climate model. CanESM2 climate model has been chosen because of its wider acceptance in many climate change impact studies in India. Further, it provides the daily predictor variables, which can be exclusively used for the SDSM model.

23.5 RESULTS AND DISCUSSION

The multiple liner regression-based statistical downscaling technique has been utilized for downscaling the monthly precipitation. However, CanESM2 climate model with three different scenarios, that is, RCP2.6, RCP4.5, and RCP8.5 have been considered for future projections. For the model calibration and validation, observed gridded rainfall data have been considered during the period 1961–2005. However, model has been calibrated with reanalysis NCEP data with the same duration. During the calibration and validation, it was found that the coefficient of determination values are 0.45 and 0.51, respectively. The graphical representation of calibration and validation between observed and estimated precipitation are shown in Figure 23.3.

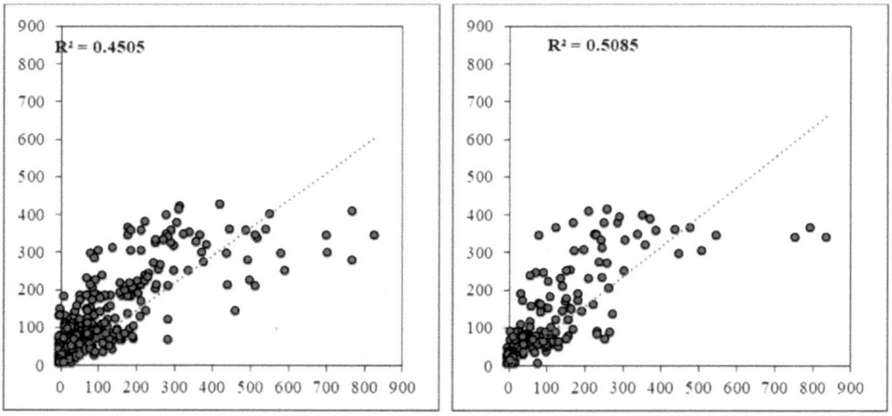

FIGURE 23.3 Scatter plot between observed and estimated precipitation during model calibration (1961–1988) and validation (1989–2005).

Further, the line graph also prepared between observed and the estimated precipitation during 1961–2005. The results present the tendency of model to under-predict the peaks which may be due to climate uncertainty and bias. The graphical representation of observed and estimated precipitation is shown in Figure 23.4. Various statistics of model performance during calibration and validation period are represented in Figure 23.5. It was also noticed that in all climate scenarios, the future precipitation trends are increasing which can be shown by box plot. The decadal future projections of RCP2.6, RCP4.5, and RCP8.5 are shown in Figure 23.6–23.8.

23.6 CONCLUSIONS

The continuous focus of IPCC on the improvement of global circulation model (GCM) which represents the precipitation and strongly relies on rigorous validation efforts for both characterization of precipitation and seasonality. Assessment of the capacity of GCM and its seasonality is very complex phenomenon which is based on historical observations of weather parameters. However, the future changes in hydroclimatic regimes are associated with climatic variability which is more complex and crucial for the realistic projection of predictand. In the recent past, water authorities, scientists, stake-holders, and policy makers were better understanding or

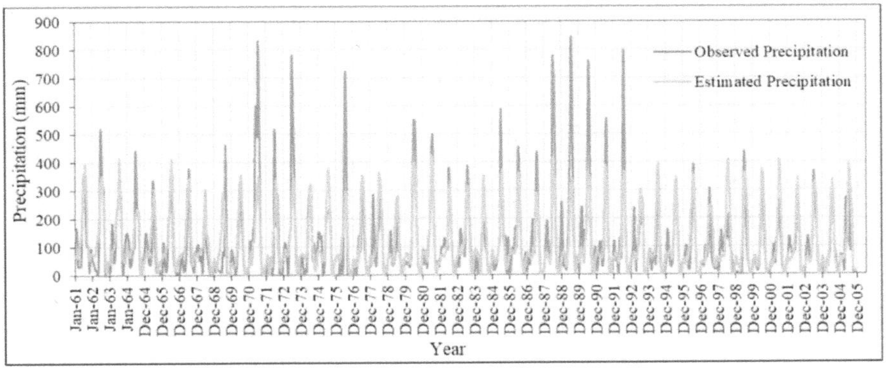

FIGURE 23.4 Observed and estimated precipitation during 1961–2005.

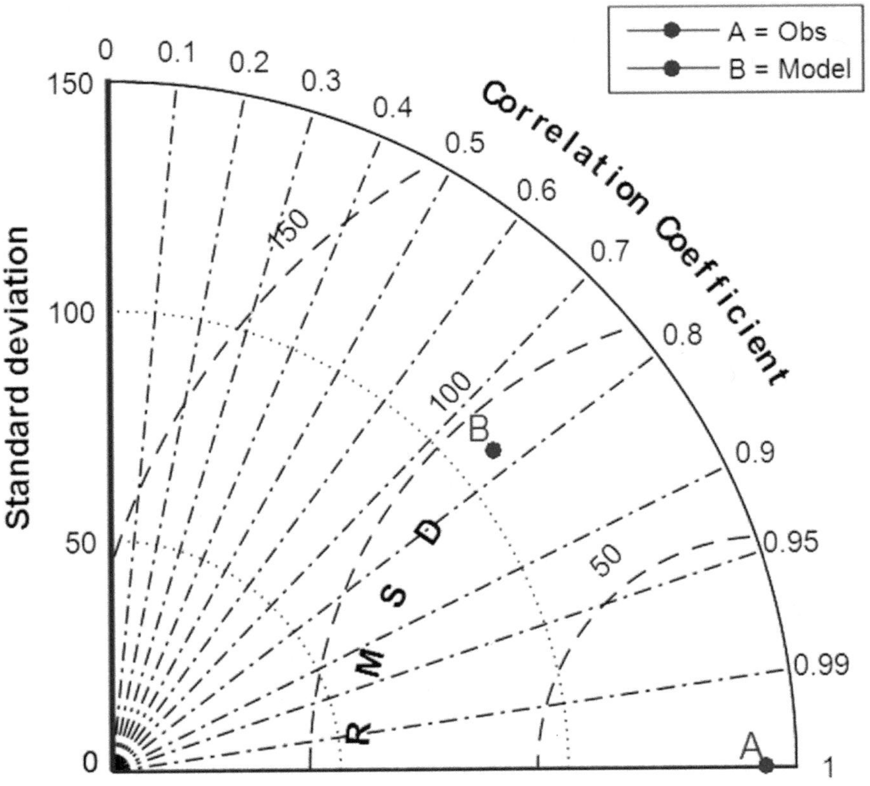

FIGURE 23.5 Taylor diagram representing efficiency statistics of model.

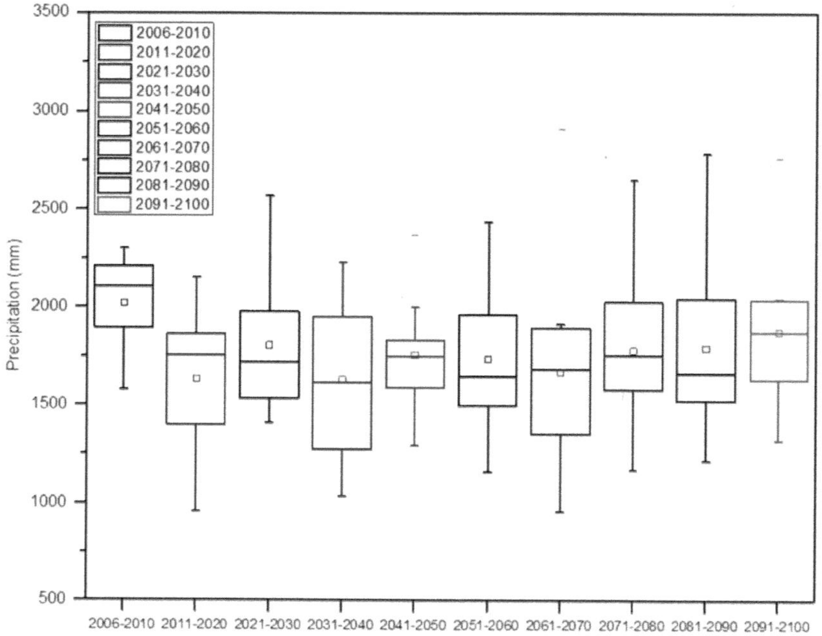

FIGURE 23.6 Projected decadal precipitation of RCP-2.6 scenario.

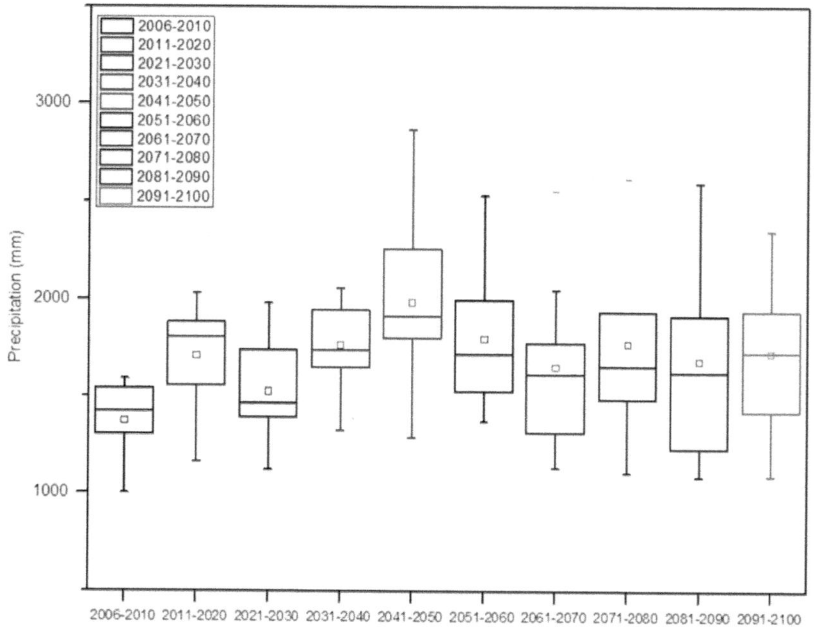

FIGURE 23.7 Projected decadal precipitation of RCP-4.5 scenario.

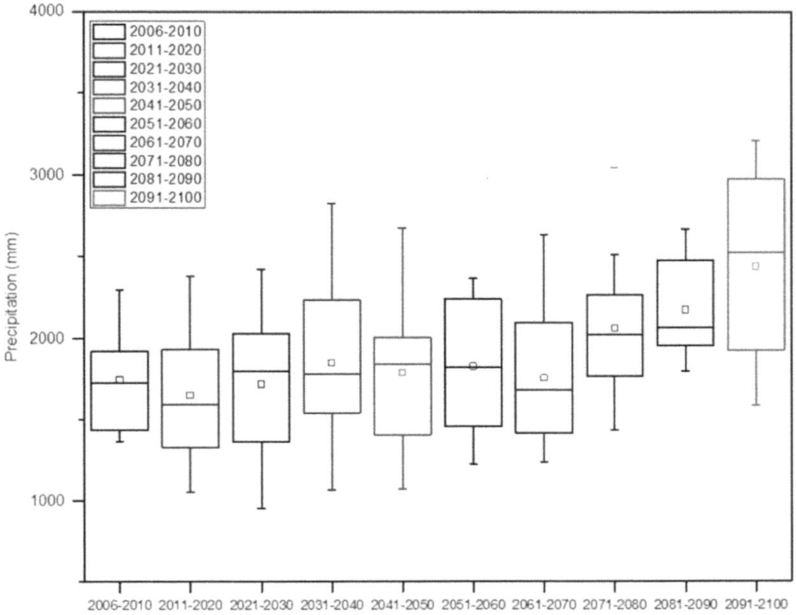

FIGURE 23.8 Projected decadal precipitation of RCP-8.5 scenario.

they anticipated the future drought, flood, and transmission condition of the climate which is directly or indirectly influencing the society. In the current study, the projection of future precipitation over Gangotri glacier at the Himalayan region has been carried out using CMIP5 climate model which can be helpful for scientific groups and society. The study highlights the future possible projections of precipitation which can be utilized for hydrological modeling for future climate in the Gangotri glacier. Additionally, positive trends have been noticed in the future precipitation of the Gangotri glacier.

KEYWORDS

- **climate change**
- **CMIP5**
- **CanESM2**
- **MLR**
- **climate scenarios**

REFERENCES

Anandhi, A.; Srinivas, V. V.; Nanjundiah, R. S.; Nagesh Kumar, D. Downscaling Precipitation to River Basin in India for IPCC SRES Scenarios Using Support Vector Machine. *Int. J. Climatol.* **2008,** *28* (3), 401–420.

Arora, M.; Kumar, R.; Kumar, N.; Malhotra, J. Assessment of Suspended Sediment Concentration and Load from a Large Himalayan Glacier. *Hydrol. Res.* **2014,** *45* (2), 292–306.

Chandniha, S. K.; Kansal, M. L. Rainfall Estimation Using Multiple Linear Regression Based Statistical Downscaling for Piperiya Watershed in Chhattisgarh. *J. Agrometeorol.* **2016,** *18* (1), 106.

Chaturvedi, R. K.; Joshi, J.; Jayaraman, M.; Bala, G.; Ravindranath, N. H. Multi-Model Climate Change Projections for India under Representative Concentration Pathways. *Curr. Sci.* **2012,** *103* (7), 791–802.

Fowler, H. J.; Blenkinsop, S.; Tebaldi, C. Linking Climate Change Modelling to Impacts Studies: Recent Advances in Downscaling Techniques for Hydrological Modelling. *Int. J. Climatol.* **2007,** *27* (12), 1547–1578.

Ghosh, S. SVM-PGSL Coupled Approach for Statistical Downscaling to Predict Rainfall from GCM Output. *J. Geophys. Res.: Atmos.* **2010,** *115* (D22).

Ghosh, S.; Mujumdar, P. P. Future Rainfall Scenario over Orissa with GCM Projections by Statistical Downscaling. *Curr. Sci.* **2006,** *90* (3), 396–404.

Kumar, K. K.; Patwardhan, S. K.; Kulkarni, A.; Kamala, K.; Rao, K. K.; Jones, R. Simulated Projections for Summer Monsoon Climate over India by a High-Resolution Regional Climate Model (PRECIS). *Curr. Sci.* **2011,** *101* (3), 312–326.

Kumar, K. R.; Sahai, A. K.; Kumar, K. K.; Patwardhan, S. K.; Mishra, P. K.; Revadekar, J. V.; Kamala, K.; Pant, G. B. High-resolution Climate Change Scenarios for India for the 21st Century. *Curr. Sci.* **2006,** *90* (3), 334–345.

Nakicenovic, N.; Swart, R. Special report on emissions scenarios. Special Report on Emissions Scenarios, Edited by Nebojsa Nakicenovic and Robert Swart, pp. 612. ISBN 0521804930. Cambridge, UK: Cambridge University Press, July 2000, 1.

Ojha, C. S. P.; Goyal, M. K.; Adeloye, A. J. Downscaling of Precipitation for Lake Catchment in Arid Region in India Using Linear Multiple Regression and Neural Networks. *Open Hydrol. J.* **2010,** *4* (1), 122–136.

Parry, M.; Cox, B. S. Technical summary. In Climate Change 2007: Impacts, Adaptation and Vulnerability. Contribution of Working Group II to the Fourth Assessment Report of the Intergovernmental Panel on Climate Change, 2007.

Solomon, S. (Ed.). Climate Change 2007-the Physical Science Basis: Working Group I Contribution to the Fourth Assessment report of the IPCC (Vol. 4). Cambridge University Press, 2007.

Taylor, K. E.; Stouffer, R. J.; Meehl, G. A. An Overview of CMIP5 and the Experimental Design. *Bull. Am. Meteorol. Soc.* **2012,** *90* (4), 485–498.

CHAPTER 24

Flow Pattern around Multiple Vane Arrangement

KARAN SOLANKI and HIMANSHU SHARMA*

Department of Civil Engineering, Institute of Infrastructure Technology Research and Management, Ahmedabad, Gujarat 380026, India

**Corresponding author. E-mail: himfce@gmail.com*

ABSTRACT

Submerged vanes are the foils placed skewed to the flow direction and are utilized to manage the sediment by altering the shear stress on the bed by generating the secondary currents on account of the pressure difference between the two sides of the vane. The range of angle of attack at which submerged vanes can be placed in the flow with respect to flow direction varies in between 10° and 40°. Since there are few studies which consider non-rectangular submerged vanes but mostly with a single tapered vane. Thus, the present study studies the optimization of tapered vanes arranged as a vane pair and multiple vanes per rows. Numerical modeling of the present study was done in ANSYS-CFX software using the K-ω turbulence closure model to simulate the vortical flow. In the present study, it was observed that due to high vortical interaction for $\delta_n = H$ &$2H$, the vanes are not able to work independently with one vane vortex suppressing another one's while for $\delta_n = 4H$, the vanes induced the vortices individually and made a more efficient system. Comparing the rectangular and tapered vane pairs showed that tapered vane induced high downwash at vane surface, which suggests that it is more proficient in generating vorticity of higher magnitude and hence, high downwash decayed suppressed vortex more rapidly in tapered

Advances in Hydrology and Climate Change: Historical Trends and New Approaches in Water Resources Management. Surendra Kumar Chandniha, Anil Kumar Lohani, Gopal Krishan, & Ajay Krishna Prabhakar (Eds.)

vane than rectangular vane. The present study showed that vorticity formation is independent of longitudinal spacing of tapered vane rows.

24.1 INTRODUCTION

The fluvial system carries a large suspension of fine sediments along with the flow. These suspended sediments are responsible for the turbidity which in turn prohibits the entrance of light in the river. Restriction of the light entrance then affects the aquatic food-chain (Van Nieuwenhuyse and LaPerriere, 1986; Davies-Colley et al., 1992). The presence of suspended sediments results in smothering of the channel bed and changes the river morphology, which may have deleterious effects on aquatic fauna and flora (Nuttall, 1972; Wright and Berrie, 1987; Doeg and Koehn, 1994). Hence, it is important to alter the bed condition in such a way that the erosion may be restricted so that they may be barred to come in the core flow region as suspended sediments.

Submerged vanes are the foils placed skewed to the flow direction and are utilized to manage the sediment by altering the shear stress on the bed by generating the secondary currents on account of pressure difference between the two sides of vane (Odgaard and Spoljaric, 1986; Odgaard and Mosconi, 1987; Odgaard and Wang, 1991a, 1991b; Wang and Odgaard, 1993). The range of angle of attack at which submerged vanes can be placed in the flow with respect to flow direction varies in between 10° and 40° (Odgaard and Kennedy, 1983; Odgaard and Wang, 1991a; Wang and Odgaard, 1993; Marelius and Sinha, 1998; Tan et al., 2005; etc.).

Submerged vanes outperform the traditional methods of sediment management such as dikes and groins. Traditional sediment management structures when placed normally to the flow redistribute eroded sediment flowing in the flow by utilizing the drag force and are less efficient in managing the sediment (Odgaard and Wang, 1991a, 1991b). While, submerged vanes minimize the skin drag acting over it and alter the flow structure by generating secondary currents (Wang and Odgaard, 1993). Several analytical and experimental studies have been reported in which the submerged vanes have been used to mitigate the problems related to sediment transport like river bank/bed erosion, intake shoaling, etc., by efficiently inducing the secondary currents (Odgaard and Wang, 1991a; Wang and Odgaard, 1993; Marelius and Sinha, 1998; Neary et al., 1999; Tan et al,. 2005; Ouyang et al., 2008). It has been observed that all the studies enlisted above consider the rectangular vane. However, there are few studies which consider non-rectangular submerged vanes (Gupta et al., 2006 2007; Ouyang, 2009; Ouyang and Lin, 2016) as shown in Figure 24.1.

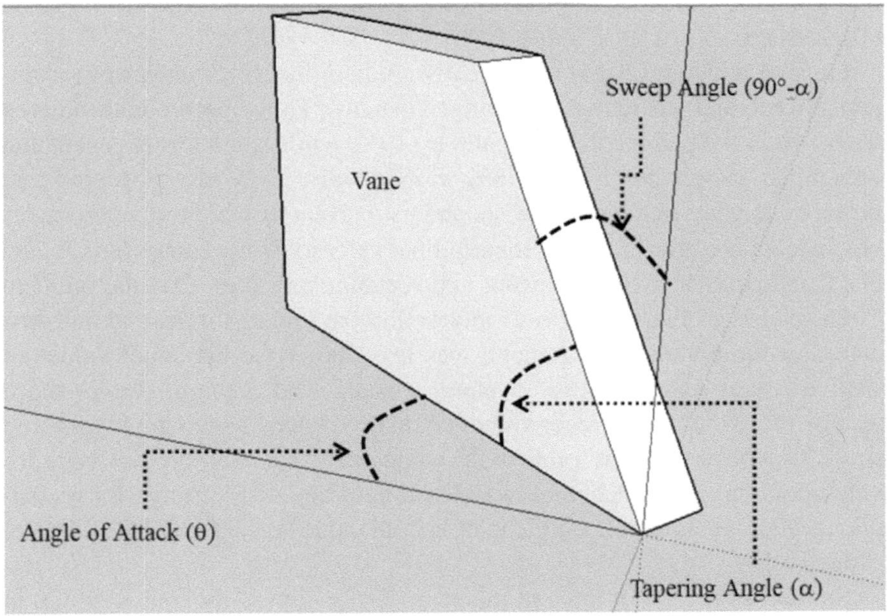

FIGURE 24.1 Parameters associated with tapered vane.

Gupta et al. (2006) experimentally studied the optimal parameters of doubly curved and J-type vane for the generation of secondary currents of maximum strength. They found that for both the vanes the optimal angle of attack is 45°. According to their observation, which wasdue to the curvature in the vanes, trailing and leading edge had the counter-rotating circulations. They also reported that the efficacy of the both type vanes were less than that of rectangular vane having same aspect ratio, in managing the sediment erosion in fluvial channel. Gupta et al. (2007) experimentally explored the optimal tapering angle of submerged tapered vane having being tapered at the leading edge in order to generate the secondary currents of maximum strength. They utilized the moment of momentum method to optimize the tapering angle and observed that the value of optimum tapering angle lies in between 33°and 45.

Ouyang (2009) numerically studied the effect of tapering vane for the sediment management in alluvial channel. He observed that for a constant surface area and height of vane, the performance of vane to manage the sediment enhanced for low tapering ratios (tapering ratio is defined as the ratio of top width of tapered vane to its base width). Ouyang (2009) also observed that by keeping the base width of tapered vane constant and increasing the vane height, efficacy of tapered vane to manage the sediment will decrease.

Ouyang and Lin (2016) numerically studied the effectiveness of rectangular, swept, and tapered vanes arranged in a row. They observed that tapered vanes raised the bed level of the bank by 19%, while same for the swept and rectangular vane were 17 and 16%, respectively. They also observed that the tapered vane generated the secondary currents of highest magnitudes which accounted for 25.8% of longitudinal velocity while same were 25 and 23.5%, respectively, for the swept and rectangular vanes. Ouyang and Lin (2016) observed that the ratio of interaction parameter for second and first vane in a three vane arrangement, was less than 1 for the small values of relative lateral spacing (ratio of lateral spacing with depth of water) and it became more than 1 for larger values of relative lateral spacing. Ouyang and Lin (2016) observed that a threshold value exists for each type of vane for which dominance of each vane was equal and they deduced that for rectangular, swept, and tapered vane, the threshold value was 0.53, 0.68, and 0.76, respectively.

Solanki et al. (2020) utilized the computational fluid dynamics (CFD) to optimize the parameters of tapered vane to generate the secondary currents of maximum strength and to evaluate the flow pattern downstream of single tapered vane. They observed that for 10°, 17° and 0.48 times depth of flow as sweep angle, angle of attack and height of vane, respectively. Tapered vane generated secondary currents of maximum strength. It was observed that in the proximity of vanes, the vorticity was maximum for a sweep angle of 35° due to the predominance of vortex lift; while for zones away from tapered vanes, the vorticity was maximum for 10° as vortex lift was dissipated under the action of viscosity and only the vortices generated by potential lift remained in the flow. While comparing, in optimal design conditions, rectangular and tapered vanes, it was observed by them that tapered vanes generated the stronger vortical structures in the flow as compared to rectangular vanes. Solanki et al. (2020) also observed that optimal rectangular vane has higher transverse velocity near the bed than a tapered vane and it can be concluded that a rectangular vane may act as a better sediment diverter than a tapered vane.

After a brief survey of literature, only Ouyang and Lin (2016) has studied the multiple vane arrangement but for a single row. The present study comprises of optimization of tapered vanes arranged in two vane per row and multiple vanes per row, with a sum total of four vane rows used, which was not studied by previous investigators.

24.2 METHODOLOGY

24.2.1 NUMERICAL MODEL BACKGROUND

Turbulence is a highly irregular and coherent phenomenon (Tennekes and Lumley, 1972; Nezu and Nakagawa, 1993). It is well-known that turbulent flow fluctuates in temporal as well as spatial domain. It makes it difficult to measure the turbulence quantities due to their randomness. While deriving Reynolds averaged Navier-Stokes (RANS) equation quantities including cross turbulence quantities product were evolved during the Reynolds averaging process which was not only difficult to measure, but also difficult to quantify these with flow parameters. Relating turbulence quantities to flow parameters were initiated by Prandtl's mixing length theory in 1936 and it gradually advanced to finally take the form of the average stress method or ASM model. In the present study, standard *K*-ω model was applied for simulation of flow. Incapability to capture the flow in separation region to over-predicting the flow made the standard *K*-ε not to be used to model the flow while *K*-ω was successfully able to capture the separation regions with efficacy (Versteeg and Malalsekera, 2007). The turbulent kinetic energy, *K*, and turbulent frequency, ω, are required to calculate the value of eddy viscosity, ξ, obtained by employing the standard *K*-ω (Wilcox, 1988) turbulence closure model. The equation of transport of *K* and ω are given as follows (where repeated indices indicate summation):

$$\frac{\partial K}{\partial t}+u_i\frac{\partial K}{\partial x_i}=\frac{\partial}{\partial x_i}\left[\frac{\left(\mu+\frac{\rho\xi}{K}\right)}{\rho}\frac{\partial K}{\partial x_i}\right]+\frac{P_k}{\rho}-\beta^* k\omega \tag{24.1a}$$

$$\frac{\partial \omega}{\partial t}+u_i\frac{\partial \omega}{\partial x_i}=\frac{\partial}{\partial x_i}\left[\nu+\frac{\xi}{\sigma_\omega}\frac{\partial \omega}{\partial x_i}\right]+\gamma_1\left[2S_{ij}.S_{ij}-\frac{2}{3}\omega\frac{\partial u_i}{\partial x_j}\delta_{ij}\right]-\beta_1\omega^2 \tag{24.1b}$$

Here, u_i = component of velocity in the *i*th direction where i = 1, 2, 3, etc.; x_i or x_j = coordinate of variable x into the *i*th or *j*th direction; μ =

dynamic viscosity of fluid; v = kinematic viscosity of fluid; S_{ij} = Strain tensor $= \frac{1}{2}\left[\frac{\partial u_i}{\partial x_j} + \frac{\partial u_j}{\partial x_i}\right]$; P_k = Production of kinetic energy $= 2\rho\xi S_{ij}.S_{ij} - \frac{2}{3}\rho K \frac{\partial u_i}{\partial x_j}\delta_{ij}$; σ_k = Coefficients in *K-ω* turbulence scheme associated with the transport of turbulent kinetic energy; σ_ω = Coefficient in Wilcox's *K-ω* turbulence scheme associated with the transport of turbulent frequency; γ_1 = Coefficient in Wilcox's *K-ω* turbulence scheme associated with the production of turbulent frequency; β_1 = Coefficient in Wilcox's *K-ω* turbulence scheme associated with the dissipation of turbulent frequency; β^* = Coefficient in Wilcox's *K-ω* turbulence scheme associated with the dissipation of turbulent kinetic energy.

24.2.2 MODEL SETUP

In the beginning, to prepare the model of a submerged vane, a volume is created of the dimensions, 10 m long, 0.910 m wide, and 0.152 m deep, which will act as a channel section in which vane will be put. At 2 m from the inlet of the prepared channel volume, a cuboidal volume is made having the dimensions as of 0.152 m long, 0.001 m thick and where the height of this volume was kept as 0.076 m. The smaller cuboidal volume was thus, trimmed for different sweep angles which ranged from 10° while the angle of attack of this volume was kept as 17°. After making the vane volume, the vane volume was duplicated and lateral spacing between the duplicated and original volume was varied from $y = H$ to $4H$ (here, H = height of vane). After developing a row, the row was copied and the distance at which the copied row will be placed was varied from $3H$ to $8H$. A total of 4 vane rows were placed (Fig. 24.2).

After finishing the geometry, process of blocking was initialized. Blocking is the process to create a block or partition around the area of interest to capture the flow features in this volume at the microscopic level. Therefore, a block or a partition-like structure was prepared around the tapered vane and was made to coincide with the vane geometry. While developing the block around the vane, the edges of vane were split by using "Split edges" option as shown in Figure 24.4. After splitting the edges, the development of the extra vertices over the surface was observed. Since the developed extra vertices are at presently unassociated, hence extra points were made over the surfaces where these vertices are to be connected over the points and hence completing the formation of blocking around the tapered vane. After having done with blocking, the block formed in the tapered vane was assigned to be "SOLID" so that block becomes impermeable to the flow occurring in the flow domain (Fig. 24.3).

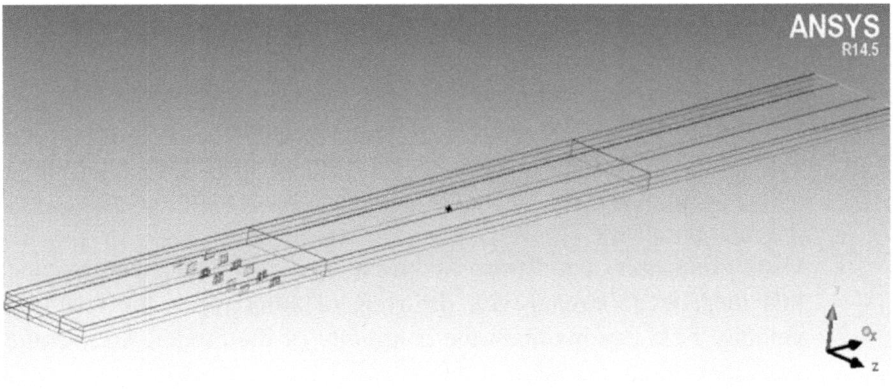

FIGURE 24.2 Model of submerged vanes generated by ICEM-CFD.

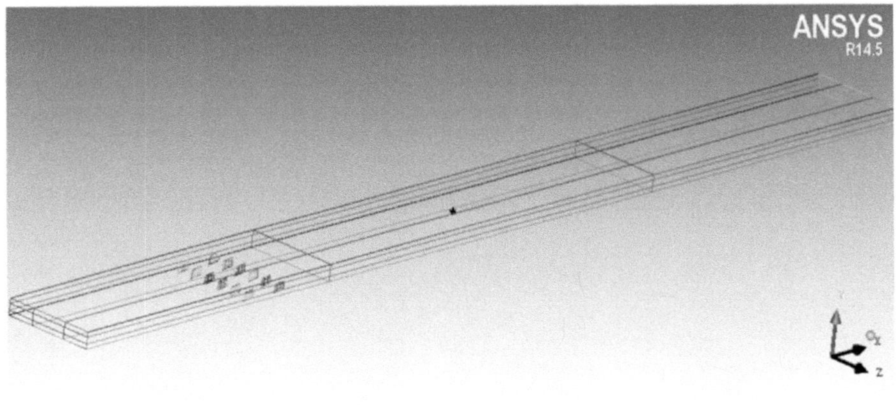

FIGURE 24.3 Blocking of submerged vanes.

The global mesh size assigned was 10 mm and in the block area, this mesh size logarithmically reduces to finer mesh in order to capture fine flow features. Mesh workability was also calculated through the mesh quality parameter. In the present study, it was observed to be within the range of

0.875–0.963 and according to Ansys CFX (2011) for a mesh to be workable; the minimum value of the mesh quality should be 0.3. After completion of all of the discussed, geometry is saved. Figure 24.5 shows the meshed geometry of channel sections along with vanes. After saving the geometry, the mesh file of geometry is imported to ANSYS CFX. After completion of geometry import, various surfaces of geometry were assigned the boundary in accordance with the role which they will serve, as below:

i. **Inlet boundary condition:** In the inlet boundary conditions, a normal flow (uniform flow) boundary condition is assumed. It is assumed that there is no lateral and perpendicular component to the longitudinal velocity. An average velocity of 0.24 m/s is set as an initial velocity at the inlet boundary.
ii. **Outlet boundary condition:** At the exit boundary plane, it is assumed that there is no stream-wise diffusion of velocity, so the resulting velocity field can maintain the continuity of the model. An average velocity of 0.24 m/s is set at the outlet.
iii. **Solid boundaries:** All the solid boundaries were entitled to "WALL" boundary conditions, which represent that the flow cannot pass through these surfaces.
iv. **Water surface:** Water surface was modeled by assuming it to be a rigid lid that has no possibility of across flow. Further, it is also assumed that the water is flowing on the surface of this rigid lid having the same velocity as assigned to the flow at the inlet boundary.

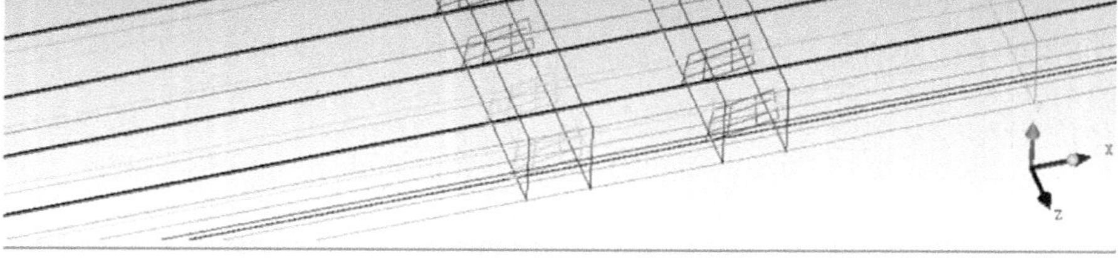

FIGURE 24.4 Meshing of submerged vanes generated by ICEM-CFD.

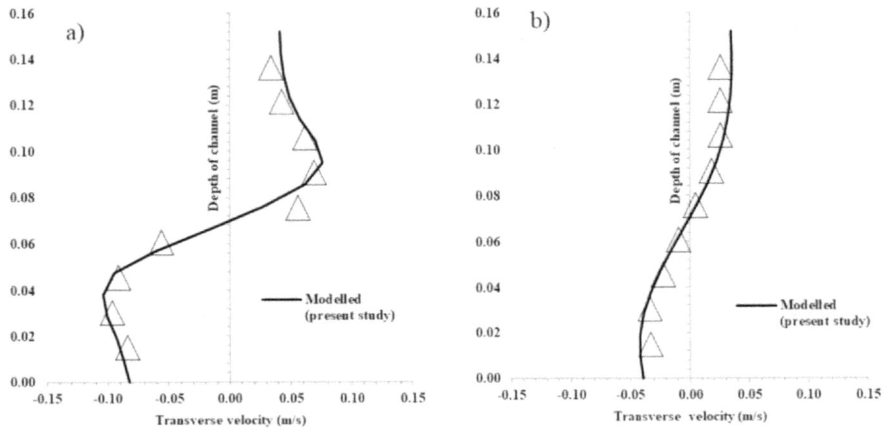

FIGURE 24.5 Validation of model for (a) $x = 2H$ and (b) $x = 20H$ (here, H = height of vane).

24.3 VALIDATION OF COMPUTATIONAL FLUID DYNAMICS MODEL

In the present study, a model of channel flume was developed and meshed in the ICEM-CFD platform and simulation was done with the help of ANSYS-CFX for an experimental study conducted by Wang and Odgaard (1993). In order to check the efficacy of the developed model, the transverse velocity computed by the model is compared with transverse velocities obtained experimentally in the study conducted by Wang and Odgaard (1993). It can be seen from Figure 24.5 that modeled transverse velocities matched well with the experimental transverse velocities of Wang and Odgaard (1993). Therefore, it may be concluded that the present model can be utilized to model the flow features in the case of multiple tapered vanes. Table 24.1 provides the parameters used in simulating the various models of tapered vanes.

Sensitivity analysis of the model was conducted to examine the effect of mesh size on the value of the transverse velocity calculated at transect. Three mesh sizes, 8, 10,and 12 mm were utilized to mesh the domain and the value of the transverse velocity profile was calculated for three longitudinal distances, $x = 2H$, $8H$, and $20H$ at mid-width of channel. Figure 24.6 shows the measured profile for all three mesh sizes at all aforesaid longitudinal distances. It can be seen that transverse velocity profiles superimpose satisfactorily over each other for all mesh sizes. It suggests that present result for the given model is independent of mesh size and hence, either of the mesh size can be used to compute the results for the present study.

TABLE 24.1 Summary of Different Model Parameters Used in the Present Study.

S. no.	Models parameters	Range
1	Length of the channel	10 m
2	Width of the channel	0.910 m
3	Depth of flow	0.152 m
4	Slope of the channel	0.000174
5	Upstream distance of first vane from inlet	2 m
6	Height of vane	0.076 m
7	Length of vane	0.152 m
8	Thickness of vane	0.001 m
9	Angle of attack of vane	17°
10	Tapered angle of vane	10° (from vertical axis)
11	Lateral spacing between two vanes in a row	0.076, 0.152, and 0.304 m
12	Longitudinal spacing between vane rows	3*H*, 5*H*, and 8*H*
13	Number of vanes per row	2–6
14	Velocity of fluid	0.240 m/s
15	Froude number	0.1967
16	Turbulence modeling use	k-ω model
17	Average mesh qualities of models	Minimum 0.902 from 1

24.4 FLOW PATTERN AND PARAMETER OPTIMIZATION OF VANE PAIR

a) *Lateral spacing between two vanes:* Wang and Odgaard (1993) observed that when vanes are put in proximity, vanes generated the secondary current that will interact with each other to reduce the vane circulation value. They observed that as the distance between the vanes was increased, each vane tends to generate circulations, individually, thus, reducing their effectiveness. Thus, they provided 2–3 times vane height as optimum range of lateral spacing which will generate the maximum strength of secondary currents. A study conducted by Ouyang et al. (2008) confirmed the findings of Wang and Odgaard (1993) and put forth that lateral spacing must range in between 2 and 3 times vane height for the generation of the secondary currents of maximum strength. Sharma et al. (2016) observed that for lateral spacing of 2.5 times height of vane, vanes-generated

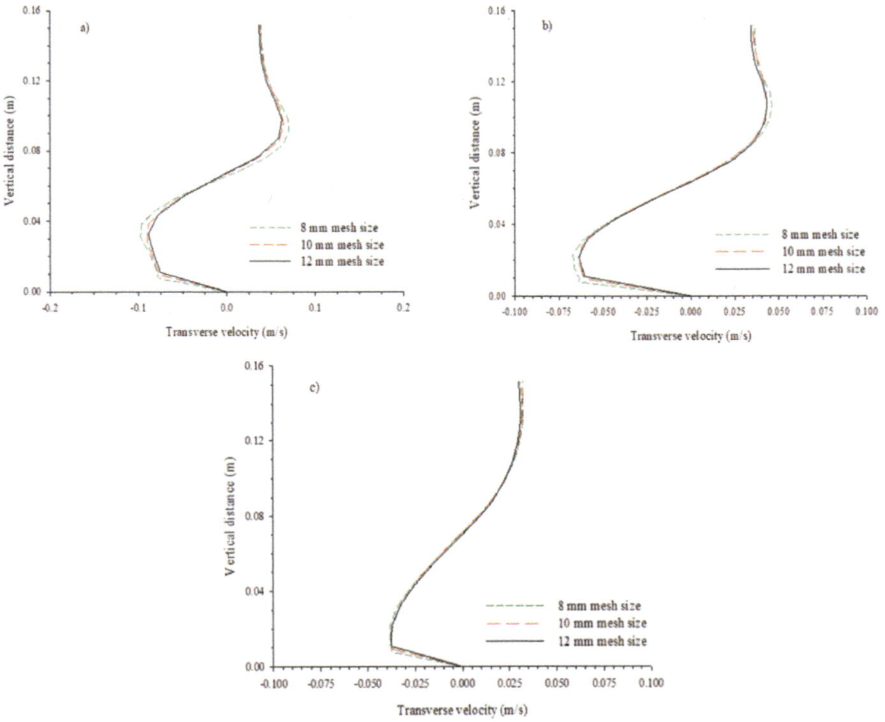

FIGURE 24.6 Sensitivity analysis of model for (a) $x = 2H$, (b) $x = 8H$, and (c) $x = 20H$.

secondary currents of maximum strength. In the present study, vanes were designed in accordance with the recommendations provided by Solanki et al. (2020). A single row of two vanes was developed in the flow domain and three lateral spacing was kept in between the vanes, which are *H*, 2*H*, and 4*H*. In Figure 24.7, one can see the contours as well as vector plot along a plane for three different vane spacing *viz.* $\delta_n = H$, 2*H*, and 4*H*. It can be seen that for $\delta_n = H$, in Figure 24.7a, the circulations generated by two vanes, started to merge with each other to make a single circulation; post separation and during mutual merger, circulation formed a single circulation of lower vorticity, which was in accordance with the Wang and Odgaard (1993). It was observed that the vortex formed by vortex then got consumed by the viscosity of fluid. In Figure 24.7b, it was observed that for $\delta_n = 2H$, the vanes generated individual vortices and these vortices were observed to exist in the flow by $x = 20H$, too. From the observation, it can be inferred that for $\delta_n = 2H$, vanes have able to generate the

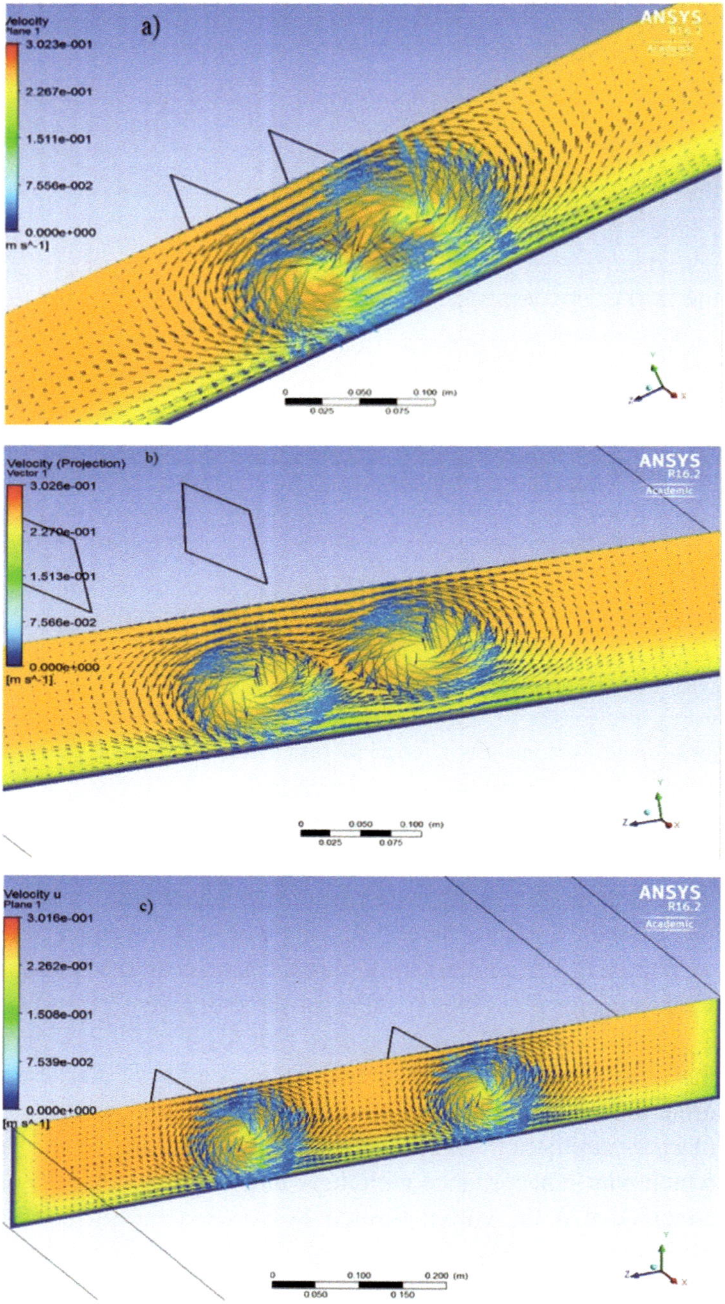

FIGURE 24.7 Flow pattern at downstream of tapered vane at $x = 2H$ for (a) $\delta_n = H$, (b) $\delta_n = 2H$, and (c) $\delta_n = 4H$ (here, δ_n = lateral spacing between the vanes).

circulations to the further downstream than $\delta_n = H$ and the circulations were able to interact with each other and merge with each other after a distance of $x = 20H$. For lateral spacing of $\delta_n = 4H$ in Figure 24.7c, it was observed that vanes have started generating vortices individually and showed that vanes have more chances to assist in the generation of vortices of equal magnitude in the flow. Figure 24.8a–c also supported that for $\delta_n = 4H$ (308 mm), the tapered vanes generated the secondary currents of higher vorticity and affected the farther downstream distance than the rest two lateral spacing. In Figure 24.8a, for $\delta_n = H$, the vorticity of circulations generated at $x = 2H$ was less, while the values of vorticity for the $\delta_n = 2H$ and $\delta_n = 4H$, were nearly of the same range. But the magnitude of circulations for $\delta_n = 4H$ was a bit higher than $\delta_n = 2H$. Now as the vortices traveled downstream, at $x = 20H$, the magnitude of vorticity was reduced for circulations generated by vanes placed at each of the spacing. It can be observed from Figure 24.8b, that for $\delta_n = H$, the circulations generated by the tapered vanes were merged and due to mutual interaction between the vortices and effect of viscosity, were reduced in a rapid manner. For $\delta_n = 2H$, it was observed that magnitude of vorticity was greater than rest of circulation while another was having lesser vorticity. Reason behind this observation can be understand by study conducted by Ouyang et al. (2008). In his study of three vane system, he showed that due to the mutual cancellation of vorticity generated by middle vane by the two outer vanes, it can be removed. Here, also the first vane is able to generate the secondary currents of higher magnitude by diverting more flow toward their generation, thus, second vane got the lesser flow for it, hence generated secondary currents of less magnitude. While for $\delta_n = 4H$, vane generated the individual circulations and magnitude of the vorticity of circulations generated by each vane were nearly equal. Thus, it can be deduced that for $\delta_n = 4H$, not only vanes will work adequately but also the effectively manage the sediment in more effective manner.

b) *Comparison of rectangular and tapered vanes:* Comparison of rectangular and tapered vane pairs was done for the same angle of attack, that is, 17°, and lateral spacing of $2H$ while tapered vane was having an angle of sweep of 10°. Figure 24.9a–c depicts the comparison of the rectangular and tapered vanes with respect to their efficacy in generating the vorticity in downstream transects. It can be seen from Figure 24.9a that for $x = 2H$, vorticity generated by

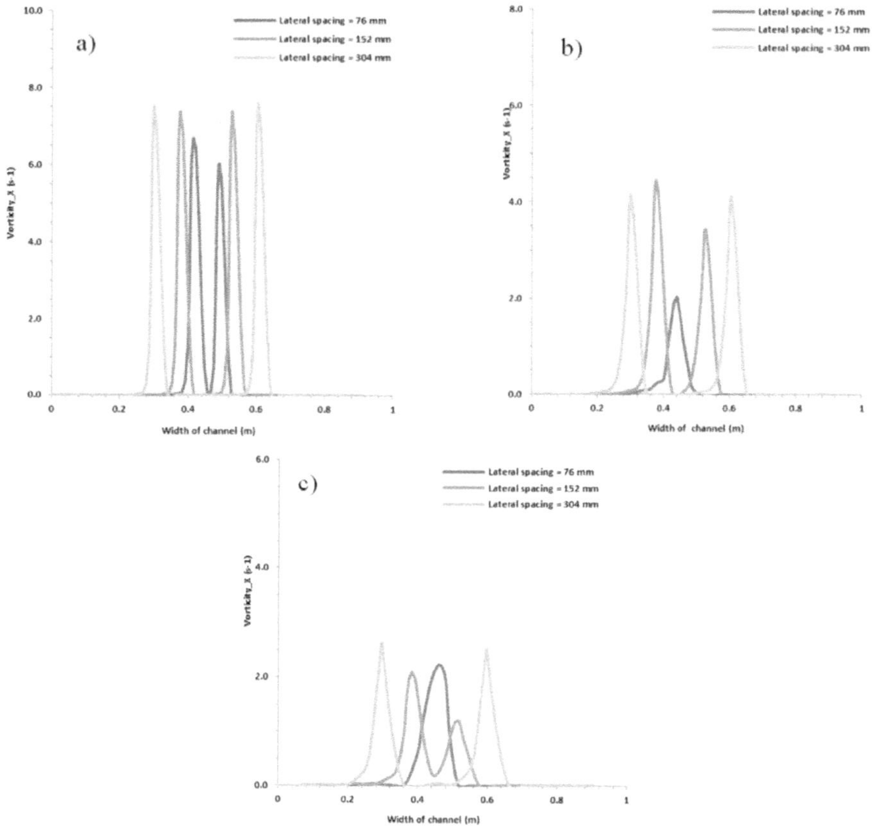

FIGURE 24.8 Variation of vorticity at downstream of tapered vane for different lateral spacing for (a) $x = 2H$, (b) $x = 8H$, and (c) $x = 20H$.

pair of rectangular and tapered vanes was of equal magnitude but as vorticity traveled further downstream, that is, at $x = 8H$ and $20H$, due to the mutual interaction one vane was observed to generate less vorticity than another. It can be observed that due to higher mutual interaction between the tapered vanes, the vorticity generated by the oppressed vane was lesser than what it was generated by rectangular vane. This can be attributed to the generation of high downwash velocity which is responsible for the more rapid decay of vorticity and higher downwash velocity is attributed to higher induced vorticity by vanes. Observations here are similar to observation made by Wang and Odgaard (1993) and Ouyang and Lin (2016). It can be seen that for ratio, $\delta_n/h = 0.5$ and 1, there was a strong vane interaction but for $\delta_n/h = 2$, the vane generated more effectively and equally, which was in accordance with the observation of Ouyang and Lin (2016) who suggested that post $\delta_n/h = 1.6$, vanes will act individually and generated circulation of equal magnitude.

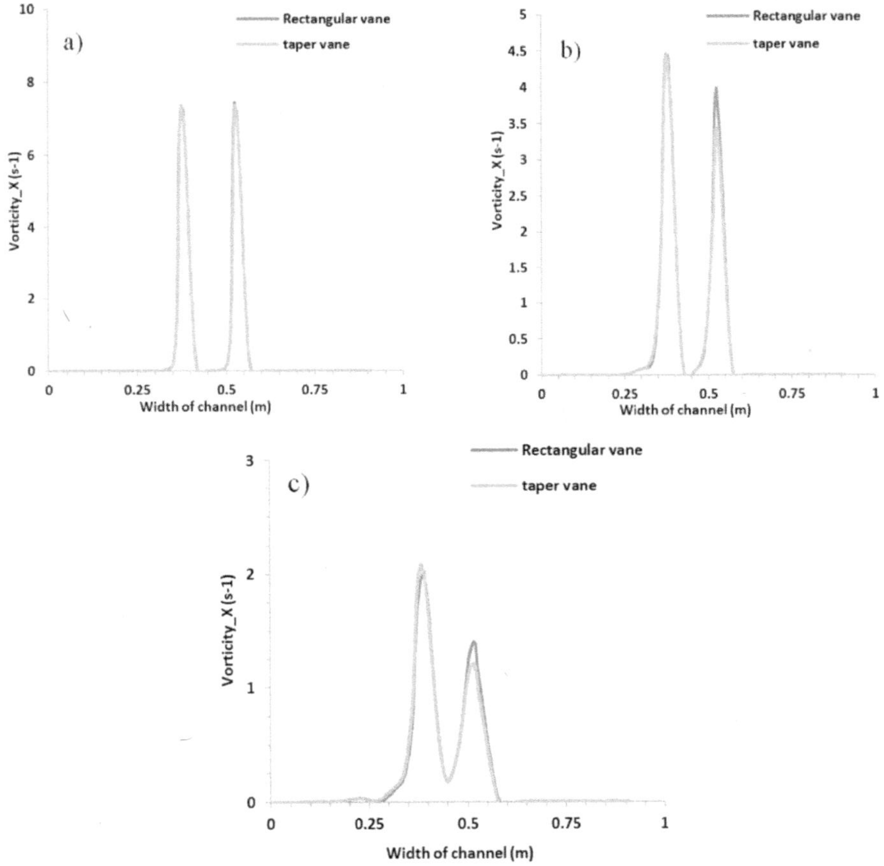

FIGURE 24.9 Comparison of rectangular and tapered vane for (a) $x = 2H$, (b) $x = 8H$, and (c) $x = 20H$.

24.5 PARAMETER OPTIMIZATION OF VANE ROWS

In order to optimize the parameter (longitudinal spacing) for the vane rows, a total of six vanes were placed in a row having a lateral spacing of $2H$. To optimize the longitudinal spacing between the two vane rows, the spacing between the rows were varied from $3H$ to $8H$ and results are presented in Figure 24.10.

Figure 24.10a–c, suggested that for all the longitudinal spacing (δ_s) specified in it, the vorticity generated by them was of equal magnitudes in the core-flow region. For the vanes, which were near to either of the side walls, the magnitude of vorticity was less than the vortices traveling in the core-flow region. The reason of suppression of vorticity in near sidewall region may be attributed to the wall-effect on the vortices, where wall assists in breaking of large vortices to vortices of smaller size. Thus, it can be said here that vorticity generation by tapered vane rows is a strong function of lateral spacing and independent of longitudinal spacing.

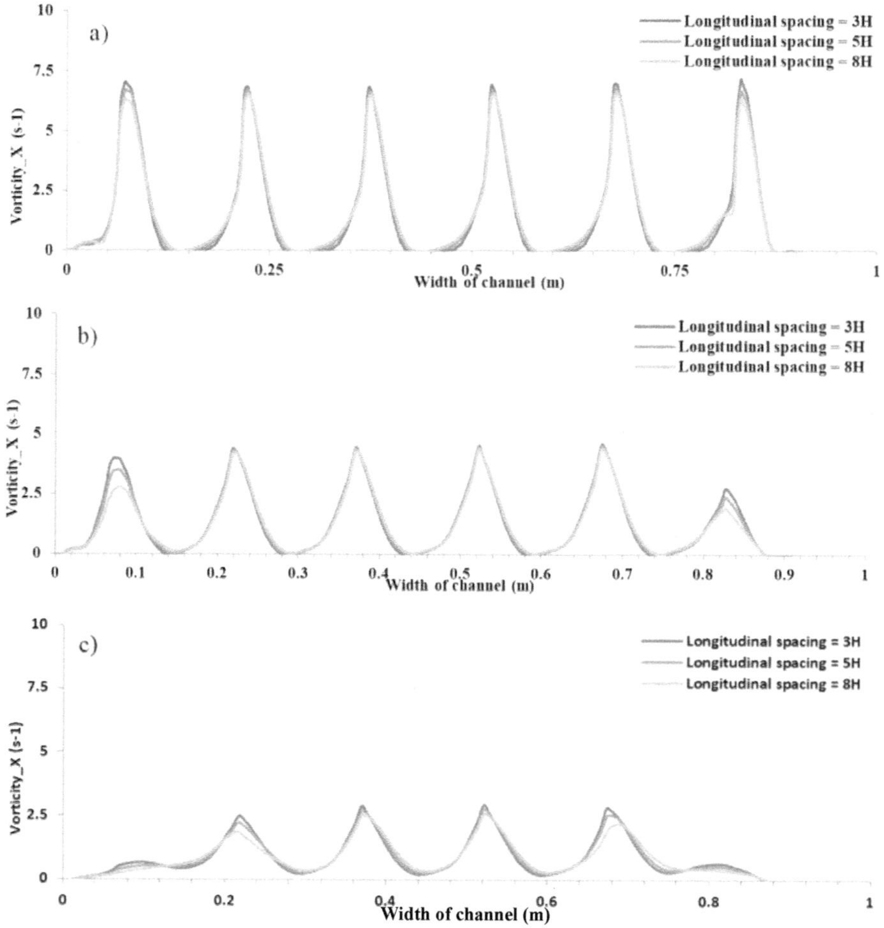

FIGURE 24.10 Comparison of vorticity generated by vane rows for different longitudinal spacing for (a) $x = 2H$, (b) $x = 8H$, and (c) $x = 20H$.

24.6 CONCLUSIONS

Thus, from the present study, following conclusions can be drawn:

a) For lateral spacing of $4H$, the vane were involved equally to generate the secondary currents in the flow for lateral spacing of H and $2H$, it was observed that the vortices start coalescing with each other to form a single vortex.
b) It was observed that for lateral spacing = H, the two individual vortices were generated at $x = 2H$ and were merged by the transect $x = 8H$ and that single vortex traveled further downstream. For lateral distance = $2H$, similar to lateral spacing = H, initially two individual vortices were formed and at $x = 8H$, again two vortices coexisted in the flow with one vortex having more vorticity than another due to interaction of vanes. At $x = 20H$, the vortices were merged for the lateral spacing = $2H$. For lateral spacing = $4H$, vane formed the individual vortices and they coexisted even after the $x = 20H$.
c) When rectangular and tapered vane pair was compared, it was observed that tapered vane interacted more with each other as compared to rectangular vane due to higher downwash generated by tapered vane which also signify the generation of higher vorticity.
d) It was observed that vorticity induced in flow were independent of the longitudinal distance between vane rows.

KEYWORDS

- **flow pattern**
- **ANSYS-CFX**
- **submerged vanes**
- **tapered vane**
- **ICEM-CDF**

REFERENCES

Ansys CFX. *Ansys ICEM CFD User's Manual*; Ansys Inc.: USA, 2011.

Davies-Colley, R. J.; Hickey, C. W.; Quinn, J. M.; Ryan, P. A. Effects of Clay Discharges on Streams: 1. Optical Properties and Epilithon. *Hydrobiologia* **1992,** *248*, 215–234.

Doeg, T. J.; Koehn, J. D. Effects of Draining and Desilting a Small Weir on Downstream Fish and Macroinvertebrates. *Regul. Rivers: Res. Manag.* **1994,** *9*, 263–278.

Gupta, U. P.; Sharma, N.; Ojha, C. S. P. Vorticity with Different Shapes of Submerged Vanes. *ISH J. Hydraulic Eng.* **2006,** *12* (1), 13–26.
Gupta, U. P.; Sharma, N.; Ojha, C. S. P. Performance Evaluation of Tapered Vane. *J. Hydraulic Res.* **2007,** *45* (4), 472–477.
Marelius, F.; Sinha, S. K. Experimental Analysis of Flow Past Submerged Vanes. *J. Hydraulic Eng.* **1998,** *124* (5), 542–545.
Neary, V. S.; Sotiropoulos, F.; Odgaard, A. J. Three-Dimensional Numerical Model of Lateral-Intake Inflows. *J. Hydraulic Eng.* **1999,** *125* (2), 126–140.
Nezu, I.; Nakagawa, N. *Turbulence in Open Channel Flows*; AA Balkema: Delft, The Netherlands, 1993.
Nuttall, P. M. The Effects of Sand Deposition upon the Macroinvertebrate Fauna of the River Camel, Cornwall. *Freshw. Biol.* **1972,** *2,* 181–186.
Odgaard, A. J.; Kennedy, J. F. River Bend Bank Protection by Submerged Vanes. *J. Hydraulic Eng.* **1983,** *109* (8), 1161–1173.
Odgaard, A. J.; Mosconi, C. E. Streambank Protection by Submerged Vanes. *J. Hydraulic Eng.* **1987,** *113* (4), 520–536.
Odgaard, A. J.; Spoljaric, A. Sediment Control by Submerged Vanes. *J. Hydraulic Eng.* **1986,** *112* (12), 1164–1181.
Odgaard, A. J.; Wang, Y. Sediment Management with Submerged Vanes. Theory: I. *J. Hydraulic Eng.* **1991a,** *117* (3), 267–283.
Odgaard, A. J.; Wang, Y. Sediment Management with Submerged Vanes. Applications: II. *J. Hydraulic Eng.* **1991b,** *117* (3), 284–302.
Ouyang, H. T. Investigation on the Dimensions and Shape of a Submerged Vane for Sediment Management in Alluvial Channels. *J. Hydraulic Eng.* **2009,** *135* (3), 209–217.
Ouyang, H. T.; Lai, J. S.; Yu, H.; Lu, C. H. Interaction between Submerged Vanes for Sediment Management. *J. Hydraulic Res.* **2008,** *46* (5), 620–627.
Ouyang, H. T.; Lin, C. P. Characteristics of Interactions among a Row of Submerged Vanes in Various Shapes. *J. Hydro-environment Res.* **2016,** *13,* 14–25.
Sharma, H.; Jain, B.; Ahmad, Z. Optimization of Submerged Vane Parameters. *Sadhana* **2016,** *41* (3), 327–336.
Solanki, K.; Sharma, H.; Joshi, N. Flow and Parameter Optimization of Tapered Vane. *J. Ecohydraulics* **2020**. Taylor and Francis (Accepted).
Tan, S. K.; Guoliang, Y.; Lim, S. Y.; Ong, M. C. Flow Structure and Sediment Motion around Submerged Vanes in Open Channel. *J. Waterway Port Coastal Ocean Eng.* **2005,** *131* (3), 132–136.
Tennekes, H.; Lumley, J. L. *A First Course in Turbulence*; MIT Press: USA, 1972.
Van Nieuwenhuyse, E. E.; LaPerriere, J. D. Effects of Placer Gold Mining on Primary Production in Subarctic Streams of Alaska. *Water Resour. Bull.* **1986,** *22*, 91–99.
Versteeg, H. K.; Malalasekera, W. *An Introduction to Computational Fluid Dynamics: The Finite Volume Method*; Longman: New Delhi, India, 1995.
Wang, Y.; Odgaard, A. J. Flow Control with Vorticity. *J. Hydraulic Res.* **1993,** *31* (4), 549–562.
Wilcox, D. C. Reassessment of Scale Determining Equation for Turbulence Models. *AIAA J.* **1988,** *26* (11), 1299–1310.
Wright, J. F.; Berrie, A. D. Ecological Effects of Groundwater Pumping and a Natural Drought on the Upper Reaches of a Chalk Stream. *Regul. Rivers: Res. Manag.* **1987,** *1*, 145–160.

Index

D

E

F

G

H

Q

R

S

T

U

V

W